Lecture Notes in Computer Science 8180

Commenced Publication in 1973
Founding and Former Series Editors:
Gerhard Goos, Juris Hartmanis, and Jan van Leeuwen

Xuemin Lin Yannis Manolopoulos
Divesh Srivastava Guangyan Huang (Eds.)

Web Information Systems Engineering – WISE 2013

14th International Conference
Nanjing, China, October 13-15, 2013
Proceedings, Part I

 Springer

Volume Editors

Xuemin Lin
The University of New South Wales, Sydney, NSW, Australia
E-mail: lxue@cse.unsw.edu.au

Yannis Manolopoulos
Aristotle University of Thessaloniki, Greece
E-mail: manolopo@csd.auth.gr

Divesh Srivastava
AT&T Labs-Research, Florham Park, NJ, USA
E-mail: divesh@research.att.com

Guangyan Huang
Victoria University, Melbourne, VIC, Australia
E-mail: guangyan.huang@vu.edu.au

ISSN 0302-9743   e-ISSN 1611-3349
ISBN 978-3-642-41229-5   e-ISBN 978-3-642-41230-1
DOI 10.1007/978-3-642-41230-1
Springer Heidelberg New York Dordrecht London

Library of Congress Control Number: 2013948675

CR Subject Classification (1998): H.3, H.4, H.5, H.2.8, J.1, D.2, I.2, C.2

LNCS Sublibrary: SL 3 – Information Systems and Application, incl. Internet/Web
and HCI

Preface

Welcome to the proceedings of the 14th International Conference on Web Information Systems Engineering (WISE 2013), held in Nanjing, China in October 2013. The series of WISE conferences aims to provide an international forum for researchers, professionals, and industrial practitioners to share their knowledge in the rapidly growing area of web technologies, methodologies, and applications. The first WISE event took place in Hong Kong, China (2000). Then the trip continued to Kyoto, Japan (2001); Singapore (2002); Rome, Italy (2003); Brisbane, Australia (2004); New York, USA (2005); Wuhan, China (2006); Nancy, France (2007); Auckland, New Zealand (2008); Poznan, Poland (2009); Hong Kong, China (2010); Sydney, Australia (2011); and Paphos, Cyprus (2012). This year, for a sixth time, WISE was held in Asia, in Nanjing to be precise.

WISE 2013 hosted four well-known keynote and invited speakers: Wen Gao of Peking University, who gave a talk on "Towards Web-Based Video Processing"; Divy Agrawal of the University of California at Santa Barbara, who gave a lecture on "Data-Driven Methodologies for Understanding, Managing, and Analyzing Online Social Networks"; Chengqi Zhang of the University of Technology Sydney, who gave a talk on "Big Data Related Research Issues and Progresses"; and Marek Rusinkiewicz of the Florida International University, who talked on "Security of Cyber-physical Systems (a Case Study)".

A total of 198 research papers were submitted to the conference for consideration, and 48 submissions were selected as full papers (with an acceptance rate of 24% approximately), plus 29 as short papers. The program also featured 10 demonstration papers and 5 challenge papers. The research papers cover the areas of web mining; web recommendation; web services; data engineering and databases; semi-structured data and modeling; web data integration and hidden web; social web; information extraction and multilingual management; networks, graphs, and web-based business processes; event processing, web monitoring and management; and innovative techniques and creations.

We wish to take this opportunity to thank the honorary general chair, Jian Lv; the industry program co-chairs, Min Wang and Lei Chen; the demo co-chairs, Hong Gao, Yoshiharu Ishikawa and Rui Zhang; the challenge program co-chairs, Weining Qian and Yabo (Arber) Xu; the panel co-chairs, Guoren Wang and Junzhou Lou; the tutorial co-chairs, Wojciech Cellary and Jeffrey Xu Yu; the workshop co-Chairs, Zhisheng Huang and Chengfei Liu; the publication chair, Guangyan Huang; the local arrangement chair, Jie Cao; the financial chair, Jing He; the publicity co-chairs, Haolan Zhang, Athena Vakali, and Wenjie Zhang; the WISE society representative, Xiaofang Zhou, and finally the webmaster, Zhiang Wu.

In addition, special thanks are due to the members of the International Program Committee and the external reviewers for a rigorous and robust reviewing process. All the papers were reviewed by at least three academic referees. In total, 727 reviews were uploaded by the members of the International Program Committee and the external reviewers.

Finally, we are also grateful to the UCAS (University of Chinese Academy of Sciences)-VU (Victoria University) Joint Lab for Social Computing and E-Health Research, to the Jiangsu Provincial Key Laboratory of E-Business at Nanjing University of Finance and Economics, and to the Nanjing Science and Technology Commission for their sponsorship of WISE 2013. We expect that the ideas that have emerged here will result in the development of further innovations for the benefit of science and society.

October 2013

Ricardo Baeza-Yates
Yanchun Zhang
Jie Cao
Xuemin Lin
Yannis Manolopoulos
Divesh Srivastava

Organization

Honorary General Chair

Jian Lv   Nanjing University, China

General Co-chairs

Ricardo Baeza-Yates   Yahoo!Research Lab, Spain
Yanchun Zhang   Victoria University, Australia
Jie Cao   Nanjing University of Finance and Economics,
  China

PC Co-chairs

Xuemin Lin   University of New South Wales, Australia
Yannis Manolopoulos   Aristotle University of Thessaloniki, Greece
Divesh Srivastava   AT&T Labs-Research, USA

Industry Program Chairs

Min Wang   Google Research, USA
Lei Chei   HKUST, Hong Kong, China

Demo Co-chairs

Yoshiharu Ishikawa   Nagoya University, Japan
Rui Zhang   University of Melbourne, Australia
Hong Gao   Harbin Institute of Technology, China

WISE Challenge Program Chairs

Weining Qian   East China Normal University, China
Yabo(Arber) Xu   Sun Yat-sen University, China

Panel Co-chairs

Guoren Wang   Northeastern University, China
Junzhou Lou   Southeastern University, China

Tutorial Co-chairs

Jeffrey Xu Yu   Chinese University of Hong Kong, China
Wojciech Cellary   Poznan University of Economics, Poland

Workshop Co-chairs

Zhisheng Huang   Vrije University of Amsterdam,
                                   The Netherlands
Chengfei Liu   Swinburne University of Technology,
                                   Australia

Publication Chair

Guangyan Huang   Victoria University, Australia

Local Arrangements Chair

Jie Cao   Nanjing University of Finance and
                                   Economics, China

Financial Chair

Jing He   Victoria University, Australia

Publicity Co-chairs

Haolan Zhang   NIT, Zhejiang University, China
Athena Vakali   Aristotle University of Thessaloniki, Greece
Wenjie Zhang   University of New South Wales, Australia

WISE Society Representative

Xiaofang Zhou   The University of Queensland, Australia

Webmaster

Zhiang Wu   Nanjing University of Finance and Economics,
                                   China

Program Committee

Alex Delis	University of Athens, Greece
Alexandros Ntoulas	Zynga, USA
Alfredo Cuzzocrea	University of Calabria, Italy
Anastasios Kementsietsidis	IBM T.J. Watson Research Center, USA
Anne Ngu	Texas State University-San Marcos, USA
Armin Haller	CSIRO, Australia
Athena Vakali	Aristotle University of Thessaloniki, Greece
Athman Bouguettaya	RMIT, Australia
Azer Bestavros	Boston University, USA
Bin Cui	Peking University, China
Birgitta König-Ries	Friedrich Schiller University of Jena, Germany
Brahim Medjahed	University of Michigan-Dearborn, USA
Bruno Martins	Instituto Superior Técnico (IST), Portugal
Chao Peng	East China Normal University, China
Chengfei Liu	Swinburne University of Technology, Australia
Chengkai Li	University of Texas at Arlington, USA
Claus Pahl	Dublin City University, Ireland
Costin Badica	University of Craiova, Romania
Dana Zhang	Google, USA
Dimitrios Katsaros	University of Thessaly, Greece
Dimitris Plexousakis	Universitat Oberta de Catalunya, Spain
Dimka Karastoyanova	University of Stuttgart, Germany
Elisa Bertino	Purdue University, USA
Evaggelia Pitoura	University of Ioannina, Greece
Fei Chen	HP Labs, USA
Fei Chiang	McMaster University, Canada
Ge Yu	Northeastern University, China
Georgia Koutrika	HP Labs, USA
Gill Dobbie	University of Auckland, New Zealand
Gong Zhang	Oracle Corporation, USA
Guangyan Huang	Victoria University, Australia
Guoliang Li	Tsinghua University, China
Hady Lauw	Singapore Management University, Singapore
Hao Yan	LinkedIn Corp, USA
Helen Karatza	Aristotle University of Thessaloniki, Greece
Hong Cheng	The Chinese University of Hong Kong, China
Hui Wang	Stevens Institute, USA
Ibrahim Kamel	University of Sharjah, UAE
Ilaria Bartolini	DEIS - University of Bologna, Italy
Ingmar Weber	Qatar Computing Research Institute, Qatar
Iraklis Varlamis	Harokopio University of Athens, Greece
Ismail Toroslu	Middle East Technical University, Turkey
Jian Yang	Macquarie University, Australia
Jiao Tao	Rensselaer Polytechnic Institute, USA

Jing He	Victoria University, Australia
Jinho Kim	Kangwon National University, South Korea
Joao Eduardo Ferreira	University of São Paulo, Brazil
John Shepherd	The University of New South Wales, Australia
Josiane Xavier Parreira	DERI - National University of Ireland, Ireland
Jun Shen	University of Wollongong, Australia
Kai-Uwe Sattler	TU Ilmenau, Germany
Karl Aberer	École Polytechnique Fédérale de Lausanne, Switzerland
Katja Hose	Aalborg University, Denmark
Kjetil Nørvåg	Norwegian University of Science and Technology, Norway
Kostas Stefanidis	Foundation for Research and Technology - Hellas, Greece
Kunal Verma	Accenture Technology Labs, USA
Ladjel Bellatreche	LIAS/ENSMA, France
Laure Berti	Institut de Recherche Pour le Développement, France
Lei Chen	Hong Kong University of Science and Technology, China
Luciano Baresi	DEI - Politecnico di Milano, Italy
Lyublena Antova	EMC Greenplum, USA
Marco Aiello	University of Groningen, The Netherlands
Mirjana Ivanovic	University of Novi Sad, Serbia
Miyuki Nakano	University of Tokyo, Japan
Mohamed Sharaf	University of Queensland, Australia
Monica Scannapieco	Istituto Nazionale di Statistica (ISTAT), Italy
Mourad Ouzzani	Qatar Computing Research Institute, Qatar
Muhammad Cheema	The University of New South Wales, Australia
Murali Mani	Flint, USA
Natwar Modani	IBM India Research Lab, India
Nick Bassiliades	Aristotle University of Thessaloniki, Greece
Oscar Corcho	Universidad Politécnica de Madrid, Spain
Panagiotis Karras	Rutgers University, USA
Panagiotis Symeonidis	Aristotle University of Thessaloniki, Greece
Panayiotis Andreou	University of Cyprus, Cyprus
Panayiotis Tsaparas	University of Ioannina, Greece
Pei Li	University of Zurich, Switzerland
Peiquan Jin	University of Science and Technology of China, China
Peter Scheuermann	Northwestern University, USA
Peter Triantafillou	University of Patras, Greece
Peter Yeh	Accenture Technology Labs, USA
Philippe Cudré-Mauroux	University of Fribourg, Switzerland
Pingpeng Yuan	Huazhong University of Science and Technology, China

Prasad Deshpande | IBM Research, India
Qi Yu | Rochester Institute of Technology, USA
Qing Liu | CSIRO, Australia
Quan Z. Sheng | The University of Adelaide, Australia
Ralf Schenkel | Saarland University, Germany
Rik Eshuis | Eindhoven University of Technology,
 | The Netherlands
Rizos Sakellariou | University of Manchester, UK
Rui Zhang | University of Melbourne, Australia
Salima Benbernou | Université Paris Descartes, France
Samir Tata | Institut Télécom, France
Sangeetha Seshadri | IBM Almaden Research Center, USA
Schahram Dustdar | TU Wien, Austria
Sebastian Michel | Saarland University, Germany
Shawn Bowers | Gonzaga University, USA
Sherif Sakr | The University of New South Wales, Australia
Solmaz Kolahi | University of British Columbia, Canada
Soon Ae Chun | The City University of New York, USA
Sourav S. Bhowmick | Nanyang Technological University, Singapore
Stelios Paparizos | Microsoft Research, USA
Tingjian Ge | University of Massachusetts Lowell, USA
Torsten Suel | Yahoo! Research, USA
Verena Kantere | Cyprus University of Technology, Cyprus
Vladimir Zadorozhny | University of Pittsburgh, USA
Walter Binder | University of Lugano, Switzerland
Wei Wang | Fudan University, China
Weiyi Meng | Binghamton University, USA
Wilfred Ng | Hong Kong University of Science and
 | Technology, China
Wojciech Cellary | Poznan University of Economics, Poland
Wolf-Tilo Balke | TU Braunschweig, Germany
X. Sean Wang | Fudan University, China
Xiangliang Zhang | King Abdullah University of Science and
 | Technology, Saudi Arabia
Xiaofang Zhou | University of Queensland, Australia
Xuemin Lin | The University of New South Wales, Australia
Yanchun Zhang | Victoria University, Australia
Ying Zhang | The University of New South Wales, Australia
Yinian Qi | Purdue University, USA
Yoshiharu Ishikawa | Nagoya University, Japan
Yuanyuan Tian | IBM Research - Almaden, USA
Yuqing Wu | Indiana University, USA
Zhisheng Huang | Vrije University Amsterdam, The Netherlands
Zhiyuan Chen | University of Maryland Baltimore County, USA
Zhongqiang Chen | Yahoo Inc!, USA
Zi Huang | The University of Queensland, Australia

Industry Program Committee

Anastasios (Tasos) Kementsietsidis	IBM Watson, USA
Bin Liu	NEC Lab America, USA
Haixun Wang	Microsoft Research Asia, China
Honesty Yang	Intel, China
Howard Ho	IBM Almaden, USA
Jingren Zhou	Microsoft, USA
Jun Rao	LinkedIn, USA
Mingxuan Yuan	Huawei Noah's Ark Lab, China
Ping Luo	HP Labs, China
Wei Fan	Huawei Noah's Ark Lab, China

External Reviewers

Ahlers, Dirk
Akritidis, Leonidas
Alvanaki, Foteini
Aly, Ahmed
Araujo, Luciano
Basaras, Pavlos
Benouaret, Karim
Bento, Carolina
Bibi, Stamatia
Bourne, Scott
Bucur, Doina
Butt, Anila Sahar
Catasta, Michele
Choi, Jae-Yong
Christoforaki, Maria
Degeler, Viktoriya
Ebaid, Amr
Fehling, Christoph
Gaaloul, Walid
Garcia-Alvarado, Carlos
Gounaris, Anastasios
Grund, Martin
Gu, Tao
Gómez Sáez, Santiago
Haupt, Florian
Huang, Jin
Huang, Xin
Huynh, Tan Dat
Jain, Prateek
Jiang, Yu

Joshi, Salil
Kim, Nam-Soo
Koloniari, Georgia
Kontopoulos, Efstratios
Kritikos, Kyriakos
Lamba, Hemank
Li, Jianxin
Liakos, Panagiotis
Liang, Yongjiang
Liu, Xiaohui
Liu, Xin
Ma, Jiangang
Meditskos, Georgios
Midi, Daniele
Narayanam, Ramasuri
Neiat, Azadeh
Nie, Tiezheng
Nizamic, Faris
Nowak, Alexander
Oikawa, Marcio K.
Papadakis, Nikolaos
Podorozhny, Rodion
Podzimek, Andrej
Prokofyev, Roman
Qi, Jianzhong
Rahman, Rameez
Rezig, Elkindi
Shang, Haichuan
Shebaro, Bilal
Shugars, David

Spicuglia, Sebastiano
Stupar, Aleksandar
Tbahriti, Salah-Eddine
Tiakas, Eleftherios
Tonon, Alberto
Tsakalozos, Konstantinos
Tzouramanis, Theodoros
Vasirani, Matteo
Vassilakopoulos, Michael
Wang, Shenlu
Wang, Xiaoyang
Warriach, Ehsan Ullah
Weiß, Andreas
Xu, Lai
Xuan, Ming
Xue, Andy Yuan
Yang, Zhenglu
Yao, Lina
Ye, Zhen
Zeginis, Chris
Zhan, Liming
Zhang, Chengyuan
Zhao, Xiang
Zheng, Wei
Zheng, Yudi
Zhong, Youliang
Zhou, Rui
Zhou, Xiangmin
Zhou, Yang

Keynotes
(Abstracts)

Towards Web-Based Video Processing

Wen Gao

Peking University
wgao@pku.edu.cn

Image and video data is becoming the majority in big data era, a reasonable improvement of video coding efficiency may get a big cost saving in video transmission and/or storage, that is why so many researchers working on the new coding technologies and standards. For example, a team under IEEE standard association works on a new standard IEEE 1857 for internet/surveillance video coding, which targets to achieves about 50% bits saving than any of existing standards. However, the video coding story will be changed in the case of web application, because so many data we can reference comparing to the case of normal video coding, in that only a few frame of images can be referenced. In this talk, the recent developments of model-based video coding will be given, special on background picture model based surveillance coding and cloud-based image coding, and a on-going project that using web image and video to enhance the efficiency of video processing will be discussed also.

Data-Driven Methodologies for Understanding, Managing, and Analyzing Online Social Networks

Divy Agrawal

University of California at Santa Barbara
agrawal@cs.ucsb.edu

Online social networks provide unprecedented amounts of information about social interactions and therefore enable the study of various problems in the context of social networks on a scale and at a level of detail that has never been possible before. In this talk, we will consider ways of systematically exploring this vast space of online social network problems. Namely, we will consider three dimensions; understanding, managing and reporting on social networks and focus on example studies relating to these dimensions. To this end, we will consider three applications: modeling adoption behavior, limiting the spread of misinformation, and trend analysis in social networks. In modeling adoption behavior, we will challenge the common use of pure local models and revive research done in the context of diffusion of innovations and demonstrate the value of this technique in two different social networks. Next, we study the notion of competing campaigns in a social network and develop protocols whose goal is to limit the spread of misinformation by identifying a subset of individuals that need to be convinced to adopt the competing (or "good") campaign so as to minimize the number of people that adopt the "bad" campaign. And finally, relating to reporting on online social networks, we explore novel trend detection mechanisms. We propose two novel structural trend definitions that use friendship information to identify topics that are discussed among clustered and unconnected users respectively. Our analyses and experiments show that structural trends provide new insights into the way people share information online.

Table of Contents – Part I

Web Mining

Web Recommendation

Web Services

Data Engineering and Database

Semi-Structured Data and Modeling

Web Data Integration and Hidden Web

Challenge

Table of Contents – Part II

Social Web

Information Extraction and Multilingual Management

Networks, Graphs and Web-Based Business Processes

Event Processing, Web Monitoring and Management

Innovative Techniques and Creations

Demo

Entity Correspondence
with Second-Order Markov Logic

Ying Xu[1,2], Zhiqiang Gao[1], Campbell Wilson[2],
Zhizheng Zhang[1], Man Zhu[1], and Qiu Ji[1]

[1] Institute of Computer Science and Engineering, Southeast University, China
[2] Faculty of Information Technology, Monash University, Australia

Abstract. Entity Correspondence seeks to find instances that refer to
the same real world entity. Usually, a fixed set of properties exists, for
each of which the similarity score is computed to support entity correspondence. However, in a knowledge base that has properties incrementally recognized, we can no longer rely only on the belief that two
instances sharing value for the same property are likely to correspond
with each other: a pair of different properties that are of hierarchies or
specific relations can also be evidential to corresponding instances. This
paper proposes the use of second-order Markov Logic to perform entity
correspondence. With second-order Markov Logic, we regard properties
as variables, explicitly define and exploit relations between properties
and enable interaction between entity correspondence and property relation discovery. We also prove that second-order Markov Logic can be
rephrased to first-order in practice. Experiments on a real world knowledge base show promising entity correspondence results, particularly in
recall.

Keywords: Entity Correspondence, Second-Order Markov Logic, Property Relation Discovery, Knowledge Base.

1 Introduction

Entity correspondence is a task that seeks to find instances that refer to the same
real world entity. This task has different names in different fields, e.g. record
linkage, entity resolution, instance matching and object reconciliation. In this
paper, we focus on entity correspondence in automatically constructed knowledge bases. Different from traditional knowledge bases that rely on experts, these
knowledge bases typically consist of probabilistic beliefs that are based on the
redundancy of the web information. Several extraction systems, including Know-
ItAll [1], Never-Ending Language Learner (NELL) [2] and REISA [3], have been
developed for this purpose. Many of these systems are based on a bootstrapping
approach that ensures automated extraction of instances and properties (including instance categories and relations between instances). NELL, in particular,
has accumulated a knowledge base of 1,829,036 asserted instances and 859 different properties after beginning with an initial set of around 800 instances and 188

X. Lin et al. (Eds.): WISE 2013, Part I, LNCS 8180, pp. 1–14, 2013.

properties[1]. This knowledge base has a locally-formatted ontology as its core for supporting information extraction, and in return, automatically extracted information can be used to enrich the knowledge base. Although it is inspiring that a machine can read the web and acquire knowledge like a human, the resulting knowledge base cannot be linked to other datasets that are available on the web, limiting the use of such a valuable knowledge base. The Semantic Web[2] is proposed to allow data to be shared and reused across application, enterprise, and community boundaries. We aim to link a knowledge base like NELL into the Semantic Web, and focus on entity correspondence as the first step to do so.

Several features of automatically constructed knowledge bases make the entity correspondence different from previous efforts. 1) The properties are incrementally recognized and thus infinite, in which case, we cannot write a fixed set of rules to correspond instances. 2) The machine may recognize properties that are of hierarchies or specific relations, which leads to hidden corresponding instance pairs. For example, in Fig. 1, if we don't know the relation between *HasWife* and *HasSpouse*, we will lose a potential corresponding pair *SameAs(Ann1, Ann2)*. 3) Due to feature 1), we cannot provide fixed property relations for entity correspondence, which requires discovering such relations at the same time with entity correspondence. Although we have the property relation *SubPropertyOf(HasWife, HasSpouse)* in Fig. 1, we may still lose the potential corresponding pair *SameAs(Bob1, Bob2)* if we cannot discover the relation between *AgentCreate* and *BookWriter*, which are newly recognized. Notice that discovering such relations and corresponding instances are in fact interactive processes, and simultaneously performing these two can help us to find more corresponding instance pairs. The Semantic Web provides the standard for relations between properties as property hierarchies *SubPropertyOf*[4], which states that all resources related by one property are also related by another, and one of the property characteristics *InverseOf*[4], which states that one property can be obtained by taking another property and changing its direction. Discovering these two kinds of relations between properties not only better supports the task of entity correspondence, but also enables us to organize the automatically recognized properties in accordance with the standard of the Semantic Web.

In this paper, we propose the use of second-order Markov Logic [4] to do entity correspondence with the help of property relation discovery. We solve the issue of incremental properties by regarding properties as variables that can be instantiated with any recognized property, which is naturally enabled by second-order Markov Logic; we solve the issue of hidden property relations by explicitly defining relations between properties using second-order logic constructors and provide discovered corresponding instances pairs as evidence for discovering property relations. And then we show how the interactive process in Fig. 1 is enabled and how entity correspondence is improved in Markov Logic. In addition, we prove that second-order Markov Logic can be rephrased to first-order in practice, which bridges these two models theoretically.

---

[1] http://rtw.ml.cmu.edu/rtw/

[2] http://www.w3.org/2001/sw/

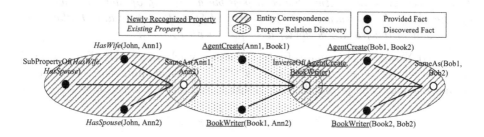

Fig. 1. Interaction between entity correspondence and property relation discovery

2 Preliminary

2.1 First-Order Markov Logic

First-order logic formulae are constructed using four types of symbols: constants, variables, functions and predicates. Constants represent objects in the domain of interests. Variable ranges over objects. Function maps tuples of objects to objects. Predicates represent relations among objects. A *term* can be any constant, variable or any function applied to a tuple of terms. An *atom* (or *atomic formula*) is a predicate symbol applied to a tuple of terms and *formulae* are recursively constructed from atoms using logical connectives and quantifiers, e.g. ¬ (negation), ∧ (conjunction), ∨ (disjunction), → (implication), ↔ (equivalence), ∀ (universal quantification) and ∃ (existential quantification). Grounding of formulae means replacing variables with constant symbols [5].

Markov network defines the joint distribution of a set of random variables $X = \{X_1, X_2, ..., X_n\} \in \mathcal{X}$. It is composed of an undirected graph $\mathcal{G}$ and a set of potential functions ϕ_k. The graph has a node for each random variable, and a potential function for each *clique*[3]. Usually, the potential functions are conveniently represented in log-linear form, where the potential function is replaced by an exponentiated weighted sum of features of a state. The probability of a state x is then given by $P(X = x) = \frac{1}{Z} \exp(\sum_{j=1}^{N} \omega_j f_j(x))$. Here, f_j is the feature function for jth clique, Z is a normalizing constant defined as $Z = \sum_{x \in \mathcal{X}} \exp(\sum_{j=1}^{N} \omega_j f_j(x))$ and N is the number of cliques [5].

Definition 1. *A Markov Logic Network (MLN) [5] L is a set of pairs (F_i, ω_i), where F_i is a formula in first-order logic and ω_i is a real number. Together with a finite set of constants $C = \{c_1, c_2, ..., c_{|C|}\}$, it defines a Markov network $M_{L,C}$ as follows:*

1. $M_{L,C}$ contains one binary node for each possible grounding of each predicate appearing in L. The value of the node is 1 if the ground predicate is true and 0 otherwise.

---

[3] If every two nodes in the node set $\mathcal{X}$ are connected by an edge, we call $\mathcal{X}$ a clique [6].

2. $M_{L,C}$ *contains one feature for each possible grounding of each formula F_i in* L. *The value of this feature is 1 if the ground formula is true and 0 otherwise. The weight of the feature is the ω_i associated with F_i in* L.

The MLNs work as templates for constructing Markov networks. Each of these networks is called a *ground Markov network*, where each clique corresponds to a ground formula. The probability distribution over state x is as follows.

$$P(\boldsymbol{X} = \boldsymbol{x}) = \frac{1}{Z} \exp(\sum_i \omega_i n_i(\boldsymbol{x})) \tag{1}$$

Here, n_i is the number of true groundings of F_i in state x.

2.2 Second-Order Markov Logic

Extending first-order Markov Logic to second-order involves grounding atoms with all possible predicate symbols as well as all constant symbols. It was first introduced by Kok and Domingos in order to perform statistical predicate invention [4]. Subsequently, Davis and Domingos reused this model to enable deep transfer among different domains [7]. Second-order Markov Logic is a probabilistic extension to second-order logic, which allows quantifications of predicates by quantifiers. For example, $\forall p \forall x(p(x) \vee \neg p(x))$ states that for each unary relation (or predicate) p of individuals and each individual x, either x is in p or it is not. This reflects a common pattern that can be shared by all unary predicates.

3 Bridge First-Order and Second-Order Markov Logic

In order to facilitate the prove of our theorem, we introduce a more general representation of higher order logic, which is called *Relational Type Theory* [8].

Definition 2 ([8]). *Types(or relational types) are defined as syntactic expressions inductively generated by: ι is a type; and if $\tau_1...\tau_k$ are types ($k \geq 0$), then $\tau = (\tau_1...\tau_k)$ is a type.*

In this paper, we intend ι to denote the type of individuals, and $(\iota_1...\iota_k)$ the type of k-ary relations between k objects of types $\iota_1...\iota_k$ respectively.

Definition 3 ([8]). *Let V be a vocabulary for the second-order formulae. A* **Henkin-V-prestructure** $\mathcal{H}$ *consists of*

- *a non-empty universe A;*
- *an interpretation in A of the V-constants; and*
- *for each type τ, a collection D^τ, where $D^\iota = A$, and $D^{(\tau_1...\tau_k)} \subseteq \mathcal{P}(D^{\tau_1} \times ... \times D^{\tau_k})$*

Example 1. For a second order formula that consists of unary atom $p(x)$, 2-ary atom of individuals $r(x,y)$, and 2-ary atom of predicates $InverseOf(p_1, p_2)$, the vocabulary V is formalized as $V = \{x, y, p, r, p_1, p_2, InverseOf\}$. According to Definition 2, x and y are of type ι; p is of type (ι); r, p_1 and p_2 are of type $(\tau_1\tau_2)$; and *InverseOf* is a V-constant of type $(\tau_1\tau_2)$ where $\tau_1 = (\tau_{11}\tau_{12})$ and $\tau_2 = (\tau_{21}\tau_{22})$. Thus, we can construct a *Henkin-V-prestructure* $\mathcal{H}$ where

- A denotes the universe;
- the V-constant *InverseOf* is interpreted as $InverseOf(A_{(\tau_{11}\tau_{12})}, A_{(\tau_{21}\tau_{22})})$;
- and the collection A is for type ι, collection $D^{(\iota)}$ for type (ι) and collection $D^{(\tau_1\tau_2)}$ for type $(\tau_1\tau_2)$.

Let V^S be an extension of V with relation-constants $T_\iota, T_{(\iota)}, T_{(\tau_1\tau_2)}, BelongsTo$ and $Triple$, of arities 1, 1, 1, 2 and 3 respectively. $T_\iota(x)$ is intended to state x is an individual, $T_{(\iota)}(x) - x$ is a set, $T_{(\tau_1\tau_2)}(x) - x$ is a 2-ary predicate, $BelongsTo(x, p)$ − x is an element of p and $Triple(x, r, y) - x$ and y have relation r.

Definition 4. *A Henkin-V-prestructure $\mathcal{H}$ can determine a unique V^S-structure $\mathcal{S} = \mathcal{S}(\mathcal{H})$ where*

- $A^S = A \cup D^{(\iota)} \cup D^{(\tau_1\tau_2)}$;
- *the interpretation in S of the V-constant, i.e. InverseOf in Example 1, is the same as their interpretation in $\mathcal{H}$; and*
- *the V^S-constant T_ι is interpreted as A, $T_{(\iota)}$ as $D^{(\iota)}$, $T_{(\tau_1\tau_2)}$ as $D^{(\tau_1\tau_2)}$, BelongsTo as $A \times D^{(\iota)}$, and Triple as $A \times A \times D^{(\tau_1\tau_2)}$*

For example 1, the second-order V-formula φ *is rephrased to* a first-order V^S-formula φ^S, which is obtained by: replacing each atom $p(x)$ with $BelongsTo(x, p)$ and $r(x, y)$ with $Triple(x, r, y)$; relativizing quantifiers over individuals to T_ι, quantifiers over sets to $T_{(\iota)}$, and quantifiers over 2-ary predicates to $T_{(\tau_1\tau_2)}$.

Lemma 1 ([8]). *A formula φ is true in $\mathcal{H}$ iff φ^S is true in $\mathcal{S}(\mathcal{H})$.*

According to Lemma 1, we can conclude the following theorem.

Theorem 1. *A second-order MLN L can determine a first-order MLN L^S that have the same probability distribution by replacing unary atom $p(x)$ with Belongs $To(x, p)$ and 2-ary atom $r(x, y)$ with $Triple(x, r, y)$.*

Proof. Let $F = \{F_i | i = 0, ..., N\}$ be the set of formulae in L and $F^S = \{F_i^S | i = 0, ..., M\}$ the set of formulae in L^S. Each F_i^S is obtained by replacing unary atom $p(x)$ with $BelongsTo(x, p)$ and 2-ary atom $r(x, y)$ with $Triple(x, r, y)$ in F_i. So, we have $N = M$. Let V and V^S be the vocabulary for F and F^S respectively. We can construct a Henkin-V-prestructure $\mathcal{H}$ for F which uniquely determines a V^S-structure $\mathcal{S}(\mathcal{H})$. According to Lemma 1, F_i in $\mathcal{H}$ and F_i^S in $\mathcal{S}(\mathcal{H})$ have the same truth value.

Let n_i be the number of true groundings of the ith formula F_i in L, and n_i^S the number of true groundings of F_i^S in L^S, we have $n_i = n_i^S$, which leads to $P(\boldsymbol{X} = \boldsymbol{x}) = \frac{1}{Z}\exp(\sum_{i=1}^N w_i n_i(\boldsymbol{x})) = \frac{1}{Z}\exp(\sum_{i=1}^M w_i n_i^S(\boldsymbol{x}))$.

4 Entity Correspondence

4.1 Basic Model

Human beings have knowledge about how to correspond instances using the properties they have. For example, it's intuitive to us that persons that are married to the same person *are likely to* refer to the same real world entity. Such knowledge can be regarded as a rule that is not certainly but probably correct, which can be perfectly modeled by a weighted formula in MLN. Accordingly, properties are mapped to predicates and property relations are mapped to predicates that take predicates as arguments. We introduce the predicate *SameAs* to correspond two instances. Then, for the above example, we can write a weighted formula like the following,

$$(\forall x_1, x_2, x_3 \quad (\text{MarriedTo}(x_1, x_3) \ \wedge \ \text{MarriedTo}(x_2, x_3)$$
$$\rightarrow \ \text{SameAs}(x_1, x_2)), \quad \omega) \tag{2}$$

where ω illustrates how strong the formula is: higher weight means greater difference between a state that satisfies the formula and the one that does not.

In an automatically constructed knowledge base, properties are recognized now and then, which results in incremental number of properties. Thanks to second-order Markov Logic, we can regard predicates as variables as well, which leads to more general formulae that can be instantiated with any predicate symbol. Formula (3) and (4) are our second-order formulae for entity correspondence, where we replace the first and second predicate-constant *MarriedTo* in Formula (2) with predicate variable p_1 and p_2, considering that a pair of different properties can be evidential to entity correspondence. And in order to provide relations between properties as evidence, we introduce predicates *CloseTo* and *InverseOf* for *SubPropertyOf* and *InverseOf* that are mentioned in Sect. 1. For the predicate *CloseTo*, we remove the constraint of *SubPropertyOf* that the direction cannot be reversed. For example, *SubPropertyOf(HasSpouse, HasWife)* has truth value *false* while *CloseTo(HasSpouse, HasWife)* has *true*. We enable this compromise because both directions of *SubPropertyOf* have the same effect on entity correspondence.

$$(\forall p_1, p_2 \quad \forall x_1, x_2, x_3 \quad \text{Similar}(x_1, x_2) \ \wedge \ p_1(x_1, x_3) \ \wedge \ p_2(x_2, x_3)$$
$$\wedge \ \text{CloseTo}(p_1, p_2) \ \rightarrow \ \text{SameAs}(x_1, x_2)), \quad \omega_1) \tag{3}$$

$$(\forall p_1, p_2 \quad \forall x_1, x_2, x_3 \quad \text{Similar}(x_1, x_2) \ \wedge \ p_1(x_1, x_3) \ \wedge \ p_2(x_3, x_2)$$
$$\wedge \ \text{InverseOf}(p_1, p_2) \ \rightarrow \ \text{SameAs}(x_1, x_2)), \quad \omega_2) \tag{4}$$

These two formulae are relatively aggressive if placed in a rule-based model as we cannot correspond instances using only one pair of properties. Some properties, e.g. *WorkFor*, maybe too weak to correspond two instances, where we need more evidence. However, from the point of view of Markov Logic, they make sense. At first, Formula (3) and (4) are just templates for generating Markov networks, where each formula may be instantiated to multiple ground formulae. More shared properties means more true ground formulae, and thus higher probability of a state. Secondly, the weighted formula itself indicates that the rule is partially correct, where each true ground formula just improves the probability of a state, but not make a certain decision. In addition, the whole Markov network generated is a connected one, where the probability of each node is influenced by many other factors, which makes the property *WorkFor* not so aggressive in corresponding entities.

Then, how can we discover property relations? Reviewing the example given in Fig. 1, we can in fact easily conclude the pattern for property relation discovery as following two weighted formulae:

$$(\forall p_1, p_2 \quad \forall x_1, x_2, x_3 \quad (\text{SameAs}(x_1, x_2) \ \wedge \ p_1(x_1, x_3) \ \wedge \ p_2(x_2, x_3)$$
$$\rightarrow \ \text{CloseTo}(p_1, p_2)), \quad \omega_3) \tag{5}$$

$$(\forall p_1, p_2 \quad \forall x_1, x_2, x_3 \quad (\text{SameAs}(x_1, x_2) \ \wedge \ p_1(x_1, x_3) \ \wedge \ p_2(x_3, x_2)$$
$$\rightarrow \ \text{InverseOf}(p_1, p_2)), \quad \omega_4) \tag{6}$$

In order to keep accordance with the symmetry and transitivity characteristics of the predicate *SameAs*, we also define symmetric and transitive *hard constraints* for *SameAs*. Additionally, *CloseTo* and *InverseOf* also have the symmetry characteristic that can refine the results of property relation discovery.

Existing tools for Markov Logic include Alchemy[4] and Tuffy[5], which are well implemented and maintained. However, none of them support second-order Markov Logic. In order to reuse these tools, we reformulate Formula (3) to (6) to first-order according to Theorem 1, replacing all 2-ary atoms $p(x, y)$ with $Triple(x, p, y)$. In the rest of this paper, we intend Formula (3)' to(6)' to denote the first-order form of Formula (3) to (6).

4.2   Decreasing the Number of Candidate Corresponding Pairs

The basic idea of *canopy* [9] is to find some measure to separate candidates that are obviously unmatched so as to decrease the number of candidate corresponding pairs. We use two measures, *name similarities* and *instance categories*, to decrease the number of candidate corresponding pairs.

1. We *only* compare instances with similar name labels. The similarity score is computed by the Overlap Coefficient[11], $\text{Overlap}(X, Y) = \frac{|X \cap Y|}{\min(|X|, |Y|)}$, where X and Y are two string sets. We introduce the predicate *Similar* for this measure and provide it as an atom of the conjunctive antecedents of Formulae (3)' and (4)'. Thus, only similar names can activate them.
2. We *only* compare instances that are in the same category. Considering the case of entity correspondence within one category, we divide the collection of instances into two subsets, one for the target category, and one for the rest. As a result, the predicate *Triple* should be replaced by $Triple_1$ and $Triple_2$, which indicates that the target instance is the first and second argument respectively. Corresponding changes should be taken to other predicates.

4.3   Weight Learning

Up to this point, we have introduced the model for entity correspondence and property relation discovery. In practice, weights can be specified manually or

---

[4] http://alchemy.cs.washington.edu/
[5] http://hazy.cs.wisc.edu/hazy/tuffy/

learnt from training data. We use discriminative learning where conditional probability of query atoms are computed given the evidence nodes. According to Equation (1), the partial derivation of the conditional log-likelihood of Markov networks is given by $\frac{\partial}{\partial \omega_i} \log P_\omega(\boldsymbol{y}|\boldsymbol{x}) = n_i(\boldsymbol{x}, \boldsymbol{y}) - E_\omega[n_i(\boldsymbol{x}, \boldsymbol{y})]$, where $\boldsymbol{x}$ is the vector of the evidence atoms' truth values, $\boldsymbol{y}$ is the truth values of the query atoms and $n_i(\boldsymbol{x}, \boldsymbol{y})$ is the number of true groundings of ith formula [12]. Thus, we can use the standard gradient-based optimization methods to learn weights.

4.4   Inference

Inference allows for query about ground atoms of one or more predicates. We aim to capture dependencies between entity correspondence and property relation discovery, so we query *SameAs*, *CloseTo* and *InverseOf* at the same time. The most widely used algorithms for inference in Markov Logic include (Loopy) Belief Propagation, Simulated Tempering, MC-SAT and Gibbs Sampling. As we have circles in our model, loopy belief propagation may not converge at last. Simulated tempering is much slower than the others as it includes a process of slow cooling to extend the solution space. According to our experiments on small datasets, Gibbs Sampling is relatively faster than MC-SAT, while, at the same time, gives equally satisfying results. So we choose Gibbs Sampling for the conditional probabilistic query. This kind of inference tells us how likely two instances correspond with each other given the evidences and we can set different thresholds to filter the results.

4.5   An Example of Bi-directional Joint Inference

In this section, we demonstrate how entity correspondence is improved in Markov Logic, taking property relation discovery into consideration. Reviewing Fig. 1, if we query the facts at white nodes and provide the black ones as evidence, Fig. 2 illustrates the Markov network that is generated after grounding. Notice that *Similar* atoms are provided additionally and instance relations are rephrase to *Triple* atoms. Here, Fig. 2 (a) is a clique that corresponds to Formula (3)', Fig. 2 (b) is a clique that corresponds to Formula (6)' and Fig. 2 (c) is a clique that corresponds to Formula (4)'. Given the evidence nodes in Fig. 2 (a), the query node *SameAs(Ann1, Ann2)* has high probability to be assigned true, which results in high probability of node *InverseOf(AgentCreate, BookWriter)* in Fig. 2 (b) and further affects the probability of node *SameAs(Bob1, Bob2)* in Fig. 2 (c). Fig. 2 (d) demonstrates the directions of message propagation between entity correspondence and property relation discovery. Fig. 2 (a) and Fig. 2 (b) propagate massage from entity correspondence to property relation discovery, which we call *single-directional* message propagation, and Fig. 2 (b) and Fig. 2 (c) propagate message from property relation discovery to entity correspondence in return, which completes the *bi-directional* message propagation loop. Notice that the above process is completed within one query, and the assignment to each query node will be optimized in a global view. As a result, entity correspondence and property relation discovery are collectively improved.

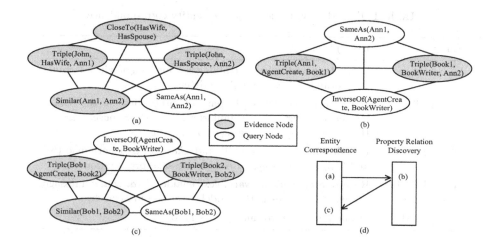

Fig. 2. Message propagation between entity correspondence and property relation discovery

5   Experiments and Analysis

5.1   Dataset

The dataset we use for experiments is the iteration 690 of the NELL knowledge base[6]. This downloadable part of NELL knowledge base contains 1,850,160 beliefs in total, among which 1,795,281 have high confidence greater than 0.9. For experiments, we focus on entity correspondence within person category. So we get 285,793 person instances and extracts those have at least one relation with other instances. The resulting candidate instances we can use is only 10,550 of them. We manually annotate the corresponding pairs among the 10,550 instances, and randomly choose 300 of them that have at least one *SameAs* relation with others. The final *SameAs* matrix is of size 300 × 300, and the number of true entries in the *SameAs* matrix is 566. Notice that grounding involves replacing all variables with all possible constants, so the generated Markov networks for the above 300 instances contain tens of thousands of nodes in practice, which limits the scale of the datasets. Nevertheless, we build a benchmark that can be used to evaluate entity correspondence with automatically recognized properties.

5.2   Experiment Settings

We conduct four groups of experiments to demonstrate four different aspects of our model. At first, We incrementally extend the original model to enable it to perform entity correspondence and property relation discovery collectively. The five models given in Table 1 are all based on Markov Logic and follow a similar

---

[6] http://rtw.ml.cmu.edu/resources/results/08m/NELL.08m.690.esv.csv.gz

Table 1. Five Markov Logic models for entity correspondence

Name	Discription
MLN+Orig	Extract properties in the training set, for each of which formula like Formula (2) is manually written.
MLN+PP	Write formulae for each pair of properties in the training set, e.g. $(\forall x_1, x_2, x_3(\text{Similar}(x_1, x_2) \land \text{HasSpouse}(x_1, x_3) \land \text{HasWife}(x_2, x_3) \to \text{SameAs}(x_1, x_2), \quad \omega)$
MLN+PR	Include predicate *InverseOf* and *CloseTo*; add Formulae (5)' and (6)'; enables *single-direction* message propagation.
MLN+PR+PV	Regard properties as variables; enables *bi-direction* message propagation. (This is the model described in Sect. 4.)
MLN+PRH+PV	Define Formulae (5)' and (6)' as hard.

nomenclature to that used in [10]. For this group, we aim to show the difference after taking property relation discovery into consideration. Secondly, we compare our approach with first-order Markov Logic [10] and the Fellegi-Sunter pairwise approach [13]. The first-order Markov Logic approach is exactly the MLN+Orig mentioned above. The Fellegi-Sunter approach is the very original pairwise model, which regards entity correspondence as a classification problem where a vector of similarity scores is given as feature. For this group, we aim to show the difference between pairwise approach and collective approach. Thirdly, as we already have property relation in training set before inference, we compare situations where we provide these property relations as evidence for test and not do so. Fourthly, we compare among MLN+PR, MLN+PR+PV and MLN+PRH+PV at discovering property relations. The correctness of these discovered property relations are checked manually.

We perform 5-fold cross validation to the whole dataset. For each iteration, we use four folds for learning and one for test. And we perform precision, recall and F1 evaluations against the benchmark we build for entity correspondence.

5.3   Experimental Results

In Figure 3, we have two thresholds. The similarity threshold is used to decide the truth value of *Similar* atoms, which are assigned *true* if the similarity scores are greater than the threshold. The probability threshold is used to decide how probable two instances correspond with each other should they be considered a positive example of corresponding instances. There is no obvious difference between the results of similarity threshold 0.5 and 0. However, the precision in Fig. 3 (a) does have a little improvement compared to that in Fig. 3 (d), which is to our intuition that higher similarity threshold can keep out more noise, and thus gives higher precision. The results within each subgraph show that with the increase of probability threshold, precisions increase and recalls decrease.

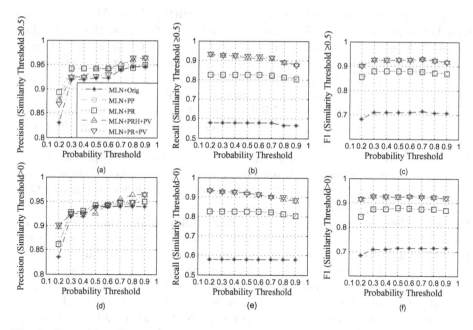

Fig. 3. Comparison between different Markov Logic models for entity correspondence

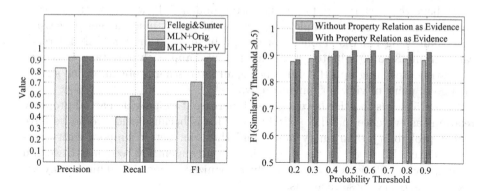

Fig. 4. Comparison between different approaches for entity correspondence (probability threshold=0.5)

Fig. 5. Entity correspondence before and after providing partial property relations as evidence in MLN+PRH+PV

Table 2. Results for property relation discovery

	Sim. Threshold ≥ 0.5		Sim. Threshold > 0	
	Avg. Prec.	# Correct	Avg. Prec.	# Correct
MLN+PR	90.39	53	88.06	54
MLN+PR+PV	90.60	**62**	**90.43**	**60**
MLN+PRH+PV	**90.77**	60	87.92	60

Figure 3 demonstrates that our models (MLN+PRH+PV and MLN+PR+PV) have better precisions (96.29%) than the others (94.90%) when the probability threshold is greater than 0.5. More obvious improvement is in recall, which is to our expectation that after taking property relation discovery into consideration, we can find more corresponding instances. MLN+Orig has the worst recalls as it loses a lot of potential corresponding pairs that are evidenced by related properties. MLN+PP and MLN+PR have almost the same results, except that MLN+PR can be used to discover new property relations. However, the newly discovered property relations will not affect entity correspondence in return as the entity correspondence formulae are fixed by pre-specified properties. In our models, bi-directional message propagation is enabled by defining properties as variables, increasing recall from 82.38% to 92.22% (probability threshold=0.5).

Figure 4 demonstrates the results for Fellegi-Sunter, MLN+Orig and MLN+PR+PV. Their precisions are 82.67%, 92.20% and 92.51%, and their recalls are 39.91%, 57.97% and 92.22% respectively. These obvious improvements come from the limitation of pairwise approaches and the first-order Markov Logic. In datasets with complex relations, pairwise approaches may lose many valuable dependencies among multiple entity correspondence decisions, for example, $SameAs(A,B)$ and $SameAs(B,C)$ can lead to $SameAs(A,C)$. First-order MLN captures such dependencies, but does not capture dependencies between entity correspondence and property relation discovery, as it is not capable of defining relations between properties. Our model enables both of the above dependencies, so we have the best results for entity correspondence. However, as the Markov network inference is #P-complete [5], we would expect much longer processing time than using pairwise model. For our problem, the time for MLN inference varies from 48 to 80 minutes, while the pairwise approach only take 4 seconds for testing. Speeding up MLN inference is an interesting problem, which should be a good point of view for our future works.

Figure 5 shows that F1 have a general increase if we provide partial property relations as evidence in MLN+PRH+PV. This is because evidence nodes have higher probability than query nodes and thus can better support entity correspondence. However, as the properties are incrementally recognized, we can never provide relations between newly recognized properties. This is why we need to discovery these relations.

Table 2 demonstrates the results for property relation discovery. The results show that MLN+PR+PV and MLN+PRH+PV can discover more related properties, which, we believe, is the result of their higher recalls for entity correspondence. It implies that dependencies between entity correspondence and property relations discovery can improve the results of both tasks.

6   Related Works

Entity Correspondence has been received a wide range of research in different fields. It is regarded as a classification problem [13] where a pair of candidates is classified as 'Match' or 'Not Match' given a vector of similarity scores. However, recent works focus on capturing dependencies among multiple decisions.

For example, Singla and Domingos[10], Parag and Domingos [14] try to model dependencies among paper correspondence and author correspondence for citation matching, as one instance may appear in multiple candidate matching pairs. Brocheler [15] propose Probabilistic Similarity Logic for reasoning about entity resolution. Bhattacharya and Getoor [16] propose collective entity resolution to jointly determine corresponding entities for co-occuring references. More recent works include joint entity resolution [17], whereby the result of resolving one dataset may benefit the resolution of another dataset.

Another kind of dependency comes from entity correspondence and other tasks. Poon and Domingos [18], Singh [19] allow dependencies between entity correspondence and segmentation, which improves the result for both tasks. Haas [20] perform data integration by leveraging information about both schema and data to improve the integrated results. Niepert [21] tries to map two ontologies by using schema information to exclude logically inconsistent corresponding pairs. Whang [22] does not explicitly capture dependencies among multiple tasks, but it tries to adapt entity correspondence rules to the changing data, where rule refinement and entity correspondence are collectively performed. Our work tries to capture the dependency between entity correspondence and property relation discovery, which is enabled by bi-directional joint inference in Markov Logic.

We use the same graphical model with [10], but we extend the first-order Markov Logic to second-order to enable discovering property realtions. We focus on automatically constructed knowledge bases that are full of probabilistic facts, while in [20][17][22], they have reliable data sources saved in databases. We have to discover relations between properties within the same knowledge base, while in [21], there are well-defined ontologies on both sides. In addition, as we regard property as variable, our model can be applied to any set of properties.

7   Conclusions

In this paper, we propose to perform entity correspondence with second-order Markov Logic, which enables us to support entity correspondence by property relation discovery. As evidenced by the experimental results, these two tasks can strengthen each other when joint inference of Markov Logic is performed. We compare among five different models, extending incrementally from the original model to our models, demonstrating consequent improvements in both precision and recall for entity correspondence. We also compare among three different approaches to illustrate the importance of capturing dependencies among multiple decisions and tasks. In addition, results for property relation discovery show an improvement when entity correspondence is improved.

Acknowledgement. We gratefully acknowledge funding from the National Science Foundation of China under grants 61170165.

References

1. Etzioni, O., Banko, M., Soderland, S., Weld, D.S.: Open Information Extraction from the Web. Communications of the ACM 51(12), 68–74 (2008)

2. Carlson, A., Betteridge, J., Kisiel, B., Hruschka Jr., E.R., Mitchell, T.M.: Toward an Architecture for Never-Ending Language Learning. In: 24th AAAI, vol. 2(4), pp. 1306–1313 (2010)

3. Mrabet, Y., Bennacer, N., Pernelle, N.: Controlled Knowledge Base Enrichment from Web Documents. In: Wang, X.S., Cruz, I., Delis, A., Huang, G. (eds.) WISE 2012. LNCS, vol. 7651, pp. 312–325. Springer, Heidelberg (2012)

4. Kok, S., Domingos, P.: Statistical Predicate Invention. In: 24th Annual International Conference on Machine Learning, pp. 433–440. ACM (2007)

5. Richardson, M., Domingos, P.: Markov Logic Networks. Machine Learning 62(1-2), 107–136 (2006)

6. Koller, D., Friedman, N.: Probabilistic Graphical Models: Principles and Techniques. MIT Press (2009)

7. Davis, J., Domingos, P.: Deep Transfer via Second-order Markov logic. In: 26th Annual International Conference on Machine Learning, pp. 217–224. ACM (2009)

8. Leivant, D.: Higher Order Logic. In: Handbook of Logic in Artificial Intelligence and Logic Programming, pp. 229–321 (1994)

9. McCallum, A., Nigam, K., Ungar, L.H.: Efficient Clustering of High-dimensional Data Sets with Application to Reference Matching. In: 6th ACM SIGKDD, pp. 169–178. ACM (2000)

10. Singla, P., Domingos, P.: Entity Resolution with Markov Logic. In: 6th International Conference on Data Mining, pp. 572–582. IEEE (2006)

11. Manning, C.D., Schütze, H.: Foundations of Statistical Natural Language Processing. MIT Press (1999)

12. Singla, P., Domingos, P.: Discriminative Training of Markov Logic Networks. In: 20th AAAI, vol. 5, pp. 868–873. AAAI Press (2005)

13. Fellegi, I.P., Sunter, A.B.: A Theory for Record Linkage. Journal of the American Statistical Association 64(328), 1183–1210 (1969)

14. Domingos, P.: Multi-Relational Record Linkage. In: Proceedings of the KDD 2004 Workshop on Multi-Relational Data Mining, pp. 31–48 (2004)

15. Brocheler, M., Mihalkova, L., Getoor, L.: Probabilistic Similarity Logic. Technical report, University of Maryland, College Park (2010)

16. Bhattacharya, I., Getoor, L.: Collective Entity Resolution in Relational Data. TKDD 1(1), 1–35 (2007)

17. Whang, S.E., Garcia-Molina, H.: Joint Entity Resolution. In: 28th International Conference on Data Engineering (ICDE). IEEE (2012)

18. Poon, H., Domingos, P.: Joint Inference in Information Extraction. In: Proceedings of the National Conference on Artificial Intelligence (2007)

19. Singh, S., Schultz, K., McCallum, A.: Bi-directional Joint Inference for Entity Resolution and Segmentation using Imperatively-defined Factor Graphs. In: Buntine, W., Grobelnik, M., Mladenić, D., Shawe-Taylor, J. (eds.) ECML PKDD 2009, Part II. LNCS, vol. 5782, pp. 414–429. Springer, Heidelberg (2009)

20. Haas, L.M., Hentschel, M., Kossmann, D., Miller, R.J.: Schema and Data: A Holistic Approach to Mapping, Resolution and Fusion in Information Integration. In: Laender, A.H.F., Castano, S., Dayal, U., Casati, F., de Oliveira, J.P.M. (eds.) ER 2009. LNCS, vol. 5829, pp. 27–40. Springer, Heidelberg (2009)

21. Niepert, M., Noessner, J., Meilicke, C., Stuckenschmidt, H.: Probabilistic-Logical Web Data Integration. In: Polleres, A., d'Amato, C., Arenas, M., Handschuh, S., Kroner, P., Ossowski, S., Patel-Schneider, P. (eds.) Reasoning Web 2011. LNCS, vol. 6848, pp. 504–533. Springer, Heidelberg (2011)

22. Whang, S.E., Garcia-Molina, H.: Entity Resolution with Evolving Rules. Proceedings of the VLDB Endowment 3(1-2), 1326–1337 (2010)

Learning Social Relationship Strength via Matrix Co-Factorization with Multiple Kernels

Youliang Zhong, Lan Du, and Jian Yang

Macquarie University, NSW 2109, Australia
{youliang.zhong,lan.du,jian.yang}@mq.edu.au

Abstract. In recent years the research on measuring relationship strength among the people in a social network has gained attention due to its potential applications of social network analysis. The challenge is how we can learn social relationship strength based on various resources such as user profiles and social interactions. In this paper we propose a *KPMCF* model to learn social relationship strength based on users' latent features inferred from both profile and interaction information. The proposed model takes an uniformed approach of integrating *Matrix Co-Factorization* with *Multiple Kernels*. We conduct experiments on real-world data sets for typical web mining applications, showing that the proposed model produces better relationship strength measurement in comparison with other social factors.

Keywords: Social networks, Relationship strength, Matrix co-factorization, Kernel learning.

1 Introduction

Social relationship strength plays an important role in our life. For instance, people usually tend to use "reliable channel" to transmit sensitive information, whereas the study in social theory indicated that extremely valuable information was often strewed through other people with "weak ties" [6]. Social relationship strength has been applied in a wide range of applications such as friendship identification, group recommendation, community discovering, etc.

Social networks here refer to the networks of people connected through the internet, helping people make friends and share information. An interesting question arises: how can we evaluate the strength of relationships among the people in a social network? If we know how the relationship strength is measured, we can then incorporate it in recommendation, apply it in community discovery. The handle to the relationship strength is twofolds: (1) user profiles; (2) people interactions. Web is an open platform, people's partial profile information becomes visible to some extent, such as location, education, skills and groups. In the meantime, it has become easier than ever for web users to interact with others in virtual space, for example posting comments, setting up tags, following twitters, sharing reviews or photos, and so on.

X. Lin et al. (Eds.): WISE 2013, Part I, LNCS 8180, pp. 15–28, 2013.

It has become quite normal for a user to have a large number of virtual "friends", and so do her/his "friends". The so-called "friends" in virtual space range from actual close friends to casual acquaintances on the web. For instance, [22] reported that some users with Flickr web site had more than 25,000 "friends". As the result, the relationship strengths in social networks have become more complex than those in a face-to-face setting, where the relationship strengths are mostly measured by explicit connections with descriptive forms.

Much work has been done on measuring social relationship strength [13,5,9,23,26,24]. In particular, [13] discussed the quantitative measurement of relationship strength using multiple indicator techniques. [5,9] measured relationship strength using statistics and graph theories. [23,26,24] predicted social relationship strength for online social networks by means of probabilistic matrix factorization, kernel learning and joint probabilities techniques. In these work, either one type of a profile or interaction data was used, or a single method was employed to handle all different forms of data sources. Therefore, restricted users' overall relationship strengths learnt from comprehensive social information.

Several recent studies indicated that users' social interactions affect their behaviors on the web, and such information could improve the accuracy of social network analysis, such as recommendation task [12]. Motivated by these studies, we believe we need an integrated approach to deal with profile and interaction information which are presented in diverse forms. In this paper, we present a new kernelized probabilistic matrix co-factorization model to learn social relationship strength among users. This model simultaneously captures users' profiles and social interactions by coding the interaction information through multiple kernel matrices. Thus, we call it *Kernelized Probabilistic Matrix Co-Factorization* (KPMCF). The main contributions of this paper are summarized as follows:

- We propose a new probabilistic matrix co-factorization model with kernels, and develop a gradient decent algorithm to estimate the maximum likelihood of the factorization. The model incorporates multiple interaction information directly into the factorization process. Therefore, the resulting latent matrices depend on all the input resources.
- We formulate users' social relationship strength in the users' latent feature space that is shared among various data sources, in our case, the user's profile matrices.
- We conduct experiments for *Friend Identification* on real-word datasets extracted from Flickr web site. The preliminary results demonstrate that the leaned relationship strength is superior to other social factors.

The rest of the paper is organized as follows. In section 2 we describe various explicit and implicit information extracted from Flickr web site. Then, we discuss the key techniques of matrix factorization and kernel methods in section 3. Section 4 elaborates our *KPMCF* model and the inference algorithm. Experimental results on Flickr datasets are reported in section 5. The related work is reviewed in section 6, and a brief remark is concluded in last section.

2 User's Profile Characteristics and Interaction Behavior

In this paper we make use of the data from Flickr web site (http://www.flickr. com/), provided by [14] through *Stanford Large Network Dataset Collection* (http:// snap.stanford.edu/data/web-flickr.html). Flickr helps users post and share photos with others, it also encourages users make contact list, join interesting groups, and add comments on photos. The Flickr users have formed extensive social networks that involve a great amount information with regard to users' profile characteristics and interaction behavior.

Throughout the paper we denote matrices by capital-italic font (and may with superscript), for instance M or M^t. An entry of M at row i and column j is denoted as m_{ij} or $(M)_{ij}$. Moreover, $m_{i:}$ stands for the i-th row, whereas $m_{:j}$ for the j-th column of the matrix.

From the original Flickr dataset, we are able to further extract a number of explicit and implicit information. In particular, three matrices are derived as profile-relevant data: *user-country (P^c)*, *user-label (P^b)* and *user-tag (P^t)*, and other four as interaction-relevant ones: *user-friend (X^f)*, *mutual-tag (X^t)*, *mutual-comment (X^c)* and *mutual-group (X^g)*. The details of these matrices are specified as follows.

user-country (P^c) - A user-country matrix P^c associates users with countries, which are converted from the locations in the original dataset. The rows of P^c stands for users, and the columns for countries from which users post photos. As one user may post a number of photos from a same country, we set each entry p^c_{ij} by the number of the photos posted by a user i from a country j.

user-label (P^b) - Labeling in Flickr is essentially a facility of classification provided by image sources, therefore the labels in some degree indicate users' preferences for photos. Of the user-label matrix P^b, the rows stand for users, and the columns for labels. Each entry p^b_{ij} is set to the count of a same label j received by a user i.

user-tag (P^t) - In Flickr, each author is able to assign tags to her/his photos, thus the tags can be read as a sign of users' intention. We calculate a *tag-vector* for each user in a tag space developed by *Bag-of-Words* model [2]. In another word, each row in P^t specifies a user's preferences for tags.

user-friend (X^f) - The user-friend matrix X^f is an interaction matrix, of which both the rows and columns are users. Friends often share many photos, so an entry x^f_{ik} is set by the numbers of the photos posted by both users i and k. Instinctively, the more shared photos, the stronger relationship of the two users.

mutual-tag (X^t) - While the user-friend matrix specifies explicit connections, mutual-tag matrix X^t represents implicit interactions among users by observing how many same tags posted by user-pairs. Each entry of X^t is set by the accumulated count of the same tags posted by the corresponding pair.

mutual-comment (X^c) - Similar to mutual-tag matrix X^t, the mutual-comment matrix X^c also represents the implicit interactions among users. Each entry of X^c is the accumulated count of the comments posted by two corresponding users on same photos.

mutual-group (X^g) - Intuitively, if two users join same groups, then the two users probably have strong social ties. The mutual-group matrix X^g is constructed by specifying the numbers of the same groups in which two users post their photos.

Thus, given the various profile and interaction matrices, how can the users' overall social relationship strength be uncovered without loosing the information buried in them?

In the subsequent sections, we denote the *user-label* matrix P^b as P to represent the profile matrices, and refer $PR = \{P^c, P^b, P^t\}$ to all the profile matrices, n_p to the size of PR. Similarly, we denote the *user-friend* matrix X^f as X to stand for the interaction matrices, and $XR = \{X^f, X^t, X^g, X^c\}$ for all the interaction matrices, n_x for the size of XR.

3   Background

In this section we briefly review the major techniques used in our model: probabilistic matrix factorization and kernelized probabilistic matrix factorization.

3.1   Probabilistic Matrix Factorization

Matrix factorization is to factorize a matrix (called *target matrix*) into a product of two or more matrices that are usually called *latent matrices*. In the case of user-label matrix $P \in \mathbb{R}^{n_u \times n_b}$ given by n_u users with regard to n_b labels, it might be decomposed into two lower-dimension matrices $U \in \mathbb{R}^{n_u \times n_d}$ and $V \in \mathbb{R}^{n_b \times n_d}$ such that $P \approx UV^T$, where n_d is the dimension of the latent vectors. The study of matrix factorization in social network analysis has shown that users' preferences or attitudes (towards *labels*, in this user-label example) can be revealed by the users' latent matrix U. This inspires our interest in modeling social relationship strength by means of matrix factorization method.

In order to handle very large datasets and deal with users having very few ratings, [17] proposed *Probabilistic Matrix Factorization* (PMF) model. A key feature of *PMF* is that, it defines a conditional distribution over the target matrix (P), and sets zero-mean Gaussian priors on latent matrices (U and V):

$$p(P|U, V, \sigma^2) = \prod_{i=1}^{n_u} \prod_{j=1}^{n_b} [\mathcal{N}(p_{ij}|u_i v_j^T, \sigma^2)]^{I(i,j)} \ , \tag{1}$$

$$p(U|\sigma_U^2) = \prod_{i=1}^{n_u} \mathcal{N}(u_i|0, \sigma_U^2 \boldsymbol{I}) \ , \ p(V|\sigma_V^2) = \prod_{j=1}^{n_b} \mathcal{N}(v_j|0, \sigma_V^2 \boldsymbol{I}) \ , \tag{2}$$

where $\mathcal{N}(x|\mu, \sigma^2)$ is a probability density function of Gaussian distribution with mean μ and variance σ^2, and $I(i,j)$ is an indicator function, which equals to 1 if a user i receives a label j, otherwise equals to 0.

Following Bayesian learning theory, the above probabilistic model can be learnt by minimizing a sum-of-squared-error objective function E as follows, with regard to latent matrices U and V.

$$E = \frac{1}{2} \sum_{i=1}^{n_u} \sum_{j=1}^{n_b} I(i,j)(p_{ij} - u_i v_j^T)^2 + \sum_{i=1}^{n_u} \lambda_U \|u_i\|_{Fro}^2 + \sum_{j=1}^{n_b} \lambda_V \|v_j\|_{Fro}^2 . \quad (3)$$

where $\lambda_U = \frac{\sigma^2}{\sigma_U^2}$, $\lambda_V = \frac{\sigma^2}{\sigma_V^2}$, and $\|.\|_{Fro}^2$ denotes Frobenius norm.

3.2 Kernelized Probabilistic Matrix Factorization

Kernel learning [18,8] offers a natural framework to study the relationships between structured objects. The kernels are used to map observations from a limited dimensional space into a much higher or possibly infinite dimensional space without explicitly computing the mapping. In particular, *Regularized Laplacian* kernel [21] is a graph kernel that effectively specify the similarities between the nodes in graphs. Let $X \in \mathbb{R}^{n_u \times n_u}$ stand for the user-friend interaction matrix, then a Regularized Laplacian kernel matrix $K = (k(i,k))_{ik}$ can be used to stipulate the correlation between any pair of two users i and k participating in X, where n_u is the number of the users in X.

Utilizing the kernel methods, some recent research proposed *kernelized probabilistic matrix factorization* to improve recommendation performance [1,25]. Especially, the *KPMF* model proposed in [25] sets kernels to the priors of latent matrices. Furthermore, *KPMF* samples latent matrices in a *column-wise* manner against the general *row-wise* approach. As the result, *KPMF* model shows great promise in exploiting side information in recommendation process.

Provided that we have user-label matrix P (representing user profiles) and user-friend matrix X (representing social interactions) under consideration, *KPMF* model might be used to "guess" the missing values in P, in order to recommend those labels which users may be interested in though yet have experienced. We outline the process of *KPMF* as follows.

(1) Use the interaction matrix X to construct a kernel matrix K_U, specifying the correlation between *users*. Likewise, construct a kernel matrix K_V for *labels*. With the kernels K_U and K_V, use Gaussian Processes [16] to initialize the prior distribution of U and V,

$$\text{Generate } u_{:d} \sim \mathcal{GP}(0, K_U), \ v_{:d} \sim \mathcal{GP}(0, K_V), \ d = 1, ..., n_d , \quad (4)$$

where $u_{:d}$ and $v_{:d}$ stand for the columns of U and V respectively, n_d is the dimensionality of the latent vectors.

(2) Set the likelihood over the target matrix P given U and V, and also assign the priors over U and V as follows.

$$p(P|U,V,\sigma^2) = \prod_{i=1}^{n_u} \prod_{j=1}^{n_b} [\mathcal{N}(p_{ij}|u_{i:}v_{j:}^T, \sigma^2)]^{I(i,j)} , \quad (5)$$

$$p(U|K_U) = \prod_{d=1}^{n_d} \mathcal{GP}(u_{:d}|0, K_U), \ p(V|K_V) = \prod_{d=1}^{n_d} \mathcal{GP}(v_{:d}|0, K_V) . \quad (6)$$

(3) Through Bayesian inference, the log-posterior over U and V can be formulated as follows, in which A is the number of non-missing entries in P, $|K_U|$ and $|K_V|$ the determinant of K_U and K_V, and C an independent constant.

$$log\ p(U,V|P,\sigma^2,K_U,K_V) = -\frac{1}{2\sigma^2}\sum_{i=1}^{n_u}\sum_{j=1}^{n_b}I(i,j)(p_{ij}-u_{i:}v_{j:}^T)^2 \qquad (7)$$

$$-\frac{1}{2}\sum_{d=1}^{n_d}u_{:d}^TK_U^{-1}u_{:d} - \frac{1}{2}\sum_{d=1}^{n_d}v_{:d}^TK_V^{-1}v_{:d}$$

$$-A\ log(\sigma^2) - \frac{n_d}{2}(log(|K_U|)+log(|K_V|)) + C\ ,$$

(4) The latent matrices U and V can be learnt by minimizing the following objective function E, using for instance gradient descent algorithm [25].

$$E = \frac{1}{2\sigma^2}\sum_{i=1}^{n_u}\sum_{j=1}^{n_b}I(i,j)(p_{ij}-u_{i:}v_{j:}^T)^2 \qquad (8)$$

$$+\frac{1}{2}\sum_{d=1}^{n_d}u_{:d}^TK_U^{-1}u_{:d} + \frac{1}{2}\sum_{d=1}^{n_d}v_{:d}^TK_V^{-1}v_{:d}\ .$$

4   Learning Users' Relationship Strength

In this section, we describe the *KPMCF* model by four parts: graphical model, multiple kernels of social interactions, matrix co-factorization over user profiles, and social relationship strength measurement.

4.1   Graphical Model

Figure (Fig. 1) shows the graphical model of *KPMCF*, following the plate notation of graph models [3]. Generally speaking, a plate indicates that the inside variables, indexed by subscripts, are repeated. The times of repetition is denoted at the corner of a plate. When a variable is embedded in two or more plates, that means it involves multiple layers of repetitions, according to the relevant plates.

As shown in the graphical model, the proposed *KPMCF* model is essentially a matrix co-factorization process, with kernels for the priors of latent vectors. In the middle of the picture, p_{ij}^c, p_{ik}^b and p_{is}^t stand for the observed values in profile matrices $\{P^c, P^b, P^t\}$, whereas c_j, b_k, t_s and u_i for the latent vectors of *countries*, *labels*, *tags* and *users* respectively. At the corners of the plates, n_c, n_b, n_t and u_u specify the dimensions of the corresponding latent vectors.

Out of the plates are input parameters. σ_{p^c}, σ_{p^b} and σ_{p^t} represent the variances of Gaussian distributions to be set over the profile matrices. On the other hand, K_C, K_B, K_T and K_U are the kernels for the priors of corresponding latent vectors. Especially, K_U is a combination of multiple kernels derived from interaction matrices: $user-friend\ (X^f)$, $mutual-tag\ (X^t)$, $mutual-group\ (X^g)$ and $mutual-comment\ (X^c)$.

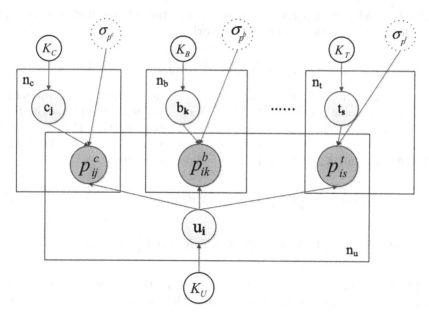

Fig. 1. Graphical model of *KPMCF*

4.2   Multiple Kernels of Social Interactions

To well match the natures of various interaction information, we construct individual kernel K^X for each interaction matrix $X \in \{X^f, X^t, X^g, X^c\}$ as follows, in which $\gamma > 0$ is a constant.

$$K^X = (\mathbf{I} + \gamma \tilde{L}^W)^{-1} \ , \text{ where } \tilde{L}^W = D^{-\frac{1}{2}}(D - X)D^{-\frac{1}{2}}, \text{ and} \qquad (9)$$

$$D \text{ is a degree matrix with } d_{ii} = \sum_{k=1}^{n_u} x_{ik}, \ i = 1, ..., n_u \ ,$$

For the prior distribution of latent matrix U, we make a linear combination K_U of all the kernels K^X. Intuitively, an entry $(K_U)_{ik}$ specifies the overall correlation between two users i and k with regard to all the given interaction information in X^f, X^t, X^g and X^c.

$$K_U = (k(i, k))_{ik}, \ k(i, k) = \frac{1}{n_x} \sum_{X \in \{X^f, X^t, X^g, X^c\}} \alpha^X (K^X)_{ik}, \text{ where} \qquad (10)$$

$$i, k \in \{1, ..., n_u\}, \ n_x = |\{X^f, X^t, X^g, X^c\}|, \text{ and } \sum \alpha^X = 1 \ .$$

Besides the latent matrix U of *users*, factorizing three profile matrices P^c, P^b and P^t will produce other three latent matrices C, B, and T. We denote $LF = \{C, B, T\}$ as the set of these matrices, and L a representative of them. At this moment, we simply set a diagonal kernel $K_L = \sigma_L^2 \mathbf{I}$ for $L \in \{C, B, T\}$, where $\mathbf{I}$ is an identity matrix. We leave the question of how finding matching kernels for profile matrices in our future work.

4.3   Matrix Co-Factoring over User Profiles

With the aforementioned kernels K_U and K_L, we now perform co-factorization over multiple profile matrices P^c, P^b, and P^t. For each profile matrix $P \in \{P^c, P^b, P^t\}$, we set the likelihood over P given latent matrices U of $users$ and L of the corresponding profile characteristic, where $L \in \{C, B, T\}$. We assign the kernels K_U and K_L to the priors of U and L respectively.

$$p(P|U, L, \sigma_P^2) = \prod_{i=1}^{n_u} \prod_{j=1}^{n_l} [\mathcal{N}(p_{ij}|u_{i:}l_{j:}^T, \sigma_P^2)]^{I(i,j)} , \qquad (11)$$

$$p(U|K_U) = \prod_{d=1}^{n_d} \mathcal{GP}(u_{:d}|0, K_U), \ p(L|K_L) = \prod_{d=1}^{n_d} \mathcal{GP}(l_{:d}|0, K_L) . \qquad (12)$$

Apply Bayesian inference to the above distributions, maximizing the log-posterior over U and L is equivalent to minimizing the following object function $\tilde{E}$,

$$\tilde{E} = \sum_{P \in PR}^{P \in \mathbb{R}^{n_u \times n_l}} \frac{1}{2\sigma_P^2} \sum_{i=1}^{n_u} \sum_{j=1}^{n_l} I(i, j)(p_{ij} - u_{i:}l_{j:}^T)^2 \qquad (13)$$

$$+ \frac{1}{2} \sum_{n_p=|PR|} \sum_{d=1}^{n_d} u_{:d}^T K_U^{-1} u_{:d} + \frac{1}{2} \sum_{L \in \mathbb{R}^{n_l \times n_d}} \sum_{d=1}^{n_d} l_{:d}^T K_L^{-1} l_{:d} .$$

A local minimum of the objective function $\tilde{E}$ can be found by executing gradient descent over each row of U and $L \in \{C, B, T\}$ alternatively.

$$\frac{\partial \tilde{E}}{\partial u_{id}} = \sum_{P \in PR}^{P \in \mathbb{R}^{n_u \times n_l}} (\frac{1}{\sigma_P^2} \sum_{j=1}^{n_l} I(i, j)(u_{i:}l_{j:}^T - p_{ij})l_{dj} + \epsilon(u)^T K_U^{-1} u_{:d}) , \qquad (14)$$

$$\frac{\partial \tilde{E}}{\partial l_{jd}} = \sum_{P \in PR}^{P \in \mathbb{R}^{n_u \times n_l}} (\frac{1}{\sigma_P^2} \sum_{i=1}^{n_u} I(i, j)(u_{i:}l_{j:}^T - p_{ij})u_{di} + \epsilon(l)^T K_L^{-1} l_{:d}) , \qquad (15)$$

where $\epsilon(k)$ denotes a k-dimensional unit vector with the k-th component being one and others being zero.

4.4   Measuring Social Relationship Strength by Latent Features

Learnt from the matrix co-factorization, the latent matrix U potentially reveals users' overall attitudes towards the inclusive profile and interaction information. The principle of homophily in social networks suggests that, people tend to build connections with other people having similar characteristics, and the stronger the relationship, the higher the likelihood that more interactions occur between the people [6,15]. Based on these assumptions, we measure the users' social relationship strengths as follows.

Firstly, by applying Pearson correlation coefficient (pcc) function to the latent matrix $U \in \mathbb{R}^{n_u \times n_d}$, we calculate the pair-wise similarities between all the pairs of the users in U. Then, for every pair of two users i and j, we find out the other users who interact with both users, named as "mutual peers" and denoted as M^{ij}. Finally, we add up the similarity between the two users and the similarities between the pair and these "mutual peers" to get the social relationship strength s_{ij} between i and j. Formally,

$$s_{ij} = pcc(i,j) + \sum_{m \in M^{ij}} (pcc(i,m) + pcc(m,j)), \text{ where}$$

$$pcc(x,y) = \frac{\sum_{d=1}^{n_d} (u_{xd} - \bar{u_x})(u_{yd} - \bar{u_y})}{\sqrt{\sum_{d=1}^{n_d} (u_{xd} - \bar{u_x})^2 \sum_{d=1}^{n_d} (u_{yd} - \bar{u_y})^2}} . \tag{16}$$

5 Experiment and Evaluation

In this section, we describe the experiment settings, and then exhibit the evaluation on an experimental web mining application: *Friend Identification*. In the evaluation, we select those user-pairs who participate in all relevant matrices. We use *ROC/AUC* as evaluation metrics.

To conduct our experiments, we extract a subset (Subset-3) from the original Flickr-PASCAL dataset [14]. The subset includes 3 profile matrices (user-country, user-label and user-group) and 4 interaction matrices (user-friend, mutual-tag, mutual-group and mutual-comment). The statistics of the original dataset and Subset-3 are listed in the following table (Table 1).

Table 1. The statistics of original and experimental datasets

Dataset	User	Country	Label	Tag	Friend	Comment	Group
PASCAL	8,698	1,222	20	27,250	—	16,669	6,951
Subset-3	1,324	42	28	1,309	1,324	1,932	562

We process the evaluation as follows. Firstly, we use a combination of users' profile and interaction matrices to compute the *KPMCF* relationship strengths (*KPMCF-strength*) among these users. Next, we calculate the relationship values for other 3 social factors: *Mutual Group*, *Mutual Tag* and *Mutual Comment*. Then, we prepare a test data with the labels of "friend" or "not friend" for all the potential user-pairs involved in the test data, according to the information in $user - friend$ matrix. Finally, we perform a ROC/AUC testing over the *KPMCF-strength* and other three sets of relationship values. The following figure (Fig. 2) shows the ROC (Receiver Operating Characteristic) curves of the testing

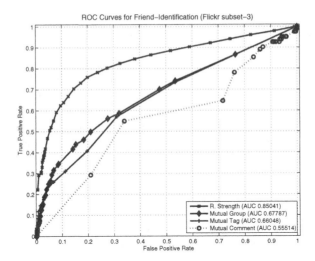

Fig. 2. ROC curves of *KPMCF*, MG, MT and MC

results, where the x- and y-coordinates represent False-Positive-Rate and True-Positive-Rate respectively. The corresponding AUC values are 0.85041 (*KPMCF-strength*), 0.67839 (*Mutual Group*), 0.66508 (*Mutual Tag*) and 0.5553 (*Mutual Comment*) respectively.

As the results demonstrate, the *KPMCF-Strength* shows superior to all other social factors in the case of *Friend-Identification*. In particular, the *KPMCF-Strength* earns the highest AUC value at 0.85, which is about 10% higher than those of *Mutual Group* and *Mutual Tag*, and 30% higher than that of *Mutual Comment*. On the other hand, the *Mutual Comment* relationship gets the lowest AUC value at 0.55. That means, whereas the relationship strength learnt from *KPMCF* would be helpful for identifying friendship between users, the number of the comments commonly posted by user-pairs seems not a meaningful factor in friendship-judgment.

6   Related Work

In this section, we concentrate on two research domains: social relationship strength and kernelized probabilistic matrix factorization, both are closely related to our proposed *KPMCF* model.

6.1   Social Relationship Strength

Social relationship strength is one of the most important research topics in social network analysis, measuring how strong or weak the relationships are among the users in a social network. In the literature, social relationship strength is also referred to *tie strength*. A theory of "The Strength of Weak Ties" was initially introduced in [6,7]. In line with the theory, [13] discussed the quantitative measurement of social relationship strength using multiple indicator techniques and friendship data.

Previous research of social relationship strength mainly focused on face-to-face social networks, where one node may connect tens of other nodes; and the relationship strengths are mostly of descriptive forms such as being friends or not, strong or weak. However, the situation on the web has been significantly changed. Usually an average user has several hundreds of connections, and some individuals may have much more than usual [22]. Because of the easily-established internet connections, some recent efforts have been made to refine the granularity of relationship strength among the users in online social networks [23,26].

When the web grows, it has become even possible to have not only users' interaction data but also users' profile information. This brings greater opportunities of learning social relationship strength. On one hand, numerous studies mainly used interaction data to predict descriptive relationship strength by statistical model [13,5] or graph theory [9]. On the other hand, some recent work exploited profile or interaction information to estimate the relationship strength for online social networks [23,26,24].

In particular, [23] executed matrix factorization over both of users' interaction activities and profile similarities. [26] utilized similarity-based kernels to estimate the relationship strength. Both of the studies used a single method to deal with profile or/and interaction information. Contrastingly, *KPMCF* model applies matrix co-factorization over profile characteristics, while employs kernel technique to capture social interactions. Such an integration approach better serves the different natures of profile characteristics and interaction behavior.

The framework proposed by [24] predicted relationship strength on various activity fields by calculating a joint distribution of *profile strength* and *interaction strength*. The so-called *interaction strength* was determined by a *relatedness* value of target resources (documents). Therefore, the measured relationship strength would be constrained by the application resources. Whereas, our *KPMCF* model defines relationship strength by users' latent features, which are inferred from users' profile and interaction information, regardless whatever the target resources are. This makes *KPMCF* model more universal than others.

6.2   Kernelized Probabilistic Matrix Factorization

Latent variable model or *matrix factorization* has been considered as one of the state-of-the-art techniques of information retrieval [20,10]. Distinctively, *Probabilistic Matrix Factorization (PMF)* [17] used probabilistic approach to model the target matrix and latent vectors such that "scales linearly with the number of observations and performs well on very sparse and imbalanced data sets" [17].

Expanded from single matrix factorization, *matrix co-factorization* was proposed to deal with the situation where some entities participate in more than one matrix [19,11,12,4]. The most common application domains of matrix co-factorization are entity clustering and prediction, based on the assumption that users' preferences are determined by the unobserved latent features. This inspires our interest in learning social relationship strength by means of users' latent features derived from matrix co-factorization.

In recent years, *Kernel learning* [18,8] is becoming increasingly popular, used in various applications such as Bayesian inference, computational biology and link analysis. Especially, some recent research employed kernel methods in matrix factorization to improve the quality of recommendation [1,25]. The *PMA* model in [1] utilized kernels to capture the covariance for rows and columns of a target matrix, then additively combined the kernels to generate prediction matrix. Very similar to but slightly different from *PMA*, the *KPMF* model proposed in [25] assigned kernels to the priors of latent matrices, and inferred prediction matrix from the product of the latent matrices.

In particular, *KPMF* model developed two advanced features on the top of *PMF*. Firstly, *KPMF* sets the prior distribution of latent features by kernels, thus exploits users' side information in matrix factorization. And secondly, *KPMF* samples the latent matrices by a "column-wise" approach instead of the conventional "row-wise" manner. The "column-wise" approach addresses an embarrassing situation in matrix factorization where entire rows or columns are missing in target matrices. Owing to the nature of recommendation, *KPMF* focused on a single rating matrix and a single kernel of side information. Contrastingly, in order to learn users' overall attitudes towards inclusive information in social networks, *KPMCF* model leverages *KPMF* by two important extensions. One is the multiple kernels of various social interactions, and the other is the co-factorization over multiple profile matrices.

The framework *SoRec* proposed in [12] is also closely related to *KPMCF* model. Both are extensions of matrix co-factorization for incorporating social information into factorization process, and both deal with interaction information. In particular, for the purpose of item-recommendation, *SoRec* places Gaussian priors on the latent vectors of users, assuming an independence among the users. In the meantime, *SoRec* introduces a social-related term c_{ik}^* in the conditional distribution over social "factor" matrix. As the result, social information is finally fused in the generated "missing values" for better recommendation. Different from *SoRec*, *KPMCF* aims to learn users' overall attitudes towards inclusive profile characteristics and interaction behavior. In particular, *KPMCF* assigns the kernels of social interactions directly to the priors of users' latent vectors, taking care of the correlations among the users in various interaction graphs. Therefore, *KPMCF* becomes able to measure users' overall social relationship strength from the users' latent features.

7   Conclusion and Future Work

In this paper we present a *KPMCF* model to learn users' relationship strength in a social network. The proposed model integrates *Matrix Co-Factorization* and *Multiple Kernels* techniques in an uniformed framework, simultaneously capturing users' profile characteristics and social interaction behavior.

The proposed *KPMCF* model provides three advantageous features: (1) employing multiple kernels to incorporate users' interactions in relationship strength measurement; (2) performing matrix co-factorization over users' multiple profile matrices to discover the underlying homophily in social networks; (3) learning social relationship strength from users' latent features derived from matrix co-factorization. The experiments have been conducted on the real data sets from *Flickr* web site, showing that the proposed model produces better relationship measurement than using other social factors.

There remain many issues and tasks for our future work. As the *KPMCF* model involves so many parameters that parameter-optimization becomes a key to deliver a higher quality of the relationship strengths. To this end, integrating *KPMCF* with Bayesian inference is under investigation. Although graph kernels are the natural facilities in incorporating social interactions into relationship measurement, how to deal with the latent features of other profile characteristics is still a great challenge. Meanwhile, we are planning to carry on more comprehensive experiments to compare our model with other state-of-the-art matrix co-factorization and kernel learning methods using large-scale datasets in terms of the size of users, profile characteristics and interaction types.

Acknowledgments.   The authors would like to thank all the anonymous reviewers for their valuable comments. The work described in this paper was partially supported by the grants of Australia Macquarie University Research Excellence Scholarship scheme and Australian Research Council's Linkage grant (LP120200231). Lan Du was supported under Australian Research Council's Discovery Projects funding scheme (project numbers DP110102506 and DP110102593).

References

1. Agovic, A., Banerjee, A., Chatterjee, S.: Probabilistic matrix addition. In: Proceedings of the Twenty-Eighth International Conference on Machine Learning (2011)
2. Baeza-Yates, R., Ribeiro-Neto, B., et al.: Modern information retrieval, vol. 463. ACM Press, New York (1999)
3. Buntine, W.L.: Operations for learning with graphical models. arXiv preprint cs/9412102 (1994)
4. Fang, Y., Si, L.: Matrix co-factorization for recommendation with rich side information and implicit feedback. In: Proceedings of the 2nd International Workshop on Information Heterogeneity and Fusion in Recommender Systems, pp. 65–69. ACM (2011)
5. Gilbert, E., Karahalios, K.: Predicting tie strength with social media. In: Proceedings of the SIGCHI Conference on Human Factors in Computing Systems, pp. 211–220. ACM (2009)
6. Granovetter, M.: The strength of weak ties. American Journal of Sociology, 1360–1380 (1973)
7. Granovetter, M.: The strength of weak ties: A network theory revisited. Sociological Theory 1(1), 201–233 (1983)
8. Hofmann, T., Schölkopf, B., Smola, A.J.: Kernel methods in machine learning. The Annals of Statistics, 1171–1220 (2008)

9. Kahanda, I., Neville, J.: Using transactional information to predict link strength in online social networks. In: Proceedings of the Third International Conference on Weblogs and Social Media, ICWSM (2009)
10. Koren, Y., Bell, R., Volinsky, C.: Matrix factorization techniques for recommender systems. Computer 42(8), 30–37 (2009)
11. Lippert, C., Weber, S.H., Huang, Y., Tresp, V., Schubert, M., Kriegel, H.P.: Relation prediction in multi-relational domains using matrix factorization. In: Proceedings of the NIPS 2008 Workshop: Structured Input-Structured Output, Vancouver, Canada (2008)
12. Ma, H., Yang, H., Lyu, M.R., King, I.: Sorec: Social recommendation using probabilistic matrix factorization. In: Proceeding of CIKM 2008, pp. 931–940. ACM (2008)
13. Marsden, P., Campbell, K.: Measuring tie strength. Social Forces 63(2), 482–501 (1984)
14. McAuley, J., Leskovec, J.: Image labeling on a network: Using social-network metadata for image classification. In: Fitzgibbon, A., Lazebnik, S., Perona, P., Sato, Y., Schmid, C. (eds.) ECCV 2012, Part IV. LNCS, vol. 7575, pp. 828–841. Springer, Heidelberg (2012)
15. McPherson, M., Smith-Lovin, L., Cook, J.M.: Birds of a feather: Homophily in social networks. Annual Review of Sociology, 415–444 (2001)
16. Rasmussen, C., Williams, C.: Gaussian processes for machine learning, vol. 1. MIT Press, Cambridge (2006)
17. Salakhutdinov, R., Mnih, A.: Probabilistic matrix factorization. In: Advances in Neural Information Processing Systems, vol. 20, pp. 1257–1264 (2008)
18. Scholkopf, B., Smola, A.: Learning with kernels: support vector machines, regularization, optimization, and beyond, vol. 1. MIT Press, Cambridge (2002)
19. Singh, A.P., Gordon, G.J.: Relational learning via collective matrix factorization. In: Proceeding of the 14th ACM SIGKDD International Conference on Knowledge Discovery and Data Mining, pp. 650–658. ACM (2008)
20. Singh, A.P., Gordon, G.J.: A unified view of matrix factorization models. In: Daelemans, W., Goethals, B., Morik, K. (eds.) ECML PKDD 2008, Part II. LNCS (LNAI), vol. 5212, pp. 358–373. Springer, Heidelberg (2008)
21. Smola, A.J., Kondor, R.: Kernels and regularization on graphs. In: Schölkopf, B., Warmuth, M.K. (eds.) COLT/Kernel 2003. LNCS (LNAI), vol. 2777, pp. 144–158. Springer, Heidelberg (2003)
22. Van House, N.A.: Flickr and public image-sharing: distant closeness and photo exhibition. In: Proceedings of CHI 2007, pp. 2717–2722. ACM (2007)
23. Xiang, R., Neville, J., Rogati, M.: Modeling relationship strength in online social networks. In: Proceedings of the 19th International Conference on World Wide Web, pp. 981–990. ACM (2010)
24. Zhao, X., Li, G., Yuan, J., Chen, X., Li, Z.: Relationship strength estimation for online social networks with the study on facebook. Neurocomputing (2012)
25. Zhou, T., Shan, H., Banerjee, A., Sapiro, G.: Kernelized probabilistic matrix factorization: Exploiting graphs and side information. In: Proceedings of SDM 2012, pp. 403–414. SIAM (2012)
26. Zhuang, J., Mei, T., Hoi, S., Hua, X., Li, S.: Modeling social strength in social media community via kernel-based learning. In: Proceedings of the 19th ACM International Conference on Multimedia, pp. 113–122. ACM (2011)

NEXIR: A Novel Web Extraction Rule Language toward a Three-Stage Web Data Extraction Model

Shengsheng Shi, Wu Wei, Yulong Liu, Haitao Wang,
Lei Luo, Chunfeng Yuan, and Yihua Huang

Department of Computer Science and Technology, Nanjing University
National Key Laboratory for Novel Software Technology, Nanjing University
Nanjing 210023, China
{s_sh_sheng,liuyulong_001}@126.com,
{weiwunju,whtacm,luolei.nju}@gmail.com,
{cfyuan,yhuang}@nju.edu.cn

Abstract. As the most popular information publishing platform, the Web contains a lot of valued data information of interests to users or applications. Nowadays, although a lot of data mining or analysis techniques have been studied in last decade, there are still not many easy-to-use web data mining tools available for users to extract useful data information from the Web. The web information extraction is a whole process involving web page navigation, data extraction and data integration. Unfortunately most of existing studies or systems lack of sufficient consideration toward the three-stage process. Also most of them lack the powerful rules to express the flexible extraction logic to extract data records with complicate structure. In this paper, we propose a novel web data extraction language, NEXIR, toward a three-stage web data extraction model. First of all, the language can define rules for system to automate the navigation process of the web pages, including deep web pages that need interactions from users. Then the language allows users to define flexible and complicated rules to extract data records from web pages and integrate extracted data into a pre-defined structure. A language engine and a prototype extraction system have been implemented based on the proposed language. The experimental results show that our language and system work effective and powerful compared with existing data extraction approaches.

Keywords: Web data extraction, Extraction Rule language, Data record, Web page navigation, Web data integration.

1 Introduction

The Web has become the most popular information publishing platform with a lot of valued data information contained in the huge number of web pages. To acquire the useful data information of interests to users or applications from the Web, many data mining or analysis techniques have been studied in last decade. However, there are still not many easy-to-use web data mining tools available for users to extract useful data information from the Web. The major reason is that the web information

X. Lin et al. (Eds.): WISE 2013, Part I, LNCS 8180, pp. 29–42, 2013.

extraction is a three-stage whole process that involves web page navigation, data extraction and data integration but most of existing studies or systems only focused on how to extract data from existing web pages without the ability to perform the whole process of web data extraction.

Before extracting data from web pages, the first question is how the system can automate the web page navigation process just like a user's browsing to find web pages of interests. Today, web applications usually use sessions to maintain certain status or transactions across web pages or whole web application. Thus directly jumping into a web page without step-by-step web page navigation will not acquire valid web pages. Also most of web applications today come in the form of deep web pages. Thus instead of simply clicking a hyperlink or button to jump into next page, a user may need to perform more complicated interactions with web pages (e.g., type in a search keyword or check a checkbox). For above reasons, we need to provide a powerful rule language to automate the web page navigation process.

After we acquire raw data items from web pages, we have to integrate them into a well-structured data entity by performing certain transformation and mapping process. This integration stage is especially important when the data items of a data entity are scattered on multiple web pages.

In addition, to describe complicated navigation, extraction and integration logic, a powerful extraction rule language is needed. However, most existing web data extraction studies or systems used simple data extraction rules and lack the powerful rules to express the flexible logic to extract data records with complicate structure.

To deal with the limitations of existing studies or systems, this paper proposes a three-stage web data extraction model and a novel web data extraction language, NEXIR (Navigation, EXtraction and Integration Rules). The proposed language can put together web page navigation, data extraction and data integration to describe the whole process of web data extraction. It can also define powerful data extraction rules to express complicated extraction logic to deal with a variety of data records in web pages.

The experimental results from extracting typical web page examples show that our language and system work effective and powerful compared with existing data extraction approaches.

The rest of the paper is organized as follows. Section 2 reviews related work. Section 3 describes the models related to web data extraction. Section 4 presents the design of our language. Section 5 gives the implementation of our language. Section 6 concludes the paper and outlines future work.

2   Related Work

Many web data extraction systems or approaches have been developed in recent decades. Most typical approaches are surveyed in [1][2][3]. According to the techniques used, existing web data extraction approaches can be mainly divided into three categories: language-based approaches, wrapper induction approaches and pattern mining approaches.

Language-based approaches: Such approaches provide special-designed languages which are used to define data extraction rules. Typical approaches of this category include   TSIMMIS[4], Minerva[5], WebOQL[6], Lixto[7][8], Wargo[9], OX-Path[10]. Among these language-based approaches, only Lixto, Wargo and OXPath support the three stage processing introduced in section 1. However, these three approaches lack the ability to extract data records with complicate structures (the column-based and grid-based data records which will be introduced in the paper) or with different numbers of DOM nodes.

Wrapper induction approaches: Such approaches try to learn data extraction rules from one or more labeled web pages. To use such approaches, the users need to label the data of interest on one or more web pages. Typical approaches of this category include SRV[11], RAPIER[12], WIEN[13], Softmealy[14], Stalker[15], DEByE[16], Vertex[17] etc.

Pattern mining approaches: Such approaches try to generate extraction rules automatically by mining patterns from one or more web pages. In recent years, many studies focused on such approaches because such approaches can reduce human labor in the largest extent. Typical approaches of this category include RoadRunner[18], EXALG[19], FiVaTech[20], IEPAD[21], DeLa[22], MDR[23], NET[24], ViNTs[25], MSE[26], DEPTA[27], G-STM[28], ViDE[29], CTVS[30] etc.

Although the wrapper induction and the pattern mining approaches can reduce human labor in large extent, they only focus on the second stage of the three stages introduced in section 1. Furthermore, these approaches generate rules without human intervention; thus the users do not have the chance to help generate more accurate extraction rules.

This paper proposes an approach of rule language toward three-stage web data extraction model and can define complicate rules to extract data records with complex structures.

3   Models Related to Web Data Extraction

3.1   Three-Stage Web Data Extraction Model

Most of existing studies or systems focused on how to automatically analyze the web pages and generate the extraction rules to complete the extraction tasks. However, a complete web data extraction process involves three stages from web page navigation, data extraction to data integration. Thus, we propose a three-stage model for a complete web data extraction process as shown in Fig. 1.

Web page navigation is an automated process during web data extraction process to simulate the browsing behaviors of users in a browser to reach and acquire the web pages containing data of interest to users. It will allow system to start from an initial page on a web site and approach the required web pages through a series of navigation actions on web pages.

Fig. 1. Three-stage web data extraction model

Data extraction stage extracts raw data from web pages acquired from the first stage. The data of interest on a web page usually appear in a data region that contains one or more data records each of which contains a set of data items. We will define hierarchical rules to extract data regions, data records and data items hierarchically.

Data integration stage integrates the data extracted in the second stage into a predefined target data structure. It needs to maintain the correct relationships between raw data extracted from different web pages.

3.2   Web Page Model

To describe the logic of web navigation, data extraction and integration, we define a page model which holds rules for web navigation, data extraction and data integration. We will define a page model for each accessed web page.

3.3   Navigation Link Model

To simulate user's interaction actions on web pages to browse web pages, a web data extraction system needs to provide mechanism to record user's interaction actions on sample web pages. Then, these recorded actions will be replayed and executed at runtime in the browser to reach valid web pages. Here, the objects associated with interaction actions are called the controls of the web page, and each interaction action is called a control action.

To describe the link relationships among web pages as   **Fig. 2.** Navigation link
well as web page models, we define a navigation link model.

A navigation link consists of a sequence of control actions. When performing web page navigation process, system will execute the actions within the navigation link to jump to a new web page. In addition, a navigation link also contains a target page model that will tell the system which page model will be applied to the new web page. Fig. 2 shows the model of navigation link.

3.4   Web Data Record Model

Data records might be displayed on a web page in various forms. In terms of the regularity of the structure of web data records, we divide web data records into two categories: regular data records and irregular data records (see Fig. 3).

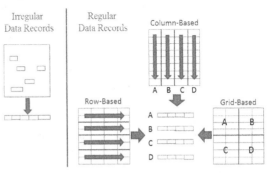

Fig. 3. Web data record model

Irregular Data Records

Data items in an irregular data record are often scattered on a web page randomly. Thus usually we cannot extract the data items of the data record in an regular area of the web page. For such data records, we usually construct extraction rules based on user interaction.

Regular Data Records

Regular data records are typically displayed on a web page in a regular form. Such data records may have several variations. In terms of presentation format, regular data records can be divided into three categories: the row-based data records, the column-based data records and the grid-based data records.

The row-based data records will have the same abstract structure as their underlying DOM tree structure. This is a simplest form that can be automatically analyzed and extracted by most of automated extraction systems. The column-based data records will be vertically displayed on their DOM tree, which will increase the difficulty for automated extraction system to analyze and extract data records from a web page due to the fact that a record will consist of a set of non-sibling DOM nodes at the same DOM tree level. The most difficult one is the grid-based form that brings much more difficulty for automated extraction system to analyze and identify the data records. The varied forms of web data records also bring more issues for expressing extraction rules. To provide enough information and logic to guide system to correctly identify and extract both irregular and regular data records, we need to introduce enough attributes and patterns to define logic for different forms on extraction rules.

The language proposed in this paper can describe extraction rules for extracting irregular data records as well as all of three categories of regular data records.

3.5   Source Data Object Model

To perform web data extraction, first we try to extract source data objects presented on web pages; then we try to integrate the data contained in the source data objects into the predefined target data structure. For the first step, we propose a hierarchical source data object model that involves three types of source data objects: data region object, data record objects and data item objects. Each data region object contains one or more data record objects and each data record object contains one or more data item objects.

According to this model, we need to define extraction rules for all source data objects in different levels.

3.6   Data Integration Model

The structure of source data objects may not be consistent with the target structure. Thus, data integration process is employed to transform the data item objects into target data items and then map the transformed data items into the target data structure. For this purpose, we propose an ETI (Extraction, Transformation, and Integration) data integration model as shown in Fig. 4. The Source Data Object Extraction

Fig. 4. Data integration model

module will extract source data objects from the web pages. The Data Transformation module will transform the data item objects according to transformation rules written in a script language. The Data Integration Module maps the transformed data items into target data records each of which possesses the target data structure.

4   Rule Language Design

4.1   Page Model

The page model is a holder to describe all logic of web page navigation, data extraction and data integration. The basic structure of a page model is shown in Fig. 5. Basically a page model consists of source data objects, navigation link objects, part or whole of a target data object (as a target data record may come from more than one page). In addition, it also contains workflow scripts to control all extraction process and data integration scripts to transform and map extracted data to target data object represented in XML.

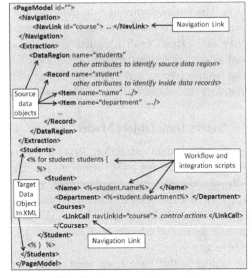

Fig. 5. A page model example

4.2   Navigation Rules

The navigation rules for a navigation link consist of two parts. The first part declares the navigation link by defining the html controls and the target page model(s) for the web page to be linked. The second part defines the link call logic that consists of the control actions to execute the navigation link. Fig. 6 shows the two parts.

In Fig. 6(1), the "navLinkId" attribute specifies the ID of a <NavLink> element. The "controlId" attribute specifies the ID of the control. The "type" attribute specifies the control type such as text box, button, seclect, and anchor. The <TargetModels>

element contains one or more <TargetModel> elements that specify candidate page model(s) for the web page to be linked.

In Fig. 6(2), the "navLinkId" attribute specifies the ID of the corresponding <Nav-Link> element to be called. Each control action is specified as a script defined between "<%" and "%>".

```
<NavLink   navLinkId ="">
   <Controls>   <Control     controlId="" type=""/> ... <Controls>
   <TargetModels>   <TargetModel   targetModelID=""/> ...   </TargetModels>
</NavLink>
```
(1) First part of navigation rule

```
<LinkCall navLinkId= ""> <% control action %> ... <% control action %> </LinkCall>
```
(2) Second part of navigation rule

Fig. 6. Navigation Rule

Table 1 lists different types of controls and their typical actions.

Table 1. Various types of controls and their typical actions

Control Type	Typical Actions
location bar	set URL, navigate with URL
textbox	set text
button	click
select	select an option
checkbox/radiobox	check/uncheck
anchor/other	click/mouseover/mouseout

4.3   Source Data Objects and Their Extraction Rules

The first level source data object is a web data region element named <DataRegion> to specify a data region that contains one or more data record elements <Record> on a web page. Each <Record> will further contain a set of data item elements <Item>. The basic description of a web data region along with its inside data records and data items is illustrated in Fig. 5. Usually a web data region can be used to wrap a set of regular data records that can be automatically analyzed and extracted by an automated data extraction process.

For system to correctly identify different source data objects from web pages, we need to adopt the extraction attributes and patterns to define complex extraction logic for this purpose. Extraction attributes can be divided into two categories: simple attributions and complex attributes.

For simple attributes, the "name" and "type" attributes are used to assign a name and type for a source data objects. For example, the web data region in Fig. 5 is assigned the name "students" and a set of student data records inside the data region can be assigned the name "student". The "type" attribute in a data region will be assigned to one of "row-based", "column-based", or "grid-based", to tell system which type of data records the data region contains and guide the system to perform associated extraction processing.

The complex attributes will work with patterns to define more complicated extraction logic. A pattern is used to define a set of attributes and one or more DOM node features to achieve specific function, such as locating, filtering and splitting the DOM tree of source data objects. The XPaths are used in a pattern and its feature(s). We synthesize the structural, content and vision-based features to constitute an XPath-based feature expression. The structural features can be a tag name, tag path, id or a class attribute that appears on the DOM tree structure. The content attribute can involve a featured text that occurs within a DOM node. The vision-based feature can be the color, font, coordinate attributes that show visual effects on a web page. Each of these features is defined as an XPath expression to choose a set of nodes. These XPath expressions will be merged if we want to combine two or more features.

4.4   Integration Rules

Our integration rules are expressed based on XML and script language. The XML is used to define the target data structure. The script language is used to define the logic to control the workflow and perform data transformation for data integration. The script language can access the data objects extracted by the extraction rules and perform fine-grained data transformation processing. All scripts are specified between "<%" and "%>".

To integrate a complicated data record that spans multiple web pages, we employ the <LinkCall> elements to link these web pages to maintain right data relationships for the data record. To do this, we will place each <LinkCall> element into suitable location among the integration rules to tell system where to insert the data extracted from linked pages.

5   Implementation of Extraction Rule Language

5.1   Execution Process of Extraction Rules

Fig. 7 illustrates the execution process of our extraction rules based on three-stage WIE model.

At the build-time, system will record all navigation actions to generate navigation logic when a user browses sample web pages. At the same time, for web pages with irregular data records, system will allow to generate extraction rules based on user's interaction. For web pages with regular data records, system will allow to adopt automated web page analysis process to generate extraction rules. Users can further add flexible scripts and integration logic into page models. All generated page models will be pre-compiled and converted to execution classes in Java language. As this paper focuses on the extraction rule language and is limited by its length, how to record navigation logic and how to generate rules based on user interaction and automated analysis will be discussed in another paper.

At the runtime, system will start the entire execution task from a start web page. By running each web page's execution class, all extraction and integration logic on the web page will be executed. By following the navigation logic defined on the page model, all subsequently linked web pages will be acquired and corresponding page models will be executed. When all execution tasks are complete, a set of target data records will be output.

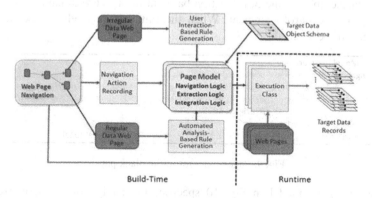

Fig. 7. Extraction process of our extraction rules based on three-stage WIE model

5.2   Extraction Rule Example and Extraction Results

Fig. 8 shows a comprehensive data extraction example with three linked web pages. The task of this extraction example is to search a particular gene name and then extract a list of literatures associated with the searched gene name. The first web page is a search page from "www.pubmed.com" website. The second web page contains a search result list returned from searching the gene name "cre". The third web page is a detail web page to describe one of listed literatures.

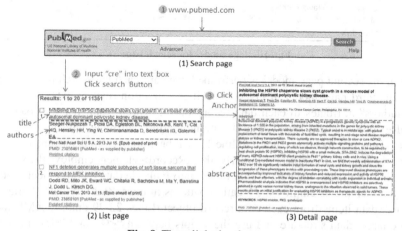

Fig. 8. Three linked web pages

In this example, we want to extract the "title" and "authors" data items in each data record on the search result list web page. In addition, we want to extract the "abstract" data item on the detail web page for each literature. To perform this task, we construct four page models as shown in Fig. 9-12. In each page model, the rules of navigation, extraction and integration are shown with different colors.

The initial page model in Fig. 9 defines navigation rules to navigate to "www.pubmed.com". In line 3, a location bar control object with name "location" is defined by the <Control> element. When the initial page model is executed, the search page will be reached.

```
1.   <PageModel id="startpage">
2.   <Navigation>   <NavLink id="urlconnection">
3.                  <Controls> <Control name="location", type="locationbar"/> </Controls>
4.                  <TargetModels> <TargetModel targetModelId="searchpage"> </TargetModels>
5.               </NavLink> </Navigation>
6.   <LinkCall navLinkId="urlconnection">
7.        <% location.navigate("www.pubmed.com"); %> </LinkCall>   </PageModel>
```

Fig. 9. Page model for initial blank page

The second page model in Fig. 10 specifies the navigation rules of executing search process. It specifies two control objects in lines 4-5. The first one is the text box control named as "textbox" while the second one is the button control named as "searchbtn". Line 7 indicates that the target page model is the one in Fig. 11 for processing search result list web page. In line 9, the text box action and button action are specified between "<%" and "%>".

```
1.   <PageModel id="searchpage">
2.     <Navigation>
3.       <NavLink id="formsubmit">
4.         <Controls> <Control name="textbox", type="textbox" xpath="//input[@id='term']" />
5.                    <Control name="searchbtn" type="button"   xpath="//button[@id='search']"/>
6.         </Controls>
7.         <TargetModels> <TargetModel targetModelId="listpage"> </TargetModels>
8.       </NavLink> </Navigation>
9.       <LinkCall navLinkId="formsubmit"> <% textbox.setText("cre"); searchbtn.click(); %> </LinkCall>
10.  </PageModel>
```

Fig. 10. Page model for search page

When the second page model is performed, we can reach the search result list web page. The third page model in Fig. 11 specifies extraction rules to extract the "title" and "authors" attributes in each data record on the search result list web page. The <DataRegion> element specifies a "locationPatternId" attribute with value "region-Parent". Thus, the location pattern in line 17 will be employed to determine the parent node of the data region object. Similarly, the <Record> element employs the split pattern in line 18 to split the data region object into multiple data record objects and the two <Item> elements employ the location patterns in lines 19-20 to locate "title" and "authors" data items from each data record object. Lines 22-29 employ XML to define the target data structure and use the scripts between "<%" and "%>" to control the loop for extracting a list of literature records. Lines 26-27 specify a <LinkCall>

element that will be executed to navigate to the detail web page. In line 27, the codes between "<%" and "%>" specify two actions. The first action specifies the real XPath of the dynamic anchor control for currently accessed literature record object; the second action is to click the anchor. For each literature data record object, its "title" data item object corresponds to dynamic anchor control and thus the XPath of the anchor control will be set as the XPath of the "title" data item object.

The page model in Fig. 12 will be employed to process each detail web page. The "abstract" data item extracted from each detail web page will be directly put into the

```
1.   <PageModel id="listpage">
2.     <Navigation>
3.       <NavLink id="anchorLink">
4.         <Controls>
5.           <Control name="anchor", type="anchor" xpath="//p[@class='title'][position()=1]/a"/>
6.         </Controls>
7.         <TargetModels>  <TargetModel targetModelId="detailpage"/>  </TargetModels>
8.       </NavLink>  </Navigation>
9.     <Extraction>
10.       <DataRegion name="geneliteratures" type="row-based" auto="false"
11.                locationPatternId="regionParent">
12.         <Record name="geneliterature" splitPatternId="record">
13.           <Item name="title" locationPatternId="titlePtn"/>
14.           <Item name="authors" locationPatternId="authorsPtn"/> </Record>
15.       </DataRegion>
16.       <Patterns>
17.         <Pattern id="regionParent" type="location" xpath="//div[./div[@class='rprt']]" />
18.         <Pattern id="record" type="split" method="head"> <Feature presentXPath="div"/>
19.         <Pattern id="titlePtn" type="location" xpath="//p[@class='title']/a" />
20.         <Pattern id="authorsPtn" type="location"  xpath="//p[@class='desc']"/> </Patterns>
21.     </Extraction>
22.   <GeneLiteratures>
23.     <% for (geneliterature : generatureLiteratures)  {%>
24.         <GeneLiterature> <Title> <%=geneLiterature.title%> </Title>
25.                  <Authors> <%=geneLiterature.authors%> </Authors>
26.           <LinkCall navLinkId="anchorLink">
27.             <% anchor.setXPath(geneLiterature.title.getXPath());   anchor.click(); %> </LinkCall>
28.         </GeneLiterature>      <%}%>
29.   </GeneLiteratures> </PageModel>
```

Fig. 11. Page model for search result list page

```
1.   <PageModel id="detailpage">
2.     <Extraction>
3.       <DataRegion name="detailregion" type="irregular" auto="false">
4.         <Record name="detailrecord">
5.           <Item name="abstract" locationPatternId="abstractPtn"/> </Record>  </DataRegion>
6.       <Patterns>
7.         <Pattern id="abstractPtn" type="location" xpath="//div[@class='abstr']/div" /> </Patterns>
8.     </Extraction>
9.   <Abstract>  <%=detailregion.detailrecord.abstract%>  </Abstract>
10. </PageModel>
```

Fig. 12. Page model for detail page

place where the <LinkCall> element resides. In this way, we can integrate the data from the detail web page with one of the literature records.

After completing the execution of the four page models, the system will output the integrated literature data in XML format. Fig. 13 shows the outline of the output, where "…" denotes the omitted texts or <GeneLiterature> elements.

```
<GeneLiteratures>
    <GeneLiterature> <Title>Inhibiting the … </Title>  <Authors> Seeger-Nukpezah T, …</Authors>
                    <Abstract>Autosomal dominant</Abstract> </GeneLiterature>
    <GeneLiterature> <Title>NF1 deletion … </Title>  <Authors>Dodd RD, …</Authors>
                    <Abstract>Soft-tissue sarcomas  …</Abstract> </GeneLiterature>
    …
</GeneLiterature>
```

Fig. 13. The outline of the result output

6   Conclusions and Future Work

In this paper, we first proposed a three-stage web data extraction model. Based on the model, we proposed a novel web data extraction language, NEXIR, to describe the extraction logic for whole process of web data extraction. To the best of our knowledge, most of existing approaches only considered the second stage. To be able to extract a variety of complicated data records on web pages, the proposed rule language adopts a variety of extraction attributes and patterns to allow users to define powerful extraction rules to extract the row-based, column-based, and grid-based data records. To the best of our knowledge, few existing web data extraction languages can define such powerful rules to extract the column-based or grid-based data records. Furthermore, when regular data records consist of different numbers of DOM nodes, existing approaches can hardly define rules to extract such data records. Our language can define rules to extract all these types of data records. Thus, compared with existing approaches, the approach and language proposed in this paper work more effectively and powerfully. To facilitate the generation of web data extraction rules, we have implemented a visual web data extraction system which provides users with GUI. Users do not need to be familiar with the language and can generate rules easily by interacting with the GUI.

In the future, we will study how to combine the user interaction approach with automated pattern mining approach to reduce human labor and generate web data extraction rules more efficiently.

Acknowledgements. The research work of this paper is sponsored by China NSF funding (#61072152) and Jiangsu Province Industry Support Program (#BE2011172).

References

1. Laender, A.H.F., Ribeiro-Neto, B.A., da Silva, A.S., Teixeira, J.S.: A Brief Survey of Web Data Extraction Tools. SIGMOD Record 31(2), 84–93 (2002)
2. Chang, C.-H., Kayed, M., Girgis, M.R., Shaalan, K.F.: A Survey of Web Information Extraction Systems. IEEE Transactions on Knowledge and Data Engineering 18(10), 1411–1428 (2006)
3. Sleiman, H., Corchuelo, R.: A Survey on Region Extractors from Web Documents. IEEE Transactions on Knowledge and Data Engineering PP(99) (2012)
4. Hammer, J., McHugh, J., Garcia-Molina, H.: Semistructured Data: The TSIMMIS Experience. In: Proceedings of the First East-European Conference on Advances in Databases and Information Systems, pp. 1–8 (1997)
5. Crescenzi, V., Mecca, G.: Grammars Have Exceptions. Information Systems 23(8), 539–565 (1998)
6. Arocena, G.O., Mendelzon, A.O.: WebOQL: Restructuring Documents, Databases and Webs. In: Proceedings of the 14th International Conference on Data Engineering, pp. 24–33 (1998)
7. Baumgartner, R., Flesca, S., Gottlob, G.: Visual Web Information Extraction with Lixto. In: Proceedings of the 27th International Conference on Very Large Data Bases, pp. 119–128 (2001)
8. Baumgartner, R., Gottlob, G., Herzog, M.: Scalable Web Data Extraction for Online Market Intelligence. Proceedings of the VLDB Endowment 2(2), 1512–1523 (2009)
9. Raposo, J., Pan, A., Álvarez, M., Hidalgo, J., Viña, A.: The Wargo System: Semi-Automatic Wrapper Generation in Presence of Complex Data Access Modes. In: Proceedings of the 13th International Workshop on Database and Expert Systems Applications, pp. 313–317 (2002)
10. Furche, T., Gottlob, G., Grasso, G., Schallhart, C., Sellers, A.: OXPath: A Language for Scalable Data Extraction, Automation, and Crawling on the Deep Web. The VLDB Journal 22(1), 47–72 (2013)
11. Freitag, D.: Information Extraction from HTML: Application of a General Machine Learning Approach. In: Proceedings of the 15th National Conference on Artificial Intelligence, pp. 517–523 (1998)
12. Califf, M.E., Mooney, R.J.: Relational Learning of Pattern-Match Rules for Information Extraction. In: Proceedings of the 16th National Conference on Artificial Intelligence, pp. 328–334 (1999)
13. Kushmerick, N.: Wrapper Induction: Efficiency and Expressiveness. Artificial Intelligence 118(1-2), 15–68 (2000)
14. Hsu, C.-N., Dung, M.-T.: Generating Finite-State Transducers for Semi-Structured Data Extraction from the Web. Information Systems 23(8), 521–538 (1998)
15. Muslea, I., Minton, S., Knoblock, C.A.: Hierarchical Wrapper Induction for Semistructured Information Sources. Autonomous Agents and Multi-Agent Systems 4(1-2), 93–114 (2001)
16. Laender, A.H.F., Ribeiro-Neto, B., da Silva, A.S.: DEByE – Data Extraction By Example. Data & Knowledge Engineering 40(2), 121–154 (2002)
17. Gulhane, P., et al.: Web-Scale Information Extraction with Vertex. In: Proceedings of the 27th International Conference on Data Engineering, pp. 1209–1220 (2011)
18. Crescenzi, V., Mecca, G., Merialdo, P.: RoadRunner: Towards Automatic Data Extraction from Large Web Sites. In: Proceedings of the 27th International Conference on Very Large Data Bases, pp. 109–118 (2001)

19. Arasu, A., Garcia-Molina, H.: Extracting Structured Data from Web Pages. In: Proceedings of the ACM SIGMOD International Conference on Management of Data, pp. 337–348 (2003)
20. Kayed, M., Chang, C.-H.: FiVaTech: Page-Level Web Data Extraction from Template Pages. IEEE Transactions on Knowledge and Data Engineering 22(2), 249–263 (2010)
21. Chang, C.-H., Lui, S.-C.: IEPAD: Information Extraction Based on Pattern Discovery. In: Proceedings of the 10th International Conference on World Wide Web, pp. 681–688 (2001)
22. Wang, J., Lochovsky, F.H.: Data Extraction and Label Assignment for Web Databases. In: Proceedings of the 12th International Conference on World Wide Web, pp. 187–196 (2003)
23. Liu, B., Grossman, R., Zhai, Y.: Mining Data Records in Web Pages. In: Proceedings of the 9th ACM SIGKDD International Conference on Knowledge Discovery and Data Mining, pp. 601–606 (2003)
24. Liu, B., Zhai, Y.: NET – A System for Extracting Web Data from Flat and Nested Data Records. In: Ngu, A.H.H., Kitsuregawa, M., Neuhold, E.J., Chung, J.-Y., Sheng, Q.Z. (eds.) WISE 2005. LNCS, vol. 3806, pp. 487–495. Springer, Heidelberg (2005)
25. Zhao, H., Meng, W., Wu, Z., Raghavan, V., Yu, C.: Fully Automatic Wrapper Generation for Search Engines. In: Proceedings of the 14th International Conference on World Wide Web, pp. 66–75 (2005)
26. Zhao, H., Meng, W., Yu, C.: Automatic Extraction of Dynamic Record Sections from Search Engine Result Pages. In: Proceedings of the 32nd International Conference on Very Large Data Bases, pp. 989–1000 (2006)
27. Zhai, Y., Liu, B.: Structured Data Extraction from the Web Based on Partial Tree Alignment. IEEE Transactions on Knowledge and Data Engineering 18(12), 1614–1628 (2006)
28. Jindal, N., Liu, B.: A Generalized Tree Matching Algorithm Considering Nested Lists for Web Data Extraction. In: Proceedings of the 10th SIAM International Conference on Data Mining, pp. 930–941 (2010)
29. Liu, W., Meng, X., Meng, W.: ViDE: A Vision-Based Approach for Deep Web Data Extraction. IEEE Transactions on Knowledge and Data Engineering 22(3), 447–460 (2010)
30. Su, W., Wang, J., Lochovsky, F.H., Liu, Y.: Combining Tag and Value Similarity for Data Extraction and Alignment. IEEE Transactions on Knowledge and Data Engineering 24(7), 1186–1200 (2012)

Heterogeneous Metric Learning for Cross-Modal Multimedia Retrieval

Jun Deng[1,2], Liang Du[1,2], and Yi-Dong Shen[1]

[1]State Key Laboratory of Computer Science, Institute of Software
Chinese Academy of Sciences, Beijing, 100190, China
[2]University of Chinese Academy of Sciences, Beijing, 100049, China
{dengj,duliang,ydshen}@ios.ac.cn

Abstract. Due to the massive explosion of multimedia content on the web, users demand a new type of information retrieval, called cross-modal multimedia retrieval where users submit queries of one media type and get results of various other media types. Performing effective retrieval of heterogeneous multimedia content brings new challenges. One essential aspect of these challenges is to learn a *heterogeneous metric* between different types of multimedia objects. In this paper, we propose a Bayesian personalized ranking based heterogeneous metric learning (BPRHML) algorithm, which optimizes for correctly ranking the retrieval results. It uses pairwise preference constraints as training data and explicitly optimizes for preserving these constraints. To further encouraging the smoothness of learning results, we integrate graph regularization with Bayesian personalized ranking. The experimental results on two publicly available datasets show the effectiveness of our method.

Keywords: Metric Learning, Heterogeneous Spaces, Multimedia.

1 Introduction

With the explosive accumulation of multimedia content on the web, cross-modal multimedia retrieval attracts much attention of industry and academia. Nowadays, the prevailing tools for retrieving multimedia content are still single-media based, e.g., search engines such as google or bing. In single-media based retrieval, the retrieval result and user query are of the same media type. For example, you type in a text query in google, then you get many textual descriptions related to the query. In fact, users demand more diversities of the retrieval result [11]. Suppose you get lost in a strange town and you want to find the way back to your hotel. By taking a photo, cross-modal multimedia retrieval is able to return all the textual materials about where you are. Cross-modal multimedia retrieval is an exciting technology and will make our life more convenient.

Cross-modal multimedia retrieval is an exciting, yet difficult problem because multimedia objects are represented in different feature spaces. Thus the traditional single-media based methods cannot directly apply to it. The main problem is how to measure the similarity between heterogeneous objects. In this paper,

X. Lin et al. (Eds.): WISE 2013, Part I, LNCS 8180, pp. 43–56, 2013.

we address the problem by automatically learning a *heterogeneous metric* over two different spaces using labeled training data. Distance metric learning is not a new research topic. Many research efforts have been devoted to it in the last decade (e.g., [1,3,5,7,8], see [9] for a comprehensive survey). Given a training dataset of pairs of similar and dissimilar objects, distance metric learning aims to learn an optimal metric that preserves the similar/dissimilar relations among the objects. Many studies have demonstrated, both empirically and theoretically, that a learned metric can significantly improve the performance in classification, clustering and retrieval tasks [9]. However, most existing algorithms focus on learning a distance metric in a single space. Few algorithm attempt to learn a metric between different spaces.

In this paper, we propose a new approach, called Bayesian personalized ranking based heterogeneous metric learning (BPRHML). Suppose that we are learning a distance metric between spaces $\mathcal{X}$ and $\mathcal{Y}$. Let x be an object from $\mathcal{X}$, y_1 and y_2 be objects from $\mathcal{Y}$. y_1 is relevant to x, while y_2 is irrelevant to x. BPRHML computes two distances d_1 and d_2 in the transformed space. d_1 is the distance between x and y_1. d_2 is the distance between x and y_2. The key idea of BPRHML is to explicitly maximize the difference between d_1 and d_2. This will encourage the relevant objects to rank in front of the irrelevant objects. To better exploit the structure information of the heterogeneous objects, we integrate homogeneous and heterogeneous graph regularization into the objective function. Homogeneous graph regularization utilizes the similarity information inside a single space while heterogeneous graph regularization use the similarity information between different spaces. By combining them together, we can preserve smoothness of the learning result in both spaces. The objective function of BPRHML mainly consists of three terms: the loss function defined on the set of pairwise preference constraints, L2 regularization and graph regularization. We derive an efficient optimization algorithm to learn the model based on gradient descent. Experiments on the Wikipedia dataset and the corel5k image dataset show that BPRHML significantly outperforms related methods.

The rest of the paper is organized as follows: Section 2 will discusses related work. In section 3, we demonstrate preliminaries and notations. Section 4 introduces our method BPRHML. Section 5 shows the experimental results. Finally, we conclude this paper in Section 6.

2   Related Work

What lies at the core of cross-modal multimedia retrieval is to learn a metric between heterogeneous multimedia objects. In distance metric learning, a distance metric is learned from labeled training data. Typically, a linear transformation is learned to transform the data into a new space. The distance metric is then defined as the Euclidean distance in the new space. In this paper, we also adopt this definition. Most existing methods learn the metric as a Mahalanobis distance which can be represented as a positive semi-definite matrix. Given pairs of similar/dissimilar objects, approaches such as [1,3,7,8] try to learn a distance metric that keeps all the data points with the same label close, while separating data points with different label far apart. For example, in [1], the authors attempted to minimize the Mahalanobis distance between similar objects while keeping a large margin between dissimilar objects. In [5], a Mahalanobis distance is learned by maximizing the posterior probability of the training data. All of these methods focus on learning a distance metric in a single space, while in this paper a distance metric is learned between two spaces.

In heterogeneous metric learning, we usually learn two transformation matrices, which transform the heterogeneous objects into a same output space. Heterogeneous metric learning is relatively a new problem. The most related approaches to our method are [4,2,6]. Canonical correlation analysis (CCA) [10] is applied in [4] to learn a heterogeneous metric. Specially, CCA attempts to maximize the correlation between same labeled objects in the transformed space. Based on the learning results of CCA, [4] further learns a high-level semantic metric by logistic regression. Another notable approach is cross-modal factor analysis (CFA) proposed in [2]. Unlike CCA, CFA adopts a criterion of minimizing the Frobenius norm between same labeled objects in the transformed space. Both CCA and CFA consider only pairs of same labeled objects as input. They do not explicitly separate different labeled objects. To overcome this problem, Wu et.al. [6] proposed to learn two orthogonal transformation matrices by minimizing the distance between same labeled objects and maximizing the distance between different labeled objects. However, all of the above methods do not optimize for correctly ranking the retrieval results, which is important for an information retrieval system. In this paper, we attempt to preserve the partial ranking information of the training data in the transformed space.

3   Preliminaries and Notations

3.1   Cross-Modal Multimedia Retrieval

In this section, we first define the problem to be addressed. Then we introduce the notations used in this paper.

Let $\mathcal{X}$ and $\mathcal{Y}$ denote two different media types such as text, image, video. Let $\mathbb{D}_{\mathcal{X}} = \{(x_1, l_1^x), (x_2, l_2^x), \cdots, (x_m, l_m^x)\}$ and $\mathbb{D}_{\mathcal{Y}} = \{(y_1, l_1^y), (y_2, l_2^y), \cdots, (y_n, l_n^y)\}$ be two sets of multimedia objects of types $\mathcal{X}$ and $\mathcal{Y}$, respectively. l_i^x and l_i^y are labels of x_i and y_i, respectively. Our goal is to retrieve relevant x in an unlabeled dataset $\mathbb{T} = \{x_1, x_2, \cdots, x_p, y_1, y_2, \cdots, y_q\}$ in response to a query y, or vice-versa.

As for the notations, we use $\mathbf{X}$ and $\mathbf{Y}$ to denote data matrices of $\mathbb{D}_{\mathcal{X}}$ and $\mathbb{D}_{\mathcal{Y}}$. Columns of $\mathbf{X}$ and $\mathbf{Y}$ correspond to objects and rows correspond to features. x_i and y_i are the i-th column vectors of $\mathbf{X}$ and $\mathbf{Y}$, respectively. $\mathbf{U}$ and $\mathbf{V}$ are the linear transformation matrices correspond to $\mathbf{X}$ and $\mathbf{Y}$. The main notations used in this paper are summarized in Table 1.

Table 1. Notations used in this paper

Notations	Explanations
$\mathbb{D}_\mathcal{X}, \mathbb{D}_\mathcal{Y}$	training object set of types $\mathcal{X}$ and $\mathcal{Y}$
$\mathbb{T}$	test object set
l_i^x, l_i^y	labels of x_i and y_i
m, n	number of objects in $\mathbb{D}_\mathcal{X}$ and $\mathbb{D}_\mathcal{Y}$
p, q	number of objects of types $\mathcal{X}$ and $\mathcal{Y}$ in $\mathbb{T}$
α, β	regularization parameters
d^x, d^y	original dimensionality of $\mathcal{X}$ and $\mathcal{Y}$
c	dimensionality of the transformed space
$\mathbf{X}$	$d^x \times m$ data matrix of objects in $\mathbb{D}_\mathcal{X}$
$\mathbf{Y}$	$d^y \times n$ data matrix of objects in $\mathbb{D}_\mathcal{Y}$
x_i	$d^x \times 1$ column vector represents i-th object in $\mathbb{D}_\mathcal{X}$
y_i	$d^y \times 1$ column vector represents i-th object in $\mathbb{D}_\mathcal{Y}$
$\mathbf{U}$	$d^x \times c$ transformation matrix for $\mathbf{X}$
$\mathbf{V}$	$d^y \times c$ transformation matrix for $\mathbf{Y}$

3.2 Bayesian Personalized Ranking

Bayesian Personalized Ranking (BPR) [12] is a famous model of recommendation system. BPR's key idea is to use partial order of items, instead of single user-item examples to train a recommendation model. It allows the interpretation of positive-only data as partial ordering of items. When we observed a positive use-item example of user u on an item i, e.g., user u viewed or purchased item i, we assume that the user prefers this item than all other non-observed items. Formally we can extract a pairwise preference dataset $\mathcal{P}_1 : U \times I \times I$ by

$$\mathcal{P}_1 := \{(u, i, j) \mid i \in I_u^+ \wedge j \in I \setminus I_u^+\} \tag{1}$$

where U is the user set, I is the item set, I_u^+ and $I \setminus I_u^+$ are the positive item set and missing set associated with user u, respectively. Each triple $(u, i, j) \in \mathcal{P}_1$ says that user u prefers item i than j. BPR optimization criterion [12] aims to find an arbitrary model class to maximize the posterior probability over these pairs. The generic optimization criterion for Bayesian personalized ranking is :

$$\text{BPR-OPT} = -\sum_{(u,i,j) \in \mathcal{P}} \ln \sigma(\hat{x}_{uij}) + \lambda_\Theta(\| \Theta \|^2) \tag{2}$$

where Θ represents the parameter vector of an arbitrary model class, λ_Θ are model specific regularization parameters, σ is the logistic sigmoid function $\sigma(x) = 1/(1 + exp(-x))$. $\hat{x}_{uij}$ is an arbitrary real-valued function of the model parameter vector Θ which captures the special relationship between user u, item i and item j.

Note that extracting pairwise preferences constraints has been widely used in learning to rank tasks [13]. The BPR optimization criterion is actually the cross entropy cost function (logistic loss) over pairs. In fact, there are also pairwise preferences in the training dataset $\mathbb{D}_{\mathcal{X}}$ and $\mathbb{D}_{\mathcal{Y}}$. Consider the following media objects: $(x, l_1) \in \mathbb{D}_{\mathcal{X}}, (y_1, l_1) \in \mathbb{D}_{\mathcal{Y}}$ and $(y_2, l_2) \in \mathbb{D}_{\mathcal{Y}}$. Let us assume that x_1 is a query. Obviously, x_1 prefers y_1 to y_2 in the retrieval result because x_1 and y_1 are same labeled, while x_1 and y_2 are different labeled. Consequently, y_1 should rank in front of y_2 in the retrieval result.

4   BPRHML

In this section, we propose Bayesian personalized ranking based heterogeneous metric learning (BPRHML) algorithm. We first briefly review the metric learning for heterogeneous data. Then we define our objective function which consists of three terms. Finally, we derive an efficient optimization strategy. Note that BPRHML is an asymmetric model. We train different models for queries from $\mathcal{X}$ and $\mathcal{Y}$. In the rest of this paper, we assume queries are of type $\mathcal{X}$ and retrieval results are of type $\mathcal{Y}$. The algorithm for the other direction is defined analogously.

4.1   Heterogeneous Metric Learning

We can construct the set of pairwise preference constraints among the heterogeneous media objects from the training dataset $\mathbb{D}_{\mathcal{X}}$ and $\mathbb{D}_{\mathcal{Y}}$:

$$\mathcal{P}_2 = \{(x_k, y_i, y_j) \mid l_k^x = l_i^y \wedge l_k^x \neq l_j^y\} \tag{3}$$

where x_k and y_i share the same label, x_k and y_j are different labeled. Each triple (x_k, y_i, y_j) is inferred from the category labels of x_k, y_i, y_j and indicates that x_k prefers y_i to y_j in the retrieval result. Consequently, y_i should rank higher than y_j in the retrieval result. Our goal is to learn a metric between $\mathcal{X}$ and $\mathcal{Y}$ that preserves the pairwise preference constraints in $\mathcal{P}_2$.

In traditional single space metric learning, the distance metric is defined as the Mahalanobis distance between objects. The Mahalanobis distance can be viewed as a linear transformation with matrix $\mathbf{L} \in \mathbb{R}^{d^x \times c}$ followed by calculating the Euclidean distance. For an object pair (x_i, x_j), the Mahalanobis distance between them is computed as follows:

$$d(x_i, x_j) = \sqrt{(\mathbf{L}^T x_i - \mathbf{L}^T x_j)^T (\mathbf{L}^T x_i - \mathbf{L}^T x_j)} \tag{4}$$

The goal of metric learning is to learn the linear transformation matrix $\mathbf{L}$ from the set of similar/dissimilar constraints. However, in heterogeneous metric learning, objects x_i and y_j are coming from two heterogeneous spaces $\mathcal{X}$ and $\mathcal{Y}$ with different features (e.g., dimensions). The similarity relation between heterogeneous data is not a metric, hence, does not fall into the standard framework of metric learning. It is not trivial to define a metric between two heterogeneous spaces. Our proposal is to learn two linear transformations, which transform heterogeneous objects into a same output space. More specially, let $\mathbf{U} \in \mathbb{R}^{d^x \times c}$ be the transformation matrix for $\mathbf{X} \in \mathbb{R}^{d^x \times m}$ and $\mathbf{V} \in \mathbb{R}^{d^y \times c}$ be the transformation matrix for $\mathbf{Y} \in \mathbb{R}^{d^y \times n}$, d^x is the original dimensionality of $\mathcal{X}$ and d^y is the original dimensionality of $\mathcal{Y}$, m and n are the number of media objects in $\mathbb{D}_\mathcal{X}$ and $\mathbb{D}_\mathcal{Y}$, c is the dimensionality of the transformed space. For an object pair (x_i, y_j), we define the heterogeneous distance as the Euclidean distance in the transformed space:

$$d(x_i, y_j) = \sqrt{(\mathbf{U}^T x_i - \mathbf{V}^T y_j)^T (\mathbf{U}^T x_i - \mathbf{V}^T y_j)} \tag{5}$$

Our formulation naturally extends conventional distance metric learning from one single space to two different spaces. Single space metric learning can be viewed as a special case of our formulation in which $\mathbf{U} = \mathbf{V}$. We aim to learn the parameter matrices $\mathbf{U}$ and $\mathbf{V}$ that preserve the pairwise preference constraints in $\mathcal{P}_2$.

4.2   Objective Function

We construct an objective function which consists of three terms for heterogeneous metric learning as follow:

$$\underset{\mathbf{U},\mathbf{V}}{\arg\min}\ l(\mathbf{U}, \mathbf{V}) + \alpha s(\mathbf{U}, \mathbf{V}) + \beta g(\mathbf{U}, \mathbf{V}) \tag{6}$$

where $l(\mathbf{U}, \mathbf{V})$ is the loss function defined on the set of pairwise preference constraints, $s(\mathbf{U}, \mathbf{V})$ is the L2 regularization, $g(\mathbf{U}, \mathbf{V})$ is the graph regularization, α and β are regularization parameters.

Loss Function. We argue that the loss function should optimize for correctly ranking the retrieval result. Minimizing the loss function will encourage relevant objects to rank in front of irrelevant objects, i.e, preserving the pairwise preference constraints in $\mathcal{P}_2$. Our proposal is to take the advantage of the Bayesian personalized ranking model introduced in section 3.2. One simple formulation of the loss function is as follows:

$$l(\mathbf{U}, \mathbf{V}) = -\frac{1}{2} \sum_{k=1}^{m} \sum_{i \in \mathbf{Y}_k^+} \sum_{j \in \mathbf{Y}_k^-} \ln \sigma(\hat{x}_{kij}) \tag{7}$$

where x_k is an object from $\mathbf{X}$, $\mathbf{Y}_k^+$ is the set of objects in $\mathbf{Y}$ that are same labeled with x_k, $\mathbf{Y}_k^-$ is the set of objects in $\mathbf{Y}$ that are different labeled with x_k, and $\hat{x}_{kij}$ is defined as follows:

$$\hat{x}_{kij} = \| \mathbf{U}^T x_k - \mathbf{V}^T y_i \|^2 - \| \mathbf{U}^T x_k - \mathbf{V}^T y_j \|^2 \tag{8}$$

where $\| \cdot \|^2$ denotes the square of L2 norm. Intuitively, for a given triple (k, i, j), minimizing $l(\mathbf{U}, \mathbf{V})$ will result in maximizing $\hat{x}_{kij}$, i.e., encouraging relevant object to rank in front of irrelevant object. However, for a training dataset which consists of m objects of type $\mathcal{X}$ and n objects of type $\mathcal{Y}$, there are possibly $O(m \times n^2)$ pairwise preference constraints in $\mathcal{P}_2$. In case of large m and n, the huge number of pairwise preference constraints in $\mathcal{P}_2$ will affect the efficiency of the optimization algorithm.

To handle the above issue, we propose to construct two *representative object* sets out of the relevant and irrelevant object sets, respectively. More specially, let

$x_k \in \mathbb{D}_{\mathcal{X}}$ be a query, $\mathbf{Y}_k^+ \subseteq \mathbb{D}_{\mathcal{Y}}$ and $\mathbf{Y}_k^- \subseteq \mathbb{D}_{\mathcal{Y}}$ be the corresponding relevant and irrelevant object sets. We construct two representative object set D_k^+ and D_k^- out of $\mathbf{Y}_k^+$ and $\mathbf{Y}_k^-$, respectively. Intuitively, objects in D_k^+ are representatives of objects in $\mathbf{Y}_k^+$ and objects in D_k^- are representatives of objects in $\mathbf{Y}_k^-$. Note that objects in $\mathbf{Y}_k^+$ share the same label, while objects in $\mathbf{Y}_k^-$ are different labeled. To construct D_k^+, we cluster the objects in $\mathbf{Y}_k^+$. Suppose that we have built M clusters $(C_1^+, \cdots, C_M^+)$ by applying a clustering algorithm such as K-means. Then we define D_k^+ to be the centroid of each cluster:

$$D_k^+ = \{cen(C_i^+) \mid 1 \le i \le M\} \tag{9}$$
$$cen(C_i^+) = \frac{1}{|C_i^+|} \sum_{y \in C_i^+} y \tag{10}$$

where $|\cdot|$ denotes set cardinality. As for D_k^-, we first divide $\mathbf{Y}_k^-$ into N clusters according to the labels, i.e., same labeled objects form a cluster. N is the number of labels in $\mathbf{Y}_k^-$. Assume that we have built N clusters $(C_1^-, \cdots, C_N^-)$. Then we also define D_k^- to be the centroid of each cluster:

$$D_k^- = \{cen(C_i^-) \mid 1 \le i \le N\} \tag{11}$$
$$cen(C_i^-) = \frac{1}{|C_i^-|} \sum_{y \in C_i^-} y \tag{12}$$

In fact, we can further perform clustering on D_k^- to reduce the number of representative objects in D_k^-. Note that we are not the first to define representative object as the centroid of corresponding object set. In [17], the authors attempted to learn latent factors by maximizing the marginal utility between user choice and the *average* of non-choices. With a slight abuse of notation, we will also denote an object from D_k^+ or D_k^- by y_i or y_j. Based on D_k^+ and D_k^-, we can construct the new set of pairwise preference constraints as follows:

$$\mathcal{P}_2' = \{(x_k, y_i, y_j) \mid 1 \le k \le m \land i \in D_k^+ \land j \in D_k^-\} \tag{13}$$

By constructing the representative objects, we reduce the number of pairwise preference constraints to $O(m \times M \times N)$, where $M \ll n$ and $N \ll n$. The loss function is defined to preserve the constraints in $\mathcal{P}_2'$:

$$l(\mathbf{U}, \mathbf{V}) = -\frac{1}{2} \sum_{k=1}^m \sum_{i \in D_k^+} \sum_{j \in D_k^-} \ln \sigma(\hat{x}_{kij}) \tag{14}$$

where $\hat{x}_{kij}$ is defined similar to (8). Intuitively, $\sigma(\hat{x}_{kij})$ defines the probability for y_i to rank in front of y_j in the retrieval result.

L2 Regularization. We define the L2 regularization as follows:

$$s(\mathbf{U}, \mathbf{V}) = \frac{1}{2}(\| \mathbf{U} \|_F^2 + \| \mathbf{V} \|_F^2) \tag{15}$$

$\| \mathbf{U} \|_F^2$ and $\| \mathbf{V} \|_F^2$ are the square of Frobenius norm of $\mathbf{U}$ and $\mathbf{V}$, respectively. L2 regularization is widely used to reduce overfitting.

Graph Regularization. Graph regularization has been widely used in dimensionality reduction [19], clustering [20] and semi-supervised learning [21]. The key assumption of graph regularization is that if two media objects are similar, they should also be close to each other in the transformed space. In heterogeneous metric learning, we have similarity constraints in single modality and across modalities. Therefore, we intend to define *homogeneous* graph regularization and *heterogeneous* graph regularization, respectively. Homogeneous graph regularization captures the similarity information inside a single modality, i.e, $\mathbf{X}$ or $\mathbf{Y}$. In the following, we define the homogeneous graph regularization for $\mathbf{X}$. The homogeneous graph regularization for $\mathbf{Y}$ is defined analogously. In homogeneous space, we can use both label information and distance information. Following [18], we define the homogeneous neighbourhood of an object x_i, denoted as $\mathcal{N}_i$, to be the k nearest neighbours, determined by Euclidean distance, that share the same label with x_i. We treat the same labeled objects outside $\mathcal{N}_i$ as outliers and ignore them. We define an undirected and symmetric data graph $G_x = (V_x, \mathbf{W}_x)$ on $\mathbf{X}$. V_x is the set of objects in $\mathbf{X}$. $\mathbf{W}_x$ is a $m \times m$ matrix and each element w_{ij} of $\mathbf{W}_x$ denotes the similarity information between the i-th media object and j-th media object of $\mathbf{X}$. Based on the homogeneous neighbourhood, $\mathbf{W}_x$ is defined as follows:

$$w_{ij} = \begin{cases} 1, & (x_j \in \mathcal{N}_i \lor x_i \in \mathcal{N}_j) \land i \neq j \\ 0, & otherwise \end{cases} \tag{16}$$

w_{ii} is set to 0 to avoid self-reinforcement. Let $\mathbf{T} = \mathbf{U}^T \mathbf{X}$ and t_i be the i-th column of $\mathbf{T}$. Intuitively, $\mathbf{T}$ represents all media objects of $\mathbf{X}$ in the transformed space. The homogeneous graph regularization is defined as follows:

$$\begin{aligned} \mathcal{O}_1 &= \tfrac{1}{4} \sum_{i=1}^{m} \sum_{j=1}^{m} w_{ij} \| \tfrac{t_i}{\sqrt{d_{ii}}} - \tfrac{t_j}{\sqrt{d_{jj}}} \|^2 \\ &= \tfrac{1}{2} tr(\mathbf{T} \mathbf{L}_x \mathbf{T}^T) \end{aligned} \tag{17}$$

where $\mathbf{L}_x = \mathbf{I} - \mathbf{D}^{-1/2} \mathbf{W}_x \mathbf{D}^{-1/2}$ is called the graph Laplacian with $\mathbf{D}$ being a diagonal matrix whose diagonal are row sums of $\mathbf{W}_x$, $d_{ii} = \sum_j w_{ij}$, $\mathbf{I}$ is an $m \times m$ identity matrix, and $tr(.)$ denotes the trace of a matrix.

Unlike in homogeneous space, we have only label information in heterogeneous spaces. So we construct the data graph $G_{xy} = (V_{xy}, \mathbf{W}_{xy})$ between $\mathbf{X}$ and $\mathbf{Y}$ from the labels. $\mathbf{W}_{xy}$ is a $m \times n$ matrix and each element w_{ij} of $\mathbf{W}_{xy}$ denotes the similarity information between the i-th media object of $\mathbf{X}$ and the j-th media object of $\mathbf{Y}$. $\mathbf{W}_{xy}$ is defined as follows:

$$w_{ij} = \begin{cases} 1, & l_x^i = l_y^j \land 1 \leq i \leq m \land 1 \leq j \leq n \\ 0, & otherwise \end{cases} \tag{18}$$

Let $\mathbf{S} = \mathbf{V}^T \mathbf{Y}$ and s_i be the i-th column of $\mathbf{S}$. $\mathbf{S}$ represents all media objects of $\mathbf{Y}$ in the transformed space. The heterogeneous graph regularization is defined as follows:

$$\begin{aligned} \mathcal{O}_2 &= \tfrac{1}{2} \sum_{i=1}^{m} \sum_{j=1}^{n} w_{ij} \| \tfrac{t_i}{\sqrt{d_{ii}^x}} - \tfrac{s_j}{\sqrt{d_{jj}^y}} \|^2 \\ &= \tfrac{1}{2} tr(\mathbf{T}\mathbf{T}^T) + \tfrac{1}{2} tr(\mathbf{S}\mathbf{S}^T) - tr(\mathbf{T}\mathbf{D}_x^{-1/2} \mathbf{W}_{xy} \mathbf{D}_y^{-1/2} \mathbf{S}^T) \end{aligned} \tag{19}$$

where $\mathbf{D}_x$ is a $m \times m$ diagonal matrix whose diagonal are row sums of $\mathbf{W}_{xy}$, $d_{ii}^x = \sum_j w_{ij}$, $\mathbf{D}_y$ is a $n \times n$ diagonal matrix whose diagonal are column sums of $\mathbf{W}_{xy}$, $d_{ii}^y = \sum_j w_{ji}$. In summary, the graph regularization $g(\mathbf{U}, \mathbf{V})$ is defined to be:

$$
\begin{aligned}
g(\mathbf{U}, \mathbf{V}) &= \tfrac{1}{2}tr(\mathbf{TL}_x\mathbf{T}^T) + \tfrac{1}{2}tr(\mathbf{SL}_y\mathbf{S}^T) + \tfrac{1}{2}tr(\mathbf{TT}^T) \\
&\quad + \tfrac{1}{2}tr(\mathbf{SS}^T) - tr(\mathbf{TD}_x^{-1/2}\mathbf{W}_{xy}\mathbf{D}_y^{-1/2}\mathbf{S}^T) \\
&= \tfrac{1}{2}tr(\mathbf{U}^T\mathbf{XL}_x'\mathbf{X}^T\mathbf{U}) + \tfrac{1}{2}tr(\mathbf{V}^T\mathbf{YL}_y'\mathbf{Y}^T\mathbf{V}) \\
&\quad - tr(\mathbf{U}^T\mathbf{XD}_x^{-1/2}\mathbf{W}_{xy}\mathbf{D}_y^{-1/2}\mathbf{Y}^T\mathbf{V})
\end{aligned}
\tag{20}
$$

where $\mathbf{L}_x' = \mathbf{L}_x + \mathbf{I}$, $\mathbf{L}_y' = \mathbf{L}_y + \mathbf{I}$, $\mathbf{L}_y$ is the graph Laplacian corresponding to $\mathbf{Y}$. Minimizing $g(\mathbf{U}, \mathbf{V})$ will encourage the smoothness of the transformation over both modalities.

4.3   Optimization Strategy

Firstly, we show how to initialize $\mathbf{U}$ and $\mathbf{V}$. Among all the related methods introduced in section 2, CFA shares the same assumption with our method. CFA also assumes there are two linear transformations that transform the heterogeneous objects into a same output space. Then the heterogeneous distance metric is defined as the Euclidean distance in the transformed space. Consequently, we propose to initialize $\mathbf{U}$ and $\mathbf{V}$ with the learning result of CFA. More specially, CFA optimizes the following objective function:

$$
\min_{\mathbf{U}, \mathbf{V}} \parallel \mathbf{U}^T\mathbf{X}_1 - \mathbf{V}^T\mathbf{Y}_1 \parallel_F^2
\tag{21}
$$

$$
s.t. \quad \mathbf{UU}^T = \mathbf{I}, \quad \mathbf{VV}^T = \mathbf{I}
$$

where $\mathbf{X}_1$ and $\mathbf{Y}_1$ are two object matrices, which consist of row-by-row coupled samples of two media types, and $\mathbf{I}$ is identity matrix of corresponding size. We can rewrite the above objective function as follows:

$$
\parallel \mathbf{U}^T\mathbf{X}_1 - \mathbf{V}^T\mathbf{Y}_1 \parallel^2 = tr(\mathbf{X}_1{}^T\mathbf{X}_1) + tr(\mathbf{Y}_1{}^T\mathbf{Y}_1) - 2tr(\mathbf{X}_1{}^T\mathbf{UV}^T\mathbf{Y}_1)
\tag{22}
$$

We can easily see from the above that matrices $\mathbf{U}$ and $\mathbf{V}$ that maximize $tr(\mathbf{X}_1{}^T\mathbf{UV}^T\mathbf{Y}_1)$ will minimize (20). It can be shown [16] that such matrices are given by singular value decomposition:

$$
\mathbf{X}_1\mathbf{Y}_1{}^T = \mathbf{U\Sigma V}^T
\tag{23}
$$

Instead of adopting stochastic gradient descent like in Bayesian personalized ranking [12], we derive our optimization algorithm based on gradient descent. There are two reasons for this choice: 1) by introducing representative object, the number of pairwise preference constraints reduces to $O(m \times M \times N)$. 2) one update on a preference triple (k, i, j) will affect all variables, i.e., $\mathbf{U}$ and $\mathbf{V}$, in the objective function. Let $\mathcal{F}(\mathbf{U}, \mathbf{V})$ denote the objective function in (6). Once the initial value of $\mathbf{U}$ and $\mathbf{V}$ are computed, we update $\mathbf{U}$ and $\mathbf{V}$ in each iteration by the following gradients:

Algorithm 1. Learning procedure of BPRHML

Input: training data matrices $\mathbf{X}$ and $\mathbf{Y}$, learning rate η, regularization parameters α and β, dimensionality K of the transformed space

Output: transformation matrices $\mathbf{U}$ and $\mathbf{V}$

1: Construct the data graph matrices $\mathbf{W}_x$, $\mathbf{W}_y$ and $\mathbf{W}_{xy}$
2: Compute the Laplacian matrices $\mathbf{L}_x$ and $\mathbf{L}_y$ based on $\mathbf{W}_x$ and $\mathbf{W}_y$
3: Compute the diagonal matrices $\mathbf{D}_x$ and $\mathbf{D}_y$ based on $\mathbf{W}_{xy}$
4: Compute the temporary matrices $\mathbf{U}'$ and $\mathbf{V}'$ by $\mathbf{X}\mathbf{Y}^T = \mathbf{U}'\mathbf{\Sigma}\mathbf{V}'^T$
5: Initialize $\mathbf{U}$ and $\mathbf{V}$ with the first K columns of $\mathbf{U}'$ and $\mathbf{V}'$, respectively
6: **repeat**
7:    update $\mathbf{U}$ by $\mathbf{U} \leftarrow \mathbf{U} - \eta\frac{\partial \mathcal{F}}{\partial \mathbf{U}}$
8:    update $\mathbf{V}$ by $\mathbf{V} \leftarrow \mathbf{V} - \eta\frac{\partial \mathcal{F}}{\partial \mathbf{V}}$
9: **until** convergence

$$\frac{\partial \mathcal{F}}{\partial \mathbf{U}} = \sum_{k=1}^{m} \sum_{i\in D_k^+} \sum_{j\in D_k^-} \frac{1}{1+exp(\hat{x}_{kij})} x_k(y_i^T - y_j^T)\mathbf{V} + \alpha\mathbf{U}$$
$$+ \beta\mathbf{X}\mathbf{L}_x'\mathbf{X}^T\mathbf{U} - \beta\mathbf{X}\mathbf{D}_x^{-1/2}\mathbf{W}_{xy}\mathbf{D}_y^{-1/2}\mathbf{Y}^T\mathbf{V} \qquad (24)$$

$$\frac{\partial \mathcal{F}}{\partial \mathbf{V}} = \sum_{k=1}^{m} \sum_{i\in D_k^+} \sum_{j\in D_k^-} \frac{1}{1+exp(\hat{x}_{kij})} ((y_i - y_j)x_k^T\mathbf{U} + (y_jy_j^T - y_iy_i^T)\mathbf{V})$$
$$+ \alpha\mathbf{V} + \beta\mathbf{Y}\mathbf{L}_y'\mathbf{Y}^T\mathbf{V} - \beta\mathbf{Y}\mathbf{D}_y^{-1/2}\mathbf{W}_{xy}^T\mathbf{D}_x^{-1/2}\mathbf{X}^T\mathbf{U} \qquad (25)$$

We use a constant learning rate to update the transformation matrices. The process of estimating $\mathbf{U}$ and $\mathbf{V}$ is described in algorithm 1.

5   Experiments

We conduct experiments on two publicly available dataset to compare the performance of our method with other state-of-the-art methods.

5.1   Datasets and Evaluation Criteria

Cross-modal multimedia retrieval is relatively a new problem. There are few publicly available benchmark datasets. To the best of our knowledge, the Wikipedia dataset proposed by Rasiwasia et.al. [4] is the only publicly available dataset specially collected for cross-modal multimedia retrieval. To further evaluate the performance of our method, we also construct experiment on the corel5k image dataset, which is widely used in image annotation [14,15]. In the following, we will introduce the above two dataset in detail.

Wikipedia dataset[1] contains documents that are selected sections from the Wikipedia's featured articles collection. This is a continually updated collection of 2700 articles that have been selected and reviewed by Wikipedia's editors since 2009. The article generally have multiple sections and pictures. Each article

---

[1] http://www.svcl.ucsd.edu/projects/crossmodal/

is split into sections based on section headings, and assign each image to the section in which it was placed by the authors. The final dataset contains a total of 2866 documents, which are text-image pairs, annotated with a label from the vocabulary of 10 semantic categories. The dataset is randomly split into a training set of 2173 documents and a test set of 693 documents.

Corel5k dataset[2] is a widely used benchmark for image annotation. The dataset consists of 4500 training images and 500 test images and there are 260 possible keywords. Each image is annotated with 1-5 key words. The average keywords per image is 3.4. We treat each keyword as a *pseudo document*. The pseudo document is relevant to an image if the corresponding keyword is used to annotate the image. The pseudo document is represented in a space in which each dimension corresponds to a keyword and the dimensionality is 260. We use co-occurrence of keywords in the annotation matrix as feature vectors. Image representation is based on the popular scale invariant feature transformation (SIFT). We perform principal component analysis on both the pseudo document matrix and the image matrix and preserve 85% variance. The final dimensionality of pseudo document matrix and image matrix is 18 and 100, respectively.

Similar to [4] and [6], we adopt Mean Average Precision (MAP) as the evaluation criteria. The MAP score is the average precision at the ranks where recall changes. It is widely used in the information retrieval literature.

5.2 Comparison Settings

In order to show the effectiveness of our approach, we compare the results with the following five baseline methods.

1. **Random:** Randomly retrieving the results.
2. **CCA:** Canonical correlation analysis is used in [4] to learn two transformation matrices that maximize the correlation between two sets of heterogeneous objects.
3. **CFA:** CFA learns two linear transformation matrices [2]. Unlike CCA, CFA optimizes for minimizing the Frobenius norm between pairwise objects in the transformed space.
4. **SCM:** SCM is proposed by Rasiwasia et.al. [4]. CCA is first applied to learn two maximally correlated subspaces. Then it applies logistic regression to learn a high-level semantic representation of the media objects.
5. **MSmethod:** MSmethod is currently state-of-the-art method [6]. It learns two orthogonal transformation matrices by minimizing the distance between relevant objects and maximizing the distance between irrelevant objects.

5.3 Experimental Results

In this section, we compare the performance of our method with the above five baseline methods on the Wikipedia dataset and the corel5k dataset.

---

[2] http://lear.inrialpes.fr/people/guillaumin/data.php

Table 2. MAP values on Wikipedia and Corel5k dataset

Task	Random	CCA	CFA	SCM	MSmethod	BPRHML
Image→Text	0.118	0.249	0.279	0.277	0.282	**0.299**
Text→Image	0.118	0.196	0.231	0.226	0.238	**0.265**
Image→Keyword	0.054	0.117	0.112	0.126	0.135	**0.179**
Keyword→Image	0.061	0.125	0.136	0.131	0.147	**0.167**
Average	0.088	0.172	0.189	0.190	0.201	**0.228**

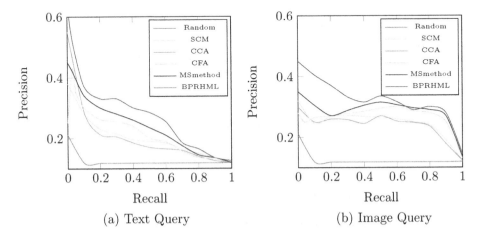

(a) Text Query   (b) Image Query

Fig.1. Precision recall curves on Wikipedia dataset

Table 2 shows the MAP values of our method and all other baseline methods. The upper part of Table 2 shows the MAP values on the Wikipedia dataset. The low part of Table 2 shows the MAP values on the corel5k dataset. The better results are shown in bold. To make a fair comparison, we tune all methods to their best according to the 5-fold cross validation on the training dataset. As for the parameters of BPRHML, we set $\alpha = 86$, $\beta = 7.1$ for the task of retrieving text by image query and set $\alpha = 1000$, $\beta = 0.001$ for the task of retrieving image by text query. For both tasks, we set $k = 5$ for K-means clustering and set $k = 50$ for computing the homogeneous neighbourhood.

Due to the factor that BPRHML explicitly optimizes for correctly ranking the retrieval result, it outperforms the compared baselines on both datasets and tasks consistently. We observe that the performance of most methods decrease on the corel5k dataset. It is reasonable since there are only keywords, rather than text in the corel5k dataset. So the corel5k dataset contains less textual information than the Wikipedia dataset, which makes it more challenge. However, BPRHML still significantly outperforms other baseline methods on this challenge dataset. It can be seen from Table 2 that the performance of MSmethod is comparable to our method. It is not surprise because MSmethod considers both similar and dissimilar information while the other baseline methods consider only similar information. CFA outperforms CCA and SMN in most of the case, this demonstrates that the definition of heterogeneous distance metric as the Euclidean distance after two linear transformations is effective for cross-modal multimedia retrieval. In addition, we observe that SCM always outperforms CCA. This suggests that we can further improve the performance of our method by learning a high-level semantic representation [4] of the heterogeneous objects. We leave it as future work. Further analysis of the results is presented in Figure 1, which shows the PR curve of all approaches for both image and text queries on the Wikipedia dataset. We can see from Figure 1 that BPRHML achieves high precision at most levels of recall.

6 Conclusion

In this paper, we propose a Bayesian personalized ranking based heterogeneous metric learning (BPRHML) algorithm to learn the distance metric between heterogeneous objects. We assume that there are two linear transformation matrices which transform the heterogeneous objects into a same output space. The heterogeneous distance metric is then defined as the Euclidean distance in the transformed space. We argue that good objective function should optimize for correctly ranking the retrieval result. So we formulate an objective function which can preserve the pairwise preference constraints in the training data. To further exploiting the structure information contained in the training data, we integrate homogeneous and heterogeneous graph regularization into the objective function. Experiments on benchmark datasets demonstrate the effectiveness of our method. The experimental datasets of this paper contain only text and images. In the future, we intend to evaluate our method on more multimedia types, such as audio and video. We will also learn a high-level semantic representation of the heterogeneous objects based on the learning result of BPRHML.

Acknowledgments. We would like to thank all anonymous reviewers for their helpful comments. This work is supported in part by NSFC grant 60970045 and China National 973 project 2013CB329305.

References

1. Xing, E., Ng, A., Jordan, M., Russell, S.: Distance metric learning with application to clustering with side-information. In: NIPS 2002, pp. 505–512 (2002)
2. Li, D., Dimitrova, N., Li, M., Sethi, I.K.: Multimedia content processing through cross-modal association. In: Proceedings of the Eleventh ACM International Conference on Multimedia, pp. 604–611 (2003)
3. Weinberger, K., Blitzer, J., Saul, L.: Distance metric learning for large margin nearest neighbour classification. In: NIPS 2006, pp. 1475–1482 (2006)
4. Rasiwasia, N., Pereira, J.C., Coviello, E., Doyle, G., Lanckriet, G.R.G., Levy, R., Vasconcelos, N.: A New Approach to Cross-Modal Multimedia Retrieval. In: Proceedings of the Eighteenth International Conference on Multimedia, pp. 251–260 (2010)

5. Liu, Y., Rong, J., Rahul, S.: Bayesian Active Distance Metric Learning. In: Proceedings of the 23rd Conference on Uncertainty in Artificial Intelligence (UAI 2007), pp. 442–449 (2007)
6. Wu, W., Xu, J., Li, H.: Learning Similarity Function between Objects in Heterogeneous Spaces. Microsoft Research Technique Report (2010)
7. Davis, J.V., Kulis, B., Jain, P., Sra, S., Dhillon, I.: Information-Theoretic Metric Learning. In: ICML 2007, pp. 209–216 (2007)
8. Hoi, S.C.H., Liu, W., Chang, F.: Semi-Supervised Distance Metric Learning for Collaborative Image Retrieval. In: CVPR 2008, pp. 1–7 (2008)
9. Liu, Y.: Distance Metric Learning: A Comprehensive Survey, School of Computer Science, Carnegie Mellon University (2006)
10. Timm, N.: Applied multivariate analysis. Springer (2002)
11. Liu, J., Xu, C.S., Lu, H.Q.: Cross-media retrieval: state-of-the-art and open issues. International Journal of Multimedia Intelligence and Security 1(1), 33–52 (2010)
12. Rendle, S., Freudenthaler, C., Gantner, Z., Schmidt-Thieme, L.: BPR: Bayesian Personalized Ranking from Implicit Feedback. In: Proceedings of the Twenty-Fifth Conference on Uncertainty in Artificial Intelligence, pp. 452–461 (2009)
13. Liu, T.: Learning to rank for information retrieval. Foundations and Trends in Information Retrieval 3(3), 225–331 (2009)
14. Guillaumin, M., Mensink, T., Verbeek, J., Schmid, C.: TagProp: Discriminative metric learning in nearest neighbor models for image auto-annotation. In: ICCV 2009, pp. 309–316 (2009)
15. Guillaumin, M., Verbeek, J., Schmid, C.: Multimodal semi-supervised learning for image classification. In: CVPR 2010, pp. 902–909 (2010)
16. Krzanowski, W.: Principles of multivariate analysis. Oxford University Press, Oxford (1988)
17. Yang, S.H., Long, B., Smola, A., Zha, H.Y., Zheng, Z.H.: Collaborative Competitive Filtering: Learning Recommender Using Context of User Choice. In: SIGIR 2011, pp. 295–304 (2011)
18. Wang, F., Sun, J., Li, T., Anerousis, N.: Two heads better than one: Metric+active learning and its applications for it service classification. In: ICDM 2009, pp. 1022–1027 (2009)
19. Belkin, M., Niyogi, P.: Laplacian eigenmaps for dimensionality reduction and data representation. Neural Computing, 1373–1396 (2003)
20. Cai, D., He, X., Han, J., Huang, T.: Graph regularized non-negative matrix factorization for data representation. IEEE Transaction on Pattern Analysis and Machine Intelligence (2010)
21. Belkin, M., Niyogi, P., Sindhwani, V.: Manifold regularization: A geometric framework for learning from labeled an unlabeled examples. The Journal of Machine Learning Research 7, 2399–2434 (2006)

Efficient Online Novelty Detection in News Streams

Margarita Karkali[1], François Rousseau[2],
Alexandros Ntoulas[3,4], and Michalis Vazirgiannis[1,2,5]

[1] Athens University of Economics and Business, Greece
[2] LIX, École Polytechnique, France
[3] National and Kapodistrian University of Athens, Greece
[4] Zynga, San Francisco
[5] Institut Mines-Télécom, Télécom ParisTech, France
{karkalimar,mvazirg}@aueb.gr, rousseau@lix.polytechnique.fr,
antoulas@di.uoa.gr

Abstract. Novelty detection in text streams is a challenging task that emerges in quite a few different scenarii, ranging from email threads to RSS news feeds on a cell phone. An efficient novelty detection algorithm can save the user a great deal of time when accessing interesting information. Most of the recent research for the detection of novel documents in text streams uses either geometric distances or distributional similarities with the former typically performing better but being slower as we need to compare an incoming document with all the previously seen ones. In this paper, we propose a new novelty detection algorithm based on the *Inverse Document Frequency (IDF)* scoring function. Computing novelty based on IDF enables us to avoid similarity comparisons with previous documents in the text stream, thus leading to faster execution times. At the same time, our proposed approach outperforms several commonly used baselines when applied on a real-world news articles dataset.

Keywords: novelty detection, inverse document frequency, news streams.

1 Introduction

A great deal of information consumption these days happens in the form of push notifications: a user specifies a general topic or stream that he is interested in watching or following and a specific service sends updates to his email, desktop or smartphone. In certain cases, the user may be interested in following *all* the stories coming from a specific source. On the other hand, some sources like Twitter, Facebook or certain news sites allow posting of variants of a given story. In such a scenario, the user might be interested in having a way of specifying that he is interested only in stories that he is not aware of, or, in other words, only in stories that are *novel*.

This problem emerges in a variety of different settings, from email threads to RSS readers on a cell phone and is commonly called *First Story Detection* (FSD)[1]. A good novelty detection algorithm can potentially save a lot of time to the user (by hiding known stories and not only previously seen articles) and can also save bandwidth, battery and storage in the mobile setting scenario.

At a high level, previous research on novelty detection consisted of the definition of a similarity (or distance) metric that is used to compare each new incoming story

---

[1] Also known as novelty detection, novelty mining, new event detection, topic initiator detection.

X. Lin et al. (Eds.): WISE 2013, Part I, LNCS 8180, pp. 57–71, 2013.

(or document) to a set of previously seen stories. If the similarity of the new incoming document is below a threshold (defined differently in each work) then the document is considered novel and therefore some relevant action is taken on the document, otherwise it is discarded. The similarity functions used in the literature range in effectiveness and complexity from simple word counts through cosine similarity to online clustering and one-class classification [3,30,12,4].

In prior work, cosine similarity has been reported to work better than most of the previously proposed approaches [3,30,4] and was shown to outperform even complex language-model-based approaches in most cases. The documents were represented as bag-of-word vectors with additional TF×IDF term weighting applied on them.

Although previous approaches have been shown to work well in most cases, they have two shortcomings. First, the *document-to-document approaches* (such as the maximum cosine similarity ones [3]) tend to be computationally expensive as we need to compare the new incoming document with all the existing previously seen documents in order to determine its novelty. If the user wishes to have a reasonably large collection of documents to compare to, this approach can prove very costly for a system supporting millions of users or, in the case of a mobile setting, may drain the phone's battery faster. On the other hand, the *document-to-summary approaches* such as the online clustering or one-class classification [22,12], where we compare the document to a summary (e.g. the centroid of a cluster) are faster and more appropriate for a mobile setting, but they were shown to perform worse than the document-to-document approaches [22,3].

To this end, we propose a document-to-summary technique that is both efficient computationally and effective in performing novelty detection. Our main idea is to maintain a summary of the collection of previously seen documents that is based on the frequency of each term. We capture the specificity of each term through its *Inverse Document Frequency* (IDF) for a given incoming document and then we show how to compute its overall specificity through the definition of a novelty score. Since our approach is document-to-summary based, we do not compare to all the previous documents and thus we can compute the novelty score faster. At the same time, we show in our experimental evaluation that our approach outperforms several commonly used baseline approaches, in certain cases by a wide margin.

The main contributions of this paper are:

- A new metric for novelty detection based on inverse document frequency that captures the difference of a document's vocabulary with regard to the past.
- An extensive experimental evaluation of our proposed method and the commonly used baselines. Our results indicate that our method outperforms previous ones in both execution time and precision in identifying novel documents.
- A novel annotated corpus that can be used as a benchmark for novelty detection in text streams extracted from a real-world news stream.[2]

2   Related Work

Novelty detection is usually described as a task in signal processing. A survey on methods for novelty detection has been published on Signal Processing Journal by Markou

---

[2] The dataset is publicly available at:
http://www.db-net.aueb.gr/GoogleNewsDataset/

and Singh. The survey is separated in two parts: statistical approaches [12] and neural networks [13]. Novelty detection is a challenging task, with many models that perform well on different data. In this survey, novelty detection in textual data was reported to be a variant of traditional text classification and it was mentioned as an alternative terminology to Topic Detection and Tracking (TDT).

In Topic Detection and Tracking (TDT) field, many papers are dealing with the problem of First Story Detection (FSD). In TDT-3 competition [1], which included a FSD task, Allan *et al.* presented a simple 1-NN approach, also known as UMass [3], that was reported to perform at least as well as the other participants. The UMass approach is constantly used as a baseline in relevant literature. An interesting report from the FSD task in the context of TDT was also published by Allan *et al.* [2], concluding that FSD based on tracking approaches bounds its performance. In our approach we do not rely on model tracking and thus such limitations do not apply.

An interesting work by Yang *et al.* [28] used topic clustering, Named Entities (NE) and topic specific stopword removal for the task of novelty detection on news. In [30], novelty detection at a document level was used in adaptive filtering. The measures tested were separated between geometric distance and language model measures. The results show that the simple approach of maximum cosine distance, introduced by Allan *et al.* in [3], work as well as complex language model measures. A recent work by Verheij et al [27] presents a comparison study of different novelty detection methods evaluated on news article from Yahoo! News Archive where language model based methods perform better than cosine similarity based ones.

Except from the TDT competition, novelty detection was also present in TREC 2002-2004 [7,21,20]. Novelty detection was examined at sentence level and the general goal of the track was to highlight sentences that contain both relevant and novel information in a short, topical document stream. A paper by Sobboroff and Harman [22] reported the significant problem in evaluating such tasks, by highlighting problems in the construction of a ground truth dataset.

Based on TREC novelty track, a significant amount of work was published on novelty detection at sentence level [4,9,8,26]. Allan *et al.* [4] evaluated seven measures for novelty detection separating them in word count measures and language model measures. The results again showed that the simple approach of maximum cosine similarity between a sentence and a number of previously seen ones, works as well as complex language model measures. The Meiji University experiments in TREC 2003 [15] proposed a linear combination of the maximum cosine similarity measure with a metric that aggregates the *TF-IDF* scores of the terms in a sentence. This metric is similar to the one presented here, but it is tested for sentence level novelty detection which is a different task from the one we tackle in the current work.

Lately the interest in novelty detection and mainly in FSD is focused at reducing the computation time as FSD is an online task, and the prevalent 1-NN approach uses exhaustive document to document similarity computation. Petrovic *et al.* [16] approximate 1-NN with Locality Sensitive Hashing (LSH). Zhang *et al.* [29] also target in improving the efficiency of novelty detection systems introducing a news indexing-tree. [10] presents a framework for online new event detection used in a real application that focuses on improving system efficiency using indices, parallel processing etc. Our method also manages to increase the efficiency of novelty detection by avoiding exhaustive comparisons (see next section).

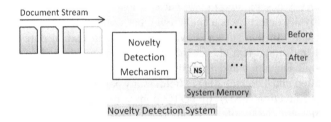

Fig. 1. The process for Novelty Detection

Benchmark Datasets for Novelty Detection. Novelty detection in text streams is usually evaluated in news applications since this is the most common form of text streams and the task of finding novel news articles makes perfect sense. Most of the work on novelty and first story detection use the TDT datasets for evaluation [1]. The most recent TDT benchmark collection (TDT5) is sparsely labeled: it includes 278,109 English news articles but only around 4,500 are annotated with 100 topics. The TREC novelty track dataset is another benchmark dataset, mainly used for sentence-level novelty detection. It is not suitable for the purpose of this paper as it contains novelty judgments per sentence and not per document. [25] is the only work using it at a document level considering the number of novel sentences per document but we believe that such an assumption cannot lead to safe conclusions. Other works [28,30] use available news article collections and apply sampling and manual labeling using well-known events in a specific time span. Details for these datasets are also available in [24].

All the above evaluation datasets are manually annotated using predefined events. Thus there is always the issue of human subjective judgment that introduces a degree of uncertainty. In addition, only a small proportion of the stream is annotated.

3   Novelty Scoring Methods

We consider a system that monitors a stream of documents. New documents reach the system at different times. We assume that documents arrive ordered by their creating time (timestamp). Each document d^t, with a timestamp t, is represented using a *bag-of-word* approach, as $< (q_1, w_1^C), (q_2, w_2^C), ..., (q_{|d^t|}, w_{|d^t|}^C) >$, where q_i is the ith unique term in document d^t and w_i^C is the corresponding weight computed with regard to a corpus C. When a new document d^t arrives in the system, the previous N ones are already stored and indexed. We use the terms *memory* and *corpus* for this set of documents interchangeably in the paper. Assuming the corpus C, for each new document d^t, a novelty score $NS(d^t, C)$ is computed, indicating the novelty of this document for the given corpus. d^t is then stored in memory and the oldest document is flushed. This process is illustrated in Figure 1.

We define the Novelty Detection (ND) problem as the characterization of an incoming document as novel with respect to a predefined window in the past. In the described context, we declare novel a document d^t when the corresponding novelty score $NS(d^t, C)$ is higher than a given threshold θ.

3.1   Baselines

Document-to-Document Using Vector Space. As mentioned earlier, methods based on cosine similarity are proved to work better in similar tasks and are frequently used as a baseline in Novelty Detection. As for the *Max Cosine Similarity* baseline, it was introduced by Allan *et al* at TDT3 in [3] and is also known as the UMass. This method is used as a baseline also by [16,29,30] and it is considered the traditional method for document novelty detection. The intuition of this metric is that if a new document is very similar to another in the corpus, the information it contains was seen before and thus the document cannot be considered as novel. We also introduce a second baseline using the *Mean Cosine Similarity*. Similarly, a document is marked as novel if its mean similarity to the documents in the corpus is below a threshold.

Assuming the cosine similarity between two documents d and d' is defined as:

$$CS(d, d') = \frac{\sum_{k=1}^{m} w_k(d) w_k(d')}{\sqrt{\sum_{k=1}^{|d|} w_k(d)^2 \sum_{k=1}^{|d'|} w_k(d')^2}} \tag{1}$$

where $w_k(d)$ the weight of the term k in document d and m the number of common terms among the two documents, then the respective similarity formulas for the aforementioned metrics are as follows:

$$MaxCS(d^t, C) = \max_{1 \leq i \leq |C|} CS(d^t, d_i) \tag{2}$$

$$MeanCS(d^t, C) = \frac{\sum_{i=1}^{|C|} CS(d^t, d_i)}{|C|} \tag{3}$$

Both approaches are simple to implement but their computational complexity depends on the length of the corpus used. In the worst case the complexity is $O(|d^t| \times |C|)$.

Document-to-Document Using Language Models. A common method to measure the similarity between two documents is using language models. A recent comparison study by Verheij et al [27], where a number of methods were used for novelty detection, reports that the best performing method was document-to-document distance based on language models.

We use minimum Kullback-Leibler (KL) divergence as a baseline approach based on LMs. We implemented the method as described in [27]. Thus, assuming the KL divergence of a document d given a document d' is as follows:

$$KL(\Theta_d, \Theta_{d'}) = \sum_{q \in d} \Theta_d(q) \log \frac{\Theta_d(q)}{\Theta_{d'}(q)} \tag{4}$$

where Θ_d is the unigram language model on document d and $\Theta_d(q)$ is the probability of term q in document d, then the respective novelty scoring formula is as follows:

$$MinKL(d^t, C) = \min_{1 \leq i \leq |C|} KL(\Theta_{d^t}, \Theta_{d_i}) \tag{5}$$

In order to avoid the problem of zero probabilities we use linear interpolation smoothing, where document weights are smoothed against the set of the documents in the corpus. Then the probabilities are defined as $\Theta_{d^t}(q) = \lambda \times \Theta_{d^t}(q) + (1-\lambda) \times \Theta_{d^1...d^{t-1}}(q)$, where $\lambda \in [0, 1]$ is the smoothing parameter and $\Theta_{d^1...d^{t-1}}$ the probability of term q in the corpus C. In our experiments, λ was set to 0.9 based on the experiments in [27].

Table 1. Extended SMART notations that include BM25 components

Notation	Term frequency	Notation	IDF	Notation	Normalization
b (boolean)	$\begin{cases} 1 & \text{if } tf > 0 \\ 0 & \text{otherwise} \end{cases}$	t (idf)	$\log \frac{N}{df}$	n (none)	1
n (natural)	tf	p (prob. idf)	$\log \frac{N-df}{df}$	u (# unique terms)	$\lvert d \rvert$
l (logarithm)	$1 + \log tf$			d (L^1 norm)	dl
k (BM25)	$\frac{(k_1+1)\cdot tf}{k_1 \times (1-b+b \times \frac{dl}{avdl})+tf}$	b (BM25)	$\log \frac{N-df+0.5}{df+0.5}$	c (L^2 norm)	$\sqrt{\sum tf^2}$
				p (pivot)	$1-b+b \times dl/avdl$

Document-to-Summary Using Vector Space. Alternatively, we can maintain a summary of the previously seen documents and compare the new one only to this summary, avoiding computationally expensive comparisons with all the past documents.

To have a complete set of baselines for the evaluation of our method, we also include a document-to-summary approach based on vector space as the document representation and cosine similarity as the novelty metric. The corpus summary is defined based on [27], as the concatenation of all the documents in corpus, i.e. $D_C = \bigcup_{d \in C} d$. Then the novelty scoring formula for the document-to-summary baseline can be defined as follows:

$$SumCS(d^t, C) = CS(d^t, D_C) \tag{6}$$

3.2   Inverse Document Frequency for Novelty

Design Principles. In this paper, we introduce a novelty score that *does not use any similarity or distance measure*. This novelty score can be considered as a way to compare a document to a corpus, which is the essence of a novelty detection task.

To do so, we capitalize on the *Inverse Document Frequency* (IDF) measure introduced in [23]. IDF is a heuristic measure for term specificity and is a function of term use. More generally, by aggregating all the IDF of the terms of a document, IDF can be seen as a function of the vocabulary use at the document level. Hence, our idea to use it as an estimator of novelty – a novel document being more likely to use a different vocabulary than the ones in the previous documents. In a way, a document is novel if its terms are also novel – i.e. previously unseen. This implies that the terms of a novel document have a generally high specificity and therefore high IDF values.

IDF was initially defined as $idf(q, C) = \log \frac{N}{df_q}$. where q is the considered term, C the collection, df_q the document frequency of the term q across C and N the size of C, i.e. the number of documents. There exists a slightly different definition known as *probabilistic* IDF used in particular in BM25 [18] where the IDF is interpreted in a probabilistic way as the odds of the term appearing if the document is irrelevant to a given information need and defined as $idf_{prob.}(q, C) = \log \frac{N-df_q}{df_q}$. Note that this IDF definition can lead to negative values if the term q appears in more than half of the documents as discussed in [17]. For ad-hoc information retrieval, it has been claimed that it violates a set of formal constraints that any scoring function should meet [5] but for novelty detection, this property could be of importance as we want to penalize the use of terms appearing in previously seen documents. We will test both versions in our experiments. Both versions also have smoothed variants for extreme cases where the document frequency could be null or equal to the size of the collection (by usually adding 0.5 to both numerator and denominator). These are the ones that we will use in practice since the collection is pretty small (memory of the last 100 documents for example) and thus subject to sparseness in the vocabulary.

Novelty Score Definition and Properties. It seems then natural to define our novelty score as a TF×IDF weighting model since we are relying on a *bag-of-word* representation and a *vector space* model. The task here is more of *filtering* than ad-hoc IR, hence the TF component needs not to be concave and pivot document length normalized as in BM25. We explored indeed a great variety of combinations for TF and IDF that we will present following the SMART notations (the historical ones defined in [19] and additional ones that include BM25 components). In general, the novelty score of a new document d for a collection C can be defined as follows:

$$NS(d, C) = \frac{1}{norm(d)} \sum_{q \in d} tf(q, d) \times idf(q, C) \qquad (7)$$

where tf, idf and $norm$ can be any of the functions presented in Table 1, ranging from a standalone IDF (btn) to a BM25 score (kbn) using the SMART triplet notation. Note that because of the way BM25 is designed, the length normalization is already included in the TF component ($k\_\_$) for a slop parameter b greater than 0. Therefore, BM25 is denoted by kbn.

The aggregation (through the sum operation) of the term scores to obtain a document score reduces the impact of synonymy which is a common problem when using *bag-of-word* representation and *vector space* model. Indeed, a document that would have terms synonymous with the ones in the other documents would probably be detected as novel since its terms have high IDF values. Nevertheless, it is very unlikely that all its terms are synonymous and overall, its score should not be as high as the one of a novel document.

Unlike the approaches described in 3.1, this measure is not related to the size N of the corpus used. Its complexity is $O(|d|)$. In addition, no document vector needs to be retrieved (and a fortiori stored in an inverted index except for d) for the computation of NS. The index is only used after the score has been assigned in order to decrease the *document frequency* of the terms occurring in the oldest document (the one being flushed). Thus, the response time of the system is not affected.

4   Experiments

4.1   Datasets

Google News Dataset. We wanted a dataset with ground truth judgments regarding the first story of each news cluster. Towards this direction, we worked for the construction of an annotated dataset from a real world news stream. We used the RSS feeds provided by the Google News aggregator.

The method for creating the *Google News dataset* was the following: we periodically collected all articles from the RSS, offered by Google News, for the category "Technology" published in the time period July 12 to August 12, 2012. All articles are from the English news stream. Each news unit consists of the article title, a small description (snippet), the URL for the article, the publication date and a cluster id, assigned by the aggregator, clustering threads of news. In addition, we used an open source script for main content extraction from news websites[3] to get the main content of the articles from the article URLs. We applied two standard preprocessings: stopword removal and Porter's stemming. Then for each article we store the set of unigrams and their corresponding local frequency (*TF*) for the article snippet and content separately.

Annotation Process: We take advantage of the cluster information provided by the Google News to create the ground truth dataset for our experiments. Thus, the goal set is to identify as novel the *first article* in each news cluster. Unfortunately, as the clustering in Google News is carried out via an automated mechanism, there is no guarantee that the articles in a single cluster refer to the same real world event. To have a reliable ground truth dataset, we assign to human annotators the task of *correcting* the clusters retrieved from Google News RSS. The annotators have to assign one of the following labels to the cluster: *clean, separable, part of an existing one* or *mixed*. A clean cluster contains articles that refer to the same event (e.g. *Release of iPhone5*). A *separable* cluster contains articles from more than one event that can easily be detected and annotated. An example of such cluster contained 22 articles for the *Antitrust investigation of Microsoft by EU* and 11 articles for *Windows 8 release on October 26*. For each *separable*, the corresponding number of new clean clusters was created. When a cluster is declared as *part of an existing one*, the two clusters are merged. If the cluster mixes too many events that could not be easily distinguished by the annotator, the cluster is marked as *mixed* and it is not considered for evaluation. We do not consider *mixed* clusters for evaluation because such clusters contain more than one article that should be considered as novel.

The dataset we produced has some advantages over the other benchmark datasets such as TDT5. In those datasets (some in the scale of 10^5 articles) it is the human annotators that decide the similarity among news articles and therefore clustering before they identify the first occurrence of the cluster. Apparently the result is introduction of noise and errors with very high probability due to the diverse background of the annotators and the chance that some articles - due to human error or negligence - are left out of the thematic news clusters. In our case the dataset contains already ground truth in grouping the articles into clusters - and the annotators only improve the few (compared to the documents) clusters. In any case there is no doubt on the first article per cluster as it is the temporally first in the clusters. Thus the probability for errors and more importantly missing the first article on a cluster is much smaller. Finally, the introduced dataset does not suffer from sparseness as all of the previously used ones.

The annotation process reduced the initial data set of 3300 articles/673 clusters to 2006 articles/555 clusters. The cluster size distribution is biased as about half of the clusters (247) consist of only one article while another 261 have between 1 and 10 articles and only 47 have more than 10 articles in each cluster. Also the topics of the news clusters are quite characteristic, we refer to Table 2 for a list of the topics and sizes of the larger clusters.

Note that we use the *actual stream* including *all articles* published during the predefined one month period. We exclude *mixed* clusters only from the final evaluation of the detection task.

---

[3] http://goo.gl/LKahS

Table 2. Sample of Topics and Cluster sizes

Cluster Topic	Size	Cluster Topic	Size
Apple Considered Investing in Twitter	20	Google Nexus 7 tablet goes on sale in US	21
VMware buys Nicira for $1.05 billion	21	Google unveils price for Gbit Internet service	21
Digg acquired by Betaworks	23	Microsoft Reboots Hotmail As Outlook	27
FTC Fines Google for Safari Privacy Violations	27	Nokia cuts Lumia 900 price in half to $50	30
Apple Brings Products Back Into EPEAT Circle	31	Yahoo confirms 400k account hacks	45

Twitter Dataset. To examine the potential of our method for very small documents, we used a second dataset consisting of real tweets. This synthetic dataset was constructed using the annotated proportion of the one described in [16]. The dataset contains 27 events of various lengths, from 2 to 837 tweets. The whole dataset consist of 2600 tweets. Events include "*Death of Amy Winehouse*", "*Earthquake in Virginia*" and "*Riots break out in Tottenham*". The stream created uses the actual temporal order of these tweets. Most of the events are well separated from each other with eight of them having a small overlap in time. Here, again we consider as novel only the first story in time for each event. The dataset is available from the website of the CROSS project[4] in the context of which it was created.

4.2   Evaluation Methodology

Detection Errors. The performance of a Novelty Detection algorithm is defined in terms of the missed detection and false alarm error probabilities as defined in [6]. A signal detection model, variation of ROC curves, is often used for evaluation; the Detection Error Trade-off (DET) curve [14], which illustrates the trade-off between missed detections and false alarms. On the x-axis is the miss rate and on the y-axis is the false alarm rate. A system is considered to perform best when it has its curve towards the lower-left of the graph. The axes of the DET curve are on a Gaussian scale.

For the detection systems evaluation, these error probabilities are usually linearly combined into a single detection cost, C_{Det} [6,11] defined as:

$$C_{Det} = (C_{Miss} \times P_{Miss} \times P_{Target} + C_{Fa} \times P_{Fa} \times (1 - P_{Target})) \qquad (8)$$

where P_{Miss} is the number of missed detections divided by the number of target articles, P_{Fa} the number of False Alarms divided by the number of non-targets, C_{Miss} and C_{Fa} the costs of a missed detection and a false alarm respectively, P_{Miss} and P_{Fa} the probabilities of a missed detection and a false alarm respectively and P_{Target} the a priori probability for finding a target. For our experiments we set the same cost for missed detections and false alarms ($C_{Miss} = C_{Fa} = 1$) and the same probability for finding a target and a non-target ($P_{Target} = 0.5$) assuming no prior knowledge for the probability of targets.

Cross-Validation. Since the goal of a detection task is to minimize the detection cost C_{DET}, $minC_{DET}$ is used to define the optimum threshold, i.e. the threshold that gets the lowest C_{DET} value is the best to use for this detection model. The $minC_{DET}$

---

[4] http://demeter.inf.ed.ac.uk/cross/

also corresponds to a certain point on the DET Curve, as the DET curve illustrates the different operating points of a detection system (i.e. the detection errors for different thresholds). In order to avoid an overfitting effect over our datasets, we used 5-fold cross validation in our experiments. We computed the $minC_{DET}$ and the corresponding threshold on the training part and then computed C_{DET} on the testing part. We will report the average C_{DET} in test for all our experiments.

Baselines. We use the ground truth information of each dataset to evaluate the performance of novelty detection for our method and in comparison against the four *baseline approaches*. These methods take into account the similarity/divergence among the document under evaluation and the previous N documents or their summary and rate it as novel based on a threshold. For the experiments, the weighting model used for the baselines is BM25 (kbn in SMART notation), which is the one used in [3].

Weighting Models. As mentioned in section 3, we are using a variety of TF×IDF weighting models that we will refer to using the SMART notations presented in Table 1. For TF, we used the variants b (boolean term representation), n (plain term frequency), l (logarithmic saturation) and k ($BM25$ saturation) and for IDF, we considered the following variants: t (the plain IDF value) and b (the form used in $BM25$). Finally we tested three different options for document length normalization: n (none), d (the document length) and u (the number of unique terms).

5   Results

In this section we present and review the results of the experiments on the datasets mentioned in the previous sections and for all the combinations of measures and parameters values mentioned.

Google News Dataset. We present here the average detection cost ($avgC_{DET}$) for the cleaned dataset with memory size (i.e. length of the corpus) N ranging between 20 and 180 with step 40 for a variety of meaningful combinations of the variants of term frequency, IDF and normalization. We report these results for the snippets and the full articles versions of the dataset (Table 3).

The result table is organized in blocks of lines based on the normalization method. The top block (model SMART acronym ending in d) corresponds to normalization based on the document length, the mid block (SMART acronym ending in u) to normalization based on the number of unique terms in the document and the third one (SMART acronym ending in n) is for the case where no normalization takes place. The last four rows of the table represent the results of the baseline methods (*MaxCS, MeanCS, MaxKL, SumCS*).

The values appearing in the cells represent the average detection cost in test (computed using 5-fold cross validation) for each combination of parameters. We excluded some combinations of the above parameters as they introduce normalization twice (ktd-b=0.75, kbd-b=0.75, ktu-b=0.75, kbu-b=0.75). This is because TF variation k, used in BM25, introduces a length normalization prior to the saturation in its formula for $b > 0$.

Given the above hints, we notice that almost all methods best results are obtained for memory size (N) either 60 or 100 thus we focus our further comments on the respective results columns.

Table 3. Average C_{DET} using 5-fold cross validation on Snippets and Content

N	Snippet					Content				
	20	60	100	140	180	20	60	100	140	180
btd	0.439	**0.407**	0.408	0.411	0.397	0.434	0.429	0.436	0.433	**0.418**
bbd	0.432	**0.404**	0.405	0.411	0.398	0.433	0.423	0.431	**0.416**	0.416
ntd	0.287	**0.266**	0.284	0.285	0.297	0.391	0.366	**0.362**	0.396	0.386
nbd	0.294	**0.267**	0.291	0.284	0.307	0.4	**0.367**	0.373	0.4	0.394
ltd	0.413	0.382	**0.359**	0.371	0.368	0.429	0.425	0.412	**0.405**	0.413
lbd	0.393	0.381	**0.36**	0.374	0.373	0.422	0.414	**0.412**	0.419	0.415
ktd-b=0	0.394	0.367	**0.354**	0.362	0.365	0.429	0.428	0.411	**0.404**	0.412
kbd-b=0	0.392	0.368	**0.346**	0.36	0.366	0.424	0.415	**0.413**	0.419	0.422
btu	0.307	0.299	**0.293**	0.296	0.311	0.447	0.444	0.451	0.434	**0.429**
bbu	0.313	0.299	**0.294**	0.297	0.307	0.44	0.452	0.45	**0.425**	0.429
ntu	0.319	**0.293**	0.294	0.317	0.303	0.464	**0.441**	0.447	0.442	0.46
nbu	0.324	**0.288**	0.298	0.302	0.308	0.458	**0.437**	0.44	0.44	0.449
ltu	0.298	0.283	**0.281**	0.283	0.299	0.455	**0.423**	0.424	0.436	0.461
lbu	0.301	**0.276**	0.281	0.288	0.299	0.455	**0.429**	0.438	0.436	0.447
ktu-b=0	0.295	0.282	**0.268**	0.28	0.296	0.452	**0.42**	0.427	0.446	0.458
kbu-b=0	0.298	0.283	**0.279**	0.279	0.302	0.458	**0.422**	0.439	0.438	0.45
btn	0.429	0.41	0.391	**0.389**	0.398	**0.503**	0.504	0.504	0.504	0.504
bbn	0.417	0.402	**0.385**	0.388	0.403	**0.496**	0.504	0.502	0.505	0.507
ntn	0.371	**0.332**	0.337	0.336	0.346	**0.483**	0.499	0.499	0.497	0.499
nbn	0.366	0.333	**0.33**	0.34	0.339	**0.492**	0.502	0.501	0.508	0.509
ltn	0.401	0.375	**0.364**	0.37	0.371	0.502	0.509	**0.502**	0.504	0.502
lbn	0.399	0.372	0.374	**0.364**	0.371	0.498	0.515	**0.501**	0.503	0.505
ktn-b=0	0.395	0.368	0.372	0.364	**0.36**	0.499	0.51	0.503	0.508	0.505
ktn-b=0.75	0.385	0.342	**0.337**	0.348	0.343	0.461	**0.432**	0.434	0.439	0.447
kbn-b=0	0.39	0.363	0.363	**0.36**	0.367	0.498	0.516	0.501	**0.5**	0.505
kbn-b=0.75	0.38	0.343	**0.335**	0.347	0.343	0.441	**0.428**	0.445	0.451	0.446
maxCS	0.375	0.369	0.365	0.368	**0.353**	0.492	0.465	0.457	0.456	**0.45**
meanCS	0.368	0.362	0.344	**0.339**	0.355	0.471	0.461	0.444	0.44	**0.439**
maxKL	0.448	0.438	0.423	**0.421**	0.43	0.49	0.458	0.449	**0.434**	0.471
CSAgg	0.451	0.425	0.424	0.425	**0.423**	0.494	0.48	0.476	**0.47**	0.489

It is evident that the proposed Novelty Scoring measure outperforms the all baselines with the best performance achieved by a L^1 normalized TF-IDF (raw TF and classic IDF – ntd) narrowly followed by the nbd model (same except for the IDF component inherited from BM25). Very good performance is achieved by the u normalization (number of unique terms) especially for the ltu and kbu models. Absence of length normalization yields to the worst results as it can be expected with documents of varying length. Nevertheless it still outperforms the best baseline results when we consider the snippets where the variation in length is limited. Considering the content of the articles for novelty detection the no normalization weighting schemes perform much worst even than the baselines. The difference in performance of this group in comparison to the one on snippets originates at the greater differences in document lengths when the full article is taken into account (snippets tend to have a constant length – around 25 terms). Note that $ktn\text{-}b = 0.75$ and $kbn\text{-}b = 0.75$ perform better than the rest of the block. This can be easily explained as both methods use the BM25 variant of TF which includes a pivot length normalization for parameter $b > 0$. We chose to display them in that block just to be consistent in terms of SMART notations.

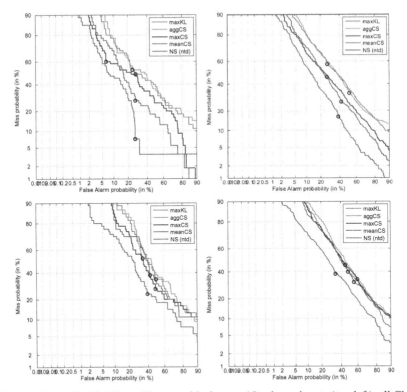

Fig. 2. DET Curves for N=100 on Clusters with size$\geq$ 10 using snippets (top-left), all Clusters using snippets (top-right), size$\geq$ 10 using content (bottom-left) and all Clusters using content (bottom-right)

*DET Curves:*We plotted DET curves showing the evolution of performance with regards to the Miss and the False Alarm probability. These diagrams indicate the evolution of the detection cost for the best performing model (*ntd*) and all the baselines. They also depict the point on each curve that corresponds to the optimum threshold, having the $minC_{DET}$.

In figure 2 we plot the DET Curves for memory N=100 on four versions of the Google News dataset, large Clusters with size$\geq$ 10 using snippets (top-right), size$\geq$ 10 using content (bottom-left) and all Clusters using content (bottom-right). We compare all baseline methods and our method using the best performing weighting schema, ntd(see table 3). It is clear that overall the ntd method outperforms the others. The baseline based on maximum KL-divergence and document-to-summary baseline perform worst. The same applies for the case of *all Clusters* data set. In addition, comparing the corresponding snippet and content DET curves we confirm again our previous claims that using the full content of an article instead of a simple summary as the first few sentences of the article introduces significant noise and makes it harder to detect the first stories.

Twitter Dataset. We report in Figure 3 the results on the Twitter dataset described in section 4.1. We again compare all baseline methods and our method using the best

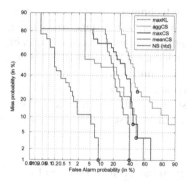

Fig. 3. DET Curve for N=100 on Twitter dataset

performing weighting schema, ntd (see table 3) for $N = 100$. As mentioned earlier, we exhaustively examine the performance of our method for the *Twitter Dataset* as well. Due to lack of space, we present the results of the best performing weighting scheme and N value using the DET Curves as a more concise means. The results are very encouraging, as our method outperforms by far all the baselines and manages to have a zero miss probability while maintains false alarm probability below 10%.

Execution Time. We compared our method in terms of execution time with the best performing method from the baselines, *MeanCS*. We ran experiments for different values of N, using content. We used the whole stream of news. The results are shown in Table 4. The values reported correspond to the average time needed to process and assign a novelty score to an article in the dataset. The time cost for database connection and communication, indexing and index updating is not considered.

It is clear that our method is considerably faster than the document-to-document competing ones as it is at least seven times faster than *MeanCS*. The difference among the methods increases as the corpus length increases, since *MeanCS*, as any document-to-document method, have to be executed on the entire corpus to compute the similarity between all documents.

Table 4. Execution times per article in microseconds for different N values, using content

	N=20	N=60	N=100	N=140	N=180
NS	124.44	154.46	128.50	200.54	134.96
meanCS	704.06	1798.30	2372.72	3156.27	3923.91

6   Conclusion

Novelty detection is an important topic in modern text retrieval systems. In this paper we proposed a new method for the novelty detection task in document streams that is accurate (i.e. performing better than several dominant baselines). We conducted extensive experiments on a real world dataset (from a news stream) where our method clearly outperforms the four baseline techniques used in the relevant literature. Moreover, as our method does not use any similarity or distance measure among documents but only stream statistics kept in memory, it is much faster and scalable than the others.

These results give strong evidence that stream statistics, such as IDF in our case, can alone be used to detect novel documents from streams. IDF is a simple yet effective indicator of both *term specificity* and *document novelty*. The first property has been known since 1972 and our work just showed the second one. In large-scale streaming, such as on Twitter that recently sparked interest in the research community, this observation may be of great importance.

Acknowledgments. M. Karkali has been co-financed by the EU (ESF) and Greek national funds through the Operational Program "Education and Lifelong Learning" of the NSRF - Heracleitus II. This work was also supported by PIRG06-GA-2009-256603.

References

1. Allan, J.: Introduction to topic detection and tracking. In: Allan, J. (ed.) Topic Detection and Tracking. The Information Retrieval Series, vol. 12, pp. 1–16. Springer, US (2002)
2. Allan, J., Lavrenko, V., Jin, H.: First story detection in tdt is hard. In: CIKM 2000, pp. 374–381. ACM (2000)
3. Allan, J., Lavrenko, V., Malin, D., Swan, R.: Detections, bounds, and timelines: Umass and tdt-3. In: Topic Detection and Tracking Workshop, TDT-3 (2000)
4. Allan, J., Wade, C., Bolivar, A.: Retrieval and novelty detection at the sentence level. In: SIGIR 2003, pp. 314–321. ACM (2003)
5. Fang, H., Tao, T., Zhai, C.: A formal study of information retrieval heuristics. In: SIGIR 2004, pp. 49–56. ACM (2004)
6. Fiscus, J.G., Doddington, G.R.: Topic detection and tracking. In: Allan, J. (ed.) Topic Detection and Tracking, ch. 1, pp. 17–31. Kluwer Academic Publishers (2002)
7. Harman, D.: Overview of the trec 2002 novelty track. In: TREC 2002, pp. 46–55. NIST Special Publication 500-251 (2002)
8. Kwee, A.T., Tsai, F.S., Tang, W.: Sentence-level novelty detection in English and Malay. In: Theeramunkong, T., Kijsirikul, B., Cercone, N., Ho, T.-B. (eds.) PAKDD 2009. LNCS, vol. 5476, pp. 40–51. Springer, Heidelberg (2009)
9. Li, X., Croft, W.B.: Novelty detection based on sentence level patterns. In: CIKM 2005, pp. 744–751. ACM (2005)
10. Luo, G., Tang, C., Yu, P.S.: Resource-adaptive real-time new event detection. In: SIGMOD 2007, pp. 497–508. ACM (2007)
11. Manmatha, R., Feng, A., Allan, J.: A critical examination of tdt's cost function. In: SIGIR 2002, pp. 403–404. ACM (2002)
12. Markou, M., Singh, S.: Novelty detection a review–part 1: statistical approaches. Signal Process. 83(12), 2481–2497 (2003)
13. Markou, M., Singh, S.: Novelty detection a review-part 2: neural network based approaches. Signal Process. 83(12), 2499–2521 (2003)
14. Martin, A., Doddington, G., Kamm, T., Ordowski, M., Przybocki, M.: The det curve in assessment of detection task performance. In: 5th European Conference on Speech Communication and Technology, pp. 1895–1898 (1997)
15. Ohgaya, R., Shimmura, A., Takagi, T., Aizawa, A.N.: Meiji university web and novelty track experiments at trec 2003. In: TREC 2003, pp. 399–407 (2003)
16. Petrović, S., Osborne, M., Lavrenko, V.: Streaming first story detection with application to twitter. In: HLT 2010, pp. 181–189. ACL (2010)

17. Robertson, S.E., Walker, S.: On relevance weights with little relevance information. SIGIR Forum 31(SI), 16–24 (1997)
18. Robertson, S.E., Walker, S., Sparck Jones, K., Hancock-Beaulieu, M., Gatford, M.: Okapi at TREC-3. In: TREC-3, pp. 109–126 (1994)
19. Singhal, A., Salton, G., Buckley, C.: Length normalization in degraded text collections. Technical report, Cornell University, Ithaca, NY, USA (1995)
20. Soboroff, I.: Overview of the trec 2004 novelty track. In: TREC 2004. NIST Special Publication, pp. 500–251 (2004)
21. Soboroff, I., Harman, D.: Overview of the trec 2003 novelty track. In: TREC 2003. NIST Special Publication, pp. 500–251 (2003)
22. Soboroff, I., Harman, D.: Novelty detection: the trec experience. In: HLT 2005, pp. 105–112. ACL (2005)
23. Sparck Jones, K.: A statistical interpretation of term specificity and its application in retrieval. Journal of Documentation 28(1), 11–20 (1972)
24. Tsai, F.S.: Review of techniques for intelligent novelty mining. Information Technology Journal 9, 1255–1261 (2010)
25. Tsai, F.S., Kwee, A.T.: Experiments in term weighting for novelty mining. Expert Systems with Applications 38(11), 14094–14101 (2011)
26. Tsai, F.S., Tang, W., Chan, K.L.: Evaluation of novelty metrics for sentence-level novelty mining. Inf. Sci. 180(12), 2359–2374 (2010)
27. Verheij, A., Kleijn, A., Frasincar, F., Hogenboom, F.: A comparison study for novelty control mechanisms applied to web news stories. In: WI 2012, pp. 431–436. IEEE Computer Society (2012)
28. Yang, Y., Zhang, J., Carbonell, J., Jin, C.: Topic-conditioned novelty detection. In: KDD 2002, pp. 688–693. ACM (2002)
29. Zhang, K., Zi, J., Wu, L.G.: New event detection based on indexing-tree and named entity. In: SIGIR 2007, pp. 215–222. ACM, New York (2007)
30. Zhang, Y., Callan, J., Minka, T.: Novelty and redundancy detection in adaptive filtering. In: SIGIR 2002, pp. 81–88. ACM (2002)

Detecting Opinion Drift from Chinese Web Comments Based on Sentiment Distribution Computing[*]

Daling Wang[1,2], Shi Feng[1,2], Dong Wang[1], and Ge Yu[1,2]

[1] School of Information Science and Engineering, Northeastern University
[2] Key Laboratory of Medical Image Computing (Northeastern University),
Ministry of Education, Shenyang 110819, P.R. China
{wangdaling,fengshi,yuge}@ise.neu.edu.cn, hrdxwangd@gmail.com

Abstract. Opinion drift is regarded as the change of sentiment distribution in this paper. In opinion and sentiment mining, how to detect opinion drifts and analyze their reasons, is an important problem for Web public opinion analysis. To tackle this problem, an approach of opinion drift detection for Chinese Web comment is proposed. For a comment set during a long time about a hot event, the proposed approach first determines possible drift timestamps according to the change of comment number, computes different sentiment orientations and their distributions at these timestamps, detects opinion drift according to the distribution changes, and analyzes the influences of related events occurring in the timestamps. Extensive experiments were conducted in a real comment set of Chinese forum. The results show that drift timestamps determined and opinion drifts detected correspond to the real event, so the approach proposed in this paper is feasible and effective in the application of Web public opinion analysis.

Keywords: Opinion Drift Detection, Sentiment Distribution, Opinion Mining.

1 Introduction

More and more users have been accustomed to publish their comments and opinions about hot events on the Web. Many facts indicate that after a hot event occurred, the users' attention to the event would vary over time. Especially, with more information about the event been issued, users' sentiment hidden in their comments may change. In the beginning, perhaps users' sentiments focus on positive (support), but when more details or related information about the event are known, users' sentiment may change from positive to negative, or vice versa by their further thinking and judging.

In machine learning, the change of a concept over time is regarded as "concept drift" [13]. Inspired by the idea, in this paper, we call the change of sentiment distribution as "opinion drift". In sentiment mining, how to detect opinion drifts and analyze their reasons, is an important problem for Web public opinion analysis.

---

[*] Project supported by the State Key Development Program for Basic Research of China (Grant No. 2011CB302200-G), State Key Program of National Natural Science of China (Grant No. 61033007), National Natural Science Foundation of China (Grant No. 61100026, 60973019), and the Fundamental Research Funds for the Central Universities (N100704001, N120404007).

X. Lin et al. (Eds.): WISE 2013, Part I, LNCS 8180, pp. 72–81, 2013.

As one of the techniques for Web public opinion analysis, opinion and sentiment mining analyzes users' opinions, appraisals, attitudes, and emotions toward entities, individuals, issues, events, topics, and their attributes [8]. The current work includes opinion extraction, opinion summary, opinion tracking, opinion retrieval based on sentiment orientation, and so on. However, for a hot event, only a few studies detect users' opinion or sentiment change over time, especially for Chinese websites. Above opinion tracking aims to track users' common sentiment [3] or content [4] change, but not every sentiment distribution change, i.e. opinion drift in this paper. For public opinion analysis, detecting users' different sentiment distribution change is important as the same as tracking users' common sentiment change.

Based on above analysis, in this paper, for a hot event, we track the comments during a long term, compute every sentiment distribution, detect opinion drift, and analyze the reason resulting in the drift. We propose an opinion drift detection model and its modeling approach, and show the effectiveness of the proposed model.

The rest of the paper is organized as follows. Section 2 introduces related work. Section 3 describes our opinion drift detection model. Section 4 presents the modeling process in detail. Section 5 shows and discusses the experiment results, and the conclusion is given in Section 6.

2   Related Work

Inspired by the concept drift, in this paper we call the change of sentiment orientation distribution as opinion drift. In some previous work, opinion drift is also called as opinion change, sentiment change, emotion change, and so on.

In this domain, [1] classified tweets in real-time and detected sentiment change by comparing the average values in two windows and analyzing the sentiment changed. [2] used association rule mining to find out opinions about products in an e-commerce website, and detected changes in patterns between two datasets from different time periods. [11] built an opinion formation model for detecting where in time and how long this change persisted and how big this change was. [7] adopted the probabilistic topic model and language grammar based sentiment analysis techniques for mining opinions on a certain topic, and detected significant changes of sentiment of the opinions on the topic. Furthermore, they identified possible reasons causing each such change. [10] researched the changes in referents and emotions over time in election-related social networking dialog by analyzing week-by-week content of Facebook wall postings by candidates' fans. [12] regarded Twitter real-time messages as streams, and researched sentiment stream classification for tracking sentiment drift and providing up-to-date sentiment analysis.

Our work is different from above related work. Firstly, the related work mainly aimed at English or Japanese, but our work is for Chinese. Because of different syntax and grammar structures, the sentiment analysis will apply different methods. Secondly, the related work focused on microblog [1, 12], product review [2], blog [7, 11], and other users generating content [10, 11]. Our work analyzes users' sentiment from the comments of hot events. There is the difference in topic number and text length between them.

3   Building Opinion Drift Detection Model

For a hot event E on the Web, let users' comment set be C, the opinion drift detection model OD for E be described as $OD=<Class, TS(Class), Drift(Class), SE(TS)>$. Here $Class=<pos, neu, neg>$ is sentiment orientation, the triplet $<pos, neu, neg>$ indicates three classes of sentiment, i.e. positive (approve), neutral, and negative (oppose), respectively. $TS(Class)$ is the timestamp set. Every $ts \in TS$ is a timestamp. Every timestamp in OD model is the time of opinion drift occurring, and called opinion drift timestamp. $Drift(Class)$ is the set of value of quantized opinion drift. $SE(TS)$ is the event set resulting in opinion drift occurring in TS. According to the description, OD can be expressed as that users' sentiment $Class$ occurs opinion drift $Drift$ due to the influence of event SE at the timestamp TS.

Obviously, opinion drift detection concerns with timestamps, i.e. when the drift occurs. For further describing and modeling OD, we give the following definitions.

Definition 1 (Timestamp). A timestamp is a time period, which can be one hour, one day, or several days. We denote a timestamp as ts.

For describing the relation of comments and timestamps, suppose that we obtained comment set C of a hot event in time period T, T can be partitioned into m shorter proximate sub-periods by timestamps ts_1, ts_2, ..., ts_m. Accordingly, C can also be partitioned into m subsets C_1, C_2, ..., C_m, and the comment number $|C|=|C_1|+|C_2|+...+|C_m|=D(ts_1)+D(ts_2)+...+D(ts_m)$. Here $D(ts_i)=|C_i|$ ($i=1, 2, ..., m$) is the number of comments in timestamp ts_i. In this case, T can be regarded as a longer timestamp.

Definition 2 (Sentiment Distribution). For a class of sentiment orientation o in comment set C, its sentiment distribution is defined as the percentage of comment number with o class in $|C|$, we denote it as $Dist_C(o)$, $o \in \{pos, neu, neg\}$.

Definition 3 (Opinion Drift). In comment set C, for two comment subset C_i, $C_j \subseteq C$ corresponding to two proximate timestamps ts_i, $ts_j \subseteq T$ and a given threshold γ, if a class of sentiment o satisfies $|Dist_{C_i}(o)-Dist_{C_j}(o)| \geq \gamma$, then the opinion drift occurs in C at timestamp ts_j. In this case, the timestamp ts_j and the value of $|Dist_{C_i}(o)-Dist_{C_j}(o)|$ are the elements of TS and $Drift$ set in OD model, respectively.

According to Definition 3, opinion drift detection for C can be done by calculating $|Dist_{C_i}(o)-Dist_{C_j}(o)|$, i.e. the sentiment distribution of two proximate comment subset C_i and C_j, so we call the timestamp ts_j which corresponds to C_j as a drift timestamp.

Definition 4 (Drift Timestamp). For the timestamp ts in T corresponding to comment set C, if the opinion drift occurs in C at ts, then ts is defined as a drift timestamp and represented by dts.

Obviously, a dts must be a ts, but not all tss are $dtss$. Moreover, only $dtss$ are the elements of TS in OD model.

Based on above definitions, modeling OD will concern the following problems. 1) Download the related comment data with a hot event. 2) Set timestamps and further determine (candidate) drift timestamps from the given timestamps. 3) Compute the sentiment distribution for every class of sentiments and detect the opinion drift at every determined (candidate) drift timestamp. 4) Analyze the influence of related events in every real drift timestamp occurring opinion drift. For problem 2) and 3), we detail the approach in the next section.

4   Implementing Opinion Drift Detection Function

4.1   Determining Detection Time and Partitioning Comment Data

The comments of a hot event will continue for a long time from the start to the end of the event, and there is different number of comments in the different timestamp over the increment of related information with the event. In intuition, the sentiment distribution may change when the comment number increases significantly. So the scale and significant increase of comment number should be two pivotal metrics for drift timestamp.

[7] found that a sentiment change of a topic was usually accompanied by hot discussions related to the topic by analyzing blog search results, so an increase of the popularity of a topic suggested a possible sentiment change to the topic. Inspired by the idea of [7], we think that similar to blogs, for the comments related with a hot event, opinion drift also accompanies number increase of the comments. Comparing with our above two metrics of drift timestamp, for a timestamp ts_j, "significant increase of comment number" can be regarded as "$D(ts_j)$ is much larger than $D(ts_i)$", and "scale of comment number" can be regarded as "$D(ts_j)$ itself is relatively large". So we apply the approach proposed by [7] for partitioning time period.

According to the idea of [7], for two proximate timestamps ts_i and ts_j $(i<j)$, Formula (1) computes a *BoundaryScore(ts_j)*.

$$BoundaryScore(ts_j) = \frac{D(ts_j)}{D(ts_i)}\log(D(ts_j)) \tag{1}$$

Moreover, [7] proposed that for timestamp ts_j, if the document number of both the split time intervals before and after it were greater than a predefined threshold, it was a qualified boundary. The authors of [7] considered the accumulation of comment scale mentioned by our paper. For our application, we detail the process as follows.

For the timestamps ts_1, ts_2, ..., ts_i, ts_j, ..., ts_m in time period T corresponding to comment set C, if ts_j is a qualified boundary, both $D(ts_1)+D(ts_2)+...+D(ts_i)+D(ts_j)$ and $D(ts_j)+...+D(ts_m)$ should satisfy a threshold, which is set to be *MinCommCnt*. That means:

$$D(ts_1) + \cdots + D(ts_i) + D(ts_j) \geq MinCommCnt \ \& \ D(ts_j) + \cdots D(ts_m) \geq MinCommCnt \tag{2}$$

We compute *BoundaryScore(ts_j)* if $D(ts_j)$ satisfies Formula (2). From $D(ts_1)$ to $D(ts_m)$, we obtain a *ts* with the largest *BoundaryScore(ts)* as the first candidate *dts*, denoted by *cdts*, namely,

$$cdts = \arg\max_{ts \subseteq T} BoundaryScore(ts) \mid ts \text{ satisfies } ts_j \text{ in Formula (2)} \tag{3}$$

If the *cdts* is a real *dts*, according to the real significance of *dts*, it should be a start of new related event and opinion drift. So based on the first *cdts* (suppose it is ts_j), we continue to find new *cdts*s in ts_1~ts_i and ts_j~ts_m. The process runs recursively until no new *cdts* is found.

4.2   Analyzing Sentiment Orientation of Comment Subset in Every Timestamp

There is the difference of length among different comments in Chinese comments. We analyze sentiment orientation of comments from word, sentence, and comment, to comment subset. According to the characteristics of Chinese, in a comment, we partition sentence with period (.), exclamatory mark (!), semicolon (;), and question mark (?), i.e. these marks are regarded as the end of a sentence. In detail, the process of computing the sentiment orientation of a comment subset is in the following.

1) **Extract sentiment words from every sentence.** We first partition every sentence into words based on two famous Chinese sentiment lexicons HowNet [5] and NTUSD [9], merge the two lexicons, and add some new popular network sentiment words such as "顶 (agree)" and "囧 (embarrassed)".

2) **Compute the orientation of every sentiment word.** We consider the orientation of every sentiment word, and utilize the approach of computing sentiment orientation based on HowNet as Formula (4) [14].

$$Orientation(w) = \sum_{i=1}^{k_1} Sim(posseed_i, w) - \sum_{j=1}^{k_2} Sim(negseed_j, w) \qquad (4)$$

where *posseed* and *negseed* are positive and negative seed words, respectively, and the numbers of two classes of seed words are k_1 and k_2 respectively. Moreover, $Sim(w_1, w_2)$ is the similarity between w_1 with w_2 computed by using HowNet.

3) **Compute sentiment orientation of every sentence.** Based on the orientation of sentiment word in a sentence, we use Formula (5) for determining the sentiment orientation of the sentence.

$$Orientation(s) = \sum_{w \in s} Orientation(w) \qquad (5)$$

where w is a sentiment word in sentence s. Moreover, we give two thresholds α and β ($\alpha > \beta$). For a sentence s, if $Orientation(s) \geq \alpha$, sentence s is defined as positive sentiment. If $Orientation(s) < \beta$, the sentence s is defined as negative sentiment. If $\alpha > Orientation(s) \geq \beta$, the sentence s is defined as neutral sentiment.

4) **Compute the sentiment distribution of every comment.** We use a triplet <pos, neu, neg> to express the sentiment distribution of a comment. Based on Formula (5), we can get the sentence sets with positive, neutral, and negative sentiment respectively from a comment consisting of some sentences. We denote them as S_{pos}, S_{neu}, and S_{neg} respectively. For a comment consisting of sentence set $S = S_{pos} + S_{neu} + S_{neg}$, its sentiment distribution is calculated with Formula (6).

$$pos = \sum_{s \in S_{pos}} |s| \Big/ \sum_{s \in S} |s| \qquad neu = \sum_{s \in S_{neu}} |s| \Big/ \sum_{s \in S} |s| \qquad neg = \sum_{s \in S_{neg}} |s| \Big/ \sum_{s \in S} |s| \qquad (6)$$

That means the sentiment distribution of a comment is the percentage of positive, negative, and neutral sentences in the comment, and the distribution value is between 0 and 1. Especially, if a comment only contains one sentence and the sentiment orientation of the sentence is the positive, the sentiment distribution of the comment is <1, 0, 0>. If its sentiment orientation is the negative, the sentiment distribution of the comment is <0, 0, 1>.

5) **Compute sentiment distribution of comment subset.** Based on the detection time determined in Section 4.1, the time period T corresponding to the comment set C

can be partitioned into some *cdts*s, and *C* can be partitioned into some related subsets. According to Definition 2 in Section 3, for every comment subset, we compute the percentage of the comments with positive, neutral, and negative sentiments in the subset respectively as the sentiment distribution of the subset.

According to Formula (6), for a comment subset $\hat{C} \subseteq C$ corresponding to a *cdts* in *T*, we can obtain the sentiment distribution *<pos, neu, neg>* of every comment $c \in \hat{C}$. Then we sum *pos*, *neu*, and *neg* for all $c \in \hat{C}$, denote the results as $\hat{C}(pos)$, $\hat{C}(neu)$, and $\hat{C}(neg)$ respectively, and compute the sentiment distribution with Formula (7).

$$Dist_{\hat{C}}(o) = < o(pos), o(neu), o(neg) > \tag{7}$$

where $o(pos) = \hat{C}(pos)/|\hat{C}|$ $o(neu) = \hat{C}(neu)/|\hat{C}|$ $o(neg) = \hat{C}(neg)/|\hat{C}|$.

To sum up the above five steps, for a comment subset $\hat{C}$, the algorithm of analyzing the sentiment distribution of $\hat{C}$ is described as Algorithm1.

Algorithm 1. Sentiment Distribution Analysis of Comment Subset

---

Input: Comment subset $\hat{C}$ corresponding to a candidate drift timestamp
Output: Sentiment distribution $Dist_C(o)$=<o(pos), o(neu), o(neg)>
  1) For every comment $c \in \hat{C}$
  2)   {partition *c* into sentence set *S*;
  3)   For every sentence $s \in S$
  4)   {extract sentiment words of *s* and construct set *W*;
  5)   For every sentiment word $w \in W$
  6)   compute sentiment orientation of *w* with Formula (4);
  7)   compute sentiment orientation of *s* with Formula (5);}
  8)   compute sentiment distribution *<pos, neu, neg>* of *c* with Formula (6);}
  9) compute $Dist_C(o)$=<o(pos), o(neu), o(neg)> with Formula (7);

4.3 Opinion Drift Detection

In this paper, opinion drift means the change of sentiment orientation distribution of users' comments about a hot event. In detail, for a comment set *C* about a hot event, its corresponding time period is *T*, we can partition *T* into some timestamps and obtain some candidate drift timestamps from them using the approach in Section 4.1. For any one candidate drift timestamp *cdts'* and its precursor timestamp *cdts"*, if the comment set in *cdts'* and *cdts"* are *C'* and *C"* respectively, we judge whether opinion drift occurs in *cdts'* by comparing $|Dist_{C'}(o) - Dist_{C''}(o)|$ with a given threshold γ according to Definition 3. If opinion drift occurs, *cdts'* is a real drift timestamp. Moreover, we analyze related events occurring in *cdts'*. Based on above analysis, the algorithm of opinion drift detection is described in Algorithm2.

In line 1) of Algorithm2, we get all candidate drift timestamps using the approach in Section 4.1. In line 4) and line 5), we call Algorithm1 for computing the sentiment distributions of timestamp *cdts'* and *cdts"* respectively. Then from line 6) to line 11), we judge whether opinion drift occurs in *cdts'*. If occurring, we continue to analyze the influences of related events in line 12).

Algorithm 2. Opinion Drift Detection

Input: Comment set C and its time period T about a hot event
Output: drift timestamp set TS, drift value set $Drift$ of every class of sentiment

   1) find candidate drift timestamp set $CDTS$;

   2) $TS=Drift=\{\}$;

   3) For every $cdts' \in CDTS$ and its precursor timestamp $cdts''$

   4)   \{compute $Dist_C(o)=<o(pos), o(neu), o(neg)>$ with Algorithm1;

                   //C' is comment subset in $cdts'$

   5)   compute $Dist_{C''}(o)=<o(pos), o(neu), o(neg)>$ with Algorithm1;

                   //C'' is comment subset in $cdts''$

   6) $Diff(pos)=|Dist_{C'}(o(pos))-Dist_{C''}(o(pos))|$;

   7) $Diff(neu)=|Dist_{C'}(o(neu))-Dist_{C''}(o(neu))|$;

   8) $Diff(neg)=|Dist_{C'}(o(neg))-Dist_{C''}(o(neg))|$;

   9)   If $Diff(pos)>\gamma$ or $Diff(neu)>\gamma$ or $Diff(neg)>\gamma$

  10)   Then $\{TS=TS+\{cdts'\}$;

  11) $Drift=Drift+\{Diff(pos), Diff(neu), Diff(neg)\}$;

  12)   analyze related event occurring in timestamp $cdts'\}\}$

Recalling the model $OD=<Class, TS(Class), Drift(Class), SE(TS)>$ in Section 3, now we can model OD by finding drift timestamps, computing sentiment distribution, detecting opinion drift, and analyze the influence of related events in Section 4.

5   Experiments

5.1   Experiment Corpora and Lexicon

The event of Huangyan Island confrontation between Chinese and Philippines is one of the hot events in 2012. We collect related comments with the event from Netease[1] for detecting opinion drift.

From Netease, we obtain comment set about the event of Huangyan Island confrontation between Chinese and Philippines in time period 2012.4.10~2012.5.14 (Huangyan Island Event for short). We first purify the comment set by deleting ads and other irrelated information with the event, and the result includes 85,447 comments. The comment number distribution is shown as Figure 1.

In Figure 1, we set timestamp be one day, so the number of timestamps is 35 from 2012.4.10 to 2012.5.14. Moreover, for partitioning every comment into words, we apply ICTCLAS 5.1 [6] from Chinese Academy of Sciences. For extracting sentiment words and analyzing sentiment orientation of comments, we apply HowNet [5] and NTUSD [9] lexicon, and add some new words such as people and place names, popular network words obtained by manual. We merge the two lexicons and above new words, eliminate the duplicates, and form a final sentiment lexicon.

---

[1] http://comment.news.163.com/news_guoji2_bbs/SPEC00017V807LPO.html

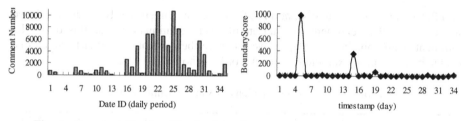

Fig. 1. Comment Number Distribution of Huangyan Island Event

Fig. 2. *BoundaryScore* Distribution

5.2   Experiment Results and Analysis

Because we have not found the related work of opinion drift detection for comments in Chinese, here we show our experiment results conforming to the facts for proving the feasibility and availability of our work.

1) Finding Candidate Drift Timestamps. Based on the comment set of Huangyan Island Event in Figure 1, we use the approach in Section 4.1, let the threshold *MinCommCnt* be 50, and obtain 3 timestamps with the most *BoundaryScore*. The *BoundaryScore* distribution is shown as Figure 2.

From Figure 2, we see timestamp 5, 15, and 19 are detected as drift timestamps. They correspond to date ID 6, 16, and 20 in Figure 1 respectively. In fact, 3 related events occur in these dates, they are "Chinese Fishery Boat 310 leaves Huangyan island sea area", "Philippines says the United States will tripling the military aid for Philippine troops", and "Philippines intends to clear the Huangyan island and Chinese logo". So it is reasonable for us to regard the 3 timestamps as candidate drift timestamps.

2) Analyzing Sentiment Orientation of Comment. For the comment set about Huangyan Island Event, we analyze the sentiment distribution by analyzing sentiment words, sentences, every comment, and comment subset according to timestamps. According to the approach in Section 4.2, our experiment process is shown as follows.

Step 1. Partition every comment into sentences and every sentence into words. Here we take the comment subset of timestamp 22 in Figure 3 as an example for describing the process. Figure 3 give an example of a comment in this timestamp.

Fig. 3. An Example of a Comment

We partition the comment into 12 sentences (The vertical line in Figure 3), then partition every sentence into words, and extract sentiment words. For example, the sentiment words extracted from second sentence include反复无常(capricious), 出尔反尔(contradict oneself), 戏弄(make fun of), 幕后(backstage), and 坏(bad).

Step 2. Compute sentiment distribution of every comment subset. By Algorithm1, we first compute the sentiment orientation of every sentiment word. Based on the sentiment words obtained in Step 1, we compute their orientation with Formula (4), and the orientation of example sentiment words is shown as Table 1.

Table 1. Sentiment Orientation of Example Sentiment Words

Sentiment Words	反复无常 capricious	出尔反尔 contradict oneself	果断 resolute	纸老虎 paper tiger	智慧 wise
Sentiment Orientation	-0.9062	-0.9117	0.8861	-0.8036	0.8636

Moreover, we compute the sentiment distribution of every sentence with Formula (5) and compute the sentiment distribution of every comment with Formula (6). For example, for the comment shown in Figure 3, the sentiment distribution is *<pos, neu, neg>=<0.286, 0.286, 0.428>*. Based on the distribution, we continue to compute the sentiment distribution of the comment subset in timestamp 22 $Dist_C(o)=<o(pos)$, $o(neu)$, $o(neg)>=<0.121, 0.148, 0.731>$.

3) Detecting Opinion Drift. With Algorithm1, we can compute the sentiment distribution of any timestamp. In this paper, we assume that by only computing the sentiment distribution of all candidate drift timestamps and their precursor timestamps, we can detect opinion drift with Algorithm2. However, for proving our assumption, we compute the sentiment distribution of every timestamp and judge whether the opinion drift occurs or not in candidate drift timestamp by calculating *Diff(<pos, neu, neg>)* according to line 6), line 7), and line 8) in Algorithm2 and Definition 3. Figure 4 shows the sentiment distribution of timestamp 10~20, and Figure 5 indicates the sentiment distribution in all timestamps of Huangyan Island Event.

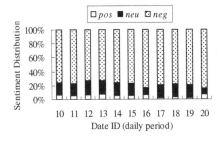

Fig. 4. An Example of Sentiment Distribution in 11 timestamps

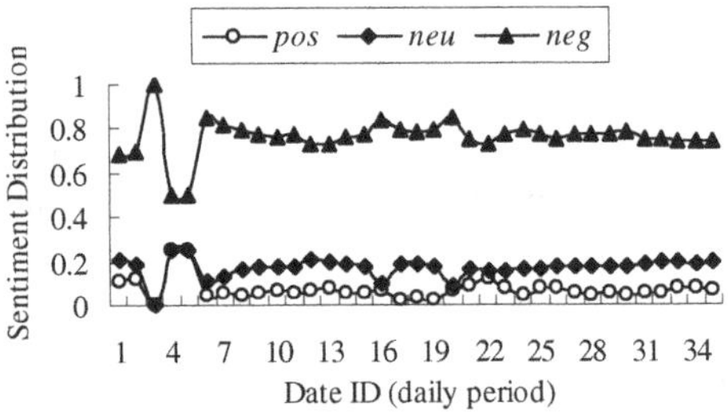

Fig. 5. Sentiment Distribution in All Timestamps of Huangyan Island Event

From Figure 5, we see that in date ID 3, 4, and 5, the sentiment distribution is undulate, but the number of comment is not enough, so it indicates that users had not paid attention to the event in this time period, the sentiment orientation is not significant. In fact, they are not detected as drift timestamp when setting threshold *MinCommCnt* 50. In date ID 6, negative sentiment distribution (opinion summarization by manual is "should solve with force") becomes larger, because the event of "Chinese Fishery Boat 310 leaves Huangyan island sea area" occurs in this day, and results in users' distrust to the government's behavior. Users think the government should take strong measures to Philippines. From date ID 6 to 16, the negative sentiment distribution diminishes gradually, and indicates government's behavior was being understood gradually. But in date ID 16, the negative sentiment distribution increases again, because another related event "Philippines says the United States will tripling the military aid for the Philippine troops" occurs in this day, users hope government's strong measure again. Similarly, in date ID 20, negative sentiment distribution increases again accompanied by the event "Philippines intend to clear the Huangyan Island and Chinese logo". The event initiates users' indignation again. Accordingly, another two sentiment distributions have the similar drift because of above 3 related events.

6   Conclusion

In this paper, we define the concept of opinion drift and propose an approach of opinion drift detection for Chinese web comment about hot events. For a comment set during a long time about a hot event, our approach first determines candidate drift timestamps according to the number change of the comments, computes sentiment distribution at the candidate drift timestamps and their precursor timestamp, detects opinion drift at the candidate drift timestamps, and analyzes the influences of related events. In the future, we will further explore more effective sentiment orientation computing methods and threshold setting in sentiment distribution and opinion drift detection, and extend the technique to more user-generated contents.

References

1. Bifet, A., Holmes, G., et al.: Detecting Sentiment Change in Twitter Streaming Data. Journal of Machine Learning Research - Proceedings Track (JMLR) 17, 5–11 (2011)
2. Cheng, L., Ke, Z., Shiue, B.: Detecting changes of opinion from customer reviews. In: FSKD 2011, pp. 1798–1802 (2011)
3. Choi, D., Kim, P.: Sentiment Analysis for Tracking Breaking Events: A Case Study on Twitter. In: Selamat, A., Nguyen, N.T., Haron, H. (eds.) ACIIDS 2013, Part II. LNCS, vol. 7803, pp. 285–294. Springer, Heidelberg (2013)
4. He, Y., Lin, C., et al.: Tracking Sentiment and Topic Dynamics from Social Media. In: ICWSM 2012 (2012)
5. HowNet, http://www.keenage.com/
6. ICTCLAS, http://www.ictclas.org
7. Jiang, Y., Meng, W., Yu, C.: Topic Sentiment Change Analysis. In: Perner, P. (ed.) MLDM 2011. LNCS, vol. 6871, pp. 443–457. Springer, Heidelberg (2011)
8. Liu, B.: Sentiment Analysis and Opinion Mining. Morgan & Claypool Publishers (2012)
9. NTUSD, http://www.datatang.com/data/11837
10. Robertson, S.: Changes in Referents and Emotions over Time in Election-Related Social Networking Dialog. In: HICSS 2011, pp. 1–9 (2011)
11. Saito, K., Kimura, M., Ohara, K., Motoda, H.: Detecting Changes in Opinion Value Distribution for Voter Model. In: Salerno, J., Yang, S.J., Nau, D., Chai, S.-K. (eds.) SBP 2011. LNCS, vol. 6589, pp. 89–96. Springer, Heidelberg (2011)
12. Silva, I., Gomide, J., Veloso, A., et al.: Effective sentiment stream analysis with self-augmenting training and demand-driven projection. In: SIGIR 2011, pp. 475–484 (2011)
13. Widmer, G., Kubat, M.: Learning in the Presence of Concept Drift and Hidden Contexts. Machine Learning (ML) 23(1), 69–101 (1996)
14. Zhu, Y., Min, J., et al.: Semantic Orientation Computing Based on HowNet. Journal of Chinese Information Processing 20(1), 14–20 (2006) (in Chinese)

Feature Extraction from Micro-blogs for Comparison of Products and Services

Peng Zhao, Xue Li, and Ke Wang

Nanjing University, 210093 China,
The University of Queensland, QLD 4072 Australia,
Simon Fraser University, Vancouver, Canada
zp10@software.nju.edu.cn,
xueli@itee.uq.edu.au,
wangk@cs.sfu.ca

Abstract. Social networks are a popular place for people to express their opinions about products and services. One question would be that for two similar products (e.g., two different brands of mobile phones), can we make them comparable to each other? In this paper, we show our system namely *OpinionAnalyzer*, a novel social network analyser designed to collect opinions from Twitter micro-blogs about two given similar products for an effective comparison between them. The system outcome is a structure of features for the given products that people have expressed opinions about. Then the corresponding sentiment analysis on those features is performed. Our system can be used to understand user's preference to a certain product and show the reasons why users prefer this product. The experiments are evaluated based on accuracy, precision/recall, and F-score. Our experimental results show that the system is effective and efficient.

Keywords: social network, feature extraction, opinion mining, sentiment analysis.

1 Introduction

In recent years, social networks are becoming a popular place for people to express their opinions about social-economical issues or on products and services. Twitter, as a representative of social networks, is a microblogging service [16]. A tweet is a post or status update on Twitter. There has been a significant increase of daily posted tweets in recent years: 50 million in 2010, 200 million in 2011, and 400 million in 2012.

The direct implication of this trend is that it is difficult for consumers to compare two similar products or services that offer similar functions but with different properties. Many studies have been done on analyzing consumer reviews from forums [6] [5] [10] [13]. Bing Liu *et al* [9] explored opinion features and they analyzed certain types of customers' reviews, but their paper didn't go further to conduct sentiment analysis for the features of the products and services. Popescu and Etzioni in [13] conducted the product feature extraction and the sentiment

X. Lin et al. (Eds.): WISE 2013, Part I, LNCS 8180, pp. 82–91, 2013.

analysis. However, their approach is based on the known consumer reports that are written by experts and this kind of second-hand information can be biased or misleading. They did not use the features discussed by the first-hand consumers in social networks. Both [6] and [13] worked on consumer reviews, their work would provide only qualitative assessment rather than quantitative analysis on consumer opinions toward those product features. To the best of our knowledge, our work is the first of this kind that directly compare and contrast two similar products or services based on the social network analysis.

In this paper, we approach the feature extraction problem based on the text in tweets for the given products or services. There are three questions to be answered, 1) what features can we extract from tweets about the given products? 2) what are people's opinions about the products based on the features extracted? and 3) how do we make recommendations based on the sentiment analysis of those extracted features? For the first question, the products are made up with some features that people would like to comment, so the first question turns out to be a feature extraction problem. For the second question, people would like, love, or even hate the products because of some features of the products. People's emotion regarding the features of products can be divided into three kinds: *positive, negative,* or *neutral.* Thus the second question becomes a sentiment analysis problem. As for the third question, recommendations can be made if the overall statistical information can be made available for all of those features identified over the given products. So we conduct comparisons on the given products not only by offering recommendations but also by explaining why people like them, based on the sentiment analysis on the extracted features.

It should be pointed out that the physical features of a product can be easily obtained from product specifications where a product is advertised. Here, we do not want to get product features in this way because there are many features that consumers do not care and there are features that might not be listed in product specifications, for which consumers may like to comment. Therefore for a particular product, features should be chosen by consumers in anyway they prefer. We call the user-preferred features as *hot features.*

There are many feature extraction algorithms such as those given in Weka [17] like Principal Component Analysis (PCA), Linear Discriminant Analysis (LDA) [7], and Latent Dirichlet Allocation (LDA) [1]. An insightful review on feature extraction from text can be found in Hu and Liu [6] and an overview on effective feature extraction approaches can be found in Guyon *et al* [4]. Sentiment analysis is an approach used to categorize the overall attitude of a sentence or a paragraph towards a certain subject [13] [20].

In this paper, a new algorithm based on the *formal concept analysis* [3] is proposed to solve the feature extraction problem in our *OpinionAnalyzer* system. The difference between our feature extraction algorithm and the other feature extraction algorithms is that the feature space of a product or service is often hierarchically structured and we need an approach that will extract features in a lattice structure with a partial order, so to make features comparable to each other. As far as we know, this is the first work that extracts features and

analyzes sentiments from social networks in this way. Another challenging issue
is on processing social network data that is noisy and sparse in terms of the variety of words used by tweets (a tweet has about 28 words on average). Our system
has shown a good performance on those data. A multi-node tree is constructed
with the proposed formal concept analysis (FCA) algorithm. Consumers' opinions are analyzed through several algorithms which have been evaluated with
precision/recall, as well as F-score. The evaluation shows that *OpinionAnalyzer*
is effective and efficient.

The rest of the paper is organized as follows: Section 2 shows the related
work. Section 3 discusses the approaches including feature extraction, sentiment
analysis, and product recommendation. Section 4 discusses the results of our
experiments. Section 5 presents conclusions of our work.

2   Related Work

Our approach is mostly relevant to Hu and Liu [6] and Popescu and Etzioni [13].
In [6], they use Part-of-Speech (POS) tagging to collect nouns or noun phrases
since features are nouns mostly. They produced POS tags for each word (whether
the word is a noun or a verb etc). After that, association rule mining is applied
to filter out the frequent feature itemsets. The result of their research shows a
good performance in analysing electronic products like DVD player, MP3 player,
digital camera and cellular phone. Obviously, our research is related but different from theirs in many ways. POS tagging and association rules mainly focused
on noun features which may skip some words of their inputs that can imply
features. For instance, there are some smart phones that people prefer white
ones rather than other colors. In such condition, people may talk about their
preference about "white phone" when they refer to appearance. But "white" is
adjective in those sentences. Which means it would be filtered off when they try
to sum up all the features. Our system based on the feature extraction does not
have this problem. We did not remove words by part of speech. Instead, we comprehensively analyze input words from both frequency and relationship between
different words. Moreover, they use comments on products from e-commerce
Web sites as input. While we use data from social networks that have a large
number of short text with sparse words, which makes association rules not applicable. They demonstrated their algorithm with a small data set (500 records),
while we tested our algorithm with more than 8,200 records. Our work is also
different from the feature extraction method in [13], that they perform mining
of consumer reviews and sentiment classification without comparing the pair of
user-specified products based on the corresponding product features.

Based on the above reasons this paper will provide the effectiveness and efficiency studies for our pioneer work.

2.1   Feature Extraction Methods

When dealing with a large volume of text, we should scan through the text only
once and generate a list of features or properties that can best represent the

content of text. In doing so, we consider to extract the meaningful keywords and calculate their TFIDF [14] to represent text as feature vectors for the computational purpose. Feature extraction is an important step in text processing to transform input text into feature vectors. Guyon et al [4] provided a comprehensive review on feature extraction from text data and relevant applications. Insightful discussions can also be found in [6] and [13]. Weka has provided open source tools for feature extractions [17].

The task of feature extraction in this paper is to transform text data into a feature space that can best describe the interests of social network users who comment on the products or services. In brief, our feature extraction is to extract only *product features* [6] [12] that have appeared in the social networks. In the feature extraction process we need to firstly search for the relevant text from tweets where the given products are mentioned, then we apply the feature extraction algorithms on the text to derive the features for those specific products.

In order to make products comparable to each other, the output product features need to be constructed as a tree structure which can be transformed from a concept lattice where some features are general and some features are specific. This requirement especially matches with the idea of discovering concept hierarchy by formal concept analysis (FCA) approaches [3].

2.2 Formal Concept Analysis

The classical Formal Concept Analysis (FCA) [3] builds up a concept hierarchy by comparing the subset relationships amongst the associated terms of a concept. In FCA a concept can be associated with a single term or a set of terms. A term is a regarded as a meaningful word not appearing in the stop word list. When a term is used in describing a concept it is considered as an attribute of that concept. All the attributes that are associated with all concepts can be organized in a two dimensional matrix: one dimension (columns) is to list all attributes and the other (rows) is to list all the concepts. Then FCA algorithm will check the columns that corresponding to the matrix and form a lattice from that. It has been proven that there is a one-to-one mapping between each matrix and its corresponding lattice [3]. It can be seen that the critical step in FCA algorithm is to generate the attribute matrix for every concept by scanning the text only once.

3 Our Approach

As an example, given a smart phone product, one might be curious about how well it is received by customers. Does it have a *good* battery life? Or does it have a *good* camera? Does it look pretty? In order to answer those questions, we might need to know people's opinions towards these features. But what features would people care about mostly? How do people like it? To give answers, firstly, we need to know what features people talk mostly about. Secondly, we need to know whether people like these features or not, or in other words, the opinions of people towards the product features.

The proposed system has three main steps: (1) feature extraction, (2) sentiment analysis, and (3) recommendations. The system is initially given a set of pre-defined keywords (e.g., brands of smart phones) to search at Twitter Web site [16]. The output of the system is the counting of people's opinions towards the extracted features of the given products and the recommendations we offer.

Given a pair of products, the system firstly crawls the tweets related to the given keywords and processes them before storing them in data set. Then the feature extraction function, which is the main contribution of this paper, will extract the features that people talk about frequently. The sentiment analysis is then applied to analyze people's opinions towards the given products and classifies the opinions into three categories: *positive, negative*, and *neutral*, with respect to each feature. The recommendation process will provide conclusive comments on the products as the product that has overall positive feedback on its relevant features. It can be seen that the recommendation is evidence-based and is not only qualitative but also quantitative. In following subsections, we discuss these steps in turn.

3.1   Feature Extraction

Before the product features are identified, we firstly pre-process the tweets that we crawled. Since people express opinions casually within social networks, there may have either explicit and complete sentences [5], which we can easily know what they mean; or there may have implicit sentences that are incomplete sentences or just some phrases. For example, an implicit sentence in following is difficult for identifying its feature: "This mobile phone works for a long time". In this case, it is difficult to tell whether this sentence is referring the battery life or not. Such sentences would have several different ways to express the same meaning which makes it even more difficult to find the patterns of features. Fortunately, we observed that those implicit sentences do not appear much in our data set (with less than 15% of the sentences). So we can focus on explicit statements in this paper and leave the process of implicit sentences to the future work.

Our algorithm (i.e., Algorithm 1) can filter those words that are popular but not regarded as product features. It analyzes the processed tweets and finds out the hierarchy of the high TFIDF words.

In processing tweets, the stop words, *url, user name, date*, and non-English characters are removed. The smart phone brand names are replaced by the code names for the privacy preservation. Different words were changed to their original form with the help of *lucene snowball* [11].

Suppose there are two random words in tweets: w_1, w_2. The tweets set that contains all the appearance of word w_1 is namely set c_1. Similarly, the tweets set that contains all the appearance of word w_2 is namely set c_2. If set c_1 is a superset of set c_2, then more likely, w_2 is a sub-concept of w_1. A tree structure

Algorithm 1. Feature Extraction

Input:
    Tw: Crawled and pre-processed tweets
Output:
    T: Concept hierarchy of product features
Description:

```
begin
   Count tf of each word(wi) in Tw
   for(each word wi)  (filter infrequent words appear in Tw)
      If tf>0.01 (set this by experiments, words' tf lower than 0.01 are not meaningful)
         add wi to word set W
   for(wi1 in W) (the double loop analyzes connections between any two words)
      for(wi2 in W)
         if(wi2 != wi1) {
             tweets collection c1, every element of it contains wi1
             tweets collection c2, every element of it contains wi2
             if (c1 \(\supset\) c2)
                wi2 is a sub-concept of wi1
                T add (wi1,wi2)
                for(any win that wi2 is a sub-concept of win)
                   if (wi1 \(\supset\) win)
                      win is a sub-concept of wi1
                      wi2 is a sub-concept of win
                   else if (win \(\supset\) wi1)
                         wi1 is a sub-concept of win
                         wi2 is a sub-concept of wi1

         }
      return T
end
```

is used to express the hierarchy like w_1, w_2 instead of a lattice as it can be seen from Figure 1.

3.2 Sentiment Analysis

This step is to explore people's opinions about the *hot* product features (features extracted in Section 3.1). People's opinions about products can be divided into three types, *positive, negative* and *neutral*. People use certain conventional words to express their feelings. Here are two examples:

"I am so glad that I have chosen this new cell phone with such a great camera."
"I love this white case."

Both sentences express positive feelings through words. One sentence uses the adjective "glad" to express the feeling toward a smart phone camera. The other uses a sentiment word "love" to express the feeling. All these words are essential words that can show consumer's sentiments. But there are also some words that people may use in social networks but have no contribution to analysis like "am". So we delete such kind of words same as that was done in [21] (However, we did some changes by removing sentimental verbs from the list). Then, we narrow down these input words again by using WordNet [18] to eliminate the words that are seldom used. We also delete the none-existing words. By tagging the existing words, a bitmap is established (listing all those existing words, tagging the existing words appeared in a tweet with value 1 the others with 0). Also tagging the orientation of each sentence is based on the sentimental orientation like *positive, negative,* or *neutral*. We show the result in the evaluation section.

Besides, people's emotion can be divided into more types. WordNet [18] has divided some sentiment words into six types including *disgust, anger, fear, sad, surprise* and *joy*. Each of these six types shows different levels of emotions which may make the analysis more sophisticated. We calculated the precision and recall of those experiments as shown in Table 1.

The taxonomy of product features provides an overview of *hot features* as well as the results of sentiment analysis of those features (Figure 1).

3.3   Recommendation

This step is to produce recommendations based on both qualitative and quantitative analysis on people's preferred features. Products that have strong similarities, like both smart phones have highly satisfied cameras that can take high quality pictures and both have good performances in battery lives. If one customer likes one of these products, probably he would like the other one as well. We use a simple tree-similarity comparison algorithm based on the results of the sentiment analysis of the *hot features*. For the limitation of the space, the recommendation algorithm is not given in this paper.

Based on our random data collection from Twitter with 8,200 tweets, we have an example summary of brands namely *htc* and *Samsung*. Both of them received good responses from customers about their applications and systems. We only used 100 of each to plot the pictures in Figure 1.

In this given set of tweets, people talked about *Samsung* appearance and seem to like it more than *htc*. While there are more people thinking *htc* is easier to use while others think *Samsung* is a little bit complex to use. Price of *htc* seems to be a problem to this band. While unlock and some fancy things about *Samsung* are both liked and disliked by customers as it can be seen from the plotted dots in Figure 1.

Apparently the above opinions are summarized based on a small data set that has only 8,200 tweets and are obviously biased and by no means representative or comprehensive in judging products. However, this does not prevent us to show the potential of our method for its application on a large scale when a large volume of tweets becomes available. In that case, the both quantitative and qualitative analysis on two similar and competitive products can be conducted significantly and precisely over social networks by using our approach.

4   Experiments and Evaluation

Our system is called *OpinionAnalyzer* and is implemented in Java. We gathered about 8,200 tweets from [16] as the original data. The time complexity of the system is in linear growth. We gathered 103 tweets in 54 seconds and for each 1,000 in ten minutes (despite the waiting time due to Twitter's limited rate which means we can only have 100 or so tweets in an hour).

We conducted the feature extraction experiments with data related to three brands of cell phones from Twitter. Original tweets include information such as

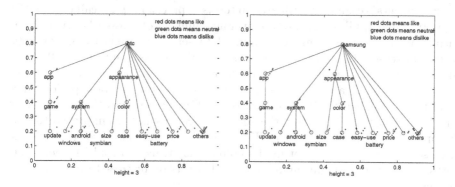

Fig. 1. Hot Features and their Sentiment Analysis for htc and Samsung

url, twitter id, post time. We deleted *url* and pre-processed crawled tweets before storing them in our data set. After that, we tagged the data for training purpose with different numbers for different classifiers and received the sentiment analysis results as *Case 1* in Table 1.

There are some words that are not frequently used in general. Reducing such kind of words would not affect the judgement of sentiment of sentences. We identified those infrequent words with the help of WordNet [18]. "Reducing infrequent words" improvements in our system are shown in *Case 2* in Table 1 as well. Compared to data for Case 1, about 10% words were removed in *Case 2*.

Also, different kinds of words might show different levels of sentiment. Mostly, "love" would show a higher preference than "like", though both of them mean "preferable". Similarly, many words would show different levels of sentiment. A "sentiment corpus" is a corpus that provides with levels of words related to emotions. In the third case, we show the results of experiment including "sentiment corpus" as *Case 3* in Table 1.

The accuracy of our system in feature extraction is above 71% which far prevails other classification algorithms, like BayesNet (33.01%), Random Forest (43.69%) and J48 (50.49%). Primarily, removing noisy data helps us filter off words not related to our topic. In addition, the feature taxonomy we derived from our algorithm helps us locate the feature. For instance, if one tweet is talking about "android" while another one is talking about "symbian". Since both of them are child-nodes of "system" in our taxonomy, we would know both of them are discussing about one thing: "system". But for any other feature extraction algorithms, the result of that would be three different unrelated features "android", "symbian", and "system". In our approach, words in the tweet show that the tweet might be talking about "android", but it mentioned "system" instead, the tagging would applied to "system" in this case. In fact, all those tweets were talking about one generic concept: "system".

Experimental results are evaluated in terms of precision (p), recall (r), and $F1$ score. In Table 1, all results are the average numbers of cross-validation experiments. With the increasing size of our data set, from 1,000 tweets to 8,000 tweets, increased 1,000 tweets every time, $F1$ score is stabilized at 0.75 for all algorithms. Among all applied classifiers, with all of our experiments with different sizes of data sets, Random Forest classifier has the best performance at all time. "Reducing infrequent words" as *Case 2* improves the performance a little in every classifier. But the sentiment corpus has no improvements as shown in *Case 3* for two classifiers. The reason might be that in social networks, those sentiment words are not used so often and straight. People in social networks have invented many more colloquial words or phrases to express their feelings.

As a result of these experiments, we would be able to recommend products to people with the reasoning based on the sentiment analysis of those specific product features. We had observed that Twitter often posted the news about a product several weeks in advance before actual official announcement of the product. For instance, we found out that the new release of *htc butterfly* had gained a good reputation among customers two weeks before the product official announcement released. This might indicate a new research issue on the authentication of the sentiment analysis.

Table 1. Evaluation of Sentiment Analysis with Different Data Pre-processing

	BayesNet			J48			Random Forest		
	p	r	f1	p	r	f1	p	r	f1
Case 1	0.741	0.752	0.745	0.675	0.676	0.638	0.798	0.8	0.77
Case 2	0.75	0.757	0.752	0.701	0.724	0.667	0.829	0.805	0.768
Case 3	0.75	0.757	0.752	0.701	0.724	0.667	0.789	0.781	0.733

5   Conclusions

When people comment on products or services online, they may choose any features they prefer. Then their feelings towards those features are expressed. For different products to be compared to each other, we need to extract those commonly discussed features to make them comparable. Those product features can subsume each other in a hierarchical structure. This paper has proposed a new algorithm based on the formal concept analysis (FCA) to have derived a taxonomy of *hot features* of the products. Then sentiment analysis is performed on those features. Based on the analysis of those features, an overview on how people like or dislike these products can be presented.

In future research, we will improve the effectiveness and scalability of our method for mining social opinions on a wide range of products and services.

References

1. Blei, D.M., Ng, A.Y., Jordan, M.I.: Latent dirichlet allocation. Journal of Machchine Learning Research 3, 993–1022 (2003)
2. Breiman, L.: Random forests. Machine Learning 45(1), 5–32 (2001)
3. Ganter, B., Wille, R.: Applied lattice theory: Formal concept analysis. In: Grätzer, G. (ed.) General Lattice Theory. Birkhäuser (1997)
4. Guyon, I., et al. (eds.): Feature Extraction: Foundations and Applications. Springer (2006)
5. Hu, M., Liu, B.: Mining and Summarizing Customer Reviews. In: Proceedings of the 2004 ACM SIGKDD International Conference on Knowledge Discovery and Data Mining, pp. 168–177. ACM Press, New York (2004)
6. Hu, M., Liu, B.: Mining opinion features in customer reviews. In: Proceedings of the National Conference on Artificial Intelligence, pp. 755–760. AAAI Press, MIT Press, Menlo Park, Cambridge (2004)
7. Lachenbruch, P.A., Goldstein, M.: Discriminant analysis. Biometrics, 69–85 (1979)
8. Linden, G., Smith, B., York, J.: Amazon.com recommendations: Item-to-item collaborative filtering. IEEE Internet Computing 7(1), 76–80 (2003)
9. Liu, B., Hu, M., Cheng, J.: Opinion observer: Analyzing and comparing opinions on the web. In: Proceedings of the 14th International Conference on World Wide Web, pp. 342–351. ACM (2005)
10. Liu, B.: Sentiment analysis and opinion mining. Synthesis Lectures on Human Language Technologies 5(1), 1–167 (2012)
11. Lucene snowball, http://lucene.apache.org/core/old_versioned_docs /versions/3_0_0/api/contrib-snowball/
12. Pang, B., Lee, L., Vaithyanathan, S.: Thumbs up? Sentiment classification using machine learning techniques. In: Proceedings of the ACL 2002 Conference on Empirical Methods in Natural Language Processing, vol. 10, pp. 79–86. Association for Computational Linguistics (2002)
13. Popescu, A.M., Etzioni, O.: Extracting Product Features and Opinions from Reviews. In: Natural Language Processing and Text Mining, pp. 9–28 (2007)
14. Salton, G., Buckley, C.: Term-weighting approaches in automatic text retrieval. Information Processing & Management 24(5), 513–523 (1988)
15. Sarwar, B., et al.: Item-based collaborative filtering recommendation algorithms. In: Proceedings of the 10th International Conference on World Wide Web. ACM (2001)
16. Twitter, https://twitter.com/
17. WEKA The University of Waikato, http://www.cs.waikato.ac.nz/ml/weka/
18. WordNet: An Electronic Lexical Database. MIT Press, Cambridge, http://wordnet.princeton.edu
19. NLProcessor-Text Analysis Toolkit (2000), http://www.inforgistics.com/textanalysis.html
20. Yi, J., Nasukawa, T., Bunescu, R., Niblack, W.: Sentiment Analyzer: Extracting Sentiments about a Given Topic Using Natural Language Processing Techniques. In: Procs. of ICDM. 2003, pp. 1073–1083 (2003)
21. http://jmlr.org/papers/volume5/lewis04a/a11-smart-stop-list/ english.stop

Directional Context Helps: Guiding Semantic Relatedness Computation by Asymmetric Word Associations

Shahida Jabeen, Xiaoying Gao, and Peter Andreae

School of Engineering and Computer Science
Victoria University of Wellington, PO Box 600, Wellington, New Zealand
{shahidarao,xgao,peter.andreae}@ecs.vuw.ac.nz

Abstract. Semantic relatedness computation is the task of measuring the degree of relatedness of two concepts. It is a well known problem with applications ranging from computational linguistics to cognitive psychology. In all existing approaches, relatedness is assumed to be symmetric i.e. the relatedness of terms t_i and term t_j is considered the same as the relatedness of terms t_j and t_i. However, there are tasks such as free word association, where the association strength assumed to be asymmetric. In free word association, the given term determines the context in which the association strength must be computed. Based on this key observation, the paper presents a new approach to computing term relatedness guided by asymmetric association. The focus of this paper is on using Wikipedia for extracting directional context of each given term and computing the association of input term pair in this context. The proposed approach is generic enough to deal with both symmetric as well as asymmetric relatedness computation problems. Empirical evaluation on multiple benchmark datasets shows encouraging results when our automatically computed relatedness scores are correlated with human judgments.

Keywords: Semantic Relatedness, Free Word Associations, Asymmetric Term Associations, Symmetric Relatedness, Directional Context.

1 Introduction

Semantic relatedness computation is a core issue in many Natural Language Processing (NLP) applications such as clustering, topic identification and word sense disambiguation. In conventional semantic relatedness computation approaches, the relatedness is assumed to be symmetric and is analogous to a commutative operation, such that $rel(a, b) \Leftrightarrow rel(b, a)$. This assumption essentially disregards the relatedness complexity of two terms by ignoring the heterogeneity of their context and by considering their relation in isolation. In real scenarios, such as in web documents where the context is quite diverse and changes rapidly, computing realistic scores for a term pair according to the context in which it appears is critical. Hence, context identification is a critical factor in computing

X. Lin et al. (Eds.): WISE 2013, Part I, LNCS 8180, pp. 92–101, 2013.

relatedness scores. In this research, we will focus on guiding term relatedness computation by directional contexts. The idea of directional context is actually borrowed from free word association task, where given a cue or stimulus word, the goal is to identify the most strongly associated response word. For this kind of problem, the given stimulus word provides the context in which association of two terms is computed. This association is assumed to be directional as it changes if the order of stimulus and response words is changed. When response word takes the role of stimulus word, the corresponding context for computing association of both terms changes entirely.

Inspired by the idea of word associations, this paper proposes a novel approach to computing semantic relatedness guided by the directional context. Our approach is unique because it adopts the concept of asymmetric association of terms to compute semantic relatedness, which (to the best of our knowledge) has not been done before. When it comes to finding association of two terms, getting the correct context is critically important issue. To extract the directional context corresponding to each input term, we opt to use Wikipedia as a knowledge source. Context mining is based on Wikipedia hyperlink network which encodes not only many lexical relations but also more sophisticated semantical relations such as cause and effect and functional associations. The proposed approach computes the semantic association of a given term pair in the individual context of each input term separately and then combines these directional associations to compute final relatedness scores. It is worth mentioning that we computed the asymmetric associations of terms differently from free word association task. The difference lies in the use of semantically rich context rather than just distributional or lexical statistics based approaches to word association computation.

2   Related Work

In this paper, we focused on approaches for semantic relatedness computation based on Wikipedia. Strube and Ponzetto [1] were the first to use Wikipedia for computing semantic relatedness. Their approach *WikiRelates* used Wikipedia category network and proposed a technique based on statistical and structure based measures of relatedness. Gabrilovich and Markovich proposed *Explicit Semantic Analysis (ESA)* [2], inspired by vector space model but rather than vector of words they generated vector of Wikipedia concepts to relate terms. Milne and Witten proposed *Wikipedia Link based relatedness Measure* (WLM) [3] for computing semantic relatedness based on Wikipedia hyperlink structure. Their approach is based on Normalized Google Distance (NGD) and tfidf inspired weighting of hyperlink vectors. Zesch and Gurevych [4] adopted the path based and information content based measures of WordNet on Wikipedia using its category network, redirects and a list of articles. Radinsky et al. [23] proposed Temporal Semantic Analysis (TSA) by incorporating temporal information in ESA. Hassan and Mihalceae [5] proposed an approach to cross-lingual semantic relatedness computation based on inter-language links of different versions of Wikipedia. Navigli and Ponzetto [6], following Hassan and Mihalceae, proposed a cross-lingual approach to computing semantic relatedness using Wikipedia. Yazdani and Popescu-Belis [7] constructed concept graphs from Wikipedia and used random walk over it to compute the relatedness scores. Halawi et al. [8] proposed *CLEAR*, an approach to large scale learning of word relatedness with constraints, where constraints are imposed by known pair of related words.

There are two main limitations of existing approaches: first, existing approaches are computationally expensive (either preprocessing the whole Wikipedia dump before actually computing the semantic relatedness or to construct a huge semantic graph or network which also requires huge amount of processing time and memory); and second, existing measures are focused on one or more aspects of relatedness such as lexical relations (synonyms, hypernyms and hyponyms), distributional relations (terms frequently occurring on the same page or functional associations) and co-locational relations (term co-occurrences) but did not represent a generalized approach for relatedness computation that covers all aspects of relatedness. In this paper, we have addressed the first limitation by proposing a new measure of relatedness, which does not require any preprocessing hence, is computationally inexpensive while still correlating well with human judgments.

3   Guiding Semantic Relatedness Computation

We propose an approach to computing semantic relatedness of terms inspired by the idea of directional context borrowed from word association. The approach is divided into three phases: In first phase, we extract the directional contexts of an input term pair. This phase makes use of Wikipedia as a knowledge source to identify corresponding context of each input term. The second phase constructs an inverted index of the extracted context. In the third phase, we compute bidirectional strengths of a term pair and combine them to get the final relatedness scores.

When we think about relating two terms, we intrinsically relate their corresponding concepts in a wider context rather then just relating two words. Since, context mining is based on Wikipedia, the context of a term consists of all Wikipedia concepts that are semantically related to the corresponding concept of that term. In Wikipedia, each concept is linked to many other semantically related concepts resulting in a huge network of hyperlinks. This hyperlink structure is implicitly semantically rich and handles various semantical and lexical relations such as synonymy, polysemy and associative relations. This research makes effective use of Wikipedia hyperlink structure for context mining. Given an input term pair (t_i, t_j), we map both terms to their corresponding Wikipedia concepts c_i and c_j respectively. For each matched Wikipedia concept, we extract its inlinks and outlinks and construct a context vector. Each element of this context vector is a Wikipedia article, also called concept. This context vector represents the respective context of an input term.

To have a faster access to various lexical and statistical features, we construct an inverted index of each concept list (representing the context of each input term). Before indexing, all concepts in each concept vector are preprocessed. This preprocessing involves a series of steps to convert the MediaWiki format of each Wikipedia concept to plain text. All redirect are mapped to their target concepts. Overly specific concepts having less than 100 non stop words and less then 5 inlinks and outlinks are also discarded. Note that we have not done any stemming or stop word removal at this point. To speed up semantic computation process, we build an inverted index of preprocessed concepts of context vector.

For a given term pair (t_i, t_j), we compute its relatedness score based on the directional association of input terms in two different contexts. When we compute the association of input term pair t_i and t_j in the context of term t_i, we call it *Forward Association Strength* (FAS). It signifies the association of the term pair in the forward direction i.e. from t_i to t_j. Similarly, when the relatedness of the same term pair is computed in the context of term t_j, it is called *Backward Association Strength* (BAS). It signifies the association of the term pair in the backward direction i.e. from t_j to t_i. Hence, the directional association of the same term pair assumes different strengths depending on the context of term in consideration.

From the inverted index, all concepts $|C|$ having both input terms t_i and t_j occurring within a reference window of size less than ϵ are extracted. In our experiments, we have set ϵ to 20. For each concept $c_a \in |C|$, its semantic strength with corresponding concept c_i and c_j of each input term t_i and t_j is computed as follows:

$$Strength(c_a) = rel(c_a, c_i) \times rel(c_a, c_j) \tag{1}$$

where, $rel(c_a, c_i)$ and $rel(c_a, c_j)$ are computed using Wikipedia Link based Measure (WLM) proposed by [3]. To relate the given two terms t_i and t_j, their directional association strength is computed as follows:

$$Association\_Strength(t_i, t_j) = \frac{\sum_{a=1}^{n} Strength(c_a)}{|T|} \tag{2}$$

where, n represents the size of $|C|$ and $|T|$ represents sum of total concepts in the context vectors of both terms.

This association strength indicates the directional association of two input terms in the forward and backward contexts. Both Forward Association Strength (FAS) and Backward Association Strength (BAS) are linearly combined to get the final Overlapping Strength based Relatedness (OSR) score for the input term pair as follows:

$$OSR(t_i, t_j) = \lambda FAS + (1 - \lambda) BAS \tag{3}$$

where, λ is a context scoring parameter and its value is set to 0.5, so as to give equal importance to both forward and backward association strengths.

4   Evaluation

For context mining of the given term pair, we have used the version of Wikipedia released in July 2011 [1]. This contains more than 3.5 million articles and occupies more than 30GB of memory space. For performance evaluation of proposed approach, we used two correlation measures to compare our results with human judgments on all datasets: Spearman's correlation coefficient (ρ) and Pearson's Correlation Coefficient (r). Spearman's correlation coefficient is essentially Pearson's correlation coefficient on ranked variables. Following are the datasets used in our experiments:

- **Miller and Charles Dataset:** Also known as M&C dataset [9], is a general English based dataset consisting of 30 noun term pairs which are rated by 38 human judges. These judgments for each term pair were averaged to get a single score on a scale of 0 (unrelated) to 4 (synonyms).
- **Rubenstein and Goodenough Dataset:** Also known as R&G dataset [10], is a collection of 65 English terms pairs sorted in ascending order of relatedness. Scores of 51 human judges were averaged to get final scores on a scale of 0-4 where 0 means *unrelated* and 4 means *synonyms*.
- **WordSimilarity-353 Dataset:** Also known as Finkelstein-353 [11], consists of 353 word pairs scored by 13 human judges on a scale of 0-10 (unsorted).
- **MeSH Dataset:** A biomedical dataset [12] consisting of 36 term pairs derived from MeSH ontology which is a taxonomic hierarchy of medical concepts. Scores of 8 human experts on 36 MeSH term pairs were averaged to provide similarity scores on a scale of 0-4.
- **MiniMayoSRS Dataset:** A domain specific dataset consisting of 30 medical term pairs annotated by physicians and medical index experts [13] on a scale of 0-4.
- **MayoMeSH Dataset:** Both previously mentioned domain specific datasets are combined to generate this dataset of 65 medical term pairs. The human scores for both datasets are normalized on a scale of 0-4.

5   Results and Discussions

We evaluated the performance of three variants of presented approach: OSR (all), OSR+WLM (all) and OSR (Non-Missing). First variant, OSR (all) represents applying OSR on all term pairs. Second variant, OSR+WLM (all) indicates using WLM measure [3] for computing the scores of only missing term pairs. Third variant, OSR (Non-Missing) represents OSR on dataset without missing term pairs.

For M&C, R&G and WordSimilarity-353 datasets, the performance of proposed approach when compared to human judgment is shown in Table 1. Clearly, the correlation of proposed approach (OSR) with human judgment on all three datasets is low when applied to datasets with all term pairs. This low correlation

---

[1] http://download.wikipedia.org

Table 1. Performance comparison of proposed approach with existing approaches on domain independent datasets

Method	M&C		R&G		WS-353	
	ρ	r	ρ	r	ρ	r
OSR (All)	**0.58**	**0.61**	**0.60**	**0.71**	**0.61**	**0.50**
OSR+WLM (All)	**0.84**	**0.77**	**0.77**	**0.63**	**0.69**	**0.60**
OSR (Non-Missing)	**0.85**	**0.73**	**0.84**	**0.73**	**0.70**	**0.61**
Strube and Ponzetto[1]	N/A	0.47	N/A	0.52	N/A	0.49
Milne and Witten[3]	0.70	N/A	0.64	N/A	0.69	N/A
Gabrilovich and Markovich[2]	0.72	N/A	0.81	N/A	0.75	N/A
Yazdani and Popescu[7]	0.69	0.50	N/A	N/A	N/A	N/A
Navigli and Ponzetto[6]	0.90	0.89	N/A	N/A	0.65	0.59
Hassan and Mihalceae[14]	0.87	0.80	0.80	0.80	0.67	0.55

is justified because OSR assigns zero score to those term pairs in which either or both terms which don't have corresponding concepts in Wikipedia. To avoid this problem, we used OSR+WLM approach which improved the results significantly. To judge the actual performance of proposed approach, we compared the results on a subset of all three datasets excluding missing term pairs and the results showed a significant improvement in both correlation values on all datasets as indicated by Table 1.

Comparison with Existing Approaches: Table 1 indicates the performance comparison of proposed approach with other existing approaches on both Spearman and Pearson correlations. For comparison with existing approaches, we have considered OSR+WLM approach, as it performs optimal on datasets with all term pairs. OSR is the third best approach on both correlation values after Navigli [6] and Hassan [14] on M&C dataset and Gabrilovich [2] and Hassan [14] on R&G dataset. On the largest dataset WordSimilarity-353, our approach outperformed all other approaches (including the best performing approaches on M&C and R&G datasets) except Gabrilovich and Markovich on Spearman's correlation. Their approach does not match input terms to corresponding Wikipedia concepts which is their advantage and our limitation but their approach is computationally quite expensive as it requires preprocessing of the whole Wikipedia corpus. Navigli and Ponzetto's approach make an efficient use of both Wikipedia and WordNet to construct a very large multi-lingual knowledge source and outperformed our approach on both M&C and R&G datasets but could not do better than our approach on WordSimilarity-353 dataset. From the results it is clear that not every approach performed good on all datasets because of their limitations as well as the implicit complexity in judging the relatedness in these datasets. If missing terms in previously mentioned datasets are considered, our approach does not perform well but if we ignore those pairs, the results are quite encouraging. The nature of Wikipedia, being continuously growing knowledge source, supports the assertion that this limitation will not be a big issue in the future. We also believe that by extending the context vector with more related concepts and avoiding the matching of input terms to corresponding Wikipedia concepts, the performance of proposed approach can be significantly improved.

5.1   Evaluation on Domain Specific Datasets

The purpose of this evaluation was two fold: First analyzing the performance of our approach on domain specific data; and second testing the semantic richness and coverage of Wikipedia on domain specific knowledge. We tested performance of our approach on three domain specific datasets: MeSH dataset, MiniMayoSRS and MayoMeSH dataset. OSR+WLM is not considered here because it does not improve OSR on all term pairs, at all. Since there are no missing term pairs in MeSH dataset, the performance of OSR (all) and OSR(non-Missing) are the same, as shown in Table 2.

Table 2. Performance of proposed approach on domain specific datasets

Dataset	OSR (All)		OSR (Non-Missing)	
	ρ	r	ρ	r
MeSH	0.80	0.78	0.80	0.78
MiniMayo(Physicians)	0.68	0.71	0.73	0.62
MiniMayo(Experts)	0.74	0.79	0.80	0.80
MayoMeSH(Physicians)	0.76	0.74	0.76	0.75
MayoMeSH(Experts)	0.78	0.76	0.76	0.76

We compared our results with following approaches for medical term relatedness computation: Hliaoutakis [12], Hisham Al-Mubaid et al. [15], Pederson et al. [13], Lin [16], Jiang and Conarth [17], Path [18], Leacock and Chodorow [19], Resnik [20]. Table 3 shows the performance comparison of proposed approach with other existing approaches. It is worth mentioning, that all these approaches have used domain specific taxonomies and knowledge sources for their relatedness computation. Since, the expert's inter-rater correlation was higher than physicians, hence we consider comparison with experts judgment more valuable than judgment of physicians. On both MeSH and MiniMayo (Expert) datasets, our approach performed second best after Hisham et al. whereas, on MiniMayo (Physician) dataset, our approach was again second best after Pederson et al. shown in Table 3. Both the best performing approaches make an efficient use of multiple biomedical ontologies whereas, our approach is based on Wikipedia, which is a generic all-purpose knowledge source and is already proven to have excellent knowledge coverage of agricultural domain [21].

The proposed approach has shown that it performed better than many other approaches which are purely based on domain specific knowledge sources. The results of experiments also showed that Wikipedia is semantically rich not only for general concepts but also for domain specific concepts and could be used as an alternative knowledge source for handling domain specific problems.

Table 3. Performance comparison of proposed approach with existing approaches on domain specific datasets

Method	Mesh	MiniMayo (Experts)	MiniMayo (Phycisians)
	r	r	r
OSR (All)	**0.78**	**0.79**	**0.71**
OSR (Non-Missing)	**N/A**	**0.80**	**0.62**
Hliaoutakis[12]	0.69	N/A	N/A
Hisham Al-Mubaid et al.[15]	0.82	0.86	N/A
Pederson et al. [13]	N/A	0.75	0.84
Lin[16]	0.50	0.75	0.60
Jiang and Conarth[17]	N/A	0.62	0.45
Path[18]	0.76	0.51	0.36
Leacock and Chodorow[19]	0.68	0.29	0.49
Resnik[20]	N/A	0.62	0.45

5.2 Asymmetric Relatedness Evaluation

We run another set of experiments on a subset of Free Association Norms (FAN) dataset [22]. FAN is a database of free word association. For asymmetric relatedness evaluation, we used a noun subset of FAN dataset consisting of 43 noun term pairs. This subset consists of forward and backward associations of five stimulus words with a number of response words for each. The number of responses of these stimulus words ranges from 6-12. Equation 2 proposed in section 2.3 is used to compute asymmetric association of each term pair in the dataset. The results are reported in Figure 1. Overall, Spearman's correlation on forward strength of FAN subset is 0.51 and Pearson's correlation is 0.26 whereas, the Spearman's correlation is 0.62 and Pearson's correlation is 0.39 on backward strength. We also computed the correlation of individual stimulus word based subset of noun dataset of FAN with human judgments. We compared our FAS and BAS with human judgments on individual stimulus words subsets using both Spearman's correlation and Pearson's Correlation. There is a clear variation on both correlation values particularly on stimulus word *Dentist*, where the Pearson's correlation ranges from -.007 (slightly visible in the graph) on Backward relatedness to 0.96 on forward relatedness. The Pearson's correlation for stimulus word *Defrost* is 0 (not visible in the graph). The results of this experiment highlighted the definitive difference of association and relatedness. Association is directional and usually based on usage and co-occurrences whereas, relatedness not only covers lexical and distributional properties of terms but also considers fine grained semantics which are not covered by the associations based measures. Human usually assign more weights to those term pairs which frequently co-occur or are lexically used together. Since, the proposed measure is not entirely based on distributional or lexical statistics that is why the overall correlation is not high. When individual term pairs were analyzed it was found that human have

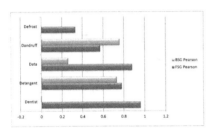   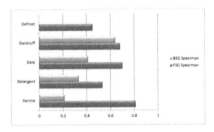

(a) Pearson's Correlation   (b) Spearman's Correlation

Fig. 1. A comparison of Spearman's correlation and Pearson's Correlation on stimulus words of FAN subset

scored the term pair (*dentist, pain*) higher than (*dentist, orthodontics*) whereas, the our measure assigns higher score to the second pair. Based on this experiment, we make an observation that semantic relatedness measures, other than ones based on lexical and distributional measures, do not correlate well with human judgment on association based task. Clearly, the reason is the perspective from which such judgments are made.

6   Conclusions

This paper addresses the problem of computing semantic relatedness of terms. There are two main contributions of this research: a new approach to semantic relatedness computation based on directional contexts extracted from Wikipedia; and a new measure of semantic relatedness based on asymmetric associations of terms. Empirical evaluations on several domain independent datasets as well as domain specific datasets have proven the effectiveness of presented approach. Results have also proven that Wikipedia can be used as a reasonable alternative knowledge source for domain specific relatedness computation. By improving the context extraction and avoiding the term matching with corresponding Wikipedia concepts, this approach can be improved further. This will be an intuitive avenue to proceed in future.

References

1. Strube, M., Ponzetto, S.P.: Wikirelate! computing semantic relatedness using wikipedia. In: Proceedings of Association for the Advancement of Artificial Intelligence (AAAI) (2006)
2. Gabrilovich, E., Markovitch, S.: Computing semantic relatedness using wikipedia-based explicit semantic analysis. In: Proceedings of the 20th International Joint Conference on Artificial Intelligence, pp. 1606–1611 (2007)
3. Milne, D., Witten, I.H.: An effective, low-cost measure of semantic relatedness obtained from wikipedia links. In: Proceeding of AAAI Workshop on Wikipedia and Artificial Intelligence: An Evolving Synergy, pp. 25–30 (July 2008)
4. Zesch, T., Gurevych, I.: Wisdom of crowds versus wisdom of linguists, measuring the semantic relatedness of words. Nat. Lang. Eng. 16(1), 25–59 (2010)

5. Hassan, S., Banea, C., Mihalcea, R.: Measuring semantic relatedness using multi-lingual representations. In: Proceedings of the First Joint Conference on Lexical and Computational Semantics, Proceedings of the Main Conference and the Shared Task, Proceedings of the Sixth International Workshop on Semantic Evaluation, SemEval 2012, vol. 1, vol. 2, pp. 20–29. Association for Computational Linguistics (2012)

6. Navigli, R., Ponzetto, S.P.: Babelrelate! a joint multilingual approach to computing semantic relatedness. In: AAAI (2012)

7. Yazdani, M., Popescu-Belis, A.: Computing text semantic relatedness using the contents and links of a hypertext encyclopedia. Artif. Intell. 194, 176–202 (2013)

8. Halawi, G., Dror, G., Gabrilovich, E., Koren, Y.: Large-scale learning of word related-ness with constraints. In: Proceedings of the 18th ACM SIGKDD International Con-ference on Knowledge Discovery and Data Mining, KDD 2012, pp. 1406–1414 (2012)

9. Miller, G.A., Charles, W.G.: Contextual correlates of semantic similarity. Language and Cognitive Processes 6(1), 1–28 (1991)

10. Rubenstein, H., Goodenough, J.B.: Contextual correlates of synonymy. Commun. ACM 8, 627–633 (1965)

11. Finkelstein, L., Gabrilovich, E., Matias, Y., Rivlin, E., Solan, Z., Wolfman, G., Ruppin, E.: Placing search in context: the concept revisited. ACM Trans. Inf. Syst. 20(1), 116–131 (2002)

12. Hliaoutakis, A.: Semantic similarity measures in mesh ontology and their applica-tion to information retrieval on medline. Master's thesis (2011)

13. Pedersen, T., Pakhomov, S.V.S., Patwardhan, S., Chute, C.G.: Measures of seman-tic similarity and relatedness in the biomedical domain. J. of Biomedical Informat-ics 40, 288–299 (2007)

14. Hassan, S., Mihalcea, R.: Semantic relatedness using salient semantic analysis. In: AAAI (2011)

15. Al-Mubaid, H., Nguyen, H.A.: Measuring semantic similarity between biomedical concepts within multiple ontologies. IEEE Transactions on Systems, Man, and Cybernetics, Part C 39(4), 389–398 (2009)

16. Lin, D.: An information-theoretic definition of similarity. In: Proceedings of the 15th International Conference on Machine Learning, pp. 296–304. Morgan Kauf-mann (1998)

17. Jiang, J., Conrath, D.W.: Semantic similarity based on corpus statistics and lexical taxonomy. In: Proceedings of International Conference Research on Computational Linguistics (ROCLING), pp. 19–33 (1997)

18. Rada, R., Mili, H., Bicknell, E., Blettner, M.: Development and application of a metric on semantic nets. IEEE Transactions on Systems, Man and Cybernet-ics 19(1), 17–30 (1989)

19. Leacock, C., Chodorow, M.: ch. 11, Combining Local Context and WordNet Simi-larity for Word Sense Identification, pp. 265–283. The MIT Press (1998)

20. Resnik, P.: Using information content to evaluate semantic similarity in a tax-onomy. In: Proceedings of the 14th International Joint Conference on Artificial Intelligence (AAAI), pp. 448–453 (1995)

21. Medelyan, O., Milne, D., Legg, C., Witten, I.H.: Mining meaning from wikipedia. International Journel of Human Computer Studies 67, 716–754 (2009)

22. Nelson, D.L., Mcevoy, C.L., Schreiber, T.A.: The University of South Florida free association, rhyme, and word fragment norms. Behavior Research Methods, Instru-ments, & Computers, 402–407 (2004)

23. Radinsky, K., Agichtein, E., Gabrilovich, E., Markovitch, S.: A word at a time: com-puting word relatedness using temporal semantic analysis. In: Proceedings of the 20th International Conference on World Wide Web, WWW 2011, pp. 337–346 (2011)

The Heterogeneous Cluster Ensemble Method Using Hubness for Clustering Text Documents

Jun Hou and Richi Nayak

School of Electrical Engineering and Computer Science,
Queensland University of Technology, Brisbane, Australia
jun.hou@student.qut.edu.au, r.nayak@qut.edu.au

Abstract. We propose a cluster ensemble method to map the corpus documents into the semantic space embedded in Wikipedia and group them using multiple types of feature space. A heterogeneous cluster ensemble is constructed with multiple types of relations i.e. document-term, document-concept and document-category. A final clustering solution is obtained by exploiting associations between document pairs and *hubness* of the documents. Empirical analysis with various real data sets reveals that the proposed method outperforms state-of-the-art text clustering approaches.

Keywords: Text Clustering, Document Representation, Cluster Ensemble.

1 Introduction

Grouping of text documents into clusters is an elementary step in many applications such as Indexing, Retrieval and Mining of data on the Web. In traditional Vector space model (VSM), a document is represented as a feature vector which consists of weighted terms. A disadvantage of VSM representation is that it is not able to capture the semantic relations among terms [2]. Researchers have introduced two approaches of enriching document representation: (1) topic modelling, such as pLSI [13] and LDA [14]; and (2) embedding external knowledge into data representation model [1][2][15]. The semantic relations between the terms discovered by these methods are limited in either original document space or concepts represented by Wikipedia articles. They fail to capture other useful semantic knowledge, e.g. Wikipedia category that contains much meaningful information in the form of a hierarchical ontology [11]. In addition, these methods failed to model and cluster documents represented with multiple feature space (or relations).

In this paper, we propose a novel unsupervised cluster ensemble learning method, entitled Cluster Ensemble based Sequential Clustering using Hubness (*CESC-H*), for enriching document representation with multiple feature space and utilising them in document clustering. We propose to enhance data model by using semantic information derived from external knowledge i.e. Wikipedia concepts and categories. We construct a heterogeneous cluster ensemble using different types of feature spaces

X. Lin et al. (Eds.): WISE 2013, Part I, LNCS 8180, pp. 102–110, 2013.

selected to maximize the diversity of the cluster ensemble. In order to build an accurate cluster ensemble learner, (1) we learn consistent clusters that hold same documents, and (2) utilize the phenomenon of high dimensional data, hubs, to represent cluster center and sequentially join inconsistent documents into consistent clusters to deliver a final clustering solution.

2   Related Work

Our work is related to document representation enrichment techniques that incorporate semantic information from external knowledge, i.e., Wikipedia into Vector space model (VSM) [1][2][15]. [1] maps Wikipedia *concepts* to documents based on the content overlap between each document and Wikipedia articles. [2][15] represented documents as Wikipedia concepts by mapping candidate phrases of each document to anchor text in Wikipedia articles. However, in these works, only Wikipedia articles are considered and our proposed cluster ensemble framework incorporates Wikipedia category and concepts both into document representation, thereby introducing more semantic features into the clustering process.

Another related work is cluster ensemble learning. Cluster ensemble learning is a process of determining robust and accurate clustering solution from an ensemble of weak clustering results. Researchers have approached this problem by finding the most optimizing partition, for instance, hypergraph cutting-based optimization [5] and probabilistic model with finite mixture of multinomial distributions [6]. Finding a consensus clustering solution has also been approached by learning ensemble information from the co-association matrix of documents such as fixed cutting threshold [7], agglomerative clustering [8] and weighted co-association matrix [10]. Our proposed cluster ensemble learning not only models documents with multiple feature spaces but also provides accurate clustering result by differentiating consistent clusters and inconsistent documents and utilizing *hubness* of document for grouping.

3   The Proposed Cluster Ensemble based Sequential Clustering using Hubness (CESC-H) Method

Figure 1 illustrates the process of clustering text documents. Firstly, each document is mapped to Wikipedia articles (concepts) and categories, and cluster ensemble matrices are formed. A heterogeneous cluster ensemble construct is obtained by applying the (same) clustering algorithm on each matrix separately. The Affinity Matrix is proposed for identifying consistent clusters and inconsistent documents based on documents consistency in the cluster ensemble. Using the concept of representative hubs, the final clustering solution is delivered by placing all inconsistent documents to the most similar consistent cluster.

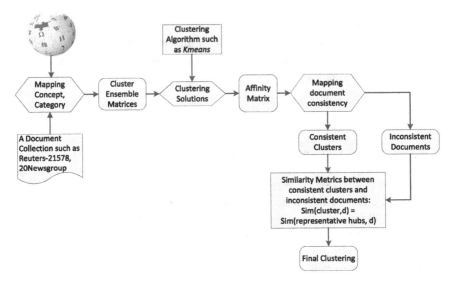

Fig. 1. The Cluster Ensemble based Sequential Clustering using Hubness Method

3.1   Cluster Ensemble Matrices

In this section, we discuss how to construct a heterogeneous cluster ensemble with different cluster ensemble matrices.

Document-Term (D-T) Matrix. A document d_j is represented as a term vector $T_{d_j} = \{t_1, t_2, \dots t_{|T|}\}$ with term set T. Each value of the vector represents the equivalent weight (TF*IDF) value of the term,

$$W_{(d_j, t_i)} = W_{TF*IDF} = tf(d_j, t_i) \times log \frac{N}{df(t_i)} \tag{1}$$

where $tf(d_j, t_i)$ is term frequency and $df(t_i)$ denotes inverse document frequency for N documents.

Document-Concept (D-C) Matrix. In the D-C matrix, a document d_j is represented by a concept vector $C_{d_j} = \{c_1, c_2, \dots c_{|C|}\}$ where C is the total number of concepts, and each value of C_{d_j} is the concept salience $SAL(d_j, c_i)$, calculated as in equation 3. If the anchor text of a Wikipedia article appears in a document, the document is mapped to the Wikipedia article. However, an anchor text, for example "tree", can appear in many Wikipedia articles, "tree (the plant)" or "tree (data structure)". Therefore, we find the most related Wikipedia article for an ambiguous anchor text by calculating the sum of the relatedness score between unambiguous Wikipedia articles and candidate Wikipedia articles. The relatedness of each pair of Wikipedia concepts (c_i, c_j), is measured by computing the overlap of sets of hyperlinks in them [9]:

$$Rel(c_i, c_j) = 1 - \frac{max(log|c_i|, log|c_j|) - log\ |c_i \cap c_j|}{W - min\ (log|c_i|, log|c_j|)} \tag{2}$$

where W is the total number of Wikipedia articles. In order to punish irrelevant concepts and highlight document topic related concepts, the concept salience [18] is applied as the weight of a concept c_i integrating local syntactic weight and semantic relatedness with other concepts c_l in document d_j:

$$SAL(d_j, c_i) = W_{(t_i, d_j)} * Rel(t_i, c_i | d_j) \qquad (3)$$

where $W_{(t_i, d_j)}$ is the syntactic weight in equation (1) of the corresponding term t_i. $Rel(t_i, c_i | d_j)$ is the sum of relatedness of concept c_i with other concepts that the rest of terms in document d_j map to:

$$Rel(t_i, c_i | d_j) = \sum_{t_l \in d_j \,\&\, t_l \neq t_i \,\&\, t_l \rightarrow c_l} Rel(c_i, c_l) \qquad (4)$$

where $Rel(c_i, c_l)$ is obtained using equation (2). If a concept is mapped to a *n-gram* (n>=2) phrase, the syntactic weight in (3) is the sum of the weight of each uni-gram term.

Document-Category(D-Ca) Matrix. A document d_j is represented by a category vector $E_{d_j} = \{E_{c_1}, E_{c_2} \dots E_{c_{|C|}}\}$ where E_{c_i} is a vector $\{e_1, e_2 \dots e_k\}$ that contains parent categories assigned for the concept c_i. The weight of a category is the weight of the corresponding concept. If a category is assigned to more than one concept, the sum of the weight of these concepts is the category's weight.

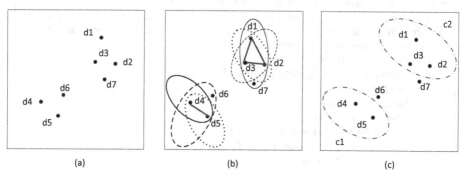

(a)   (b)   (c)

Fig. 2. The Process of Identifying Consistent Clusters and Inconsistent Documents. (a) A document collection (b) Three Component clustering results (ovals with solid line, dotted line and dashed line) and documents which have larger value than threshold θ (connected with bold line) (c) Consistent clusters (c1 and c2) and inconsistent documents (d_6 and d_7).

3.2   Affinity Matrix

In this section, we discuss how to construct an ensemble learner based on Affinity Matrix to identify consistent clusters. Figure 2 illustrates the process in simpler manner. Affinity Matrix (AM) is constructed by identifying document pairs which co-locate in the same partition of all component solutions (steps 1-3 in Figure 3). Let Cluster ensemble $\Pi = \{\Pi_1 \Pi_2, \dots, \Pi_H\}$ contains all component clustering solution. A component clustering solution $\Pi_i = \{\pi_1, \pi_2, \dots, \pi_K\}$ contains partitions for the specific feature space. For a document d_k, function $f_i(\cdot)$ searches through each clustering partition space π_m in each Π_i to identify which partition d_k belonging to:

$$f_i(d_k) = \begin{cases} m, & if\ d_k \in \pi_m \\ 0, & otherwise \end{cases}, 0 < m < K+1 \qquad (5)$$

where K is the number of partitions and i is the identifier of the component clustering result. The consensus function $con(\cdot)$ then accumulates the total co-occurrence of a document pair (d_x, d_y) in all component clustering solutions using $f_i(\cdot)$ in (5):

$$con(d_x, d_y) = \sum_{i=1}^{H} \delta\left(f_i(d_x), f_i(d_y)\right), x \neq y, \delta(a,b) = \begin{cases} 1, & if\ a = b \\ 0, & if\ a \neq b \end{cases} \qquad (6)$$

Affinity Matrix ($N \times N$ symmetric matrix where N is the number of documents) contains consistency degree ($con(d_x, d_y)$) for each pair of documents (d_x, d_y) in the corpus. The proposed method differentiates consistent clusters and inconsistent documents by setting a threshold (θ) on the values in AM and, consequently, is able to obtain reliable consistent clusters with setting a high threshold (steps 4-15 in Figure 3). Documents whose consistency degree is above θ are combined to form consistent clusters (c1 and c2 in Figure 2). Documents with lower consistency degree are dropped to a waiting list Z as inconsistent documents (d_6 and d_7 in Figure 2). The next section shows the process of joining inconsistent documents to consistent clusters.

Input: Cluster ensemble $\Pi = \{\Pi_1 \Pi_2, \dots, \Pi_H\}$ for document collection D with N documents on H types of feature space; consistency degree threshold θ; number of nearest-neighbour k and representative hub threshold η.
Output: Final document partition $C = \{C_1, C_2, \cdots C_K\}$ and K is the number of clusters.
Initialization: Set the affinity matrix AM as a null $N \times N$ matrix, and set C and Z (inconsistent document set) as empty.

```
1:   for i = 1 to H do
2:       AM(d_x, d_y) ← Π_i using equation (5) and (6)
3:   end for
4:   for x = 1:|D| do
5:       for y = 1:|D| do
6:           if AM(d_x, d_y) > θ and d_y ≠ d_x
7:               if d_x ∈ C_i && d_y ∈ C_i && C_i ∈ C
8:                   C_i ← (d_x, d_y)
9:               else
10:                  C_{i+1} ← (d_x, d_y); C ← C_{i+1}
11:          else
12:              Z ← d_x
13:          end if
14:      end for
15:  end for
16:  for z = 1 to |Z| do
17:      (C_m ∈ C) ← d_z using equation (8)
18:      Re(d_z) and Up(H_m) using equation (7)
19:  end for
20:  Return final partition C
```

Fig. 3. The Unsupervised Cluster Ensemble Learning Algorithm

3.3 Hubness of Document

The traditional centroid based partitional methods usually fail to distinguish clusters due to the curse of high dimensionality [4]. In this paper, in order to join inconsistent documents to the consistent clusters, we propose to use hubs as representation of the cluster center instead of centroids (step 16-19 in Figure 3). Let d_x be a document in a consistent cluster and d_y be a documents in the document collection D. Let $D_k(d_x, d_y)$ denote the set of documents, where document d_x is among the $k-$ nearest-neighbour list of document d_y and $d_y \neq d_x$. The hubness score of d_x, $N_k(d_x)$, is defined as:

$$N_k(d_x) = |D_k(d_x, d_y)| \tag{7}$$

The hubness score of d_x depends on the $distance(d_x, d_y)$ and the k−nearest-neighbour at data point d_y. We make use of top-η (top η proportion of) documents ranked by hubness score as hubs. Let H_m be the set of representative hubs for cluster $C_m (C_m \in C)$. For an inconsistent document d_z, we find the most similar consistent cluster C_m whose representative hub set H_m is most close to d_z where K is the number of clusters:

$$(C_m \in C) = \underset{K}{argmin}(\|H_K - d_z\|) \tag{8}$$

4   Experiments and Evaluation

4.1 Experimental Setup

As shown in table 1, two subsets from the 20Newsgroups data set are extracted in the same line as [2]: Multi5 and Multi10. Other two subsets were created from the Reuters-21578 data set as [2]: R-Min20-Max200 and R-Top10. For each data set, Wikipedia concepts and categories are mapped to the terms of documents via methods discussed in the section 3.1.

Table 1. Data set summary

Data Set	Description	#Classes	#Documents	#Terms	#Wikipedia Concepts	#Wikipedia Categories
D1	Multi5	5	500	2000	1667	4528
D2	Multi10	10	500	2000	1658	4519
D3	R-Min20Max200	25	1413	2904	2450	5676
D4	R-Top10	10	8023	5146	4109	9045

The proposed approach (*CESC-H*) is benchmarked with following approaches: *Single Feature Space.* This is a vector-space-model-based Bisecting K-means approach [3] based on each feature space: term (D-T), concept (D-C), caetegory (D-Ca) which are represented as a, b and c in Tables 2.

Linear Combination of Feature Space. This approach clusters do-cuments based on linearly combined syntactic and semantic feature space: term and concept (D-(T+C)), term and category (D-(T+Ca)) and term, concept and category (D-(T+C+Ca)) which are d, e and f, respectively in Tables 2.

HOCO [3]. This is a High-Order Co-Clustering method using the consistency information theory to simultaneously cluster documents, terms and concepts.

Cluster Ensemble Based Methods. The *CSPA, HGPA* and *MCLA* are hypergraph-based methods [5] whereas *EAC* uses evidence accumulation [8].

CESC. Variation of the proposed method CESC-H but computing similarity between consistent clusters and inconsistent documents using the cluster centroid instead of hubs.

4.2   Experimental Results

Table 2 presents the clustering performance in *FScore* and *NMI* on each data set and method respectively. *CSPA* was not able to work on data sets D3 and D4 due to computation complexity.

Table 2. Fscore *(F)* and *NMI* for Each Data Set and Method

Data Set	D1		D2		D3		D4		Ave	
	F	NMI	F	NMI	F	NMI	F	NMI	F	NMI
a	0.771	0.664	0.580	0.484	0.624	0.692	0.593	0.548	0.642	0.597
b	0.752	0.659	0.61	0.54	0.622	0.698	0.576	0.54	0.64	0.609
c	0.734	0.636	0.557	0.481	0.62	0.69	0.567	0.532	0.633	0.584
d	0.771	0.662	0.601	0.506	0.609	0.695	0.564	0.543	0.636	0.601
e	0.766	0.648	0.597	0.501	0.615	0.702	0.581	0.544	0.639	0.598
f	0.774	0.665	0.613	0.548	0.632	0.708	0.595	0.555	0.653	0.619
HOCO	0.972	0.917	0.705	0.603	0.692	0.779	0.601	0.569	0.742	0.717
CSPA	0.950	0.855	0.321	0.313	-	-	-	-	0.635	0.584
HGPA	0.797	0.601	0.466	0.61	0.648	0.702	0.509	0.179	0.605	0.523
MCLA	0.822	0.692	0.286	0.583	0.691	0.747	0.73	0.467	0.632	0.622
EAC	0.323	0.319	0.714	0.625	0.722	0.746	0.806	0.546	0.641	0.559
CESC	**0.982**	**0.922**	0.791	0.68	0.758	0.782	0.814	0.587	0.836	0.743
CESC-H	**0.982**	0.921	**0.799**	**0.691**	**0.771**	**0.797**	**0.822**	**0.591**	**0.844**	**0.75**

We can see that the proposed methods (*CESC-H* and *CESC*) and *HOCO* get significantly better performance than clustering with linear combination of feature space. More importantly, *CESC-H* (and *CESC*) outperforms *HOCO* as it uses heterogeneous cluster ensemble with additional external knowledge (i.e. Wikipedia category). The *CESC* and *CESC-H* approach consistently outperforms other cluster ensemble methods, *CSPA, HGPA, MCLA* and *EAC*. Different from our proposed method, these ensemble methods do not differentiate consistent clusters and inconsistent documents. Moreover, *CESC-H* performs better than *CESC* on each data set, as hubness scores of clusters can better represent the cluster than the cluster centriod, thereby improving the accuracy of joining inconsistent documents to consistent clusters.

4.3   Sensitivity Analysis

As shown in Figure 4, when θ increases, performance of *CESC-H* is improved. The larger θ value compels a document pair to be grouped in the same partition by more cluster ensemble components. Similarly with the higher value of η ($\eta = 0.85$), *CESC-H* achieves the best result. When the neighbourhood size is large enough ($k = 15$), representative hubs are stable and accurate clustering solution is obtained.

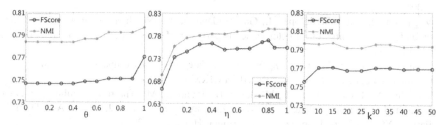

Fig. 4. FScore/NMI as function of different trade-off parameters: consistency degree shold θ; representative hub threshold η; and number of nearest-neighbour k

5   Conclusions and Future Work

The proposed novel Cluster Ensemble based Sequential Clustering using Hubness (*CESC-H*) method, integrating unsupervised cluster ensemble learning and hubness of documents, has the capability of clustering documents represented with multiple feature space (or relations). *CESC-H* is able to introduce and model more external knowledge for document representation and maximize the diversity of cluster ensemble. With the support of *hubness* of documents, *CESC-H* learns accurate final clustering solution by joining inconsistent documents into consistent clusters. Experiments on four data sets demonstrate that the proposed approach outperforms many start-of-the-art clustering methods. In future, we will investigate other cluster ensemble learning approaches with more features.

References

1. Hu, X., Zhang, X., Lu, C., Park, E., Zhou, X.: Exploiting wikipedia as external knowledge for document clustering. In: Proc. of the 15th ACM SIGKDD, pp. 389–396 (2009)
2. Jing, L., Yun, J., Yu, J., Huang, J.: High-Order Co-clustering Text Data on Semantics-Based Representation Model. In: Huang, J.Z., Cao, L., Srivastava, J. (eds.) PAKDD 2011, Part I. LNCS, vol. 6634, pp. 171–182. Springer, Heidelberg (2011)

3. Steinbach, S., Karypis, G., Kumar, V.: A comparison of document clustering techniques. In: Proc. of the Workshop on Text Mining at ACM SIGKDD (2000)
4. Agrawal, R., Gehrke, J., Gunopulos, D., Raghavan, P.: Automatic subspace clustering of high dimensional data for data mining applications. In: Proc. of the 1998 ACM SIGMOD, pp. 94–105 (1998)
5. Strehl, A., Ghosh, J.: Cluster Ensembles – A Knowledge Reuse Framework for Combining Multiple Partitions. Journal of Machine Learning Research 3, 583–617 (2003)
6. Topchy, A., Jain, A., Punch, W.: A mixture model for clustering ensembles. In: Proceedings of the SIAM International Conference on Data Mining, pp. 331–338 (2004)
7. Fred, A.: Finding consistent clusters in data partitions. In: Kittler, J., Roli, F. (eds.) MCS 2001. LNCS, vol. 2096, pp. 309–318. Springer, Heidelberg (2001)
8. Fred, A.N., Jain, A.K.: Combining multiple clusterings using evidence accumulation. IEEE Transactions on Pattern Analysis and Machine Intelligence 27(6), 835–850 (2005)
9. Medelyan, O., Witten, I., Milne, D.: Topic indexing with wikipedia. In: Proc. of AAAI (2008)
10. Vega-Pons, S., Ruiz-Shulcloper, J., Guerra-Gandón, A.: Weighted association based methods for the combination of heterogeneous partitions. Pattern Recognition Letters 32(16), 2163–2170 (2011)
11. Köhncke, B., Balke, W.-T.: Using Wikipedia Categories for Compact Representations of Chemical Documents. In: Proc. of the ACM CIKM (2010)
12. Tomašev, N., Radovanović, M., Mladenić, D., Ivanović, M.: The role of hubness in clustering high-dimensional data. In: Huang, J.Z., Cao, L., Srivastava, J. (eds.) PAKDD 2011, Part I. LNCS, vol. 6634, pp. 183–195. Springer, Heidelberg (2011)
13. Deerwester, S.C., Dumais, S.T., Landauer, T.K., Furnas, G.W., Harshman, R.A.: Indexing bylatent semantic analysis. Journal of the American Society of Information Science 41(6), 391–407 (1990)
14. Blei, D., Ng, A., Jordan, M.: Latent Dirichlet allocation. Journal of Machine Learning Research 3, 993–1022 (2003)
15. Huang, A., Milne, D., Frank, E., Witten, I.H.: Clustering documents using a wikipedia-based concept representation. In: Theeramunkong, T., Kijsirikul, B., Cercone, N., Ho, T.-B. (eds.) PAKDD 2009. LNCS, vol. 5476, pp. 628–636. Springer, Heidelberg (2009)

Exploiting User Queries for Search Result Clustering

Abdul Wahid, Xiaoying Gao, and Peter Andreae

School of Engineering and Computer Science,
Victoria University of Wellington,
19 Kelburn Parade 6012. Wellington, New Zealand
{abdul.wahid,xgao,pondy}@ecs.vuw.ac.nz
http://ecs.victoria.ac.nz

Abstract. Search Result Clustering (SRC) groups the results of a user query in such a way that each cluster represents a set of related results. To be useful to the user, the different cluster should contain the results corresponding to different possible meanings of the user query and the cluster labels should reflect these meanings. However, existing SRC algorithms often ignore the user query and group the results based just on the similarity of search results. This can lead to two problems: *low quality cluster*, where the results within a single cluster are related to different meanings of the query; and *poor cluster labels*, where the label of the cluster does not reflect the query meaning associated with the results in the cluster.

This paper presents a new SRC algorithm called QSC that exploits the user query and uses both syntactic and semantic features of the search results to construct clusters and labels. Experiments show that the query senses are good candidates for the cluster labels and the algorithm can lead to high quality cluster and more semantically meaningful labels than other state-of-the-art algorithms.

Keywords: Web Clustering Engine, Search Result Clustering, Query Senses, Document Clustering.

1 Introduction

The goal of Search Result Clustering is not only to cluster search results but also to provide semantically meaningful cluster labels. A Cluster label is a one-phrase description of all the documents in a cluster enabling users to decide whether to browse the list of documents in a cluster by looking at the cluster label. It is a common practice to use the most common keywords shared by all the documents in a cluster as a cluster label. Documents can have common keywords that might represent either more than one sense or might not represent any sense of the user query. Therefore, cluster labels based on common keywords are not always useful to the user. Also the clusters will be more useful to the user if all the documents in a cluster represent only one particular sense of the user query.

X. Lin et al. (Eds.): WISE 2013, Part I, LNCS 8180, pp. 111–120, 2013.

Traditional Search Result Clustering algorithms which ignore the user query are more vulnerable to the problems of *low quality cluster* and *poor cluster labels*. *Low quality cluster* is having documents in a cluster that represent more than one senses of the user query and *poor cluster labels* are cluster labels that do not represent any senses of the user query.

The similarity between two documents is often measured using word frequency. Such similarity measures are regarded as syntactic measures because they only consider counts of words. In order to minimize the problem of *low quality cluster*, this paper uses both syntactic and semantic features (topics) of the documents.

This paper presents a new algorithm Query Sense Clustering (QSC) that exploits the user query and combines semantic and syntactic features of a document for the clustering solution. The paper is organized as follows: section 2 highlights the related work; section 3 discusses the representation and similarity measures of the documents and the query senses; section 4 describes the algorithm; section 5 focuses on the evaluation and analysis of the results and section 6 concludes the paper.

2   Related Work

Search Result Clustering (SRC) methods can be classified into three categories: data-centric, description-aware and description-centric [4].

The data-centric category contains traditional clustering algorithms (hierarchical, partitioning) and the focus is on the clustering process. The Scatter/Gather algorithm [10,19] is the pioneer example of the data-centric category. The main drawback of this category is the *poor cluster labels* which are often generated from the text and are often meaningless.

The description-aware methods carefully select one or more features to construct meaningful cluster labels. Suffix Tree Clustering (STC) [23] was the first algorithm that used suffix trees to build cluster labels and perform clustering on search results. The issue with description-aware methods is that the cluster labeling procedure dominates the clustering process and the overall quality of the clusters is compromised.

The description-centric methods are specialized clustering methods that not only focus on cluster labels but also try to provide quality clusters. Examples in this category include LINGO [17]. Our algorithm QSC also belongs to this category.

3   Representation and Similarity Measure

This work uses query senses to generate initial clusters and then uses a new document similarity measure to refine the initial clusters. The new document similarity measure is based on a new document representation using both syntactic and semantic features (topics). The following subsections introduce the new document representation, the document similarity measure, the query sense representation and the sense similarity measure. The algorithm is presented in section 4.

3.1 Document Representation and Document Similarity Measure

The traditional bag-of-words model is widely used in document clustering to represent documents in Vector Space. Terms are commonly weighted using the tf-idf weighting scheme [21]. A document d in term-space is represented as

$$Tm(d) = \{tfidf(t_1, d), tfidf(t_2, d), tfidf(t_3, d), ..., tfidf(t_n, d)\} \qquad (1)$$

where n is the total number of terms and $tfidf$ is the tf-idf function defined as

$$tfidf(t, d) = tf(t, d) \times \log \frac{|D|}{df(t)} \qquad (2)$$

where $tf(t, d)$ is the frequency of term t in the document d, $|D|$ is the total number of documents and $df(t)$ is the number of documents containing term t. A criticism of this model is that it only uses a syntactic representation of the document and ignores semantic representation of the document. One semantic representation is based on topics representing the subjects or concepts that a document is about. If we can identify all the topics of a documents, then we can represent a document as a vector in topic space with weights for each topic representing the importance of the topic to the document. We propose a new document representation in which a document d containing topics $\tau_1...\tau_m$ in topic-space is represented as

$$Tp(d) = \{w(\tau_1, d), w(\tau_2, d), w(\tau_3, d), ..., w(\tau_m, d)\} \qquad (3)$$

where m is total number of topics and $w(\tau, d)$ is a weight of a topic τ, generated using topic detector of Wikiminer Toolkit[1] [15], in document d.

The most common and well known similarity measure for comparing documents is cosine similarity function [18]. We define the combined cosine similarity that includes semantic and syntactic features of document d_i and d_j as

$$Sim(d_i, d_j) = \lambda Cosine(Tp(d_i), Tp(d_j)) + (1 - \lambda)Cosine(Tm(d_i), Tm(d_j)) \qquad (4)$$

where λ is a scaling variable and the value of λ is 0.1 based on the preliminary experiments, $Tp(d)$ is document vector in topic-space and $Tm(d)$ is document vector in term space.

3.2 Query Sense Representation and Sense Similarity Measure

We represent a query using a set of senses $S = \{s_1, s_2, s_3...s_n\}$ of the query which is generated using Wikiminer[2] [15] word disambiguation. These raw senses are

---

[1] The topic detector is comparable to state-of-the-art LDA based topic detectors.

[2] Wikiminer parses Wikipedia disambiguation pages to get different senses of a word.

filtered and noise is removed by using tokenization, stemming and stop word removal techniques. Tokens generated from these senses are mostly bi-grams such as *jaguar car, sepecat jaguar, fender jaguar, mac os*. Other examples of senses are *panthera* and *south alabama jaguar football*.

We define the similarity score between a document d_i and a sense s_j as a weighted sum of six different criteria:

$$SimSense(d_i, s_j) = \frac{|s_j|}{|d_i|} \sum_{k=1}^{6} w_k \cdot cmp_k(d_i, s_j) \qquad (5)$$

The six criteria for cmp are exact sequence matching, semantic matching, partial matching in both term space and topic space of the document d_i for sense s_j. The exact sequence matching counts the number of occurrence; the semantic matching counts overlap of either exact or synonyms; and partial matching counts overlap of individual words in sense s_j and document d_i.

4   QSC Algorithm

We had developed a new algorithm called QSC[3] that uses our new document representation and similarity measures. It includes three main steps: the first step is to group all the documents according to their similarity to the different senses of the user query; the second step is to iteratively optimize clusters by relocating documents from one cluster to another cluster based on the similarity between documents and the clusters; the third step is to rank the documents and clusters based on similarity with the user query.

Step 1: Initial Cluster Generation. The initial clusters are formed by calculating the similarity of each document with each user query sense and assigning each document to each cluster associated with the maximally similar sense. Each cluster is labeled with its associated sense. Documents that are not sufficiently similar to any sense are placed in a cluster labeled *general*. The set of initial clusters C consists of all the clusters that contains at least one document.

Step 2: Cluster Optimization. Initial clusters were based on the similarity between documents and the senses. Base cluster labels can provide quality labeling of clusters. However the clusters, especially the general cluster may contain a mixed group of documents that might not be similar. We developed an iterative method to reassign some documents in order to improve cluster quality by increasing intra-cluster coherence and inter-cluster distinctiveness.

Step 3: Cluster Ranking. Users are interested in only those documents that are most closely related to the query. Therefore the ranking of clusters and documents are computed with respect to the query.

---

[3] The full paper contains the pseudo code with more details and is available on request.

All the clusters were sorted, by calculating the relatedness score between the user query and the cluster label, using the term similarity measure $WikiSim$ [12]. The $WikiSim$ is Wikipedia based similarity measure that computes relatedness between two terms. Documents in its own cluster were also sorted by calculating the similarity of a document to its mean in its own cluster. The ranked result list is then sent to user for browsing.

5   Results

The QSC was evaluated on two datasets, AMBIENT [5] and MORESQUE [16].

Comparison 1: The results on the larger dataset, which consist of all queries of AMBIENT and MORESQUE, based on purity and entropy were not given in [14]. However we found another recent paper [8] that compared nine algorithms using F1-measure on this large dataset. Therefore we compared our algorithm QSC with these nine algorithms using F1-measure calculated by taking the harmonic mean of precision and recall of the cluster [6]. The comparison was made between STC, LINGO, KeySRC [1], Curvature [9], SquaT++ [16,8], B-MST [7], HyperLex [22], Chinese Whispers [2] and QSC.

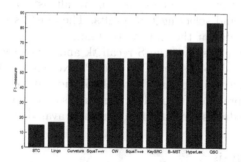

Fig. 1. Comparison of SRC methods

Figure 1 shows the percentage values of F1-measure of 10 methods on combined dataset of AMBIENT and MORESQUE taken from the paper [8] and the computed value of QSC. Clearly the QSC performed significantly better than others and have the highest value 83.62 (percentage) of F1-measure.

Other evaluation criteria Adjusted Rand Index(ARI) and Jaccard Index(JI) are also used for comparing the clustering algorithms in paper [8]. However we believe that they are not suitable for these two datasets. More discussion is provided in the last part of section 5.4.

Comparison 2: The search results needs to be diverse and top ranked results should represent different senses of the user query. In order to determine the diversification of this work, the search results were evaluated based on **S-recall@K (Subtopic recall at rank K)** and **S-precision@r (Subtopic precision at**

recall) [24] on combined dataset of AMBIENT and MORESQUE. The former evaluates the performance of the system based on K top-ranked results for number of topics of query q. S-precision@r measures the ratio of subtopics covered by minimum set of results at given recall r.

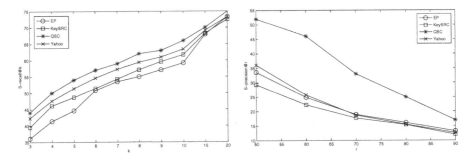

Fig. 2. S-recall@k on all queries   **Fig. 3.** S-precision@r on all queries

These two measures are used to compare search engines (Yahoo! and Essential Pages) that return ranked list of search results. The results returned by QSC were compared by flattening the clusters. The result list was formed by iterating through clusters and selecting top results. The clusters that only had one document were appended at the end to avoid noise.

Figure 2 and Figure 3 shows the S-recall@k and S-precision@r respectively for search results of Yahoo, Essential Pages(EP), KeySRC and QSC. The QSC performs relatively better in terms of S-recall@k and significantly outperformed others in terms of S-precision@r for the given values of k and r. This shows that QSC produced more diverse results than currently available search engines.

5.1   Further Analysis

The detailed analysis consists of three sub sections: the first discusses the cluster labels; the second discusses the processing time of the QSC, and the third discusses the cluster numbers and some observations about final clusters.

Cluster Label Analysis: The goal of the QSC algorithm is to generate a useful set of distinct clusters with informative labels.

Table 1 shows the cluster labels of the clusters generated for the query *Jaguar* by STC, LINGO and QSC (the cluster labels are not in ranked order). The labels for STC and LINGO were generated using the Carrot2 framework by adjusting the parameter of maximum clusters number to 8. Table 1 shows that the cluster labels generated by QSC provides more precise, intuitive and distinct labels than the cluster labels from STC and LINGO.

Processing Time: The QSC was evaluated on standalone workstation using Linux (64 bit) with Intel(R) Core(TM) i7-3770 CPU @ 3.40GHZ, 8GB RAM and

Table 1. Cluster labels of STC, LINGO and QSC of the query Jaguar

STC	LINGO	QSC
Jaguar Car	Auto Show	Jaguar Car
S-Type, Used Jaguar	Jaguar Parts	Jaguar E-Type
XK, 2006 2007, Price-Quotes and Reviews	Dealer Price Quotes and-Reviews	Jaguar XK
Ford Motor Company-Division	Ford Motor Company-Division	
Jaguar Cars		
Jaguar Panthera Onca	Jaguar Panthera Onca	Panthera
Jaguar Animal		
	Website of Fender Musical-Instrument	Fender Jaguar
Information	Jaguar Video	Mac OS X
New		SEPECAT Jaguar
		South Alabama Jaguar-Football

1TB HD. Figure 4 shows the processing time of all the queries in AMBIENT Dataset. The average time required for processing the query is under 1.0 second for both AMBIENT and MORESQUE datasets. Most of the queries were processed under one second with few exceptional cases. The maximum processing time was 6.3 seconds on a query *jaguar* because it had 54 senses to be processed. This processing time was reduced to 1 second by eliminating overlapping senses and processing only 10 distinct senses.

Strictly speaking, we cannot directly compare the processing time of other algorithms due to different machines and platforms. However we would like to give indications that word sense induction based algorithms (Curvature, Squat++, B-MST, HyperLex and Chinese Whispers) need to construct the graph to identify the senses from the huge corpus, whereas QSC extract the senses from the Wikipedia. Therefore the word sense induction based algorithms might require more processing time than QSC. The processing time of clustering, without considering the time spent on graph construction, for all algorithms is under 1 second except for SquaT++ algorithm. The SquaT++v and SquaT++e spent around 28 and 21 seconds respectively for clustering results as described in their paper [8].

Cluster Analysis: The average number of clusters for all queries in the AMBIENT dataset was 7.84 i.e on average 7-8 clusters are formed for each query. The average number of clusters for all queries in the AMBIENT and MORESQUE datasets was 5.4. There were a few queries with a high number of clusters and the maximum number of clusters was 18 for the query *Monte Carlo*. In contrast the

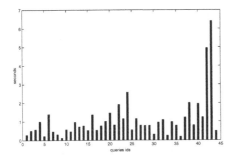
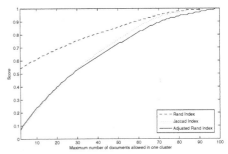

Fig. 4. Processing Time for All Queries   **Fig. 5.** RI, ARI and JI Analysis

query *Life on Mars* just had 1 big cluster. The reason for having many clusters was the large number of distinct query senses. The query *Life on Mars* had very few senses and they were overlapping with each other, e.g. *Life on Mars (TV series)*, *Life on Mars (U.S TV series)*, that causes single cluster for the query.

The QSC provided a more fine-grained clustering solution than the gold standard (manually labeled search results). The gold standard for the query *jaguar* had 7 clusters but QSC solution provided 9 clusters. The three clusters *jaguar car*, *jaguar e-type* and *jaguar xk* in QSC were sub clusters of gold standard *jaguar car*.

The QSC was not compared with other algorithms using index based evaluation measures (ARI and JI) because these measures have many issues [20,13]. One of the problems is that they do not handle fine-grained clustering solutions. If a gold standard G has a cluster g_i that contains 90 documents and clustering C has clusters c_j, c_{j+1} and c_{j+2} that contain all 90 documents then ARI and JI will penalize the clustering solution heavily. However the fine-grain clustering solution is consistent with the coarser solution and should not be penalized heavily. In fact it may even be better solution because it provides the distinctiveness that are not provided by the gold standard. ARI and JI do not measure this.

Figure 5 shows the phenomena of heavy penalty of ARI and JI as compared to Rand Index (RI) [11] on sub clusters. This experiment was performed on the AMBIENT and MORESQUE datasets by evaluating the perfect sub-clusters that gradually increased the limit of the maximum number of documents allowed in a cluster from 2 to 98. All the documents were perfectly assigned to the clusters and the values of RI, ARI and JI were computed at each iteration. The lowest value of RI, ARI and JI were 0.54, 0.05 and 0.04 respectively when the maximum allowed number of documents in sub clusters were 2. Figure 5 shows ARI and JI penalize small clusters and small sub-clusters heavily. The gold standard in our dataset had very unbalanced number of clusters. A few clusters were very small, and had 2 documents in a cluster and other were very large and had more than 90 documents. It was observed that the comparison based on ARI and JI is suitable only when the gold standard do not have sub clusters and all the clusters have almost the same number of members.

6 Conclusion

This paper presents a new description-centric search result clustering algorithm QSC that exploits query senses to generate meaningful cluster labels and use syntactic and semantic features of documents to generate quality clusters.

This paper shows that QSC outperforms existing algorithms. QSC is computationally inexpensive and provides better quality clusters with meaningful labels as compared to other algorithms, hence it has the potential to be applied to real time search result clustering applications.

The future direction for this work is to use Google WebIT and ukWac corpus along with Wikipedia to enhance the quality of query senses. The similarity measure and documentation representation are the key factors and a better similarity measure could bring more improvement. The greedy search in step 2 of the QSC could be improved to avoid local optima, by using the query senses in addition to document similarity. The currently used topic detection technique is not as good as state-of-the-art topic detection techniques such as LDA [3]; using LDA to detect topics from search results by considering query senses may further improve this work.

References

1. Bernardini, A., Carpineto, C., D'Amico, M.: Full-subtopic retrieval with keyphrase-based search results clustering. In: IEEE/WIC/ACM International Joint Conferences on Web Intelligence and Intelligent Agent Technologies, WI-IAT 2009, vol. 1, pp. 206–213. IET (2009)
2. Biemann, C.: Chinese whispers: an efficient graph clustering algorithm and its application to natural language processing problems. In: Proceedings of the First Workshop on Graph Based Methods for Natural Language Processing, pp. 73–80. Association for Computational Linguistics (2006)
3. Blei, D., Ng, A., Jordan, M.: Latent dirichlet allocation. The Journal of Machine Learning Research 3, 993–1022 (2003)
4. Carpineto, C., Osiński, S., Romano, G., Weiss, D.: A survey of web clustering engines. ACM Computing Surveys (CSUR) 41(3), 17 (2009)
5. Carpineto, C., Romano, G.: Ambient dataset (2008)
6. Crabtree, D., Gao, X., Andreae, P.: Improving web clustering by cluster selection. In: Proceedings of the 2005 IEEE/WIC/ACM International Conference on Web Intelligence, pp. 172–178. IEEE (2005)
7. Di Marco, A., Navigli, R.: Clustering web search results with maximum spanning trees. In: Pirrone, R., Sorbello, F. (eds.) AI*IA 2011. LNCS, vol. 6934, pp. 201–212. Springer, Heidelberg (2011)
8. Di Marco, A., Navigli, R.: Clustering and diversifying web search results with graph-based word sense induction. Computational Linguistics, 1–76 (just accepted, 2013)
9. Dorow, B., Widdows, D., Ling, K., Eckmann, J.-P., Sergi, D., Moses, E.: Using curvature and markov clustering in graphs for lexical acquisition and word sense discrimination. arXiv preprint cond-mat/0403693 (2004)

10. Hearst, M., Pedersen, J.: Reexamining the cluster hypothesis: scatter/gather on retrieval results. In: Proceedings of the 19th Annual International ACM SIGIR Conference on Research and Development in Information Retrieval, pp. 76–84. ACM (1996)

11. Hubert, L., Arabie, P.: Comparing partitions. Journal of Classification 2(1), 193–218 (1985)

12. Jabeen, S., Gao, X., Andreae, P.: Harnessing wikipedia semantics for computing contextual relatedness. In: Anthony, P., Ishizuka, M., Lukose, D. (eds.) PRICAI 2012. LNCS, vol. 7458, pp. 861–865. Springer, Heidelberg (2012)

13. Meilă, M.: Comparing clusterings–an information based distance. Journal of Multivariate Analysis 98(5), 873–895 (2007)

14. Meiyappan, Y., Iyengar, N.C.S.N., Kannan, A., Suyoto, Y.D., Suselo, T., Prasetyaningrum, T., Tlili, R., Slimani, Y., Dufreche, S., Zappi, M., et al.: Srcluster: Web clustering engine based on wikipedia. International Journal of Advanced Science and Technology 39(1), 1–18 (2012)

15. Milne, D., Witten, I.H.: An open-source toolkit for mining wikipedia. Artificial Intelligence (2012)

16. Navigli, R., Crisafulli, G.: Inducing word senses to improve web search result clustering. In: Proceedings of the 2010 Conference on Empirical Methods in Natural Language Processing, pp. 116–126. Association for Computational Linguistics (2010)

17. Osiriski, S., Stefanowski, J., Weiss, D.: Lingo: Search results clustering algorithm based on singular value decomposition. In: Intelligent Information Processing and Web Mining: Proceedings of the International IIS: IIPWM 2004 Conference held in Zakopane, Poland, p. 359 (2004)

18. Pang-Ning, T., Steinbach, M., Kumar, V.: Introduction to data mining. WP Co. (2006)

19. Pirolli, P., Schank, P., Hearst, M., Diehl, C.: Scatter/gather browsing communicates the topic structure of a very large text collection. In: Proceedings of the SIGCHI Conference on Human Factors in Computing Systems, pp. 213–220. ACM (1996)

20. Rosenberg, A., Hirschberg, J.: V-measure: A conditional entropy-based external cluster evaluation measure. In: Proceedings of the 2007 Joint Conference on Empirical Methods in Natural Language Processing and Computational Natural Language Learning (EMNLP-CoNLL), vol. 410, p. 420 (2007)

21. Salton, G., McGill, M.J.: Introduction to modern information retrieval (1986)

22. Véronis, J.: Hyperlex: lexical cartography for information retrieval. Computer Speech & Language 18(3), 223–252 (2004)

23. Zamir, O., Etzioni, O., Madani, O., Karp, R.: Fast and intuitive clustering of web documents. In: Proceedings of the 3rd International Conference on Knowledge Discovery and Data Mining, pp. 287–290. MIT Press (1997)

24. Zhai, C.X., Cohen, W.W., Lafferty, J.: Beyond independent relevance: methods and evaluation metrics for subtopic retrieval. In: Proceedings of the 26th Annual International ACM SIGIR Conference on Research and Development in Informaion Retrieval, pp. 10–17. ACM (2003)

Towards Context-Aware Social Recommendation via Trust Networks

Xin Liu

École Polytechnique Fédérale de Lausanne (EPFL)
CH-1015 Lausanne, Switzerland
x.liu@epfl.ch

Abstract. Utilizing social network information to improve recommendation quality has recently attracted much attention. However, most existing social recommendation models cannot well handle the heterogeneity and diversity of the social relationships (e.g., different friends may have different recommendations on the same items in different situations). Furthermore, few models take into account (non-social) contextual information, which has been proved to be another valuable information source for accurate recommendation. In this paper, we propose to construct trust networks on top of a social network to measure the quality of a friend's recommendations in different contexts. We employ random walk to collect the most relevant ratings based on the multi-dimensional trustworthiness of users in the trust network. Factorization machines model is then applied on the collected ratings to predict missing ratings considering various contexts. Evaluation based on a real dataset demonstrates that our approach improves the accuracy of the state-of-the-art social, context-aware and trust-aware recommendation models by at least 5.54% and 9.15% in terms of MAE and RMSE respectively.

Keywords: Recommender System, Social Network, Trust Network, Contexts.

1 Introduction

The large amount of data generated everyday on the web, on the one hand, provides rich information for users to consume, i.e., extracting knowledge, but on the other hand, also easily overloads users if no appropriate tools are provided to process such huge information for decision making. By suggesting information that is likely to interest users, recommender systems have become a promising tool to handle information overload in many online applications like e-commerce (e.g., ebay), social networks (e.g., LinkedIn), to name a few [23]. The mainstream technique for recommendation is collaborative filtering, which predicts a user's interest in an item by mining the patterns of the past rating information of other *similar* users and/or items [24]. In particular, factorization models, e.g., matrix factorization [8] have been proved to be one of the most successful collaborative filtering algorithms.

Recently, the increasing popularity of online social networks has brought an additional information source to enhance traditional rating based recommendation models. The assumption of such approaches is that users that are linked with each other in a social network tend to share similar tastes and interests, which can help improve the quality of recommendations and mitigate the data sparsity issue [12,13,5].

X. Lin et al. (Eds.): WISE 2013, Part I, LNCS 8180, pp. 121–134, 2013.
© Springer-Verlag Berlin Heidelberg 2013

However, most social recommendation models suffer from three limitations: (i) collaborative filtering algorithms are applied on all available ratings (i.e., without sophisticated information filtering). Although social information is considered, this still inevitably introduces noisy data, which damages the recommendation quality. (ii) Diversity of the social relationships is not well handled. Social connections may reflect different types of "friendships". For instance, two users are connected because of different reasons: friends, family, business partners, fans, etc. Given such heterogeneous connections, it is essential to assess the usefulness of a friend's recommendations in different situations, e.g., is a bookworm's recommendation on a video game reliable? (iii) Although proved to be important, to the best of our knowledge, few social recommendation models systematically leverage (non-social) contextual information[1], which can further refine the heterogeneous social information for better recommendation.

In order to address the three mentioned limitations, we propose to collect relevant rating information from context-aware trustworthy friends (and friends of friends, etc.) and then apply factorization machines to predict the missing preference taking into account various contexts. Specifically, the main contributions of this paper are summarized as follows: (1) We investigate and identify a variety of contextual information in recommender systems, and classify such information that will be used by different components of our social recommendation model. (2) On top of a social network, we construct trust networks based on the trustworthiness of users' friends in different contexts. That is, by mining past ratings, the trustworthiness of a user is represented by a trust vector, where each element represents the trust score of this user in the corresponding context. Given a target user, based on such context-aware trust, we apply random walk algorithm to collect a set of relevant ratings from the user's trustworthy neighbors[2] for recommendation (i.e., context-aware social similarity propagation). The relevance of a rating is determined by whether this rating is provided by a neighbor who is trustworthy in the contexts that are of interest. It is worth mentioning that since ratings are collected via trust networks, our approach can be naturally applied to a decentralized setting where the global rating information is not available. (3) With the collected ratings, we apply factorization machines [20,21] to systematically combine rating information and contextual information to predict missing ratings. The main advantage of factorization machines model is that it shares with other factorization models the high prediction accuracy with the support for integrating diverse types of contexts. (4) We conduct experiments on real dataset to evaluate the performance of our model. Experimental results show that our approach outperforms the state-of-the-art social, context-aware and trust-aware recommendation models (the least improvements are 5.54% and 9.15% in terms of MAE and RMSE respectively).

The rest of this paper is organized as follows: In Section 2, we review related social/trust-aware/context-aware recommendation models. In Section 3, we first discuss

---

[1] Note that social relationship is considered to be a type of context in some literatures. However, given its dynamism and complexity, we handle social information and contextual information separately in this work.

[2] We use *neighbor* in a trust network which is equivalent to *friend* in a social network, i.e., one-hop neighbor is equivalent to direct friend, and two-hop neighbor is equivalent to friend of friend, etc.

the contextual information in recommender systems in Section 3.1, and then present how to construct context-aware trust networks, collect relevant ratings and integrate contextual information to predict ratings in Section 3.2, 3.3 and 3.4 respectively. We report real dataset based experimental results in Section 4. In Section 5 we conclude this paper and outline future research directions.

2   Related Work

Several trust-based recommendation models have been proposed recently [11,4,25]. Based on the explicit and/or implicit trust information, the accuracy and coverage of recommendation is greatly improved compared to traditional rating based recommender systems. For instance, in [17], profile-level and item-level trust based methods are proposed to integrate into collaborative filtering algorithms. Jamali et al. [3] applied random walk to combine trust information and collaborative filtering to (1) address cold-start issue and (2) ensure high recommendation quality. However, these works focus on trust relationship, which is different from social relationship that is another useful information source for accurate recommendation and is more pervasive in practical systems. Actually, utilizing social network information to recommend items has recently become a popular trend. In [12], the authors proposed a probabilistic matrix factorization based approach to combine rating matrix and users' social network information. In [9], a neighborhood-based approach is proposed to generate social recommendations. A set of experiments were conducted to compare social based and nearest neighbor based recommendations.

Although greatly improving traditional recommendation models, most social/trust-aware recommendation models do not consider contexts when measuring the similarity between two users. For instance, even if a friend has very similar tastes with a user, her rating on a movie may be greatly influenced by other factors, for instance, her emotion, or with whom she watched the movie. Recent works have started looking at contexts when handling social network information. For instance, Xu et al. [26] proposed to cluster users and items such that like-minded users and their items are grouped. Subgroup information (i.e., a type of context) is then utilized by applying collaborative filtering to improve top-N recommendation quality. Yang et al. [27] first argued that a user may trust different subsets of friends regarding different domains, and then proposed a category specific circle-based model to make recommendation. However, these works only consider very basic contextual information, e.g., category/group. Jiang et al. [5] proposed to integrate social contexts (individual preference and interpersonal influence) into matrix factorization. However, such contextual information is only related to social relationships, so non-social contexts are largely ignored. SoCo [10] is probably the first work that systematically fuses rich contexts and social information for recommendation. By mixing diverse contextual information, the authors employed random decision trees to partition a rating matrix. Matrix factorization with social regularization is then applied to the partitioned matrices for rating prediction. The relevance and suitability of different contexts are ignored by SoCo, but this issue will be studied in this work.

There are also a lot of work dedicated to integrate contexts into a recommendation model. For instance, Adomavicius et al. [2] presented a multidimensional recommendation model based on multiple dimensions, i.e., user/item dimension as well as various contextual information. Karatzoglou et al. [7] proposed a multiverse recommendation model by modeling the data as a user-item-context N-dimensional tensor. Tucker decomposition is applied to factorize the tensor. An improvement was then proposed to reduce the computational complexity and handle different types of contexts by utilizing factorization machines [22]. However, these works only focus on context integration without taking into account any other potentially useful information, e.g., social and trust information.

In contrast, the novelty of our approach is two-fold. First, we consider both trust relationship and social relationship (i.e., the heterogeneity/diversity of friendships) to collect relevant rating information. Second, we integrate contextual information into our model in two ways: (i) categorical valued contexts are considered when constructing trust networks for relevant rating collection, and (ii) continuous valued contexts are integrated into factorization machines for model based context-aware recommendation.

3   Our Approach

In this section, we first investigate contextual information that is likely to influence users' rating behavior in Section 3.1. Trust network construction and relevant rating collection is discussed in Section 3.2 and 3.3. Factorization machines based context-aware recommendation is presented in Section 3.4.

3.1   Contextual Information in Recommender Systems

Context is a multifaceted concept that has been studied in different disciplines [23]. Despite numerous definitions that emphasize on different aspects in different scenarios, in this work, we define contextual information in recommender systems as the *auxiliary information that describes the characteristics of participants and that associates with interactions between participants*. Based on this definition, we identify contexts from three aspects: (1) Contexts of a user, obtained from her profile, such as her system age, gender, address (city/country), the number of ratings provided by this user (and the average rating), in which categories this user is active, etc. (2) Contexts of an item. This type of contexts are application dependent. For instance, in online auction, such contexts include category, price, average item price in the same category, etc.; in a movie review site, the contextual information may include a movie's actors/directors/producers, genres, in which year the movie is released, etc. (3) Contextual information of users and items is mainly collected or inferred (by applying statistical analysis or data mining, etc. [18]) from their characteristics. Besides such static information, another class of contexts can be obtained from the instantaneous information that is associated with interactions between users and items. For instance, the information about the location, time and weather when recommending a tour route; a user's emotion or accompanying persons when rating a music or movie; intent of purchasing a product (self-use versus gift), etc.

It is worth noting that not all contextual information has equal impact on recommendations, and furthermore, irrelevant contextual information can degrade the recommendation quality. In order to assess the relevance of contexts, we apply statistical tests [2,16] to investigate the interplay between contextual information and ratings. Specifically, we apply three statistical tests: (1) paired t-test (2) χ^2 test and (3) one way ANOVA. The null hypothesis is that context values and ratings are independent. Rejecting the null hypothesis (significance level p-value < 0.001) indicates that the context values and ratings are dependent so the corresponding contextual information is relevant. We finally select top-K contexts according to their rankings and p-values by the three tests.

After having the relevant contexts, we divide them into two categories: the ones with categorical values and the ones with continuous values, which will be used for determining multi-dimensional trust and incorporating into factorization machines respectively. The rationale behind this classification and usage is that when representing multi-dimensional trust (i.e., trust in different contexts), it is non-trivial to reasonably divide a continuous value into multiple intervals where each interval has a meaningful interpretation (like the discrete value of a categorical context). One may apply the methods designed to handle continuous valued attributes in decision trees [19] to discretize continuous valued contexts but this is not straightforward, difficult to interpret trust and incurs non-negligible overheads. On the other hand, factorization machines properly handle the contexts with continuous values. We will elaborate how different types of contexts are incorporated into our model in the following subsections.

3.2   Trust Network Construction

As mentioned in the introduction section, the diversity and heterogeneity of social information require sophisticated information filtering strategy to distinguish the usefulness of friends' recommendations in different contexts. To this end, we propose to construct context-aware trust networks on top of a social network (Fig. 1) to collect relevant rating information. We denote the selected contexts by $\mathcal{C} = \mathcal{C}_a \bigcup \mathcal{C}_o$, where $\mathcal{C}_a$ represents categorical valued contexts and $\mathcal{C}_o$ represents continuous valued contexts. For each user u_i, her friends list is denoted by $F_i = \{f_1, f_2, ...\}$.

Although trust is a complex concept and has several connotations [14], in this work we refer to trust as the ordinal correlation between the trustor u_i's ratings and the trustee (i.e., u_i's friend) f_j's ratings. Specifically, u_i's trust in a friend f_j is modeled as the correlation between their ratings on a set of items that f_j rated before u_i did. That is, if f_j first rated an item and then u_i rated the same, f_j' recommendation may influence u_i's choice of item[3]; and for the commonly rated items, if u_i and f_j's ratings are highly correlated, these two users have the similar tastes. In other words, our trust metric considers both the impact of a friend's recommendations that is measured by the amount of commonly rated items considering rating order and the friend's taste which is measured by the correlation between the ratings. That is, the more correlated between u_i and f_j's ordinal ratings on the commonly rated items, the more trustworthy f_j is from the perspective of u_i. Among numerous similarity measures, we apply Pearson Correlation Coefficient (PCC) to measure u_i's trust in f_j:

---

[3] We are not claiming all ordinal ratings indicate f_j's influence on u_i's choice, e.g., u_i may simply rate an item based on her interest without being influenced by her friends. However, on the high level, such ordinal ratings indeed reflect the relative impact of recommendations among friends.

$$T_{i,j} = \frac{\sum\limits_{k=1}^{n} R_{u,k}R_{f,k} - \frac{\sum\limits_{k=1}^{n} R_{u,k} \sum\limits_{k=1}^{n} R_{f,k}}{n}}{\sqrt{\sum\limits_{k=1}^{n} R_{u,k}^2 - \frac{(\sum\limits_{k=1}^{n} R_{u,k})^2}{n}}\sqrt{\sum\limits_{k=1}^{n} R_{f,k}^2 - \frac{(\sum\limits_{k=1}^{n} R_{f,k})^2}{n}}}. \tag{1}$$

Where $R_{u,k}$ and $R_{f,k}$ are the kth ratings in the rating vectors (for the commonly rated items considering rating order) of u_i and f_j respectively, and n is the rating vector size. One advantage of PCC (by considering average rating) is that it takes into account the fact that some users tend to give high ratings (e.g., 4 or 5 in five-point likert scale), while some more serious users may generally issue low ratings (e.g., 2 or 3 in five-point likert scale). Since PCC value is in the range [-1,1], we employ simple function like $f(x) = (x + 1)/2$ to bound the value into [0,1], which is easier to interpret trust.

In order to handle the diversity and heterogeneity of social relationships, we extend the single trust score to multi-dimensional trust considering the contexts C_a (i.e., categorical valued contexts). That is, u_i's trust in f_j is formulated as a trust vector: $\mathcal{T}_{i,j} =< T_{i,j}^{C_1}, T_{i,j}^{C_2}, ..., T_{i,j}^{C_{|C_a|}} >$, where $T_{i,j}^{C}$ is the trust score calculated by Eq. 1 in context C. In this way, we are able to measure the trustworthiness of different friends at a finer granularity. For instance, in Fig. 1, on top of a social network, we construct two trust networks in two contexts based u_i's context-aware trust in her friends. We set trust threshold, e.g., 0.6 such that only 'trustworthy' friends are involved in a certain trust network construction. Note that a friend may appear in multiple trust networks.

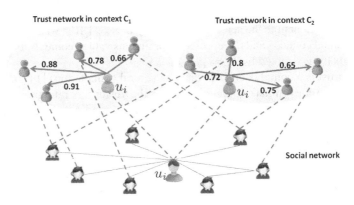

Fig. 1. Trust network construction (values at edges indicate trust scores in a certain context)

3.3   Relevant Rating Collection

Trust networks enable user u_i to explore trustworthy users to collect relevant ratings for better recommendation. Basically, u_i first finds a set of friends whose trust score is higher than a predefined threshold in certain contexts, then she collects these friends' ratings that are generated in the corresponding contexts.

However, in practice, users may have only a few friends, so they may not be able to collect sufficient relevant ratings. Such data sparsity issue greatly influences the accuracy of factorization models. Furthermore, only considering immediate friends' ratings will obviously damage the recommendation diversity [1]. In order to address these issues, we apply a random walk algorithm to explore the trust networks to obtain collect more relevant ratings from trustworthy friends of friends. The algorithm is outlined in Alg. 1. Specifically, to collect a set of relevant ratings $\mathcal{R}_i$, u_i first calculates the trustworthiness of her friends in contexts C' that are of interest (see Eq. 1), and then selects top-K most trustworthy friends[4] $\mathcal{F}_i^{C'}$ (Line 4). For each friend $f_j \in \mathcal{F}_i^{C'}$, her ratings that are generated in contexts C' are added to $\mathcal{R}_i$ (Line 6-10). If the size of $\mathcal{R}_i$ reaches the predefined threshold[5], $\mathcal{R}_i$ is returned (Line 11-13). Otherwise, starting from one of her friends, u_i explores the second hop neighbors (friend of friend) to collect more relevant ratings (Line 15-19). Note that each friend f_j is chosen to be further explored with a probability P_j that is equal to her trust score. Obviously, the more trustworthy a friend is, this friend's friends are more likely to be trustworthy to u_i (transitive trust [6]), and hence these friends will be explored with a higher probability.

Algorithm 1. Relevant rating collection (code for the user u_i)

1: Input: $\mathcal{R}_i = \Phi$, categorical valued contexts that are of interest $C' \in \mathcal{C}_a$
2: Output: filled $\mathcal{R}_i$ for user u_i
3: **function**: ratingCollection(u_i , C')
4: Calculating u_i's friends' trustworthiness in C', and selecting top-K most trustworthy friends: $\mathcal{F}_i^{C'}$.
5: **for** $f_j \in \mathcal{F}_i^{C'}$ **do**
6:     **for** each rating $R_{j,k}$ of f_j **do**
7:         **if** $R_{j,k}$ is generated in C' **then**
8:             $\mathcal{R}_i = \mathcal{R}_i \bigcup R_{j,k}$.
9:         **end if**
10:     **end for**
11:     **if** $|\mathcal{R}_i| \geq$ threshold **then**
12:         return $\mathcal{R}_i$.
13:     **end if**
14: **end for**
15: **for** $f_j \in \mathcal{F}_i^{C'}$ **do**
16:     With a probability $P_j = f_j$'s trust score, call ratingCollection(f_j , C').
17:     **if** $|\mathcal{R}_i| \geq$ threshold **then**
18:         return $\mathcal{R}_i$.
19:     **end if**
20: **end for**

---

[4] The value of K is application dependent, and can be learned based on historical records. Another option is to select friends with trust score larger than a threshold (e.g., 0.6). Experimental results show that there is no evident difference between these two options.

[5] The optimal rating matrix size for factorization machines model can be learned by conducting cross validation on existing ratings.

The trust network exploration is terminated if one of the two conditions is met: (1) the size of $\mathcal{R}_i$ reaches the predefined threshold (Line 17-19); (2) the distance (i.e., hops) between u_i and the user to be explored exceeds a limit ϵ. Such a limit is determined based on the theory of *six-degrees of separation* [15], i.e., $\epsilon = 6$. Note that the hop $h + 1$ users start to be processed only after all hop h users have been processed.

3.4 Context-aware Social Recommendation

Once relevant ratings are collected, we apply factorization machines model to predict a user's preference on an item taking into account contextual information (i.e., continuous valued contexts $\mathcal{C}_o$). Note that categorical valued contexts $\mathcal{C}_a$ has been used in trust network construction.

We describe a rating between user u and item v under a set of contexts $\mathcal{C}_o$ by a tuple $(\mathcal{X}, R)$, where $\mathcal{X}$ is the feature vector containing u, v and any context $c \in \mathcal{C}_o$, and R is the numeric rating from u to v. All such tuples construct a matrix $X \in \mathbb{R}^{N \times (|\mathcal{X}|+1)}$, where N is the number of tuples and $|\mathcal{X}|$ is the size of the feature vector. Factorization machines model all interactions (up to order d) among the $|\mathcal{X}|$ features. Specifically, if we set d to 2, i.e., factorization machines model interactions between any pair of features. The missing rating $\hat{R}$ is predicted as:

$$\hat{R} = \omega_0 + \sum_{k=1}^{|\mathcal{X}|} \omega_k x_k + \sum_{k=1}^{|\mathcal{X}|} \sum_{k'=k+1}^{|\mathcal{X}|} \left(\sum_{l=1}^{m} \tau_{k,l} \tau_{k',l} \right) x_k x_{k'}, \tag{2}$$

where m is the dimensionality of the factorization, and $\Pi = \{\omega_0, \omega_1, ..., \omega_{|\mathcal{X}|}, \tau_{1,1}, ..., \tau_{|\mathcal{X}|,m}\}$ is a set of model parameters to be learned.

According to Eq. 2, factorization machines first encode the effect of each feature $x_k \in \mathcal{X}$ on the predicted $\hat{R}$. Furthermore, factorization machines encode the effects of all interactions between two features. Although similar to linear regression model, instead of using independent weight, factorization machines employ factorized weight to model the effect of each interaction between feature x_k and $x_{k'}$:

$$W_{k,k'} = \sum_{l=1}^{m} \tau_{k,l} \tau_{k',l}. \tag{3}$$

Such a difference makes factorization machines, by using low rank approximation, reliably estimate the weights of the interactions even when few observations are available (i.e., data sparsity), which is very common in recommendation problems, while traditional models like polynomial regression require sufficient training data to learn independent weights.

It is worth noting that Eq. 2 can be easily extended to higher order ($d > 2$) factorization (see Eq. 4). However, in practice, learning high order factorization machines on a large dataset is quite expensive. So throughout this paper, we only consider 2 order factorization machines.

$$\hat{R} = \omega_0 + \sum_{k=1}^{|\mathcal{X}|} \omega_k x_k + \sum_{p=2}^{d} \sum_{k_1=1}^{|\mathcal{X}|} \dots \sum_{k_d=k_{d-1}+1}^{|\mathcal{X}|} \left(\sum_{l=1}^{m} \prod_{i=1}^{p} \tau_{k_i,l} \right) \prod_{i=1}^{p} x_{k_i}. \tag{4}$$

To predict a rating from user u to item v, model parameters Π must be reliably learned. Typically, the optimization task is to find Π to minimize the sum of the losses over the training data, where the loss is measured by the difference between a predicted rating and the corresponding real rating. We adapt least-squares loss (function) defined as follows:

$$L(R, \hat{R}) = (R - \hat{R})^2, \tag{5}$$

where R is the real rating and $\hat{R}$ is the predicted rating (see Eq. 2). The object function is then defined by summing up all losses over the training data:

$$\mathcal{L} = \arg\min_{\Pi}(\sum L(R, \hat{R})). \tag{6}$$

Since factorization machines typically involve a considerable number of model parameters, to avoid overfitting, we add regularization terms for different parameters:

$$\mathcal{L} = \arg\min_{\Pi}(\sum L(R, \hat{R}) + \sum_{\pi \in \Pi} \lambda_\pi \pi^2). \tag{7}$$

Eq. 7 can be solved using three approaches: (i) stochastic gradient descent (SGD), (ii) alternating least squares (ALS) and (iii) markov chain monte carlo (MCMC). In this paper, given its simplicity, low computational and storage complexity, we choose SGD to demonstrate how a factorization machines model is solved. The basic idea behind SGD is to iteratively (over the training data) update each model parameter $\pi \in \Pi$ by performing gradient descent in π:

$$\pi \leftarrow \pi - \eta(2(R - \hat{R})\frac{\partial \hat{R}}{\partial \pi} + 2\lambda_\pi \pi), \tag{8}$$

where η is the learning rate (empirically set to 0.0001 in our experiments). The computational complexity of SGD is $O(kS(X))$, where k is the dimensionality of the factorization and $S(X)$ is the number of nonzero elements in the user-item-contexts matrix X.

3.5 Summary

To sum up, by collecting relevant rating information, the size of the input matrix is significantly decreased, thus reducing the computational complexity. Another point that needs to be mentioned is that our rating collection strategy can be further improved by conducting finer rating selection. For instance, we can measure the similarity between items and select a rating based on the similarity with the target item. We leave as a future work a more detailed discussion on such an extension.

It is also worth mentioning that although cold-start issue is still an open question in recommender systems, addressing this issue is not the focus of this work; instead, we aim at improving the quality of recommendation by leveraging diverse information sources, e.g., rating information, social information, trust information, contextual information, etc. We next demonstrate the accuracy of our recommendation model by comparing with the state-of-the-art approaches.

4 Evaluation

4.1 Settings

We use a dataset collected from Douban (www.douban.com), one of the largest Chinese social platforms for sharing reviews and recommendations of books, movies and music. A user can provide ratings (1 to 5 stars) to indicate her preference on the items. A timestamp is associated with a rating. A user, although has not rated an item, may still express her interest by indicating "wish" (e.g., she wishes to read a book in the near future). A social network is provided, where one user can follow another user whose recommendations are considered to be helpful. Table 1 summarizes the statistics of Douban dataset[6]. Note that we only show explicit ratings, i.e., "wish" expressions are not counted.

Table 1. Statistics of Douban dataset

	# of ratings	# of users	# of items
Book	812,037	8,598	169,982
Movie	1,336,484	5,227	48,381
Music	1,387,216	23,822	185,574
All	3,535,737	25,560	403,937

For context-aware recommendation, we extract a set of contextual information from the dataset: (1) category of an item; (2) hour-of-day, i.e., which hour a rating is given; (3) day-of-week, i.e., weekday or weekend a rating is given; (4) number of "wish" on an item when a rating is given; (5, 6, 7) average rating, number of ratings, and standard deviation of ratings on an item when a new rating arrives; (8, 9, 10) average rating, number of ratings, and standard deviation of ratings of the target user when she rates a certain item. By applying statistical tests (see Section 3.1), we choose contexts (1), (3), (4), (5) and (8) for context-aware recommendation models. libFM [21] is used to implement factorization machines.

We compare our approach with three representative recommendation models that utilize trust information, social information and contextual information respectively: (1) TrustWalker [3] combines random walk (on trust network) and item based collaborative filtering to improve recommendation quality. Since Douban has no explicit trust information, a trust network is constructed as described in Section 3.2. (2) Social regularization model (SoReg) [13] is a social recommendation model. On the basis of a basic matrix factorization model, a social regularization is added to control friends' taste difference. (3) Factorization machines model, which is implemented using libFM [21].

Since the objective of our approach is to improve rating prediction accuracy, we use two standard metrics to measure the accuracy of various models, i.e., Mean Absolute Error (MAE): $MAE = \frac{1}{N} \sum_{r=1}^{N} |R - \hat{R}|$ and Root Mean Square Error (RMSE): $RMSE = \sqrt{\frac{1}{N} \sum_{r=1}^{N} (R - \hat{R})^2}$, where N is the total number of predictions, R is the real rating of an item and $\hat{R}$ is the corresponding predicted rating.

---

[6] This dataset is shared by Erheng Zhong [28].

For the factorization model based approaches, the model parameters such as latent factor dimensionality, learning rate, regularization parameter and iterations (for learning), etc. are determined using 10-folder cross-validation. Note that all comparison related results are statistically significant, proved by two-tailed, paired t-test with p-values < 0.001.

4.2   Results

In this subsection, we first demonstrate the accuracy of our approach when different design parameters are applied, and then compare the accuracy of different models.

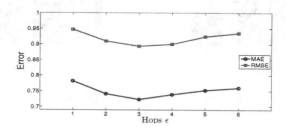

Fig. 2. Error rates with different hops ϵ

Fig. 2 shows how the accuracy of our approach varies with different hops ϵ for relevant rating collection (see Section 3.3). We observe that when $\epsilon = 1$, i.e., only immediate friends' ratings are considered, the highest MAE and RMSE are incurred. This is because for many users, only considering immediate friends cannot collect sufficient relevant ratings, thus affecting the accuracy of factorization machines. When ϵ increases, more relevant ratings are considered, and hence higher prediction accuracy is obtained. On the other hand, larger ϵ means more *less trustworthy* users' ratings are involved, which may introduce noises and hence decrease the overall accuracy. This is demonstrated in Fig 2: when $\epsilon > 3$, MAE and RMSE gradually increase. So in the following experiments we use the optimal parameter $\epsilon = 3$ for our model.

Tab. 2 summarizes the accuracy of our approach with/without context selection (see Section 3.1). Obviously, when context selection is performed (i.e., 5 contexts are selected), the accuracy is clearly better than that when all contexts are applied. This demonstrates that our statistic tests based context selection indeed helps to reduce the effects of irrelevant contextual information.

Recall that our approach handles contexts in two ways, i.e., context-aware trust network construction and factorization machines based context-aware rating prediction. In order to demonstrate the contextual information is used judiciously, we compare

Table 2. Impact of context selection

	MAE	RMSE
with context selection	0.7201	0.8892
without context selection	0.7513	0.9984

our *C*ontext-*A*ware *S*ocial *R*ecommendation model (CASR) with three variants: (1) CASR-TN: contextual information is only used in trust network construction. That is, factorization machines only takes as input user-item-rating tuples without considering any contexts. (2) CASR-FM: contextual information is only handled by factorization machines. That is, trust is established without taking into account any contexts. (3) CASR-NC: contextual information is ignored in both trust network construction and factorization machines.

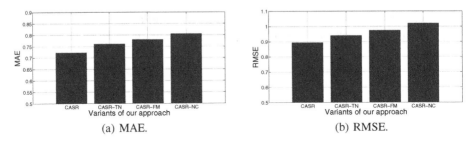

(a) MAE.   (b) RMSE.

Fig. 3. Error rates of our approach with different context configurations (variants)

Fig. 3 shows that CASR-NC is outperformed by other variants that incorporate contextual information in different ways. This demonstrates the importance of contexts in recommendation. CASR-TN incurs slightly lower MAE and RMSE than CASR-FM does, which means context-aware trust network construction (and hence relevant rating collection) plays an important role in reducing prediction errors. Finally, by integrating diverse types of contextual information into both trust establishment and factorization machines, CASR outperforms all variants, demonstrating that contexts are utilized judiciously in our approach.

We finally compare the accuracy of our approach CASR with that of three representative recommendation models: TrustWalker, SoReg and FM. To comprehensively evaluate the accuracy in different experiment settings, we randomly select 40%, 60% and 80% of all ratings respectively as training data to build recommendation models, and predict the remaining ones. From Tab. 3 we observe that more training data leads to more accurate recommendation for all models. FM incurs the highest MAE and RMSE. One possible reason is that the selected contexts (even after being statistically tested) still have limited correlation with ratings so FM cannot reliably model the interactions among users, items, contexts and ratings. This also proves that the accuracy of FM largely depends on the relevance of contexts. By considering trust information, Trust-Walker provides better results: improving the accuracy of FM by 6.03% and 3.88% in terms of MAE and RMSE respectively (80% training data). This demonstrates the advantages of trust-aware recommendation models. The fact that SoReg is slightly better than TrustWalker indicates the importance of social network information. By measuring the similarity between a user and her friends for assigning appropriate weights, and applying matrix factorization, SoReg further improves the recommendation accuracy of the trust-aware model. On the other hand, CASR consistently outperforms all other models in all settings due to three key designs: (1) Trust information is considered within a social network to study the heterogeneity of social relationships. (2) Contexts

Table 3. Accuracy comparison

metrics (training data%)	CASR	TrustWalker	SoReg	FM
MAE (80%)	0.7201	0.7900	0.7623	0.8313
RMSE (80%)	0.8892	1.0635	0.9788	1.0966
MAE (60%)	0.7262	0.7944	0.7656	0.8343
RMSE (60%)	0.8921	1.0698	0.9813	1.0990
MAE (40%)	0.7324	0.7987	0.7688	0.8385
RMSE (40%)	0.8952	1.0733	0.9855	1.1018

are utilized such that trust relationships are established at a finer granularity. (3) Factorization machines are applied to properly handle diverse types of contexts. CASR improves SoReg by 5.54% for MAE and 9.15% for RMSE (80% training data).

5   Conclusion

In this work, we impose context-aware trust on a social network to investigate the heterogeneity and diversity of social relationships, based on which relevant ratings are collected for factorization machines based recommendation. Compared to the state-of-the-art recommendation models, the advantages of our approach lie in sophisticated information filtering and proper integration of diverse types of contexts in different components. Furthermore, since data is collected along the trust paths, our approach can be easily applied to decentralized settings where global rating information is not available. Real dataset based evaluation demonstrates that our approach outperforms the representative recommendation models that also rely on trust information, social information and contextual information respectively. For the future work, we plan to apply the proposed recommendation model to real application scenarios for further performance optimization.

Acknowledgement. This work was partially supported by the grant *Reconcile: Robust Online Credibility Evaluation of Web Content* from Switzerland through the Swiss Contribution to the enlarged European Union.

References

1. Adomavicius, G., Kwon, Y.: Improving aggregate recommendation diversity using ranking-based techniques. IEEE Trans. on Knowl. and Data Eng. 24(5), 896–911 (2012)
2. Adomavicius, G., Sankaranarayanan, R., Sen, S., Tuzhilin, A.: Incorporating contextual information in recommender systems using a multidimensional approach. ACM Trans. Inf. Syst. 23(1), 103–145 (2005)
3. Jamali, M., Ester, M.: Trustwalker: a random walk model for combining trust-based and item-based recommendation. In: Proceedings of the 15th SIGKDD (2009)
4. Jamali, M., Ester, M.: A matrix factorization technique with trust propagation for recommendation in social networks. In: Proceedings of the Fourth ACM Conference on Recommender Systems (2010)

5. Jiang, M., Cui, P., Liu, R., Yang, Q., Wang, F., Zhu, W., Yang, S.: Social contextual recommendation. In: Proceedings of the 21st CIKM (2012)
6. Jøsang, A., Gray, E., Kinateder, M.: Analysing topologies of transitive trust. In: Proceedings of the Workshop of Formal Aspects of Security and Trust (2003)
7. Karatzoglou, A., Amatriain, X., Baltrunas, L., Oliver, N.: Multiverse recommendation: n-dimensional tensor factorization for context-aware collaborative filtering. In: Proceedings of the Fourth ACM Conference on Recommender Systems (2010)
8. Koren, Y., Bell, R., Volinsky, C.: Matrix factorization techniques for recommender systems. Computer 42(8), 30–37 (2009)
9. Liu, F., Lee, H.J.: Use of social network information to enhance collaborative filtering performance. Expert Systems with Applications 37(7), 4772–4778 (2010)
10. Liu, X., Aberer, K.: Soco: A social network aided context-aware recommender system. In: Proceedings of the 22nd International World Wide Web Conference, WWW (2013)
11. Ma, H., King, I., Lyu, M.R.: Learning to recommend with social trust ensemble. In: Proceedings of the 32nd SIGIR (2009)
12. Ma, H., Yang, H., Lyu, M.R., King, I.: Sorec: social recommendation using probabilistic matrix factorization. In: Proceedings of the 17th CIKM (2008)
13. Ma, H., Zhou, D., Liu, C., Lyu, M.R., King, I.: Recommender systems with social regularization. In: Proceedings of the 4th WSDM (2011)
14. Harrison Mcknight, D., Chervany, N.L.: The meanings of trust. Technical report, University of Minnesota (1996)
15. Milgram, S.: The small world problem. Psychology Today 1(1), 61–67 (1967)
16. Odic, A., Tkalcic, M., Tasic, J.F., Kosir, A.: Relevant context in a movie recommender system: Users' opinion vs. statistical detection. In: Proceedings of the 4th Workshop on Context-Aware Recommender Systems (2012)
17. O'Donovan, J., Smyth, B.: Trust in recommender systems. In: Proceedings of the 10th International Conference on Intelligent User Interfaces (2005)
18. Palmisano, C., Tuzhilin, A., Gorgoglione, M.: Using context to improve predictive modeling of customers in personalization applications. IEEE Trans. on Knowl. and Data Eng. 20(11), 1535–1549 (2008)
19. Quinlan, J.R.: Improved use of continuous attributes in c4.5. J. Artif. Int. Res. 4(1), 77–90 (1996)
20. Rendle, S.: Factorization machines. In: Proceedings of the 10th ICDM (2010)
21. Rendle, S.: Factorization machines with libFM. ACM Trans. Intell. Syst. Technol. 3(3), 57:1–57:22 (2012)
22. Rendle, S., Gantner, Z., Freudenthaler, C., Schmidt-Thieme, L.: Fast context-aware recommendations with factorization machines. In: Proceedings of the 34th SIGIR (2011)
23. Ricci, F., Rokach, L., Shapira, B., Kantor, P.: Recommender Systems Handbook. Springer (2011)
24. Su, X., Khoshgoftaar, T.M.: A survey of collaborative filtering techniques. Adv. in Artif. Intell. (2009)
25. Walter, F.E., Battiston, S., Schweitzer, F.: A model of a trust-based recommendation system on a social network. Autonomous Agents and Multi-Agent Systems 16, 57–74 (2008)
26. Xu, B., Bu, J., Chen, C., Cai, D.: An exploration of improving collaborative recommender systems via user-item subgroups. In: Proceedings of the 21st WWW (2012)
27. Yang, X., Steck, H., Liu, Y.: Circle-based recommendation in online social networks. In: Proceedings of the 18th SIGKDD (2012)
28. Zhong, E., Fan, W., Wang, J., Xiao, L., Li, Y.: Comsoc: adaptive transfer of user behaviors over composite social network. In: Proceedings of the 18th SIGKDD (2012)

Personalized Recommendation on Multi-Layer Context Graph

Weilong Yao[1], Jing He[2], Guangyan Huang[2], Jie Cao[3], and Yanchun Zhang[1,2]

[1] University of Chinese Academy of Science, Beijing, China
yaoweilong12@mails.ucas.ac.cn
[2] Centre for Applied Informatics, Victoria University, Melbourne, Australia
{Jing.He,Guangyan.Huang,Yanchun.Zhang}@vu.edu.au
[3] Nanjing University of Finance and Economics, Nanjing, China
caojie690929@163.com

Abstract. Recommender systems have been successfully dealing with the problem of information overload. A considerable amount of research has been conducted on recommender systems, but most existing approaches only focus on user and item dimensions and neglect any additional contextual information, such as time and location. In this paper, we propose a Multi-Layer Context Graph (MLCG) model which incorporates a variety of contextual information into a recommendation process and models the interactions between users and items for better recommendation. Moreover, we provide a new ranking algorithm based on Personalized PageRank for recommendation in MLCG, which captures users' preferences and current situations. The experiments on two real-world datasets demonstrate the effectiveness of our approach.

Keywords: recommender system, context, random walk, graph.

1 Introduction

Recommender systems have been successfully dealing with the problem of information overload. While a substantial amount of research has already been conducted by both industry and academia in the area of recommender systems, most approaches only recommend items to users without taking into account any additional contextual information, such as time, location, age, or genre. Thus, the conventional recommender systems operate in the two-dimensional $USER \times ITEM$ space [1].

Incorporating contextual information into a recommendation process can achieve better recommendation accuracy than considering only user and item information. For example, a vacation recommendation for a given user may depend on season, age and interest. More specifically, in summer, an elderly user would probably prefer to enjoy her/his vacation on a peaceful beach rather than a ski resort.

Generally speaking, we consider three types of context in recommender systems: 1) User context describes user's meta attributes, such as gender, age, education and social information. Users sharing more common user context tend to

X. Lin et al. (Eds.): WISE 2013, Part I, LNCS 8180, pp. 135–148, 2013.
© Springer-Verlag Berlin Heidelberg 2013

have similar tastes or preferences. 2) item context enables measuring relativity between two items. 3) The third type of contextual information we call decision context, involves context where the decision is made, such as time, location or mood. The same user with different decision context shows different preferences. For instance, the style of songs that a user listens at home on weekends may be different from the style of songs he listens in office on weekdays. Therefore, putting users and items in context can be beneficial to higher accuracy. However, there have been few methods which can incorporate all the three types of context into a recommendation process, and most approaches focus on only partial context (see Section 2).

In this paper, we propose a Multi-Layer Context Graph (MLCG) model which considers all of the three types of context into recommender systems, and models the interactions between users and items in the corresponding decision context. A personalized random walk-based ranking method on a MLCG is further presented, which captures users' preferences and decision context.

The most related work that also incorporates contextual information on graph by utilizing nodes to represent multidimensional data is proposed in [4]. This work strengthens the connections between users and items by merging different dimensions of context. The difference between MLCG and that work lies in that our method focuses on the instant situation where users interacts with items. That is, MLCG emphasizes the fact that interactions between users and items occur in a certain context.

To summarize, our main contributions are as follows:

1. We propose a Multi-Layer Context Graph (MLCG) model that incorporates all the three types of context to model the decision making by users. In particular, MLCG emphasizes the instant situation of the interactions between users and items.
2. Based on MLCG, we provide a new ranking algorithm, which extends Personalized PageRank for increasing accuracy of top-k recommendation through running on a MLCG that models the influence flow between/in layers.
3. Our experiments based on two real-world datasets demonstrate the effectiveness of our proposed method.

The rest of this paper is organized as follows. In Section 2, we cover related work on context-aware recommendation and graph-based recommendation. In Section 3, we define the problem of context-based recommendation. In Section 4, we present a MLCG construction algorithm and a recommendation method based on MLCG. In Section 5, we compare our method with counterparts on real-world datasets. Section 6 concludes this paper.

2   Related Work

In this section, we present a survey of work on context-aware recommendation and graph-based recommendation.

2.1   Context-Aware Recommendation

Early work in context-aware recommender utilized contextual information for pre-processing or post-processing. Recent work has focused on integrating contextual information with the user-item relations.

In [2], they use contextual information about the user's task to improve the recommendation accuracy. However, this method operates only within the traditional 2D $USER \times ITEM$ space. In [3], a reduction-based pre-filtering approach is proposed, which uses the user's prior item preferences to help match the current context for recommending items.

In [5], a regression-based latent factor model is proposed to incorporate features and past interactions. In [6], a context-aware factorization machine is provided to simultaneously take context into account to enhance predictions. The approach in [7] is based on tensor factorization. However, these methods are infeasible for scenarios with only implicit feedback data.

2.2   Graph-Based Recommendation

Recommendation on a graph is comprised of two steps: constructing a graph from training data and ranking item nodes for a given user.

In [8], a two-layer graph model is proposed for book recommendation based on similarity among user nodes or item nodes. In [9], a graph is constructed by connecting two item nodes rated by at least one user, then the node scoring algorithm, ItemRank, is executed to rank item nodes according to the active user's preference records. In [10], several Markov-chain model based quantities are considered, which provide similarities between any pair of nodes on a bipartite graph for recommendation. However, all the above approaches consider only $USER$ and $ITEM$ dimensions without additional contextual information.

In [11], ContextWalk algorithm is provided for movie recommendation, which uses original random walk on a graph by considering the user and item context (e.g., actor, director, genre), but it ignores the decision context where a user chooses to watch a movie. As an example in [1], a user may have significantly different preferences on the genre of movies she/he wants to see when she/he is going out to a theater with her/his boyfriend/girlfriend as opposed to going with her/his parents. In [12], a graph-based method is presented that aims to improve recommendation quality by modeling user's long-term and short-term preferences. This method can deal with time information, but can not incorporate other types of contextual information, such as location and mood. In [4], a bipartite graph is proposed to model the interactions between users and items based on a recommendation factor set, F, where each recommendation factor $f \in F$ is defined as a combination of multidimensional data. Nodes corresponding to recommendation factors are connected with item nodes. That is, the method utilizes duplicable dimensions to enhance the interactions, which may bring noises into recommendation. In addition, they do not provide principles to define recommendation factor set F, which is crucial to recommendation performance.

3   Problem Formulation

We define the context-based recommendation as follows. In recommender systems, we have a set of users $U = \{u_k | k = 1, 2, ..., |U|\}$ and a set of items $I = \{i_k | k = 1, 2, ..., |I|\}$. Formally, let $C_U = \{C_{Uk} | k = 1, 2, ..., |C_U|\}$ be the context of users, where C_{Uk} is a domain of user context (e.g., AGE, GENDER), let $C_I = \{C_{Ik} | k = 1, 2, ..., |C_I|\}$ be the context of items, where C_{Ik} is a domain of item context (e.g., GENRE, ARTIST), let $C_D = \{C_{Dk} | k = 1, 2, ..., |C_D|\}$ be the context where decisions are made, where C_{Dk} is a domain of decision context (e.g., TIME, LOCATION). A recommendation task can be expressed as: given a user $u \in U$ with user context $c_u = \{c_{uk} | c_{uk} \in C_{Uk}, k = 1, 2, ..., |C_U|\}$ in a particular decision context $c_d = \{c_{dk} | c_{dk} \in C_{Dk}, k = 1, 2, ..., |C_D|\}$, a ranked list of items $R(u, c_d) \subseteq I$ is provided as the potential items ranked by relevance scoring function $utility$, where $utility$ is defined as $utility(u, c_u, c_d, i)$ to measure the relevance between tuple $< u, c_u, c_d >$ and item $i \in I$. For example, recommending songs for a user, Ted, when he is at $home$ on $Saturday$ $night$ can be interpreted as finding the songs with top-K relevance with tuple $< u = Ted, c_u :$ $\{Gen. = M\}$, $c_d : \{DayofWeek = Sat., Time = Night, Loc. = Home\} >$. For simplicity, we assume that the contextual information is denoted by categorical values.

4   Proposed Method

In this section, we start by presenting the MLCG construction algorithm, then provide a novel random walk-based ranking method based on MLCG.

4.1   Graph Construction

Construction Algorithm. We first describe the data format. We consider users' attributes as user context, and attributes of items as item context. Hence in Table 1, C_U is $\{GEN. : \{M\}, AGE : \{[18 - 30]\}\}$, and in Table 2, C_I is $\{ARTIST : \{M.J.\}, GENRE : \{Rock\}\}$. Meanwhile, log data in Table 3 describes the song a user listened to in a certain situation. In graph-based methods, each item or user is represented as a node. Most graph-based methods describe the interaction between a user and an item by directly creating an edge between the two nodes [4, 8, 10–13]. However, these methods ignore the effect of decision context, while we argue that, in a recommender system, it is in a certain decision context where users interact with items. That is, users' preferences on items should flow though particular decision context before reaching the item nodes.

Therefore, we design a new type of node, decision node $node_{decision} =<$ $u, c_{d1}, c_{d2}, ..., c_{d|C_D|} >$, where $u \in U$ and $c_{dk} \in C_{Dk}$, to characterize the decision context and model the interactions between users and items. Note that $node_{decision}$ is a combined node with a user and a decision context. The underlying intuition of $node_{decision}$ is that decision context is a local effect and should not be shared by all users as a global effect. That is, the same decision context for

Table 1. Example of User Data

USER	GENDER	AGE
Ted	M	[18-30]
Mike	M	[18-30]

Table 2. Example of Item Data

SONG	ARTIST	GENRE
Beat It	Michael Jackson	Rock
Rock with you	Michael Jackson	Rock

Table 3. Example of Log Data

USER	DayofWeek	LOCATION	SONG
Ted	Saturday	Home	Beat it
Mike	Sunday	Office	Rock with it
Mike	Sunday	Office	Beat it
Ted	Sunday	Office	Beat it

different user have different impacts on decision making. The proposed MLCG construction algorithm is outlined in Algorithm 1.

As the Algorithm 1 shows, a MLCG is a three-layer graph that consists of a user context layer, an item context layer and a decision context layer. Figure 1 illustrates an example of MLCG constructed from Tables 1−3. The number on the edge represents the co-occurrence of two end-nodes.

Only one type of entity node is denoted as a square node in Figure 1, on each layer, such as $node_{decision}$ on decision context layer. Entity nodes are characterized by their own context nodes on the same layer. For every context/attribute of an entity, there is a corresponding edge between the entity node and the context node. As Figure 1 shows, node M describes the gender of the user nodes connected with it. Furthermore, the nodes of the same entity type interact through their context nodes. That is, influence of an entity node propagates to another entity through sharing context nodes. In the case of Figure 1, the more rock songs a user listened to, the more of the user's preferences flow to other rock songs through the node $ROCK$.

In addition to the intra-layer edges, inter-layer edges are also available to represent the interactions between layers. In our model, user nodes do not directly interact with items because we emphasize that interactions should occur in a certain situation. In this sense, the interaction is expressed as: an edge from a user node on a user layer to the corresponding $node_{decision}$ and the other edge from the $node_{decision}$ to a song node on an item layer. For an active user, songs which were listened to in the same context are connected to the same corresponding

---

Algorithm 1. Construct a Multi-Layer Context Graph

Input: Set of users U and context C_U; Set of Items I and context C_I; Decision Context C_D; Log record table $LogTable$, where each log is in the form of $< u, c_{d1}, c_{d2}, ..., c_{d|C_D|}, i >$, $u \in U$, $i \in I$ and $c_{dk} \in C_{Dk}$

Output: Multi-Layer Context Graph $\mathcal{G}$

1: Initialize a graph $\mathcal{G}$ with $USER\text{-}Layer$, $ITEM\text{-}Layer$ and $Decision\text{-}Layer$
2: $CreateLayer(C_U, U, USER\text{-}Layer)$
3: $CreateLayer(I_U, I, ITEM\text{-}Layer)$
4: **for** each context domain $c_d \in C_D$ **do**
5:     **for** each context value $v \in c_d$ **do**
6:         Create a decision context node for v on $Decision\text{-}Layer$
7:     **end for**
8: **end for**
9: **for** each log record $log \in LogTable$ **do**
10:     Create a $node_{decision}$ $v = < u, c_{d1}, c_{d2}, ..., c_{d|C_D|} >$ on Decision Layer
11:     Connect user node u and v
12:     Connect $c_{d1}, c_{d2}, ..., c_{d|C_D|}$ nodes and v
13:     Connect item node i node and v
14: **end for**
15: Return $\mathcal{G}$

Subroutine CreateLayer($C, T, Layer$)
16: **for** each context domain $c \in C$ **do**
17:     **for** each context value $v \in c$ **do**
18:         Create a context node for v on Layer
19:     **end for**
20: **end for**
21: **for** each entity $t \in T$ **do**
22:     Create a node for t on $Layer$
23:     Connect node t and its corresponding context nodes
24: **end for**

$node_{decision}$ node. In this way, the effect of current context is distributed precisely over these songs listened to in that context.

Weight Assignment. Most graph-based recommendation methods consider a recommendation process as a node ranking task on a graph, hence several random walk-based ranking measures are proposed [4, 9, 12, 13]. However, these ranking methods are performed on a homogeneous graph, ignoring the different types of edges, so they will not work for MLCG. By fusing different edge types, we transform the heterogeneous multi-layer graph to a homogeneous graph.

We denote $N(j)$ as a set of nodes connected with node j, denote $N_s(j) \subseteq N(j)$ as a set of nodes on the same layer with node j, and $N_d(j) = N(j) - N_s(j)$. Given node j and $k \in N_s(j)$ on any layer, the edge weight $w(j, k)$ is defined as follows:

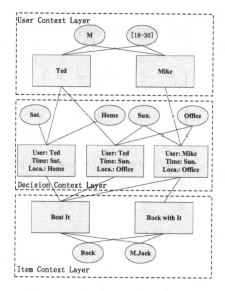

Fig. 1. An example: the MLCG constructed based on Tables 1–3

$$w(j,k) = \begin{cases} \dfrac{\alpha}{|T(N_s(j))|} \dfrac{f(j,k)}{\sum_{t \in N_s(j)} f(j,t)} & \text{if } N_d(j) > 0 \\[2ex] \dfrac{co\text{-}occu(j,k)}{\sum_{t \in N_s(j)} co\text{-}occu(j,t)} & \text{if } N_d(j) = 0 \\[2ex] 0 & otherwise, \end{cases} \qquad (1)$$

where function $f(j,k)$ denotes the importance score for node k with regard to node j, and $co\text{-}occu(j,k)$ is the co-occurrence of node j and node k. As an example in Figure 1, the user, Ted, completed two interactions in the same decision context where it is Sunday, hence the co-occurrence of node $Sunday$ and $node_{decision}$ is 2. $T(N_s(j))$ is a set of node types of nodes in $N_s(j)$.

Meanwhile, $f(j,k)$ should satisfy the following intuition-based criteria: 1) rare context/attributes are likely to be more important, whereas common context/attributes are less important. For example, thousands of people like football but only Ted and $Mike$ like PingPong, in which case node $PingPong$ carries more benefit for looking for similar user than node $Football$. 2) the more co-occurrence of two nodes, the more related they are. For example, for a user who is inclined to listen to songs at home, his preferences propagate mainly through node $Home$ to other songs. Considering the two intuitive rules, we borrow the idea of TF-IDF from information retrieval area and define $f(j,k)$ as follows.

$$f(j,k) = co\text{-}occu(j,k) \log \frac{\sum_{v \in \Omega(k)} \sum_{t \in N_s(v)} co\text{-}occu(v,t)}{\sum_{t \in N_s(k)} co\text{-}occu(k,t)}, \qquad (2)$$

where $\Omega(k)$ is a set of nodes having the same node type with node k. For example, $\Omega(Saturday) = \{Saturday, Sunday\}$ since they share a node type $DayofWeek$. Given node j and $k \in N_d(j)$, the edge weight $w(j,k)$ is calculated as:

$$w(j,k) = \begin{cases} \frac{(1-\alpha)co\text{-}occu(j,k)}{\sum_{t \in N_d(j)} co\text{-}occu(j,t)} & \text{if } N_s(j) > 0 \\ \frac{co\text{-}occu(j,k)}{\sum_{t \in N_d(j)} co\text{-}occu(j,t)} & \text{if } N_s(j) = 0 \\ 0 & otherwise. \end{cases} \quad (3)$$

Here, α controls the trade-off between intra-layer and inter-layer interactions. The larger α is, the more influence flow to nodes on the same layer, and the more effect intra-layer interactions have.

4.2   Recommendation on MLCG

Here, we consider the recommendation task as a ranking task on a graph. We then extend the Personalized PageRank [15], which is a variation of the PageRank algorithm[14]. The original PageRank score of a node is calculated as follows:

$$PR(k+1) = \alpha \cdot M \cdot PR(k) + (1-\alpha) \cdot d, \quad (4)$$

where $PR(k)$ denotes the rank value at the k-th iteration, M is a transition probability matrix, α is a damping factor that is normally given as 0.85 and d is a vector defined as $d_k = \frac{1}{n}, k = 1, 2, ..., n$, where n is the number of nodes.

As Equation 4 shows, PageRank can be considered as a Markov process with restart, and the probability of a random walker will jump to a node after a restart is equal to others. That is, the PageRank algorithm considers all nodes equally without biasing any important nodes. In [15], a topic-sensitive PageRank is proposed to introduce a personalized vector to bias users' preferences. More specifically, vector d is built as a user-specific personalized vector, where $d_k = 1$ if k-th node represents the active user, otherwise $d_k = 0$. Then the Personalized PageRank is calculated in Equation 4.

By using a hint from [9] and [12], we extend the personalized PageRank which captures both users' preferences and current decision context. For a given user $u \in U$ and current decision context $c_d = \{c_{dk} | c_{dk} \in C_{Dk}, k = 1, 2, ..., |C_D|\}$, we define $\varepsilon = \{i_k | i_k \in I, k = 1, 2, ..., |\varepsilon|\}$ as a set of items that u accessed before, then we construct $\tilde{d}$ as follows:

$$\tilde{d}_j = \begin{cases} \frac{\lambda}{|\varepsilon|} & \text{if node } j \in \varepsilon \\ \frac{1-\lambda}{|c_d|+1} & \text{if node } j \in c_d \text{ or node } j = u \\ 0 & otherwise, \end{cases} \quad (5)$$

where λ adjusts the radio of bias between users' preferences and current decision context. d is $\tilde{d}$ after L-1 normalization. Meanwhile, we consider the recommendation process as a multi-path random walk process with multiple starting points. Hence $PR(0)$ is defined as follows.

$$PR(0)_j = \begin{cases} 1 & \text{if node } j \in c_d \text{ or node } j = u \\ 0 & otherwise. \end{cases} \tag{6}$$

Then, we normalize vector $PR(0)$ to ensure that the sum of its non-negative elements is 1.

So, the extended personalized PageRank is summarized as: construct d for the active user based on Eq. 5, then run Eq. 4 until convergence with initializing $PR(0)$ based on Eq. 6. In order to make recommendations, we rank item nodes based on their PR value and keep only the top-K item nodes as recommendations.

5   Experiment

In this section, we present an experimental study which is conducted on two real-world datasets to demonstrate the effectiveness of our approach.

5.1   Dataset

We use two implicit datasets for our experiments instead of explicit rating data such as MovieLens[1], since we aim at improving the top-K recommendation other than explicit rating estimation.

The first dataset is Last.fm[2] which contains 19,150,868 music listening logs of 992 users (till May, 4th 2009). We extract the logs from April to May and remove these songs which were listened to less than 10 times, then the final dataset contains 992 users, 12,286 songs and 264,446 logs. In this case, the user context only includes domains of COUNTRY and AGE, while item context includes a domain of ARTIST. We notice that the original log tuples only consist of USER, SONG and TIMESTAMP domains. By transforming TIMESTAMP into different temporal features, we obtain several decision context domains, such as *Day of Week* and *Time Slice* (i.e., each time slice lasts for 6 hours). Finally, we use logs in April as a training set, and randomly choose 1000 logs from 1st to 4th May 2009 as a test set.

The second dataset is CiteULike[3] which provides logs on *who* posted *what* and *when* the posting occurred. By removing users who posted less than 5 papers and papers which were posted by less than 5 users, we obtain a subset of original dataset which contains 1,299 users, 5,856 items and 40,067 user-items pairs from January to May 2007. In this case, there is no user context because of the lack of user information. Similarly, we transform TIMESTAMP into several temporal features, and TAGs of papers are considered as item context. Finally, we use logs in the two final weeks as a test set, and others as a training set.

---

[1] http://www.movielen.umn.edu

[2] http://www.last.fm.com

[3] http://www.citeulike.org

5.2 Evaluation Metrics

We use HitRatio@K [16], Mean Reciprocal Rank (MRR@K) [17] and Recall@K to evaluate the performance of our top-K recommendation.

Given a test case $< u, c_d, i >$ in test set $TEST$, where a user u accessed an item i in decision context c_d, the recommendation method generates a ranked list of items $R(u, c_d)$, where $|R(u, c_d)| = K$. Then HitRatio@K is defined as follows:

$$HitRatio@K = \frac{1}{|TEST|} \sum_{<u,c_d,i>} I(i \in R(u, c_d)), \tag{7}$$

where $I(\cdot)$ is an indicator function.

In addition, we also measured the MRR@K for evaluating the rank of the target item i:

$$MRR@K = \frac{1}{|TEST|} \sum_{<u,c_d,i>} \frac{1}{rank(i)}, \tag{8}$$

where $rank(i)$ refers to the rank of target item i in $R(u, c_d)$.

Furthermore, we use Recall@K to evaluate the overall relevancy performance of recommendation methods:

$$Recall@K = \frac{1}{|TEST|} \sum_{<u,c_d>} \frac{|T(u, c_d) \cap R(u, c_d)|}{|T(u, c_d)|}, \tag{9}$$

where $T(u, c_d)$ is the set of items the user u accessed in context c_d.

5.3 Baseline Methods

We evaluate the effectiveness of our method through comparing it with other existing methods:

Frequency based (FreMax): A user independent method, which ranks the items by the times they were accessed. That is, FreMax generates a same list of items for any user.

User-based CF (UserCF): A N-neighbor user-based collaborative filtering method, which uses Pearson Correlation as the user similarity measurement. The optimal value of N is 10 in experiments on CiteULike dataset, and N is 1 on Last.fm dataset.

ItemRank [9]: A random walk-based item scoring algorithm, which is performed on an item graph. The item graph is constructed by connecting two items if they were rated by at least one user.

GFREC [4]: A contextual bipartite graph-based method, which defines a recommendation factor set F, to transform a given log table into a bipartite graph. In our experiments, we use one of the best settings of F according to [4].

We do not consider item-based CF, since the number of items is much greater than users, and user-based CF methods can provide more accurate recommendation. In addition, as [18] indicates, SVD methods only achieve slight improvement on implicit feedback datasets, thus, we do not compare our method with

SVD-based methods. Since UserCF and ItemRank operate in a $USER \times ITEM$ space, we transform our training data into a pseudo rating matrix by considering the normalized access count of a user on an item as the user's pseudo rating on the item.

5.4   Experimental Results

In this subsection, we present our experimental results in two scenarios. Note that for Last.fm dataset, our experiments provide a ranked list of songs, which might have been listened to before by the given user, since users generally listen to the same song more than once. For the experiments on CiteULike dataset, we recommend papers that have not been posted by the active user.

Here, λ in each experiment is set to 0.5. α in each layer is a tunable parameter. For Last.fm dataset, $\alpha_{user} = 0.005$, $\alpha_{decision} = 0.2$ and $\alpha_{item} = 0.01$. For CiteULike, $\alpha_{user} = 0.0$ (user context is unavailable), $\alpha_{decision} = 0.05$ and $\alpha_{item} = 0.01$.

HitRatio@K Analysis. Figure 2 illustrates the HitRatio@K of our experiments on the two datasets. For Last.fm, the methods with contextual information (i.e., GFREC and MLCG) show substantial improvement over FreMax, ItemRank and UserCF, which confirms the importance of context for recommendation. FreMax shows the lowest HitRatio, revealing the fact that users have their own preferences on songs, and would not be affected by popularity. The HitRatio of MLCG is 54.3%(K=5)-75.3%(K=15) greater than that of UserCF. Moreover, there is a gain of 35.0% over GFREC from MLCG in top-5 recommendation. Generally, on Last.fm MLCG shows close/comparable performance with GFREC over the metrics. For CiteULike, it is clear that our method significantly outperforms counterpart methods. Similar to the case on Last.fm, contextual information contributes to better performance. Among the context-based methods, MLCG achieves perceptible improvement, where HitRatio of MLCG is 3.8 times (K=5) and 42.4% (K=15) higher than that of GFREC.

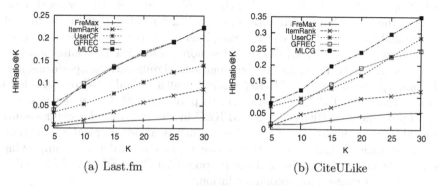

(a) Last.fm   (b) CiteULike

Fig. 2. Comparing the HitRatio@K of MLCG against Baseline Approaches

It should be noticed that Top-K recommender systems benefit more from higher accuracy when K is small, such as user experience.

MRR Analysis. As shown in Figure 3, MLCG significantly outperforms other methods in MRR, that is, MLCG generates more accurate recommendations. For Last.fm, MLCG achieves a performance gain over GFREC by 35.9% (MRR@5). We notice that in CiteULike, the MRR of GFREC is lower than that of UserCF. The reason is that superfluous combinations of multidimensional data bring connections, as well as noises. MLCG addresses this issue by grouping context into the three layers to explore their relationships, obtaining an improvement of MRR by up to 20.4% (K=15) over UserCF.

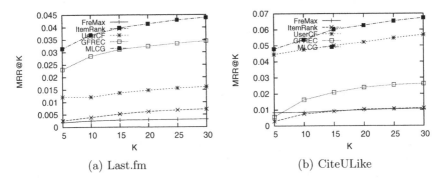

(a) Last.fm   (b) CiteULike

Fig. 3. Comparing the MRR@K of MLCG against Baseline Approaches

Recall Analysis. We give the results of recall in Figure 4. It can be seen that by incorporating contextual information, MLCG and GFREC have higher recall than other methods. And the improvements of MLCG over the 4 comparison algorithms on both datasets are still clear. More specifically, while GFREC in general performs the best of the baseline methods, MLCG outperforms it with recall of up to 32.3% (K=5) and 17.2 times (K=5) greater. Higher recall indicates that the algorithm returns most of the relevant results based on the instant situation and the active user. In this sense, having a better understanding of the context of the active user, MLCG provides more personalized recommendations than other methods. In summary, our methods significantly outperforms several counterparts in terms of MRR@K and Recall@K, and it is comparable to or better than GFREC in HitRatio@K. In particular, MLCG achieves significant improvement of HitRatio over GFREC in the scenario of recommending previously unseen items (i.e., CiteULike), which is a more typical application. These observations demonstrate that context plays a vital role in improving recommendation performance, and the proposed MLCG is effective in blending various types of context for recommendation.

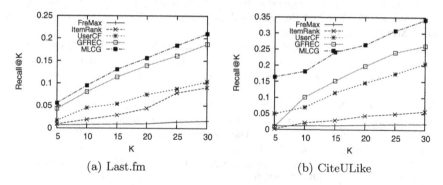

Fig. 4. Comparing the Recall@K of MLCG against Baseline Approaches

6   Conclusion

In this paper, a Multi-Layer Context Graph (MLCG) model is proposed. MLCG utilizes contextual information to construct a layer for each type of context respectively, and models the decision making by users. In particular, our model emphasizes that users interact with items in an instant context. Furthermore, we take different edge types into consideration, distinguishing not only the intra-layer from inter-layer interactions but also various contextual domains, such as *Day of Week*. Based on MLCG, we extend Personalized PageRank to rank items in MLCG which captures users' preferences and current decision context. Finally, experiments based on two real-world datasets demonstrate that the effectiveness of the proposed method exceeds other existing ones in all evaluation metrics. This work can form a basis to introduce social relationships into the user context to improve accuracy further. For example, the similarity of interest in groups/communities can be used to improve recommendation diversity and accuracy.

Acknowledgement. This work is partially supported by the National Natural Science Foundation of China (Grant No. 61272480).

References

1. Adomavicius, G., Tuzhilin, A.: Toward the Next Generation of Recommender Systems: A Survey of the State-of-the-Art and Possible Extensions. IEEE Trans. on Knowl. and Data Eng. 17(6), 734–749 (2005)
2. Herlocker, J.L., Konstan, J.A.: Content-Independent Task-Focused Recommendation. IEEE Internet Computing 5(6), 40–47 (2001)
3. Adomavicius, G., Sankaranarayanan, R., Sen, S., Tuzhilin, A.: Incorporating contextual information in recommender systems using a multidimensional approach. ACM Transactions on Information Systems (TOIS) 23(1), 103–145 (2005)
4. Lee, S., Song, S.-I., Kahng, M., Lee, D., Lee, S.-G.: Random walk based entity ranking on graph for multidimensional recommendation. In: Proceedings of the Fifth ACM Conference on Recommender Systems, pp. 93–100 (2011)

5. Agrawal, D., Chen, B.: Regression-based latent factor models. In: Proceedings of the 15th ACM SIGKDD International Conference on Knowledge Discovery and Data Mining, pp. 19–28 (2009)

6. Rendle, S., Gantner, Z., Freudenthaler, C., Schmidt-Thieme, L.: Fast context-aware recommendations with factorization machines. In: Proceedings of the 34th International Conference on Reseach and Development in Information, pp. 635–644 (2011)

7. Xiong, L., Chen, X., Huang, T.-Y., Schneider, J., Carbonell, J.: Temporal collaborative filtering with bayesian probabilistic tensor factorization. In: Proceedings of SIAM International Conference on Data Mining, pp. 211–222 (2010)

8. Huang, Z., Chung, W., Ong, T.-H., Chen, H.: A graph-based recommender system for digital library. In: Proceedings of the 2nd ACM/IEEE-CS Joint Conference on Digital Libraries, pp. 65–73 (2002)

9. Gori, M., Pucci, A., Roma, V., Siena, I.: Itemrank: A random-walk based scoring algorithm for recommender engines. In: Proceedings of the 20th International Joint Conference on Artifical Intelligence, pp. 2766–2771 (2007)

10. Fouss, F., Pirotte, A., Renders, J.-M., Saerens, M.: Random-walk computation of similarities between nodes of a graph with application to collaborative recommendation. IEEE Transactions on Knowledge and Data Engineering 19(3), 355–369 (2007)

11. Bogers, T.: Movie recommendation using random walks over the contextual graph. In: Proc. of the 2nd Intl. Workshop on Context-Aware Recommender Systems (2010)

12. Xiang, L., Yuan, Q., Zhao, S., Chen, L., Zhang, X., Yang, Q., Sun, J.: Temporal recommendation on graphs via long-and short-term preference fusion. In: Proceedings of the 16th International Conference on Knowledge Discovery and Data Mining, pp. 723–732 (2010)

13. Gori, M., Pucci, A.: Research paper recommender systems: A random-walk based approach. In: IEEE/WIC/ACM International Conference on Web Intelligence, pp. 778–781 (2006)

14. Page, L., Brin, S., Motwani, R., Winograd, T.: The PageRank Citation Ranking: Bringing Order to the Web. Technical Report (1999)

15. Haveliwala, T.H.: Topic-sensitive pagerank. In: Proceedings of the 11th International Conference on World Wide Web, pp. 517–526 (2002)

16. Karypis, G.: Evaluation of item-based top-n recommendation algorithms. In: Proceedings of the Tenth International Conference on Information and Knowledge Management, pp. 247–254 (2001)

17. Vallet, D., Cantador, I., Jose, J.M.: Personalizing web search with folksonomy-based user and document profiles. In: Gurrin, C., He, Y., Kazai, G., Kruschwitz, U., Little, S., Roelleke, T., Rüger, S., van Rijsbergen, K. (eds.) ECIR 2010. LNCS, vol. 5993, pp. 420–431. Springer, Heidelberg (2010)

18. Hu, Y., Koren, Y., Volinsky, C.: Collaborative filtering for implicit feedback datasets. In: Eighth IEEE International Conference on Data Mining, pp. 263–272 (2008)

Recommending Tripleset Interlinking through a Social Network Approach

Giseli Rabello Lopes[1], Luiz André P. Paes Leme[2], Bernardo Pereira Nunes[1,3], Marco Antonio Casanova[1], and Stefan Dietze[3]

[1] Department of Informatics, Pontifical Catholic University of Rio de Janeiro, Rio de Janeiro/RJ – Brazil, CEP 22451-900
{grlopes,bernardo,casanova}@inf.puc-rio.br
[2] Computer Science Institute, Fluminense Federal University, Niterói/RJ – Brazil, CEP 24210-240
lapaesleme@ic.uff.br
[3] L3S Research Center, Leibniz University Hannover, Appelstr. 9a, 30167 Hannover, Germany
dietze@l3s.de

Abstract. Tripleset interlinking is one of the main principles of Linked Data. However, the discovery of existing triplesets relevant to be linked with a new tripleset is a non-trivial task in the publishing process. Without prior knowledge about the entire Web of Data, a data publisher must perform an exploratory search, which demands substantial effort and may become impracticable, with the growth and dissemination of Linked Data. Aiming at alleviating this problem, this paper proposes a recommendation approach for this scenario, using a Social Network perspective. The experimental results show that the proposed approach obtains high levels of recall and reduces in up to 90% the number of triplesets to be further inspected for establishing appropriate links.

Keywords: Linked Data, Recommender Systems, Social Networks.

1 Introduction

One of the design principles of Linked Data is to include URIs linkages [1], or simply *links*, which allow the "navigation" among triplesets and the discovery of related resources and additional data [2]. Therefore, an important task in the publishing process of a tripleset t involves the selection of triplesets for which one may define links with t.

However, this is a non-trivial task. Indeed, a fully manual process requires considerable effort from the data publisher and will become impractical as the number of triplesets grows. According to Nikolov et al. [3], the selection of a tripleset u for which one may define links with t can be influenced by three factors: (i) *degree of overlap* - the number of resources of u related to resources of t; (ii) *additional information provided by the tripleset* - the amount of additional

X. Lin et al. (Eds.): WISE 2013, Part I, LNCS 8180, pp. 149–161, 2013.

information u can provide for the resources of t; and (iii) *popularity of the triple-set* - how easy it will be for t to be discovered because it has links to popular triplesets.

We refer to the problem of the discovery and selection of triplesets for which one may define links with a given tripleset as the *tripleset recommendation problem*.

In this paper, we propose to address the tripleset recommendation problem using strategies borrowed from Social Networks. We introduce a procedure that receives as input a tripleset t and a set of triplesets S, and returns a ranked list of triplesets $u \in S$ such that links from t to u are more likely to be defined for the triplesets in the beginning of the list. Therefore, the effort of creating links from t to triplesets in S would be reduce, since one would have to analyze just the first few triplesets in the ranking. The procedure we propose could be used as an initial filtering phase to other more costly recommendation techniques based, for example, on schema and ontology matching, which might be applied only to the better ranked triplesets.

To generate the ranked list, the procedure uses a recommendation function adapted from link prediction measures used in Social Networks. Informally, we say that a tripleset t is *connected* to another tripleset u iff there are at least one link between resources from t to resources in u. Basically, to adapt the link prediction measures, we interpreted the connections between triplesets as relational ties and the triplesets as the actors. In the paper, we evaluate the performance of two link prediction measures, using data obtained from the Data Hub catalogue.

In general, recommendation systems [4] alleviate problems associated with information overload [5]. Recommendation systems aim at suggesting items to users based on their interests, i.e., from the analysis of their profiles. Currently, many e-commerce Web sites use this type of system to rank suggestions of their products to potential buyers [6]. It is noteworthy that such systems not only gained prominence in e-commerce [7], but also in several application areas. Indeed, such systems have been applied to different domains such as recommendation of books [8], restaurants [9], movies [10], news [11] and social networks [12]. In particular, in the context of Social Networks, measures based on analysis of the relational ties between actors have been used to recommend links between actors [13–15].

The remainder of this paper is organized as follows. Section 2 presents related work. Section 3 details our recommendation approach. Section 4 shows an experimental evaluation and discusses the results obtained. Section 5 presents the conclusions and suggests further work.

2   Related Work

Recommendation of triplesets to be interlinked in the Linked Data domain is a research area in expansion. However, there are still few approaches developed specifically for this purpose. In this section, we briefly review the research more closely related to ours.

Nikolov et al. [3, 16] propose an approach to identify relevant triplesets for data linking. Their approach establishes two main steps: (i) searching for potentially relevant resources in other triplesets using as keywords a subset of labels in the new published tripleset; and (ii) filtering out irrelevant triplesets by measuring semantic similarities applying ontology matching techniques. In the filtering step, they consider only the triplesets with higher degrees of semantic similarity, discarding the others.

The following two references [17, 18] aim at recommending triplesets relevant to answer queries expressing the user requirements. Lóscio et al. [17] propose the recommendation of relevant triplesets that contribute for answering queries posed to an application. The authors argue that a tripleset may contribute to answer queries of an application, but the returned response may not meet the user requirements. Thus, they propose to discover triplesets relevant for applications in a specific domain using information quality (IQ) as multidimensional criteria. Their recommendation function estimates a degree of relevance of a given tripleset based on the following IQ criteria: correctness, schema completeness and data completeness.

Oliveira et al. [18] use application queries and user feedback to discover relevant triplesets in Linked Data. The application queries help filter triplesets that are potentially strong candidates to be relevant and the user feedback helps analyze the relevance of such candidates. They argue that the consideration of both queries and user feedback helps recommending triplesets related to the user requirements.

To summarize, all previous works perform an analysis at the instance or schema levels, using techniques such as filtering by keyword-based searches, schema and ontology matching, user feedback and information quality.

Our proposed approach differs from these since it considers the links among triplesets as a "high" level information and it does not require an analysis at the instance or schema levels.

Our recommendation function aims at recommending candidate triplesets $u \in S$ to a tripleset t, such that t could possibly be interlinked with u. The inputs of our approach are the previous links among the candidate triplesets and some known triplesets that t can be interlinked with. For the generation of the recommendation ranking, we propose to apply link prediction measures adopted in Social Networks to the Linked Data context. To the best of our knowledge there is no previous work that takes this approach.

3   A Recommendation Approach

3.1   Recommendation Procedure

Briefly, recall that an *RDF triple* is a triple of the form (s, p, o), where s is the *subject* of the triple, which is an RDF URI reference or a blank node, p is the *predicate* or *property* of the triple, which is an RDF URI reference, and o is the *object*, which is an RDF URI reference, a literal or a blank node.

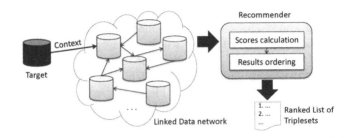

Fig. 1. Schematic description of the recommendation procedure

A *tripleset* t is a set of RDF triples. A resource identified by an RDF URI reference s is *defined* in t iff s occurs as the subject of a triple in t.

Let t and u be two triplesets. A *link* from t to u is a triple of the form (s, p, o), where s is an RDF URI reference identifying a resource defined in t and o is an RDF URI reference identifying a resource defined in u; we say that (s, p, o) *interlinks* s and o. We also say that t can be *interlinked* with u iff it is possible to define links from t to u.

A *Linked Data network* is a graph $G = (S, C)$ such that S is a set of triplesets and C contains edges (t, u), called *connection* from t to u, iff there is at least one link from t to u; we also say that t *points at* or *references* u. Note that there can be only one edge from t to u, even when there are multiple distinct links from t to u.

Let $G = (S, C)$ be a Linked Data network. The *context* of a tripleset $u \in S$, denoted C_u, is the set of all $v \in S$ such that $(u, v) \in C$; and the *inverse context* of $u \in S$, denoted C'_u, is the set of all $v \in S$ such that $(v, u) \in C$.

Our recommendation procedure analyzes a Linked Data network much in the same way as a Social Network. The inputs of our recommendation procedure are (see Figure 1):

– a *Linked Data network* $G = (S, C)$
– a *target tripleset* t not in S (intuitively the user wishes to define links from t to the triplesets in S)
– a *target context* C_t for t consisting of one (or more) triplesets u in S (intuitively the user knows that t can be interlinked with u).

The procedure outputs a list L of triplesets in S, called a *ranking*. Intuitively, the triplesets in the initial positions of the ranking have a higher probability that resources in t can be interlinked with their resources.

The procedure adds a tripleset $u \in S$ to the ranking L iff $C_t \cap C_u$ is not empty, where, we recall, C_t is the context of t (given as input to the procedure) and C_u is the context of $u \in S$ (defined from the Linked Data graph).

To order the triplesets in L, the procedure estimates score values between t and the triplesets u in L: the higher the score of a tripleset u, the topmost u will be in the ranking. Intuitively, the score of a tripleset u is a predicted value of the relevance of u with respect to the probability of defining links from t to u. As stated before, to estimate the scores, the procedure applies measures used for link prediction in Social Networks, detailed in Section 3.2, to the Linked Data network $G = (S, C)$.

Finally, we remark that the recommendation procedure may be used iteratively, considering user feedback. The user indicates a first context for a target tripleset t. The procedure then outputs a ranking of triplesets such as t could possibly be interlinked with them. The user inspects the content of the top-most ranked triplesets and includes new links in t. Then, using the connections induced by the new links, the procedure outputs a new ranking, and so on.

3.2   Adapted Measures

Among the traditional measures originated from graph theory, we chose the Jaccard and the Adamic-Adar coefficients. We selected such measures because the results reported by Liben-Nowell and Kleinberg [13], which analyze co-authorship social networks, indicate that these two measures achieve good performance. Furthermore, they estimate non-zero score values only between nodes with two degrees of separation in the graph.

In what follows, let $G = (S, C)$, t and C_t respectively be the Linked Data network, the target tripleset and the target context given as input to the recommendation procedure. Let u be a tripleset in S. Recall that the context C_u of $u \in S$ is the set of all $v \in S$ such that $(u, v) \in C$, and that the inverse context C'_w of $w \in S$ is the set of all $v \in S$ such that $(v, w) \in C$.

Jaccard Coefficient. Intuitively, the larger the cardinality of the intersection of the contexts of t and u, the greater the likelihood that the two triplesets can be connected. This effect can be measured by the Jaccard coefficient, defined as follows.

$$jc(t, u) = \frac{|C_t \cap C_u|}{|C_t \cup C_u|} \tag{1}$$

where:

- $|C_t \cap C_u|$ is the cardinality of the intersection of the contexts of t and u
- $|C_t \cup C_u|$ is the cardinality of the union of the contexts of t and u.

Adamic-Adar Coefficient. Intuitively, if two triplesets t and u point to the same tripleset w and w is also pointed by many other triplesets, then w must be a generic tripleset and, therefore, it does not necessarily suggest any possible connection between t and u. On the other hand, if there is no tripleset other than t and u which points at w, then this might be a strong indication that w is a very particular tripleset for both t and u and, therefore, a connection between t and u could as well be defined. Thus, the strength of the belief in the existence of connections between t and u increases inversely proportional to the number of triplesets, other than t and u, which points at w, i.e., depends on the popularity of w.

The Adamic-Adar coefficient aa computes a measure of belief in the connection between t and u as a summation of the inverse of the logarithm of the

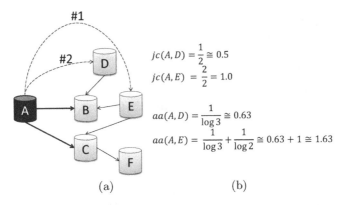

$$jc(A, D) = \frac{1}{2} \cong 0.5$$

$$jc(A, E) = \frac{2}{2} = 1.0$$

$$aa(A, D) = \frac{1}{\log 3} \cong 0.63$$

$$aa(A, E) = \frac{1}{\log 3} + \frac{1}{\log 2} \cong 0.63 + 1 \cong 1.63$$

(a)   (b)

Fig. 2. Example of (a) the inputs and the outputs of the recommendation procedure and (b) the values of the coefficients

popularity of the triplesets in the intersection of the contexts of t and u and is define as follows.

$$aa(t, u) = \sum_{w \in C_t \cap C_u} \frac{1}{\log |C'_w|} \tag{2}$$

where:

- $|C'_w|$ is the cardinality of the inverse context of w (the popularity of w).

3.3   Example

Figure 2 shows an schematic example of the computation of the coefficients, indicating: (a) the inputs and the outputs of the recommendation procedure; and (b) the values of the coefficients (using log_2). In the example depicted, the inputs are:

- The Linked Data network composed of the triplesets B, C, D, E and F and their connections, represented by solid lines
- The target tripleset A for which new connections must be recommended
- The context for A, pointed by thicker lines, consist of the triplesets B and C, which the user indicates that he can define connections from A to them.

The output of the procedure is a ranking of recommended triplesets, represented by dashed lines connecting the target to them (the number preceded by # indicates the ranking position of each recommended tripleset). Discarding the triplesets in the context of A, the recommendation technique has to rank the remainder triplesets D, E and F according to the chance of defining links from resources in A to resources in D, E and F. Adopting the Jaccard or the Adamic-Adar coefficient, the procedure will return E in the first position (#1) and D in the second (#2).

The tripleset F will not be recommended because there are no connections from F to triplesets pointed by A. Thus the score values of the Jaccard and Adamic-Adar coefficients between A and F are zero.

In this example, E points at two triplesets, B and C, which are pointed by A, whereas D points at just at one, B.

For the Jaccard coefficient, the number of triplesets in the intersection of the contexts of A and E with respect to the total number of triplesets in the union of their contexts is greater than that for the contexts of A and D.

For the Adamic-Adar coefficient between E and A, among the triplesets in the intersection of the contexts of A and E, the tripleset C is considered more important than B (just A and E points at C, while B is also pointed by D).

3.4 Interpretation of the Measures Application

The principle of our approach is that one can infer that t can be connected to u, i.e., t contains URIs that can be linked with URIs of u, iff the context of t overlaps the context of u. However, such analogy must be analyzed in order to better ground its validity.

If the two triplesets t and u share a connection to a tripleset w through the property $rdfs{:}sameAs$, then there would be triples of the form $(s_1, rdfs{:}sameAs, o_1)$, where $s_1 \in t$ and $o_1 \in w$, and $(s_2, rdfs{:}sameAs, o_2)$, where $s_2 \in u$ and $o_2 \in w$. Now, recall that $rdfs{:}sameAs$ is reflexive and transitive. Thus, if $o_1 \equiv o_2$ holds then $(s_1, rdfs{:}sameAs, s_2)$ will also hold. That is, there will be a link from t to u.

On the other hand, if the interlinking property was not $rdfs{:}sameAs$ but, for instance, $hasAuthor$ and $wasAttendedBy$, the probability that t and u share a connection would be lower, but still possible. Indeed, assume that there are triples of the form $(s_1, hasAuthor, o_1)$, where $s_1 \in t$ and $o_1 \in w$, and $(s_2, wasAttendedBy, o_2)$, where $s_2 \in u$ and $o_2 \in w$. Furthermore, assume that $o_1 \equiv o_2$. Then, we might understand s_1 as a paper presented in event s_2 and, therefore, a triple of the form $(s_1, wasPresentedIn, s_2)$ could be added to t to link t and u, provided that $wasPresentedIn$ could be added to the vocabulary of t.

To sum up, in the second case one cannot say that the analogy holds in all situations in the context of Linked Data. However, as indicated in the literature [19], the prevalence of links of type $rdfs{:}sameAs$ in the Web of Data justifies the use of the link prediction measures for the recommendation of triplesets based on their connections.

4   Experimental Evaluation

4.1 Description of the Data and the Experiment

We tested the recommendation procedure with data available in the Data Hub catalogue[1], a repository of metadata of open triplesets, in the style of Wikipedia.

---

[1] http://datahub.io

It is openly editable and is running a data cataloguing software (CKAN)[2] maintained by the Open Knowledge Foundation[3].

The description of each tripleset includes a multivalued property, called *relationships*, exposed by the REST API[4] of the catalogue, whose range is the complete set of catalogued triplesets. This property permits asserting that a tripleset t points at a tripleset u by adding the assertions $t[relationships] = \_node$ and $\_node[object] = u$ to the catalogue data. We used the property *relationships* to extract the connections between triplesets in the Data Hub catalogue. Data was gathered at the end of the 2012, adding to 797 triplesets and 15,012 connections among them. This data therefore induced a Linked Data graph $G = (S, C)$.

To evaluate the technique, we adopted the 10-fold cross validation approach. We split the Linked Data graph $G = (S, C)$ into *recommendation partitions* and *testing partitions* in ten different ways, and defined *target contexts* as follows:

- A *recommendation partition* is a subgraph $G_i = (S_i, C_i)$ of $G = (S, C)$ such that S_i is a set of triplesets to be considered for recommendation and C_i is the set of connections among the triplesets in S_i induced by the *relationships* property
- A *testing partition* is a pair $Tp_i = (T_i, aC_i)$ such that T_i is the set of triplesets in S, but not in S_i, called *recommendation targets*, and aC_i is a set of sets such that, for each $t \in T_i$, aC_i contains the set aC_t of all triplesets u in S_i such that there is a connection from t to u in C
- For each recommendation target $t \in T_i$, a *target context* C_t consists of some chosen triplesets in aC_t.

Additionally, for each different recommendation partition $G_i = (S_i, C_i)$, testing partition $Tp_i = (T_i, aC_i)$, recommendation target $t \in T_i$, with target context $C_t \in aC_i$, we define:

- the *gold standard* for t and is defined as the set $Gs_t = aC_t - C_t$ and represents the triplesets that must be recommended
- a *relevant tripleset* to be recommended for t is a tripleset in Gs_t
- a *candidate tripleset* to be recommended for t is a tripleset in $S_i - C_t$.

Unlike the traditional cross-validation approach, where partitions are used as training sets, the recommendation partitions were used as recommendation subgraphs only, since the proposed technique does not require a training step. The overall performance is taken as the average of the performances in the testing partitions.

In the experiments, the results were evaluated using traditional Information Retrieval measures [20, 21], Recall and Mean Average Precision (MAP). The *overall Recall* is the mean of the recall of each testing partition. The recall of a

---

[2] http://ckan.org
[3] http://okfn.org
[4] http://datahub.io/api/rest/tripleset/[triplesetid]

testing partition Tp_i is defined as the average of the recall values of each tripleset $t_j \in T_i$:

$$Recall(Tp_i) = \frac{\sum_{j=1}^{|T_i|} Recall(t_j)}{|T_i|} \tag{3}$$

where:

- $Recall(t_j)$ is defined as the ratio between the number of relevant triplesets that are recommended for t_j and the total number of triplesets that must be recommended $|Gs_{t_j}|$.

The *overall MAP* is defined as the mean of the MAP of each testing partition. The MAP of a testing partition Tp_i is in turn defined as the mean of the average precision scores of each tripleset $t_j \in T_i$:

$$MAP(Tp_i) = \frac{\sum_{j=1}^{|T_i|} AveP(t_j)}{|T_i|} \tag{4}$$

where:

- $AveP(t_j)$ is the average precision in the ranking of the tripleset t_j. It is computed as a average of the precision values obtained for each relevant tripleset. For this calculation, the position k in which a relevant tripleset was ranked is considered. Each precision value in position k, only the triplesets whose positions are lower or equal to k are considered, i.e., precision will be the ratio between the number of relevant triplesets recommended until the position k and this position number k. For instance, if in the tenth position was ranked the fifth relevant tripleset from a complete set of twenty relevant then the precision in p_{10} would be $p = 5/10$. For each relevant not recommend, the precision value used to calculate the $AveP$ is zero.

4.2   Evaluation and Results

To better understand the available data, Figure 3 presents the total number of triplesets calculated in function of a minimum number of connections (number of triplesets pointed by them). Figure 3 shows that most of the triplesets in the Data Hub catalogue has very few connections. The average of the number of connections per tripleset in the Data Hub catalogue was approximately 18.83.

In the experiments, we evaluated the ranking recommendations generated using the measures presented in Section 3.2. As the measures depend on both t and a tripleset s that points to at least one tripleset u which is also pointed by t, they estimate a score different from zero for the same triplesets in the *recommendation partition*. The overall recall was calculated as a function of the

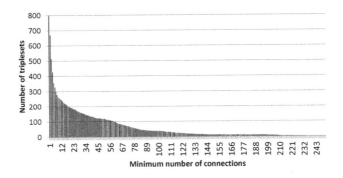

Fig. 3. Number of triplesets *vs.* minimum number of connections

cardinality of the target context (remember that the target context consists of some chosen triplesets pointed by t). Recall is used to analyze the coverage of the recommendation procedure. We considered all triplesets for which the score values are greater than zero. The results obtained showed that:

- For small target contexts, the overall recall is relatively high, being on the average greater than 75%
- For context sizes greater than 4 triplesets, the overall recall is higher than 90%.

These results show evidences that it is possible to recommend many relevant triplesets, even knowing just a few connections from the target tripleset. This is very important to validate the practical applicability of these Social Networks measures (that consider only the direct neighbors) to recommend triplesets in a Linked Data environment.

After these analyses, we evaluated the ordering of the recommendations in the rankings. For this purpose, we used the overall Mean Average Precision (MAP) to verify the accuracy of the generated ranking. Remember that the overall MAP estimation considers the gold standard induced by the choice of the context C_t, i.e., it is not defined by users. The results are presented in Figure 4 and show that the overall MAP values for Adamic-Adar are higher than those for the Jaccard coefficient, for context sizes smaller than 36, which means that, on average, for the same recall, the Adamic-Adar is more precise than Jaccard coefficient. This probably happens because the Adamic-Adar coefficient better differentiates the importance of the common triplesets pointed by the target and the candidate triplesets which tends to require less knowledge, or known triplesets in the context of the target.

In addition, we also calculated the average position of the last relevant tripleset in the ranking. These results were divided by the total number of triplesets in the corresponding recommendation partition. This analysis estimates the average percentage of the top of the ranking that needs to be verified to discover all the relevant triplesets that were recommended. Figure 5 presents these results. To better understand the results, we also calculated what would be, on the

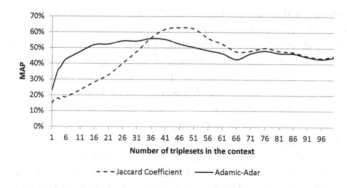

Fig. 4. Overall Mean Average Precision *vs.* context size

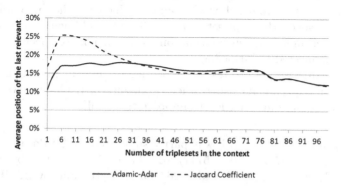

Fig. 5. Average relative position of the last relevant recommended tripleset in the ranking *vs.* context size

average, the maximum reduction possible (finding all relevant triplesets at the top positions of the ranking, for all triplesets) using a context size equal to 1. We obtained a maximum reduction value of 3.85%.

The worst performance of the Jaccard and Adamic-Adar coefficients indicates that one needs to examine, on the average, respectively, 25% and 18% of the top ranking triplesets to find all the relevant recommended triplesets. The best result, obtained using the Adamic-Adar coefficient considering only one connection in the context, indicates that one needs, on the average, to examine only 10% of the ranking to find all the relevant recommended triplesets. It shows evidences that the Adamic-Adar coefficient is more appropriate to rank the results in this scenario than the Jaccard coefficient. This also shows that is not necessary to know many triplesets in the context of target (what would otherwise invalidate the practical application of the procedure) to obtain suitable rankings.

5   Final Remarks

In this paper we proposed the use of link prediction measures to address what we called the *tripleset recommendation problem* in the Linked Data domain.

Our approach generates a ranking of triplesets to be linked with a tripleset t to be published. The ranking can be used to reduce the candidates that t can be interlinked with, thereby reducing the set of triplesets to be further inspected by other more costly techniques, if necessary. The experiments tested two different link prediction measures. The results show that such measures obtain good results, even when few triplesets in the context of t are available. Specifically, the results show that the approach can reduce up to 90% of the search space for the interlinking candidates.

We have defined the tripleset network as an unweighted graph $G = (S, C)$, thus disregarding the number of links between triplesets. This assumption favors triplesets from related information domains and penalizes generic ones. For instance, DBpedia is frequently referenced by many triplesets because it is a generic repository. Therefore, most likely, the weight of the connections to DBpedia would be very high, which would end up influencing the ranking in favour of DBpedia. However, the tripleset to be published would neither get more visibility nor unveil more hidden information from other more specific triplesets, because it is connected to DBpedia.

As further work, we plan to test other score measures for ranking generation and to perform experiments using other catalogues of triplesets. We will also consider using domain information to improve the results.

Acknowledgments. This work was partly supported by CNPq, under grants 160326/2012-5, 301497/2006-0, 475717/2011-2 and 57128/2009-9, by FAPERJ, under grants E-26/170028/2008 and E-26/103.070/2011, and by CAPES under grant PROCAD/NF 1128/2010.

References

1. Berners-Lee, T.: Linked Data - Design Issues. W3C (June 2009), `http://www.w3.org/DesignIssues/LinkedData.html` (accessed on March 2013)
2. Bizer, C., Heath, T., Berners-Lee, T.: Linked Data - The Story So Far. International Journal on Semantic Web and Information Systems (IJSWIS) 5(3), 1–22 (2009)
3. Nikolov, A., d'Aquin, M.: Identifying Relevant Sources for Data Linking using a Semantic Web Index. In: WWW 2011 Workshop on Linked Data on the Web, LDOW. CEUR Workshop Proceedings, vol. 813. CEUR-WS.org (2011)
4. Jannach, D., Zanker, M., Felfernig, A., Friedrich, G.: Recommender Systems: An Introduction. Cambridge University Press (2011)
5. Bergamaschi, S., Guerra, F., Leiba, B.: Guest editors' introduction: Information overload. IEEE Internet Computing 14(6), 10–13 (2010)
6. Linden, G., Smith, B., York, J.: Amazon.com recommendations: item-to-item collaborative filtering. IEEE Internet Computing 7(1), 76–80 (2003)
7. Schafer, J.B., Konstan, J., Riedl, J.: Recommender systems in e-commerce. In: Proceedings of the 1st ACM Conference on Electronic Commerce, EC 1999, pp. 158–166. ACM, New York (1999)
8. Mooney, R.J., Roy, L.: Content-based book recommending using learning for text categorization. In: Fifth ACM Conference on Digital Libraries, DL 2000, pp. 195–204. ACM, New York (2000)

9. Burke, R.D.: Hybrid recommender systems: Survey and experiments. User Model. User-Adapt. Interact. 12(4), 331–370 (2002)
10. Golbeck, J., Hendler, J.: Filmtrust: movie recommendations using trust in web-based social networks. In: 3rd IEEE Consumer Communications and Networking Conference, CCNC 2006, vol. 1, pp. 282–286 (2006)
11. Montaner, M., López, B., de la Rosa, J.L.: A taxonomy of recommender agents on the internet. Artif. Intell. Rev. 19(4), 285–330 (2003)
12. Meo, P.D., Nocera, A., Rosaci, D., Ursino, D.: Recommendation of reliable users, social networks and high-quality resources in a social internetworking system. AI Commun. 24(1), 31–50 (2011)
13. Liben-Nowell, D., Kleinberg, J.: The link-prediction problem for social networks. J. Am. Soc. Inf. Sci. Technol. 58(7), 1019–1031 (2007)
14. Newman, M.E.J.: The structure and function of complex networks. SIAM Review 45(2), 167–256 (2003)
15. Quercia, D., Capra, L.: Friendsensing: recommending friends using mobile phones. In: Third ACM Conference on Recommender Systems, RecSys 2009, pp. 273–276. ACM, New York (2009)
16. Nikolov, A., d'Aquin, M., Motta, E.: What should I link to? Identifying relevant sources and classes for data linking. In: Pan, J.Z., Chen, H., Kim, H.-G., Li, J., Wu, Z., Horrocks, I., Mizoguchi, R., Wu, Z. (eds.) JIST 2011. LNCS, vol. 7185, pp. 284–299. Springer, Heidelberg (2012)
17. Lóscio, B.F., Batista, M.C.M., Souza, D., Salgado, A.C.: Using information quality for the identification of relevant web data sources: a proposal. In: 14th International Conference on Information Integration and Web-based Applications & Services, IIWAS 2012, pp. 36–44. ACM, New York (2012)
18. de Oliveira, H.R., Tavares, A.T., Lóscio, B.F.: Feedback-based data set recommendation for building linked data applications. In: 8th International Conference on Semantic Systems, I-SEMANTICS 2012, pp. 49–55. ACM, New York (2012)
19. Halpin, H., Hayes, P.J.: When owl:sameAs isn't the same: An analysis of identity links on the semantic web. In: Proceedings of the WWW 2012 Workshop: Linked Data on the Web (2010)
20. Baeza-Yates, R.A., Ribeiro-Neto, B.A.: Modern Information Retrieval - the concepts and technology behind search, 2nd edn. Pearson Education Ltd., Harlow (2011)
21. Manning, C.D., Raghavan, P., Schütze, H.: Introduction to Information Retrieval. Cambridge University Press (July 2008)

Improving Rocchio Algorithm for Updating User Profile in Recommender Systems*

Chong Wang, Yao Shen, Huan Yang, and Minyi Guo

Department of Computer Science & Engineering, Shanghai Jiao Tong University
{herculeshhp,yanghuanflc}@gmail.com, {yshen,guo-my}@cs.sjtu.edu.cn

Abstract. The Rocchio algorithm is a widely used relevance feedback algorithm in Information Retrieval which helps refine queries. Rocchio algorithm is operated in the vector space model. Since in most content-based recommender systems, items and user profile are represented as vectors in a specific vector space, Rocchio algorithm is exploited for learning and updating user profile. In this paper we show how to improve the Rocchio algorithm by distinguishing recommended items from two aspects: 1) the similarity between an item and a user profile, 2) users' ratings on recommended items. We conducted experiments on Movie-Lens dataset and the results show that the improved Rocchio algorithm outperforms the original one.

Keywords: Rocchio algorithm, recommender system, user profile.

1 Introduction

Rocchio algorithm[1] is a widely used relevance feedback algorithm in Information Retrieval which helps refine queries. The algorithm classifies documents into relevant and non-relevant to users' needs. As developed using vector space model, the algorithm represents documents as vectors and modifies initial query by taking the vector sum over all relevant and non-relevant documents.

Since in most content-based recommender systems, items and user profile are represented as vectors in a specific vector space, Rocchio algorithm is exploited for learning and updating user profiles[2–4]. When updating user profiles in these systems, all the items are treated equally except that each item is labeled as interesting or uninteresting according to users' feedback. That is, all interesting item vectors are directly added to the original profile vector and all uninteresting item vectors are just directly subtracted from the original profile vector. However, we think Rocchio algorithm can be improved when applied in the following two scenarios.

---

* This work was supported in part by the 863 Program of China (No.2011AA01A202) and the Doctoral Fund of Ministry of Education of China (No. 20100073120022). Yao Shen is the corresponding author.

X. Lin et al. (Eds.): WISE 2013, Part I, LNCS 8180, pp. 162–174, 2013.

– Scenario A

Suppose there is a news recommender system. User Bob's current profile is like < "tennis"=1, "match"=1 >. Now the system recommends two pieces of news to Bob which are $news_1$ =< "tennis"=0.5, "match"=0.5 > and $news_2$ =< "tennis"=0.5, "advertisement"=1 >. Since Bob likes tennis, he labels both news as interesting. Then the system applies Rocchio algorithm to modify Bob's profile to < "tennis"=2, "match"=1.5, "advertisement"=1 >. We can see the "advertisement" in the profile is increased with a relatively high value. But it is not quite reasonable since Bob likes $news_2$ because of the "tennis" rather than "advertisement". If we use cosine similarity to measure the similarity between news and Bob's previous profile, we can get similarity($news_1$, Bob)=1 and similarity($news_2$, Bob)=0.316. It suggests that when updating a user profile, items with larger similarity to the profile should have more influence than those with lower similarity.

– Scenario B

This time the news recommender system is improved to allow users to rate recommended news with a scalar from 1 to 10. The ratings indicate the level of interest a user has on the recommended news. Bob rates 10 for $news_1$ and 7 for $news_2$. Again, we think $news_1$ should have more impact on the modified profile than $news_2$ since $news_1$ gets a higher rating which implies Bob is more interested in $news_1$.

So these two scenarios raise a question: how to determine the influence of items in order to produce more precise user profiles? This is the problem we want to solve in this paper.

In this paper, we try to improve the Rocchio algorithm for updating user profiles in some recommender systems by distinguishing recommended items from two aspects:

1. The similarity between an item and a user profile

   Intuitively, a user profile vector should be closer to the item vectors more similar to the profile and be further away from not similar item vectors. Thus, item vectors which have very high or very low similarity with the user profile vector should have more influence on the modified user profile.

2. Users' ratings on recommended items

   Ratings can indicate the level of interest a user has on the recommended items. Items with extreme ratings (very high or very low) can reflect users' preference better than those with average ratings. It is reasonable that items with extreme ratings should have more influence on the user profile.

The rest of the paper is organized as follows. In Section 2, we elaborate the improvement of Rocchio algorithm. Experimental results and evaluation are presented in Section 3. Section 4 briefly reviews the related work followed by the conclusion in Section 5.

2   Algorithm Improvement

2.1   Basic Rocchio Algorithm

The core of the algorithm is shown as Equation (1).

$$\overrightarrow{Q_m} = \alpha * \overrightarrow{Q_o} + \beta * \sum_{\overrightarrow{D_i} \in D_r} \frac{\overrightarrow{D_i}}{|D_r|} - \gamma * \sum_{\overrightarrow{D_j} \in D_{nr}} \frac{\overrightarrow{D_j}}{|D_{nr}|} \tag{1}$$

Here $\overrightarrow{Q_m}$ is the modified query vector and $\overrightarrow{Q_o}$ is the original query vector. D_r and D_{nr} mean set of relevant and non-relevant documents respectively. α, β and γ are parameters that control the influence of the original query and the two prototypes of documents on the modified query. By applying the equation, the modified query vector gets closer to the centroid of relevant documents and moves away from the centroid of non-relevant documents.

When applied in recommender systems, Rocchio algorithm can be modified as follows to update user profile:

$$\overrightarrow{P_m} = \alpha * \overrightarrow{P_o} + \beta * \sum_{\overrightarrow{I_j} \in I_{in}} \overrightarrow{I_j} - \gamma * \sum_{\overrightarrow{I_k} \in I_{nin}} \overrightarrow{I_k} \tag{2}$$

$\overrightarrow{P_m}$ and $\overrightarrow{P_o}$ are modified profile vector and original profile vector respectively. I_{in} and I_{nin} mean set of user's interested and not interested items respectively.

From Equation (2), we can see all the items are treated equally when user profiles are updated. That is, all interesting item vectors are directly added to the original profile vector and all uninteresting item vectors are just directly subtracted from the original profile vector. However, we have explained that it is more reasonable to distinguish items by considering similarity and ratings.

2.2   Improvement with Similarity

Given a user profile $\overrightarrow{P}$ and an item $\overrightarrow{I}$, we use cosine similarity to determine the similarity between $\overrightarrow{P}$ and $\overrightarrow{I}$:

$$sim(\overrightarrow{P}, \overrightarrow{I}) = \frac{\overrightarrow{P} \cdot \overrightarrow{I}}{|\overrightarrow{P}| \times |\overrightarrow{I}|} \tag{3}$$

Then we modify Equation (2) by taking into account the similarity between a user profile and an item as follows:

$$\overrightarrow{P_m} = \alpha * \overrightarrow{P_o} + \beta * \sum_{\overrightarrow{I_j} \in I_{in}} \overrightarrow{I_j} * SF(sim(\overrightarrow{P_o}, \overrightarrow{I_j})) - \gamma * \sum_{\overrightarrow{I_k} \in I_{nin}} \overrightarrow{I_k} * SF(sim(\overrightarrow{P_o}, \overrightarrow{I_k})) \tag{4}$$

Here, SF is a function that maps $sim(\overrightarrow{P}, \overrightarrow{I})$ to an influence factor. We call SF as influence factor function. We will discuss properties of SF in Sec. 2.5.

2.3   Improvement with Rating

Suppose a recommender system allows users to rate recommended items with minimum rating r_{min} and maximum rating r_{max} and a user u gives a rating r on an item $\overrightarrow{I}$, we use the following equation to measure the level of interest that u shows on $\overrightarrow{I}$:

$$int(u, \overrightarrow{I}) = r - \frac{r_{min} + r_{max}}{2} \qquad (5)$$

Here, we regard the medium rating as the separating line between user's interested and not interested items. We say u is interested in $\overrightarrow{I}$ if $int(u, \overrightarrow{I}) > 0$ and the greater $int(u, \overrightarrow{I})$ is, the more interested u is in $\overrightarrow{I}$. It is the contrary for $int(u, \overrightarrow{I}) < 0$.

Then we modify Equation (2) by taking into account the ratings u gives on items:

$$\overrightarrow{P_m} = \alpha * \overrightarrow{P_o} + \beta * \sum_{\overrightarrow{I_j} \in I_{in}} \overrightarrow{I_j} * RF(int(u, \overrightarrow{I_j})) - \gamma * \sum_{\overrightarrow{I_k} \in I_{nin}} \overrightarrow{I_k} * RF(int(u, \overrightarrow{I_k})) \quad (6)$$

Here, RF is a function similar to SF that maps $int(u, \overrightarrow{I})$ to an influence factor. We will discuss properties of RF in Sec. 2.5.

2.4   Combining Similarity and Rating

By combining similarity and rating, Equation (2) is modified as follows:

$$\begin{aligned} \overrightarrow{P_m} = \alpha * \overrightarrow{P_o} \\ + \beta * \sum_{\overrightarrow{I_j} \in I_{in}} \overrightarrow{I_j} * SF(sim(\overrightarrow{P_o}, \overrightarrow{I_j})) * RF(int(u, \overrightarrow{I_j})) \\ - \gamma * \sum_{\overrightarrow{I_k} \in I_{nin}} \overrightarrow{I_k} * SF(sim(\overrightarrow{P_o}, \overrightarrow{I_k})) * RF(int(u, \overrightarrow{I_k})) \end{aligned} \qquad (7)$$

For each item, both the influence of similarity and rating are taken into account. We multiply SF by RF to gain a final influence factor. We will evaluate the parameters α, β and γ in experiments.

2.5   Properties of SF and RF

Properties of SF. We discuss the properties of SF in two cases:

1. The user is interested in $\overrightarrow{I}$
   If $sim(\overrightarrow{P}, \overrightarrow{I}) > 0$, it implies that some features of $\overrightarrow{I}$ really appeal to the

user. As a result, values of the corresponding features in $\overrightarrow{P}$ should be increased. And the closer $sim(\overrightarrow{P}, \overrightarrow{I})$ is to 1, the more interested the user may be in $\overrightarrow{I}$, which suggests the greater $SF(sim(\overrightarrow{P}, \overrightarrow{I}))$ should be.

On the other hand, if $sim(\overrightarrow{P}, \overrightarrow{I}) < 0$, we can infer that the user may begin to like some features he/she was not interested in previously. Thus the values of corresponding features in $\overrightarrow{P}$ should be increased to reflect the drift of interest. And the closer $sim(\overrightarrow{P}, \overrightarrow{I})$ is to -1, the more values such features should be increased with, which means the greater $SF(sim(\overrightarrow{P}, \overrightarrow{I}))$ should be.

2. The user is not interested in $\overrightarrow{I}$

If $sim(\overrightarrow{P}, \overrightarrow{I}) > 0$, it implies that the user may begin to lose interest in the features of $\overrightarrow{I}$ which he/she was interested in previously. Thus the values of corresponding features in $\overrightarrow{P}$ should be decreased. And the closer $sim(\overrightarrow{P}, \overrightarrow{I})$ is to 1, the more value such features should be decreased with, which means the greater $SF(sim(\overrightarrow{P}, \overrightarrow{I}))$ should be.

On the other hand, if $sim(\overrightarrow{P}, \overrightarrow{I}) < 0$, we can infer that the user is really not interested in the item. The item vector should be subtracted from the profile vector. And the closer $sim(\overrightarrow{P}, \overrightarrow{I})$ is to -1, the greater $SF(sim(\overrightarrow{P}, \overrightarrow{I}))$ should be.

So we can conclude that the figure of function SF should be like letter "U" or "V". Figure 1 shows some possible choices for SF. Notice that the three chosen functions are symmetric which means items with higher similarity have the same influence with items whose similarity is lower. But it is not necessary for SF to be symmetric.

We can consider the original Rocchio algorithm as a special case where $SF(x) \equiv 1$.

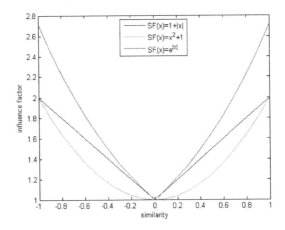

Fig. 1. Possible figures of SF

Properties of *RF*. When $int(u, \vec{I}) > 0$, the greater $int(u, \vec{I})$ is, the more interested u is in $\vec{I}$. Thus the profile vector should be modified to be closer to $\vec{I}$ which means the greater $RF(int(u, \vec{I}))$ should be. On the contrary, if $int(u, \vec{I}) < 0$, the smaller $int(u, \vec{I})$ is, the less u likes $\vec{I}$. As a result, the profile vector should be modified to be further away from $\vec{I}$ which means the greater $RF(int(u, \vec{I}))$ should be. We can notice that the figure of RF is similar to that of SF, so we set RF to be the same as SF in experiments.

3   Evaluation

3.1   Dataset

We chose MovieLens dataset[1] as the experiment dataset. The original dataset contains 10000054 ratings applied to 10681 movies by 71567 users. Each movie is assigned at least one genre. There are totally 18 different genres including "Action", "Adventure", "Crime", etc. Ratings are made on a 5-star scale, with half-star increments. We filtered out those users who made less than 100 ratings. The modified dataset contains 7073661 ratings made by 19479 users. We then split the modified dataset into one training set and five test sets. For each user, the training set and each of the test set contain at least 20 ratings from that user.

Both the user profile vector and movie vector are of 18 dimensions. Each dimension corresponds to one of the 18 genres. The value of a dimension in a user profile vector indicates the level of interest the user shows for the corresponding movie genre. For a movie vector, the value of a specific dimension is set to 1 if the movie has been assigned the corresponding genre. Otherwise, the value is 0.

We considered a user to be interested in a movie if he/she rated it for more than 3 stars and not interested if the rating is less than 3 stars.

3.2   Evaluation of Improvements

To evaluate the original Rocchio algorithm, improvement with similarity, improvement with rating and improvement by combining similarity and rating, we conducted an experiment as follows: For each of the four algorithms, the training set was utilized to learn the initial user profile. We then used this initial user profile to recommend movies contained in the first test set. After the recommendation, user profile was updated by employing the algorithm. The updated user profile was then used to recommend movies in the second test set. We repeated the process until the last test set. In this experiment, we set all parameters (α, β and γ) to be 1 and we chose the linear function as SF i.e. $SF(x) = 1 + |x|$.

On each round of recommendation, movies were ranked according to its similarity (by applying Equation 3) with a user profile. We then used MAP (Mean Average Precision) over all users to evaluate the recommendation. We used

---

[1] http://www.grouplens.org/node/12

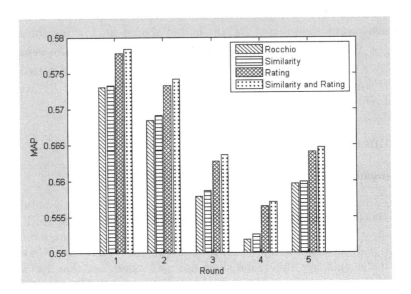

Fig. 2. MAP@$\geq$4 for evaluation of the improvement of Rocchio algorithm

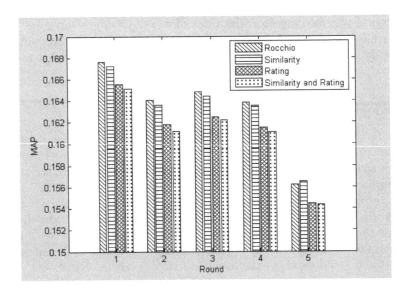

Fig. 3. MAP@$\leq$2 for evaluation of the improvement of Rocchio algorithm

MAP@$\geq$4 (MAP for movies whose ratings are greater than or equal to 4 stars) and MAP@$\leq$2 (MAP for movies whose ratings are less than or equal to 2 stars) to evaluate the quality of recommendation. A good algorithm should rank those user interested movies as high as possible and meanwhile put the movies with low rating in the end of the list. Thus, the greater MAP@$\geq$4 is, the better while it is contrary for MAP@$\leq$2. The experiment results are shown as Figure 2 and Figure 3.

The results show that improvement with similarity has slightly better performance than the original Rocchio algorithm. The rating algorithm presents a more significant improvement than the similarity algorithm which is an understandable result since ratings can reflect users' interests more precisely. By combining similarity and rating, the improved algorithm can achieve the best performance.

3.3   Evaluation of SF

We evaluated different selections of the influence factor function SF including linear function ($SF(x) = 1 + |x|$), quadratic polynomial function ($SF(x) = x^2 + 1$) and exponential function ($SF(x) = e^{|x|}$). We utilized Equation (7) to update user profile. The process was same as that in the previous experiment. Still we used MAP@$\geq$4 and MAP@$\leq$2 to evaluate recommendation quality. The result is shown as Figure 4 and Figure 5.

We can see both the quadratic polynomial and exponential SF have better performance than the linear one. Among them, the exponential SF achieves the best result. When applying the quadratic polynomial and exponential SF, items with extreme similarity or rating (i.e. very high or very low) have much more influence on the updated profile than those with moderate similarity or rating. The experiment result implies that we should pay more attention to the effect of items with extreme similarity or rating.

3.4   Evaluation of Parameters

We also conducted experiments to evaluate the influence of the three parameters α, β and γ in Equation (7). In the following experiments, we chose the exponential function as SF. The training set was used to learn the initial user profile and the first test set was utilized to make recommendations.

Firstly, we held α to 1 and changed β and γ from 0.1 to 1.0 with 0.1 increments. The result is shown as Table 1 and Table 2. MAP@$\geq$4 reached the highest value when $\beta = 0.2, \gamma = 0.9$ while MAP@$\leq$2 was lowest when $\beta = 0.1, \gamma = 0.3$, $\beta = 0.2, \gamma = 0.6$ or $\beta = 0.3, \gamma = 0.9$. Overall, the performance of the algorithm degrades with increasing β and improves with increasing γ. The algorithm achieves the best performance around $\beta = 0.2$ and $\gamma = 0.9$. The result suggests we should pay more attention on users' not interested movies when updating their profiles.

In the next experiment, we set β and γ to 0.2 and 0.9 respectively by considering the result of the previous experiment. α was changed from 0.0 to 1.0 with

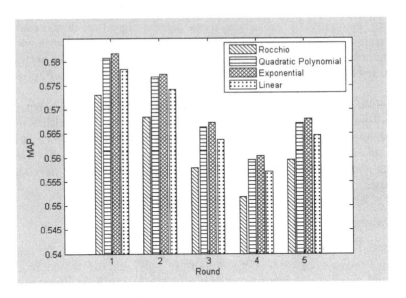

Fig. 4. MAP@$\geq$4 for evaluation of SF

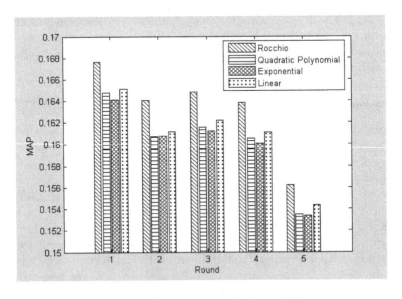

Fig. 5. MAP@$\leq$2 for evaluation of SF

Table 1. MAP@$\geq$4 for evaluation of β and γ

	γ									
β	0.1	0.2	0.3	0.4	0.5	0.6	0.7	0.8	0.9	1.0
0.1	0.5818	0.5854	0.5871	0.5877	0.5878	0.5872	0.5867	0.5863	0.586	0.5857
0.2	0.5784	0.5818	0.5839	0.5854	0.5869	0.5871	0.5875	0.5877	**0.5879**	0.5878
0.3	0.5769	0.5797	0.5818	0.5831	0.5843	0.5854	0.5864	0.5869	0.5871	0.5874
0.4	0.5764	0.5784	0.5802	0.5818	0.5828	0.5839	0.5847	0.5854	0.5861	0.5869
0.5	0.5759	0.5776	0.5793	0.5806	0.5818	0.5825	0.5834	0.5841	0.5849	0.5854
0.6	0.5756	0.5769	0.5784	0.5797	0.5809	0.5818	0.5824	0.5831	0.5839	0.5843
0.7	0.5753	0.5767	0.5778	0.579	0.58	0.581	0.5818	0.5824	0.583	0.5835
0.8	0.5751	0.5764	0.5773	0.5784	0.5794	0.5802	0.5811	0.5818	0.5823	0.5828
0.9	0.5749	0.5761	0.5769	0.5779	0.5789	0.5797	0.5804	0.5812	0.5818	0.5823
1.0	0.5748	0.5759	0.5767	0.5776	0.5784	0.5793	0.5799	0.5806	0.5813	0.5818

Table 2. MAP@$\leq$2 for evaluation of β and γ

	γ									
β	0.1	0.2	0.3	0.4	0.5	0.6	0.7	0.8	0.9	1.0
0.1	0.1641	0.1616	**0.1607**	0.1609	0.1613	0.1617	0.1628	0.1635	0.164	0.1645
0.2	0.1666	0.1641	0.1627	0.1616	0.161	**0.1607**	0.1608	0.1609	0.1609	0.1613
0.3	0.1675	0.1656	0.1641	0.1631	0.1624	0.1616	0.161	0.1608	**0.1607**	0.1609
0.4	0.1682	0.1666	0.1652	0.1641	0.1634	0.1627	0.1621	0.1616	0.161	0.161
0.5	0.1685	0.167	0.1661	0.1649	0.1641	0.1636	0.163	0.1625	0.162	0.1616
0.6	0.1688	0.1675	0.1666	0.1656	0.1648	0.1641	0.1636	0.1631	0.1627	0.1624
0.7	0.1689	0.1679	0.167	0.1663	0.1654	0.1646	0.1641	0.1637	0.1634	0.163
0.8	0.1691	0.1682	0.1672	0.1666	0.1659	0.1652	0.1645	0.1641	0.1638	0.1634
0.9	0.1691	0.1684	0.1675	0.1669	0.1663	0.1656	0.165	0.1645	0.1641	0.1638
1.0	0.1692	0.1685	0.1678	0.167	0.1666	0.1661	0.1654	0.1649	0.1644	0.1641

0.1 increments. The results is shown as Figure 6 and Figure 7. We can see overall the performance of the algorithm improves with increasing α and achieves the best when $\alpha = 0.8$ which implies that previous user profiles have a great impact on updating the profiles.

4   Related Work

Some early improvements of Rocchio algorithm include better term weighting[5, 6], query zoning[7]. [5] introduces a 2-Poisson model for term frequencies and [6] discusses methods of document length normalization. By applying these improved term weighting schemes, documents can be better represented and thereby improves the Rocchio algorithm. The algorithm proposed in [7] is somewhat similar to our similarity algorithm. By using query zoning, only the set of non-relevant documents which have high similarity with the original query is chosen for query updating but there is no discrimination among those chosen documents.

Recently, Rocchio algorithm is exploited in some text filtering and recommender systems. [2] modifies Rocchio algorithm for profile building and adaptation in

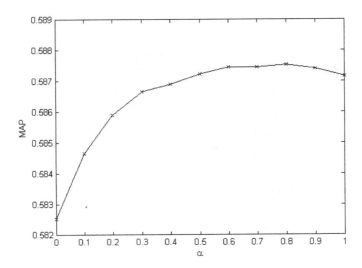

Fig. 6. MAP@$\geq$4 for evaluation of α

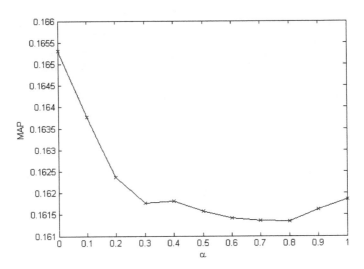

Fig. 7. MAP@$\leq$2 for evaluation of α

adaptive filtering. It does not only consider relevant and non-relevant documents but also take into account undetermined documents. Undetermined documents are documents that were not labeled interesting or not interesting by users. Those undetermined documents whose similarity with a user profile is below a threshold are regarded as pseudo-negative documents and treated as real negative documents. [8] proposes an algorithm called MIT (multiple topic tracking). MIT maintains multiple profiles to represent the multiple interests of a single user. For a given document, only the profile which has the maximum similarity with the document will be updated. [9] researches on dynamic selection of parameters in Rocchio algorithm. It shows that by dynamically learning good parameter configurations, Rocchio algorithm can adapt to differences in user behaviour among users. [10] improves Rocchio algorithm with users' implicit feedback. The implicit feedback is obtained based on the analysis of user behaviours such as click number and browsing duration.

Although these previous work improves Rocchio algorithm in different aspects from ours, we think if they are combined with our work, the Rocchio algorithm can achieve better performance.

5   Conclusion

In this paper we improve the Rocchio algorithm for updating user profiles in recommender systems through distinguishing recommended items from two aspects: 1) the similarity between an item and a user profile and 2) users' ratings on recommended items. When updating a user profile, different items may have different influence factors on the modified user profile. To evaluate the improved algorithm, we conducted experiments on MovieLens dataset. Results show that by taking into account similarity and ratings the improved algorithm can learn user profiles better than the original Rocchio algorithm. We also find that the exponential and quadratic polynomial influence factor function are better than the linear function. Influence of parameters in the algorithm are also evaluated and results show that previous user profiles and users' not interested items have much impact on the updated profiles.

Our future work will focus on evaluating the time factor i.e. distinguishing between users' long-term and short-term interests. We think users' long-term and short-term interested items should have different influence on updated user profiles. It may yield more precise user profiles by considering the difference between long-term and short-term interests.

References

1. Rocchio, J.: Relevance feedback in information retrieval. The SMART System: Experiments in Automatic Document Processing, 313–323 (1971)
2. Xu, H., Yang, Z., Wang, B., Liu, B., Cheng, J., Liu, Y., Yang, Z., Cheng, X., Bai, S.: Trec-11 experiments at cas-ict: Filtering and web. TREC11 (2002)
3. Pon, R.K., Cárdenas, A.F., Buttler, D.J., Critchlow, T.J.: iscore: Measuring the interestingness of articles in a limited user environment. In: Computational Intelligence and Data Mining, pp. 354–361. IEEE (2007)
4. Papadogiorgaki, M., Papastathis, V., Nidelkou, E., Waddington, S., Bratu, B., Ribiere, M., Kompatsiaris, I.: Two-level automatic adaptation of a distributed user profile for personalized news content delivery. International Journal of Digital Multimedia Broadcasting (2008)
5. Robertson, S.E., Walker, S.: Some simple effective approximations to the 2-poisson model for probabilistic weighted retrieval. In: Proceedings of the 17th Annual International ACM SIGIR Conference on Research and Development in Information Retrieval, pp. 232–241. Springer-Verlag, New York, Inc. (1994)

6. Singhal, A., Buckley, C., Mitra, M.: Pivoted document length normalization. In: Proceedings of the 19th Annual International ACM SIGIR Conference on Research and Development in Information Retrieval, pp. 21–29. ACM (1996)
7. Singhal, A., Mitra, M., Buckley, C.: Learning routing queries in a query zone. ACM SIGIR Forum 31, 25–32 (1997)
8. Pon, R.K., Cardenas, A.F., Buttler, D., Critchlow, T.: Tracking multiple topics for finding interesting articles. In: Proceedings of the 13th ACM SIGKDD International Conference on Knowledge Discovery and Data Mining, pp. 560–569. ACM (2007)
9. Pon, R.K., Cárdenas, A.F., Buttler, D.J.: Online selection of parameters in the rocchio algorithm for identifying interesting news articles. In: Proceedings of the 10th ACM Workshop on Web Information and Data Management, pp. 141–148. ACM (2008)
10. Qi, X.: Research on user profiling technology for personalized demands. In: Proceedings of the 2010 International Conference on Intelligent Computation Technology and Automation, pp. 198–201 (2010)

Time-Aware Travel Attraction Recommendation

Kai Wang, Richong Zhang, Xudong Liu, Xiaohui Guo,
Hailong Sun, and Jinpeng Huai

School of Computer Science and Engineering, Beihang University,
Beijing, 100191 China
{wangkai,zhangrc,liuxd,guoxh,sunhl,huaijp}@act.buaa.edu.cn

Abstract. The increasing number of tourists uploaded photos make it possible to discover attractive locations. Existing travel recommendation models make use of the geo-related information to infer possible locations that tourists may be interested in. However, the temporal information, such as the date and time when the photo was taken, associated with these photos are not taken into account by most of existing works. We advocate that this information give us a chance to discover the best visiting time period for each location. In this paper, we exploit a 3-way tensor to integrate context information for tourists visited locations. Based on this model, we propose a time-aware recommendation approach for travel destinations. In addition, a tensor factorization-based approach by maximizing the ranking performance measure is proposed for predicting the possible temporal-spatial correlations for tourists. The experimental results on the real tourists uploaded photos at Flickr.com show that our model outperforms existing approaches in terms of the prediction precision, ranking performance and diversity.

1 Introduction

With the boom of e-Tourism, there are emerging many online communities bringing convenience to travelers, and e-Business in tourism domain is becoming more and more flourishing. Meanwhile, tourists, especially with the widely spreading of the mobile devices, highly prefer to using such web facilities and mobile applications to enjoy their trips. In addition, travelers are likely to generate many media content, like photo, video, travelogue, etc., to share experiences in online social network. Thus, the travelers leave many online traces, which provides us expended opportunities to analyze the history travel data, so that to help more potential tourists in making decisions, such as acquainting some desired destinations, booking some suitable travel services, and planning the itinerary. However, without effective data processing methods, facing so many information, the decision-making is time-consuming and hard to reach a satisfying result. In this context, travel recommender system becomes increasingly popular and plays an important part in current tourism online services.

Existing studies on travel recommendation mainly focus on personalization, which aim to find the interaction of individual preference and features of attractions. Although the recommended items, such as destinations and tourism

X. Lin et al. (Eds.): WISE 2013, Part I, LNCS 8180, pp. 175–188, 2013.
© Springer-Verlag Berlin Heidelberg 2013

sites, visited by other similar tourists would make a great source of references for making the final travel decision, while the temporal information of when tourists should visit is missing. According to our statistics (presented in Section 4.1), these factors have a significant impact on the attractions choice of the tourist. For example, in late autumn, with the growing red of the maple leaves, the Fragrant Hill becomes one of the most charming places in Beijing. Then the probability that travelers would appreciate red leaves of the Fragrant Hill in late autumn should increase. Or in another case, to prove the influence of context factor, a tourist wants to know which attraction best fits his preference at night, while the recommender of ignoring the temporal context, may give an unreasonable result such as climbing Great Wall. This fact makes coherently modeling the temporal context with tourist preferences important.

The use of context-aware recommendation techniques may solve the problem of incorporating temporal context to the existing travel recommender system and building a time-aware recommendation model for predicting the most proper time period for visiting tourism sites. Several studies [1, 2] have taken into account the influence of contextual information such as the date or time for context-aware recommendations. The objective of the existing list-wise context-aware approaches is to find the best model parameters to maximize the mean average precision (MAP). However, MAP only evaluates the binary relevance of the recommended item. In practice, the preference of a user on an item is non-binary and usually evaluated by graded relevance.

To overcome the limitation of the existing context-aware recommendation approaches, in this paper, we synthesize the influence of personalization and temporal context on attraction recommendation by Tensor Factorization with a graded-degree relevance measure as the optimization objective. In particular, we exploit the nDCG, a cumulative and multilevel measure of ranking quality, as the objective function for Tensor Factorization to overcome the limitation of binary relevance measures. A gradient descent algorithm is also proposed to obtain the local optimal parameters. Moreover, we compare our proposed model with other existing matrix factorization or tensor factorization approaches on the real-life data from Flickr.com and the experimental results confirm the effectiveness of our model.

In summary, the contribution of this study is summarized as follows:

- We identify and bring to awareness the importance of temporal context for travel destination recommendation, a problem widely existing in the travel recommender systems.
- We propose the use of Tensor Factorization for temporal context-aware recommendation and utilize nDCG as the optimization objective for achieving a better ranking quality.
- We design an algorithm for predicting tourism sites that users may be interested in and the best time period for visiting. The comparative study on real-life data demonstrates the effectiveness over existing approaches.

The remainder of this paper is organized as follows. Section 2 delivers the overview of related works. Section 3 proposes the definition of our problem and the model we chose to present the interactions among travelers, attractions and temporal contexts. We also present our algorithm in this section. In Section 4, we introduce the experimental evaluation of our approach. The paper ends with brief conclusions and some discussion on future works.

2   Related Works

In order to generate appropriate recommendations and ensure the performance of recommendation systems, researchers have proposed different approaches. In this section, we present an overview of these works on travel recommendation, context-aware recommendation and learning to rank.

2.1   Travel Recommendation

With the widely development of mobile device, location-based recommendation receives more attentions from the academic circle than ever before. A large number of applications based on GPS positioning, for example, Flickr.com, a website where user can upload photos with geo-tags and timestamps, have emerged. In some studies [3–5], the timestamps of tourists uploaded photos are utilized to generate personal travel timed pathes. Then some models such as undirected graph and Markov model are adopted to recommend a travel route in given restrained conditions such as travel time and budget. Shi *et al.* [6] integrates the information of landmark categories to achieve a higher performance than basic matrix factorization and non-personalized recommendation based on popularity. However, these works do not take the context information into consideration when generating route or landmark recommendation.

Some studies claim that attractions recommendation is different from traditional collaborative filtering, such as movie or music recommendation [7, 6]. They find that almost all the travelers will visit the most popular attractions such as Forbidden City, Summer Palace when they visit Beijing. Thus, traveling histories of these attractions cannot precisely indicate their individual preferences. From another aspect, recommending less famous attractions will bring more assistance to travelers than well-known sites [6]. This characteristic of travel recommendation has also been concerned and evaluated in this study.

2.2   Context-Aware Recommendation

In [8], Schmidt *et al.* define the *context* that describes as a situation or environment a device or user is in. The intuition of context-aware recommender system (CARS) entails that, in some application scenario, user preferences are not monotonous which might leads to bad performance of context-unaware recommender systems. Based on different stages of integrating the contextual information, CARS can be classify as contextual pre-filtering, contextual post-filtering and contextual modeling [2]. Recently, Tensor Factorization, as a method of contextual modeling, arouses the attention of researchers. The effectiveness of this model has been confirmed by a number of studies [9–11]. In [12], a context-aware recommender system for mobile application discovery is proposed in this study to utilize the implicit feedback of personal usage history to form a binary tensor. In our study, we advocate that binary tensor does not carry the graded degree of users' interests. So that we make use of the nDCG, a commonly used performance measure in information retrieval to characterize the user interest level on item.

2.3   Learning to Rank

As this study aims to predict the ranking of unobserved values, our task can also be viewed as a learning to rank problem. Basically, learning to rank can be classified into three types: point-wise approach, pair-wise approach and list-wise approach [13]. Most of the earlier studies focus on point-wise and pairwise approach, such as pair-wise algorithm for tag recommendation [14]. Recently, the list-wise method, which aims to minimize a list-wise loss function defined on the prediction list, usually shows a comparative outperformance to other learning to rank methods [15]. A tendency of directly optimizing the IR metric such as MAP, MRR and nDCG, has emerged in latest research on list-wise method [11, 16, 17]. The greatest challenge of learning to rank is that the non-smoothness of these metrics are not always available [18]. In this study, we present a logistic function to approximately translate the non-differentiable measures to solve the non-smoothness problem.

3   Model

3.1   Problem Definition

For the reason of taking the temporal context into account when recommending attractions to users, we extend the traditional matrix factorization model for collaborative filtering, which only considers the interaction between users and attractions, to 3-way tensor factorization model by incorporating the temporal information as another dimension besides the users and attractions.

Specifically, we denote $X \in \mathbb{R}^{|\mathcal{U}| \times |\mathcal{S}| \times |\mathcal{T}|}$ the preferences of users to attractions in some given contexts, where $\mathcal{U}$ is the set of users with $|\mathcal{U}| = m$, $\mathcal{S}$ ($|\mathcal{S}| = n$) the set of all attractions, and $\mathcal{T}$ ($|\mathcal{T}| = l$) the set of different temporal contexts. Each entry of the tensor denoted by X_{ust} indicates the degree of user $u \in \mathcal{U}$ interested in attraction $s \in \mathcal{S}$ under some temporal context $t \in \mathcal{T}$. Actually, this tensor is incomplete and noisy in practice, i.e., only a subset $O \subseteq [m] \times [n] \times [l]$ out of the whole $m \times n \times l$ entries of X is observed, and the other unobserved entries $X_{[m] \times [n] \times [l] \setminus O}$ are set to zeros.

Inspired by the family of latent factor model, such as Matrix Factorization model, we adopt CP model [19] based tensor factorization methods to learn three f-dimensional latent factor matrixes $U \in \mathbb{R}^{m \times f}$, $S \in \mathbb{R}^{n \times f}$, and $T \in \mathbb{R}^{l \times f}$ to predict the missing values. Then the entire tensor could be fitted to $\tilde{X}$, in which each entry $\widetilde{X}_{ust}$ could be predicted through the inner product of the corresponding latent feature vectors.

$$\widetilde{X}_{ust} = \sum_{k=1}^{f} U_{uk} S_{sk} T_{tk} = < U_u, S_s, T_t > \tag{1}$$

As for the optimization objective function of these latent factors, Least Square approximations is a traditional and intuitional one, which we use as a baseline method in the experiment section, that is defined as following:

$$U^*, S^*, T^* = \arg\min_{U,S,T} \| X - \widetilde{X} \|^2. \tag{2}$$

Obviously, this loss function considers the errors for every entry of the whole tensor in an "equal odds". But for the information retrieval task, especially for the recommendation application in this work, we only highlight the precision of a few at the top of predicted results. E.g., in the Top-K recommendation which we usually show a few top-ranked items to users, the recommendation performance of the rendered items is our focus, but others with less significance.

The MAP and $nDCG$ are both rank-position sensitive and list-wise evaluation measures, but $nDCG$ reduces the contribution of the recalled items on the bottom of the returned list with a logarithmic decay factor of the position. Furthermore, different from MAP only take binary relevance, $nDCG$ can also accommodate graded, even real valued, relevance judgements.

From this point of view, we formulate our problem as to maximize the $nDCG$ values of the returned lists for all the users under all contexts. For each user $u \in \mathcal{U}$, since the value of entry $\widetilde{X}_{ust}$ represents the predicted preference of user u to attraction s under context t, the higher of this value is, the more possible the user will visit the attraction. We re-rank the attraction set according to the fitted fiber $\widetilde{X}_{u \cdot t}$ of tensor $\widetilde{X}$. Then the $nDCG$ value of the reordered list could be calculated by:

$$nDCG_{ut} = Z_{ut} \sum_{i=1}^{n} \frac{G_{uit}}{\log(1 + P_{uit})} \tag{3}$$

where P_{uit} denotes the position of ith attraction in the reordered list, and $G_{uit} = 2^{R_{uit}} - 1$ represent the gain of ith attraction, where $R_{uit} = X_{uit}$. Z_{ut} is the normalized factor related to $iDCG$. Then, the final $nDCG$ of tensor X can be computed as follows:

$$nDCG = \frac{1}{ml} \sum_{u=1}^{m} \sum_{t=1}^{l} nDCG_{ut} \tag{4}$$

3.2   NDCG Optimization Oriented Tensor Factorization

In this section, we will introduce how to optimize $nDCG$ based Tensor Factorization model. Firstly, a surrogate objective function approximating to $nDCG$ measure is proposed. Secondly, we present the corresponding optimization method. Finally, we present a Gradient Decent algorithm for inferring model parameters.

Based on Eq. (4), we average the $nDCG$ of all the above mentioned fibers of the entire tensor. Then, our optimization objective function can be written as

$$\mathcal{L}(U, S, T) = \frac{1}{ml} \sum_{u=1}^{m} \sum_{t=1}^{l} Z_{ut} \sum_{i=1}^{n} \frac{2^{X_{uit}} - 1}{\log\left(1 + (1 + \sum_{j=1, j\neq i}^{n} I(\tilde{X}_{ujt} \geq \tilde{X}_{uit})))\right)} \quad (5)$$

where

$$I(\tilde{X}_{ujt} \geq \tilde{X}_{uit}) = \begin{cases} 1, & \text{if } \tilde{X}_{ujt} \geq \tilde{X}_{uit}; \\ 0, & \text{else.} \end{cases} \quad (6)$$

Unfortunately, because of the non-smoothness of above indicator function, our objective function Eq. (5) is non-differentiable with respect to the variables U, S and T. Inspired by the work [20], we can smooth Eq. (6) with the logistic function as follows:

$$I(\tilde{X}_{ujt} \geq \tilde{X}_{uit}) \approx g(\tilde{X}_{ujt} - \tilde{X}_{uit}) = \frac{1}{1 + e^{-(\tilde{X}_{ujt} - \tilde{X}_{uit})}}$$

Furthermore, according to [21], Maximizing Eq. (5) is equivalently minimizing following reconstructed surrogate objective function:

$$\mathcal{L}'(U, S, T) = \frac{1}{ml} \sum_{u=1}^{m} \sum_{t=1}^{l} Z_{ut} \sum_{i=1}^{n} (2^{X_{uit}} - 1) \sum_{j=1, j\neq i}^{n} g(\tilde{X}_{ujt} - \tilde{X}_{uit}) \quad (7)$$

Ignoring the constant factor $\frac{1}{ml}$ and adding Frobenius norms of U, S, T to avoid over-fitting, the objective function to be minimized could be finally rewritten as below:

$$\mathcal{L}''(U, S, T) = \sum_{u=1}^{m} \sum_{t=1}^{l} Z_{ut} \sum_{i=1}^{n} (2^{X_{uit}} - 1) \sum_{j=1, j\neq i}^{n} g(\tilde{X}_{ujt} - \tilde{X}_{uit}) \quad (8)$$
$$+ \frac{\lambda}{2} (\|U\|^2 + \|S\|^2 + \|T\|^2)$$

Then, we adopt the steepest gradient descent method to minimize the regularized surrogate objective function $\mathcal{L}''$ and derive the gradients of Eq. (8) w.r.t. the variables U, S and T respectively as following:

$$\frac{\partial \mathcal{L}''}{\partial U_u} = \sum_{t=1}^{l} Z_{ut} \sum_{i=1}^{n} G_{uit} \sum_{j=1, j\neq i}^{n} g'(< U_u, S_j - S_i, T_t >)[(S_j - S_i) \circ T_t] + \lambda U_u \quad (9)$$

$$\frac{\partial \mathcal{L}''}{\partial S_s} = \sum_{u=1}^{m} \sum_{t=1}^{l} Z_{ut} \sum_{i=1, i\neq s}^{n} (G_{uit} - G_{ust}) g'(< U_u, S_s - S_i, T_t >)(U_u \circ T_t) + \lambda S_s \quad (10)$$

$$\frac{\partial \mathcal{L}''}{\partial T_t} = \sum_{u=1}^{m} Z_{ut} \sum_{i=1}^{n} G_{uit} \sum_{j=1, j\neq i}^{n} g'(< U_u, S_j - S_i, T_t >)[(S_j - S_i) \circ U_u] + \lambda T_t \quad (11)$$

where $\circ$ represents the element-wise product operation. With a randomized start point and suitable steps, following the negative gradient direction, objective function $\mathcal{L}''$ in Eq. (8) must gradually reach a local minimum. Algorithm 1 summarizes this steepest gradient decent based optimization method.

Algorithm 1.

Input: tensor X, regularization parameter λ, learning rate η, dimension of latent factors f, max iterations $itmax$, convergence parameters $\varepsilon_u, \varepsilon_s, \varepsilon_t$
Output: Matrix U, S, T
1: Initialize U, S, T with random value
2: **for** $it = 1, it <= itmax, it++$ **do**
3: **for** $u = 1, u <= m, u++$ **do**
4:    update $U_u = U_u - \eta \frac{\partial \mathcal{L}''}{\partial U_u}$ according to Eq. (9)
5: **end for**
6: **for** $s = 1, s <= n, s++$ **do**
7:    update $S_s = S_s - \eta \frac{\partial \mathcal{L}''}{\partial S_s}$ according to Eq. (10)
8: **end for**
9: **for** $t = 1, t <= l, t++$ **do**
10:   update $T_t = T_t - \eta \frac{\partial \mathcal{L}''}{\partial T_t}$ according to Eq. (11)
11: **end for**
12: **if** $(\| \frac{\partial \mathcal{L}''}{\partial U} \| \leq \varepsilon_u \&\& \| \frac{\partial \mathcal{L}''}{\partial S} \| \leq \varepsilon_s \&\& \| \frac{\partial \mathcal{L}''}{\partial T} \| \leq \varepsilon_t)$ **then**
13:    break
14: **end if**
15: **end for**
16: **return** U, S, T;

---

4   Experiments

In this section, we present a series of comparative experiments based on real data collected from Flickr. The results show the outperformance of the proposed method comparing with other state-of-the-art methods on recommendation.

4.1   Dataset Description

We use the public Flickr API to download 208,452 photos of Beijing, which are taken by 14,928 users. And we remove those photos without accurate taken time by judging whether the uploading time is equal or earlier than the taken time. Then, we use the geo-tags to map photos to attractions by coordinate matching and set the matching radius is set to 1000 meters. Finally, the left 180,467 effective photos are used to initialize the 3-way tensor, whose degree of sparseness is 1.822%.

We divide one day into four intervals of morning, afternoon, night and late night, and use those periods as temporal contexts. As the upload photos contain the taken time information, we can split photos into different temporal bins. Details are shown as Table 1.

Table 1. Temporal context

Context	Morning	Afternoon	Night	Late night
Period	$05:00 \sim 11:59$	$12:00 \sim 17:59$	$18:00 \sim 23:59$	$00:00 \sim 04:59$

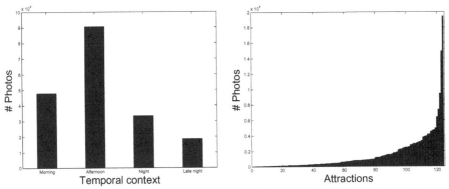

(a) photo number in each temporal context.   (b) photo number of each attractions.

Fig. 1. Preliminary statistical analysis of the dataset

We conduct a preliminary analysis on photos distribution of different periods and different attractions. Fig. 1(a), obviously, reflects the following two objective and natural facts: (1)very few people travel after the midnight; (2)the photos taken in daytime by the tourists are obviously more than nighttime, and afternoon is the prime-time for visiting and taking photos. As Fig. 1(b) demonstrated, the distribution of attractions apparently takes on a long tail effect and the popularity of attractions varies widely. The top 5 popular attractions are distinctly higher than

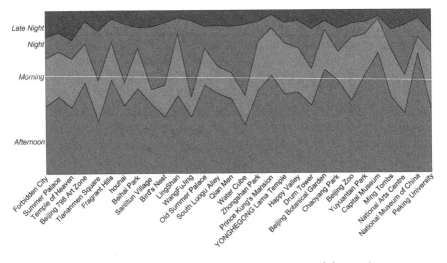

Fig. 2. The Temporal Dynamic Characteristics of Attractions

others. This phenomena conforms to the fact that hot and famous sceneries usually attract more tourists to visit. Based on this consideration, we conduct our experiments in the latter section by gradually removing those attractions to demonstrate the different characteristics between PopRec and latent factor models.

In above sections of this paper, we have claimed that the temporal distribution of visiting attractions varies dramatically. To show this fact, we investigate the proportion of photos taken at different temporal context for the most 30 famous attractive sites in Beijing. Fig.2 demonstrates that the *fraction* of uploaded photos at a specific time period over the total uploaded photos on this site is greatly influenced by the temporal contexts. This fact indicates that the different attractions show greatly diverse temporal property under the four predefined contexts. For instance, Capital Museum is closed at night, then the photos proportion under daytime contexts is sharply dominated. In addition, because Sanlitun Village is a famous Bar Street in Beijing and Water Cube usually is decorated with fancy neon lighting at night, both of them are with more photos at night context except afternoon. This statistical analysis supports our assumption that the temporal context is an important information for trave recommendation and it also confirms the motivation and the application potential of this work.

4.2   Baseline Methods

We conducted a series of comparative experiments with four baseline methods which are listed below to show the effectiveness of the proposed $nDCG$ optimization oriented tensor factorization ($TF\text{-}nDCG$).

1. **PopRec**

   Popularity based recommendation (*PopRec*) is an intuitional and effective approach in tourism domain, since the attractions generally follow a power law distribution and people usually visit some famous sceneries. So we assume the attractions with larger number of users visited under the given temporal contexts are more popular, and recommend the top-k popular unvisited attractions. We regard this method as a non-personalized approach to compare with our latent factor based personalized model.

2. **Matrix Factorization**

   In the recommendation research works, matrix factorization (*MF*) is a basic and important benchmark model from the family of latent factor models. Because it can only handle two dimensional factors, we separate the tensor into temporal context matrix slices by fixing temporal dimension indices, then employ the regularized least squares approximation to fit each individual slice.

3. **Tensor Factorization Models**

   Regularized least square optimization for tensor decomposition as defined in Eq. (2), is a commonly used method for missing value prediction, which is essentially point-wise fitting the relevance judgements. In order to perform more subtle comparison, we feed two types of implicit feedback to these models, namely binary relevance and numerical relevance (i.e., real valued graded relevance). For the binary relevance, we set the entry X_{ust} to 0 or 1 depending on whether the user upload photos of the corresponding attraction and context or not. Correspondingly, we treat the normalized proportion of photo numbers as numerical relevance. We name these two models *TF-Binary* and *TF-Numerical* respectively.

In this study, we assume that the number of user uploaded photos indicates the degree of preference a user favors some attractive sites. This measure naturally arose to measure the user preference distribution, thus we take the proportion of user uploaded photos on a specific site over the total number of photo uploaded by this user and refer this value as the graded user preferences.

4.3   Experiment Result

Accuracy and Ranking Performance. To evaluate the accuracy and ranking performance, we adopt the P@n and MAP@n as the evaluation metric respectively. We randomly choose 80% observed data of each user to form the training set, and the remaining 20% are used for evaluation. Note that we set the regularization parameter λ to be 0.01, the learning step η to be 0.01 and the number of factors d to be 5. The results are shown in Fig. 3.

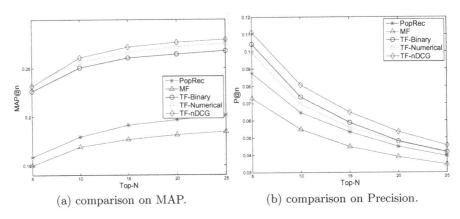

(a) comparison on MAP.   (b) comparison on Precision.

Fig. 3. Comparison on MAP@n and P@n

It can be observed that the performance of PopRec is much lower than other latent factor models based on Tensor Factorization that consider both personalization and context information. Overall, our TF-nDCG achieves the best performance in terms of MAP@n and P@n. Moreover, with integrating the normalized photo number, an improvement of 4% in MAP has been attained by TF-Numerical over TF-Binary. In addition, we notice that the performance of MF is even worse than PopRec, and TF-nDCG is slightly inferior (1.48%) to TF-Numerical in MAP@5.

Table 2. MAP@5 after removing k most popular attractions

k	PopRec	MF	TF-Binary	TF-Numerical	TF-nDCG
1	0.1345	0.1433	0.1902	0.2032	0.2141
2	0.1152	0.1305	0.1590	0.1781	0.1942
3	0.1011	0.1188	0.1390	0.1568	0.1864
4	0.0842	0.1146	0.1338	0.1459	0.1801
5	0.0784	0.1034	0.1103	0.1209	0.1698

Table 3. P@5 after removing k most popular attractions

k	PopRec	MF	TF-Binary	TF-Numerical	TF-nDCG
1	0.0773	0.0697	0.0908	0.0892	0.1065
2	0.0677	0.0634	0.0799	0.0811	0.0997
3	0.0632	0.0599	0.0728	0.0729	0.0948
4	0.0540	0.0579	0.0661	0.0672	0.0898
5	0.0486	0.0529	0.0588	0.0584	0.0829

For the characteristic of travel recommendation that travelers usually visit the most popular attractions regardless of their individual preference, the observed data are mostly concentrated on popular attractions. Past studies [7, 6] discover that travel histories in most famous attractions can not fully reflect personal preferences. Thus recommending less famous attractions is more important than recommending well-known ones. To show this less popular attraction recommendation capability, we gradually remove the most k famous attractions when generating the recommendation list of attractions.

The details of performance comparison after removing k popular attractions are described in Table 2 and Table 3. Due to the length limitation, we only present the comparison by MAP@5 and P@5. It can be observed that, with the increasing of number k, the performance of MF gradually becomes better than PopRec, e.g., 31.9% improvement in terms of MAP@5 and 8.85% in terms of P@5 when k is 5. Meanwhile, TF-nDCG improves the performance of MAP@5 and P@5 by 40.4% and 41.95% respectively over TF-Numeric when k is 5.

Inter-user Diversity. It is not enough to merely measure the performance by accuracy metrics. Other metric, such as diversity, is also important for meeting user requirements and enhancing user experiences. To illustrate the diversification, we evaluate the inter-user diversity of each compared approaches, excluding the non-personalized method PopRec, with the metric of hamming distance. The formulation of hamming distance is defined as below.

$$H_{ijt} = 1 - \frac{S_{ij}(L_t)}{L_t} \tag{12}$$

where $S_{ij}(L_t)$ represents the number of same attractions between the recommendation list in context t of user i and user j, and L_t represents the length of the recommendation list.

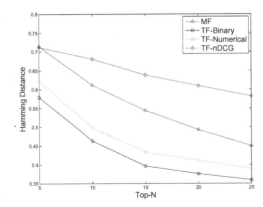

Fig. 4. Comparison on diversity by hamming distance of top-n list

We denote M the number of pairs of user, and then we can compute the overall hamming distance with Eq. (13).

$$H_{overall} = \frac{\sum_{t=1}^{T} \sum_{i=1}^{U} \sum_{j=1, j \neq i}^{U} H_{ijt}}{T \times M} \qquad (13)$$

Fig. 4 illustrates the diversification of TF-nDCG, TF-Binary, TF-Numerical and MF. It can be seen that our model outperforms other methods in terms of the inter-user diversity. For all the other compared approaches, the diversity decreases with the increasing of N, the number of items evaluated in top-N evaluation.

In this section, we have experimentally shown the superiority of our proposed model in comparison with other commonly-used factorization techniques, i.e., matrix factorization and traditional tensor factorization. In specific, we demonstrate a concrete example of the temporal distributions of visits for attractive sites in Beijing and confirms the value of our model. In addition, we evaluate the performance of TF-nDCG in terms of accuracy, ranking performance, and diversity in a real-life dataset collected at Flickr.com.

5   Conclusion and Future Work

In this paper, we proposed a novel list-wise approach based on tensor factorization and nDCG optimization to improve the performance of time-aware recommendation for attractions. We have also presented the context-aware recommendation model and algorithm to predict the items and their best corresponding temporal context. The mathematical inference validates the effectiveness of our model. We have compared our model with other state-of-the-art algorithms and our method demonstrates a significant improvement over existing context-aware recommendation algorithms on precision, MAP and diversity.

We have not yet taken the issue of better parameter selection, such as the choosing of number of latent vectors into account. We plan to include this in our future work to increase the accuracy of travel recommendations. Moreover, our future work also includes analyzing the influence of other contextual information on attractions recommendation such as season and weather and evaluating our proposed model on other datasets. Moreover, the ranking performance metric nDCG is used as the objective function for optimization. We can investigate the feasibility of applying other metric, such as MRR, into our context-aware recommendation framework.

Acknowledgments. This work was supported partly by National Natural Science Foundation of China (No. 61103031), partly by China 863 program (No. 2012AA011203), partly by the Fundamental Research Funds for the Central Universities (No. YWF-12-RHRS-016), partly by A Foundation for the Author of National Excellent Doctoral Dissertation of PR China, partly by Beijing Nova Program and partly by Program for New Century Excellent Talents in University.

References

1. Karatzoglou, A., Amatriain, X., Baltrunas, L., Oliver, N.: Multiverse recommendation: n-dimensional tensor factorization for context-aware collaborative filtering. In: Proceedings of the Fourth ACM Conference on Recommender Systems, pp. 79–86. ACM (2010)
2. Adomavicius, G., Tuzhilin, A.: Context-aware recommender systems. In: Recommender Systems Handbook, pp. 217–253. Springer (2011)
3. Kurashima, T., Iwata, T., Irie, G., Fujimura, K.: Travel route recommendation using geotags in photo sharing sites. In: Proceedings of the 19th ACM International Conference on Information and Knowledge Management, pp. 579–588. ACM (2010)
4. De Choudhury, M., Feldman, M., Amer-Yahia, S., Golbandi, N., Lempel, R., Yu, C.: Automatic construction of travel itineraries using social breadcrumbs. In: Proceedings of the 21st ACM Conference on Hypertext and Hypermedia, pp. 35–44. ACM (2010)
5. De Choudhury, M., Feldman, M., Amer-Yahia, S., Golbandi, N., Lempel, R., Yu, C.: Constructing travel itineraries from tagged geo-temporal breadcrumbs. In: Proceedings of the 19th International Conference on World Wide Web, pp. 1083–1084. ACM (2010)
6. Shi, Y., Serdyukov, P., Hanjalic, A., Larson, M.: Personalized landmark recommendation based on geotags from photo sharing sites. In: ICWSM 2011, pp. 622–625 (2011)
7. Clements, M., Serdyukov, P., de Vries, A.P., Reinders, M.J.: Using flickr geotags to predict user travel behaviour. In: Proceedings of the 33rd International ACM SIGIR Conference on Research and Development in Information Retrieval, pp. 851–852. ACM (2010)
8. Schmidt, A., Beigl, M., Gellersen, H.W.: There is more to context than location. Computers & Graphics 23(6), 893–901 (1999)
9. Hidasi, B., Tikk, D.: Fast ALS-based tensor factorization for context-aware recommendation from implicit feedback. In: Flach, P.A., De Bie, T., Cristianini, N. (eds.) ECML PKDD 2012, Part II. LNCS, vol. 7524, pp. 67–82. Springer, Heidelberg (2012)

10. Wermser, H., Rettinger, A., Tresp, V.: Modeling and learning context-aware recommendation scenarios using tensor decomposition. In: 2011 International Conference on Advances in Social Networks Analysis and Mining (ASONAM), pp. 137–144. IEEE (2011)
11. Shi, Y., Karatzoglou, A., Baltrunas, L., Larson, M., Hanjalic, A., Oliver, N.: Tfmap: Optimizing map for top-n context-aware recommendation. In: Proceedings of the 35th International ACM SIGIR Conference on Research and Development in Information Retrieval, pp. 155–164. ACM (2012)
12. Karatzoglou, A., Baltrunas, L., Church, K., Böhmer, M.: Climbing the app wall: enabling mobile app discovery through context-aware recommendations. In: Proceedings of the 21st ACM International Conference on Information and Knowledge Management, CIKM 2012, pp. 2527–2530. ACM, New York (2012)
13. Cao, Z., Qin, T., Liu, T.Y., Tsai, M.F., Li, H.: Learning to rank: from pairwise approach to listwise approach. In: Proceedings of the 24th International Conference on Machine Learning, pp. 129–136. ACM (2007)
14. Rendle, S., Schmidt-Thieme, L.: Pairwise interaction tensor factorization for personalized tag recommendation. In: Proceedings of the Third ACM International Conference on Web Search and Data Mining, pp. 81–90. ACM (2010)
15. Xia, F., Liu, T.Y., Wang, J., Zhang, W., Li, H.: Listwise approach to learning to rank: theory and algorithm. In: Proceedings of the 25th International Conference on Machine Learning, pp. 1192–1199. ACM (2008)
16. Shi, Y., Karatzoglou, A., Baltrunas, L., Larson, M., Oliver, N., Hanjalic, A.: Climf: learning to maximize reciprocal rank with collaborative less-is-more filtering. In: Proceedings of the Sixth ACM Conference on Recommender Systems, pp. 139–146. ACM (2012)
17. Taylor, M., Guiver, J., Robertson, S., Minka, T.: Softrank: optimizing non-smooth rank metrics. In: Proceedings of the International Conference on Web Search and Web Data Mining, pp. 77–86. ACM (2008)
18. Quoc, C., Le, V.: Learning to rank with nonsmooth cost functions. In: Proceedings of the Advances in Neural Information Processing Systems, vol. 19, pp. 193–200 (2007)
19. Kolda, T.G., Bader, B.W.: Tensor decompositions and applications. SIAM Review 51(3), 455–500 (2009)
20. Chapelle, O., Wu, M.: Gradient descent optimization of smoothed information retrieval metrics. Information Retrieval 13(3), 216–235 (2010)
21. Valizadegan, H., Jin, R., Zhang, R., Mao, J.: Learning to rank by optimizing ndcg measure. In: Advances in Neural Information Processing Systems, vol. 22, pp. 1883–1891 (2009)

CGMF: Coupled Group-Based Matrix Factorization for Recommender System

Fangfang Li[1,2], Guandong Xu[2], Longbing Cao[2],
Xiaozhong Fan[1], and Zhendong Niu[1]

[1] School of Computer Science and Technology, Beijing Institute of Technology, China
Fangfang.Li@student.uts.edu.au, {fxz,zniu}@bit.edu.cn
[2] Advanced Analytics Institute, University of Technology, Sydney, Australia
{Guandong.Xu,Longbing.Cao}@uts.edu.au

Abstract. With the advent of social influence, social recommender systems have become an active research topic for making recommendations based on the ratings of the users that have close social relations with the given user. The underlying assumption is that a user's taste is similar to his/her friends' in social networking. In fact, users enjoy different groups of items with different preferences. A user may be treated as trustful by his/her friends more on some specific rather than all groups. Unfortunately, most of the extant social recommender systems are not able to differentiate user's social influence in different groups, resulting in the unsatisfactory recommendation results. Moreover, most extant systems mainly rely on social relations, but overlook the influence of relations between items. In this paper, we propose an innovative coupled group-based matrix factorization model for recommender system by leveraging the user and item groups learned by topic modeling and incorporating couplings between users and items and within users and items. Experiments conducted on publicly available data sets demonstrate the effectiveness of our approach.

1 Introduction

With the advent of online social networks, more and more social information is incorporated to RS, and social RS is becoming an active area in RS [7] [3]. The main motivation behind social RS is to leverage the auxiliary friend relations of users to tackle the common challenges in RS, e.g., cold-start and sparsity. For example, for a new user to RS, it is usually difficult to find the like-minded users due to the lack of the new user's ratings. However, through the social information known from social networking, this difficulty could be partially overcome. The underlying assumption of the social recommendation approach is that a user's taste is influenced by his/her friends in social networking. Accordingly, assigning more weights to items that the friends are interested in will potentially improve the satisfaction of recommendations. However, the extant social RS treats user's friends equally, but ignores the fact that user's social interests are intrinsically multifaceted.

X. Lin et al. (Eds.): WISE 2013, Part I, LNCS 8180, pp. 189–198, 2013.

Everyone has specific preference in particular groups. This indicates that a user may trust different subsets of friends in different groups. More specially, a user may have friends working in different domains, and join in activities across different domains. This is evidenced by that the extant social networks such as Google+, Facebook, Twitter already have such mechanisms to divide users into groups for sharing different information with different groups. Undoubtedly, in social RS, utilizing such social group information will be able to provide better personalized services for users. But most of extant social Web applications such as Tweeter, Sina Weibo, Delicious etc. do not provide reliable mechanisms to allow users to differentiate social connections from individual groups. Some recent researches integrate the distinguished group information into recommendation algorithms, e.g. [13] leverages the social trust circles from item-category information for social recommendation. Despite of the superior results demonstrated from the given multi-category rating data sets, this approach has a major limitation that it relies on the explicit item category information to form user circles, upon which social recommendation is made. However, such information is not always available in existing social networks e.g., Facebook or Twitter might not have such explicit category information, resulting in difficulties in applying the proposed algorithm. In this work, we attempt to address this unknown category information problem based on hidden topic modeling.

The extant social RS mainly focused on capturing social friendships within users and mutual relations between users and items. However, just considering social friendships and mutual relations are not enough for recommendation. Actually, items are often coupled together, if item o_i is closely relevant to item o_j, the preference of user u on item o_i would be influenced by item o_j. The effectiveness of recommendation would be probably increased through analyzing these relations of items. In fact, the complex relations in social RS can be abstracted into two classes. One is inter-couplings such as user preferences or ratings on items, the other one is intra-coupling including user intra-coupling and item

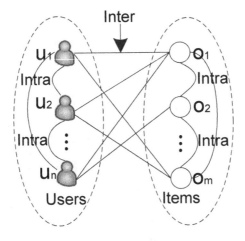

Fig. 1. Coupling Relations in Recommender Systems

intra-coupling [1], which are shown in Fig. 1. These coupling relations should be considered simultaneously and evenly in learning the recommendation model. The inter-couplings between users and items have been well studied [2] and utilized in the extant RS for enhancing the result of recommendation. However, the user intra-coupling relations are not sufficiently exploited and item intra-coupling relations are often ignored in social RS. In addition, few solutions have been proposed to integrate user-intra, item-intra and user-item inter-coupling relations into a unified manner. A complete considerations of such couplings can provide a practical mean for enhancing the effectiveness of social RS and solving the cold start and data sparsity problems. In this work, we incorporate the coupling relations and group information into the matrix factorization model.

The contributions of the paper are concluded as follows:

- We propose a Coupled Group-based Matrix Factorization model (CGMF) which incorporates couplings between/within users and items such as user intra-coupling, item intra-coupling and user-item inter-coupling.
- We apply topic modeling on item descriptions to automatically extract hidden topics of items and derive the user groups, then integrate group information into matrix factorization as an additional constraint in learning recommendation models.
- We conduct experiments to verify our algorithms and recommendation models.

The rest of the paper is organized as follows. Section 2 presents the related work. In Section 3, we analyze the coupled interactions between users and items and within users and items. Then we introduce the group formation algorithm in Section 4. After that, the CGMF model is proposed integrating user and item groups. Experimental results and analysis are presented in Section 6 followed by the conclusion.

2   Related Work

Collaborative filtering (CF)[11] is one of the most successful approaches taking advantage of user rating history data to predict users' interests. Research efforts have been invested to make use of complimentary information in order to address the cold-start and sparsity problem. Slope One is a family of algorithms used for collaborative filtering, introduced in [6]. Arguably, it is the simplest form of non-trivial item-based collaborative filtering based on ratings of another item. However, CF algorithms do not consider user intra-coupling and item intra-coupling existed in RS and the users and items are assumed to be independent and identically distributed.

Matrix factorization [4] [5] is a latent factor model which is generally effective at estimating overall structure that relates simultaneously to most or all items. The basic matrix factorization approach for RS is based on an assumption that users are independent and identically distributed. This approach ignores the social activities between users, which is not consistent with the reality that we normally ask friends for recommendations. With the advent of social network, many researchers have started to analyze social recommender systems and various models integrating social networks such as Social Recommendation (SoRec) [8], Social Trust Ensemble (STE)[7], Recommender Systems with Social Regularization [9], etc. have been proposed. Social Matrix Factorization approaches actually consider the social activities of users, but social relations are mixed together and treated equally. As a result, it is impossible to differentiate social recommendations from different friends in terms of their preferred areas. Apart from this, item intra-couplings are also ignored.

3   Coupled-Interaction Analysis

Coupled-Interaction includes intra-couplings within users and items and inter-couplings between users and items.

3.1   Inter-coupled Interaction

Users often directly interact with items, for example, some users will give their rating after they watched a movie, which is the most intuitionistic interaction between users and items. Reflecting the relations between users and items, inter-coupling between user u and item o_i can be directed computed by $\delta_{u,i}^{Ie} = P_u Q_i^T$, where each item o_i is associated with a vector $Q_i \in \mathbb{R}^d$, and each user u is associated with a vector $P_u \in \mathbb{R}^d$.

3.2   Intra-coupled Interaction

Besides inter-coupling, RS also have massive intra-couplings which contain user-intra-coupling and item-intra-coupling. Intra-coupling within users and items can be modeled by Eqn. 1

$$\delta_{u,i}^{Ia} = \sum_{v \in F(u)} S_{u,v} P_v Q_i^T + \sum_{j \in N(i)} W_{i,j} P_u Q_j^T \tag{1}$$

where the first part is user-intra-coupling and the second part is item-intra-coupling. $S_{u,v}$ is the friendship relation of users u and v, and W_{ij} is the relevance of items o_i and o_j.

Eqn. 1 not only says that user profile P_u should be similar to his friends' profile P_v, but also says if user u is interested in item o_i, he/she will also interest in item o_j which is closely relevant to item o_i.

After coupled interactions are considered, MF prediction model is modeled as follows:

$$\hat{R}_{u,i} = r_m + \delta_{u,i}^{Ie} + \delta_{u,i}^{Ia} \tag{2}$$

which $\delta_{u,i}^{Ia}$ represents the intra-couplings within users and items, $\delta_{u,i}^{Ie}$ represents the inter-coupling between users and items.

4   Group Formation

Topic modeling is a proper way to automatically divide all items into different topics which can be considered as groups. Theoretically, LDA is a probabilistic generative model for a text corpus. The basic idea of LDA is based on the hypothesis that a person has certain topics in mind when writing an article. To address a topic, the author needs to pick up a word with a certain probability from a bag of words reflecting that topic. In this manner an item is represented as random mixtures over latent topics and each topic is characterized by a set of related words with a probability distribution. In the context of social networks, the obtained topics represent the commonly shared perception of the items by collaborative users, and the words of the specic topic constitute a common vocabulary contributed to the topic. In a summary, topic modeling can be used to partition different items into different groups especially when items have corresponding text description information.

Through the LDA model, we can capture the hidden topics and item assignments to these topics. Once we get the topic probability distribution of the items, we can easily analyze the user's affiliation on such topics according to the following Eqn. 3.

$$p_r\left(u|g_k\right) = \sum_{o_i} p_r\left(u|o_i\right) p_r(o_i|g_k) \tag{3}$$

with an item $o_i = \{w_{i,n}, n = 1, \ldots, N_i\}$ is generated by picking a distribution over the topics from a Dirichlet distribution, $p_r\left(u|g_k\right)$ and $p_r\left(u|o_i\right)$ are the probality that the user u belongs to group $g_k(1 \leq k \leq K)$ and the interests on item o_i. The whole process of item group formation and user probability distribution on these groups is described as following algorithm 1.

Algorithm 1. Group Formation Algorithm

Input: Items set $\{o_1, o_2, ..., o_m\}$
Output: Item groups $\{g_1, ..., g_K\}$ and the probability distribution matrix of
           users belongs to these groups

1 Classify items set to different topic groups $\{g_1, ..., g_K\}$ in terms of significant
   text-related information by LDA topic modeling;
2 Compute user distribution on the classified topics by Eqn. 3;
3 Assign the probabilistic weight to users which indicates how much the users
   belong to the groups.

---

5   Coupled Group-Based MF Model

After considering inter-couplings and intra-couplings, we aim to integrate them with group information in a unified model, namely Coupled Group-based MF model as follows.

$$\hat{R}_{u,i}^{(g)} = r_m + \delta_{u,i}^{I_e\,(g)} + \delta_{u,i}^{I_a\,(g)} \tag{4}$$

Different from MF, the prediction task of matrix $\hat{R}$ is transferred to compute the mapping of users and items to factor matrices P, Q, coupling relations and group information. Once this mapping is completed, RS can easily predict the rating a user will give to any item in a specific group by using Eqn. 4. The computation of the mapping can be optimized by minimizing the regularized squared error on the set of observed ratings. The objective function is given as Eqn. 5.

$$
\begin{aligned}
L^{(g)} = & \frac{1}{2} \sum_{(u,i)\in K} \left(R_{ui}^{(g)} - \hat{R}_{u,i}^{(g)} \right)^2 + \\
& \frac{\lambda}{2} \left(\|Q_i^{(g)}\|^2 + \|P_u^{(g)}\|^2 + \sum_{v\in F(u)^{(g)}} \|S_{u,v}^{(g)}\|^2 + \sum_{j\in N(i)^{(g)}} \|W_{i,j}^{(g)}\|^2 \right)
\end{aligned}
\tag{5}
$$

The training process starts at randomly initiate values of $P^{(g)}$ and $Q^{(g)}$. Then it iterates to update $P^{(g)}$, $Q^{(g)}$, $S_{u,v}^{(g)}$ and $W_{ij}^{(g)}$ by the gradient decent approach on the objective function $L^{(g)}$ until convergence. After $P^{(g)}$ and $Q^{(g)}$ are learned from the training process, we can predict the ratings for user-item pairs (u, o_i) by Eqn. 4.

6   Experiments and Results

In this section, we evaluate our proposed model and compare it to the existing approaches respectively using Movielens, LastFm and DBLP citation database [12].

6.1   Data Set

MovieLens data set has been widely explored in collaborative filtering research in last decade. MovieLens 10M data set consists of 72,000 users, 10,000 movies and 10 million ratings data. MovieLens is a classic data set for evaluating recommendation models, however, this data set does not contain the friendship of users i.e. the user intra-coupling. Therefore, the following experiments on MovieLens can not show the sensitivity of user intra-couplings. But the data set actually has a special genre feature which is applied for grouping all the movies, so the results on MovieLens show the influence of inter-couplings and item intra-couplings.

Different from MovieLens, LastFm data set contains social networking, tagging, and music artist listening information involving 1892 users, 17632 artists, 12717 bi-directional user friend relations, 92834 user-listened artist relations, and 11946 tag assignments. However, the data set does not have the rating data of users on artists. We know that the listening count indirectly reflects the preference of users on the artist, therefore, we normalize the listening count for the users on artists to [0,5] to indicate the implicit preferences. The data set has tagging information which is used for group formation by topic modeling.

Therefore, the results on LastFm show the sensitivity of couplings between and within users and items, and group information after this adjusted setting.

The DBLP citation database contains 1,572,277 papers and 2,084,019 citations. Each paper is associated with title, abstract, authors, year, venue, citation number and references. The data set consists of various coupling relations such as co-author and citation relations. For the DBLP data set, in the experiments, authors and papers are separately thought as users and items. Co-author relations are taken as friendships between users since a user must have friendship with the co-authors of his/her publications. And the item intra-couplings are captured by the citation network and the text similarity based on the paper's title and abstract. The "write" and "write-by" relations are converted to "0-1" ratings representing the preference of the user to the paper.

Overall, the following experiments on MovieLens and LastFm data sets are separately used for movie and artist recommendation, while DBLP data set is explored for paper recommendation for testing our CGMF model.

6.2   Experimental Settings

The 5-fold cross validation is performed in our experiments. In each fold, we have 80% of data as the training set and the remaining 20% as testing set. Here we use Root Mean Square Error (RMSE) and Mean Absolute Error (MAE) as evaluation metrics.

RMSE and MAE are defined as follows:

$$RMSE = \sqrt{\frac{\sum_{(u,i)|R_{test}} \left(r_{u,i} - \hat{r}_{u,i}\right)^2}{|R_{test}|}} \tag{6}$$

$$MAE = \frac{\sum_{(u,i)|R_{test}} |r_{u,i} - \hat{r}_{u,i}|}{|R_{test}|} \tag{7}$$

where R_{test} is the set of all pairs (u, o_i) in the test set.

To evaluate the performance of our proposed CGMF we consider three baseline approaches:

- CF: This is the well-known item based collaborative filtering method called Slope One.
- BasicMF: This method is a probabilistic matrix factorization approach in [10] which does not take the social network into account.
- SocialMF: This is the model which just considers the social friendships but ignores the item intra-couplings and group information.

6.3   Experimental Results and Discussions

Effectiveness of Couplings and Groups. DBLP and LastFm data sets have ample couplings within users and items and between users and items, and text related information used for forming groups, so the experimental results on the

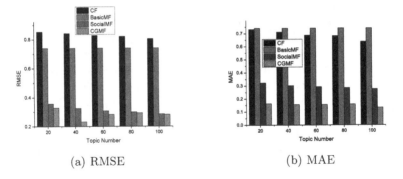

(a) RMSE   (b) MAE

Fig. 2. RMSE and MAE Comparison on DBLP with Different Number of Topics

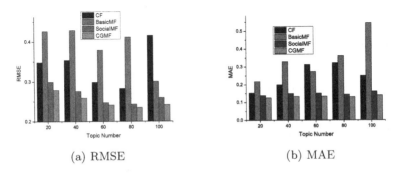

(a) RMSE   (b) MAE

Fig. 3. RMSE and MAE Comparison on LastFm with Different Number of Topics

two data sets can demonstrate the impacts of couplings and group information. We depict the effectiveness comparisons with respect to each method on DBLP and LastFm data sets in figures 2 and 3. From Fig. 2, we can clearly see that, our proposed CGMF method outperforms the counterparts in terms of RMSE and MAE. Compared to SocialMF, CGMF achieves up to 4.3% improvement on RMSE and 14.7% on MAE, while immense improvements compared to CF and Basic MF. On LastFm data sets, Fig. 3 evidences that CGMF performs much better than the benchmark methods. We can see that CGMF can reach a prominent improvements compared to CF and Basic MF approaches, which is resulted from considering complete coupling relations.

Effectiveness of Item Intra-couplings and Groups. Because the users of MovieLens data set do not have friendships which mean the user intra-couplings can not be captured, but MovieLens actually has natural genre feature which is used to form groups, the experimental results on Fig. 4 can show the performance of item intra-couplings and group information. The results indicated that our proposed CGMF can reach an average improvements of 3.5% and 3.1% on RMSE, and 1.6% and 1.8% on MAE compared to CF and Basic MF approaches.

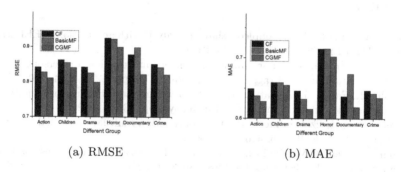

(a) RMSE   (b) MAE

Fig. 4. RMSE and MAE Comparison on MovieLens with Different Group

The biggest improvements are 4.3% on RMSE, and 3.0% on MAE compared to CF in the Dramma group. Compared to Basic MF, our proposed CGMF can improve by 7.6% on RMSE and 5.5% on MAE in the Documentary group.

Adaptiveness of Different Group Formation Methods. The experimental results on DBLP and LastFm data sets indicate that topic modeling is an effective group formation method, while results on Movielens show that category information which is used for group formation can also contribute to improving the effectiveness. That is to say, no matter which group formation methods (topic modeling or category information) are chosen, our proposed CGMF can be applied. The very significance of CGMF is topic modeling can be chosen for grouping when the data set does not have category information.

Overall, the RMSE and MAE figures on all the three data sets show the significant improvements of CGMF compared to benchmark methods. Therefore, we can conclude that by taking the couplings and group information into consideration, our approach can reach a better recommendation.

7   Conclusion

This paper proposed a coupled group-based matrix factorization model for recommender system, which incorporates coupling relations between and within users and items. CGMF first extracts items hidden groups via a Latent Dirichlet Allocation (LDA) model and derives user groups. Then coupling relations and group information are incorporated into CGMF. The experiments conducted on the real data sets demonstrated the superiority of the approach against the state-of-the-art methods.

Acknowledgements. This work is sponsored in part by Australian Research Council Discovery Grants (DP1096218 and DP130102691) and ARC Linkage Grant (LP100200774).

References

1. Cao, L., Ou, Y., Yu, P.S.: Coupled behavior analysis with applications. IEEE Trans. Knowl. Data Eng. 24(8), 1378–1392 (2012)
2. Feng, W., Wang, J.: Incorporating heterogeneous information for personalized tag recommendation in social tagging systems. In: KDD, pp. 1276–1284 (2012)
3. Jamali, M., Ester, M.: A matrix factorization technique with trust propagation for recommendation in social networks. In: RecSys, pp. 135–142 (2010)
4. Koren, Y.: Factorization meets the neighborhood: a multifaceted collaborative filtering model. In: KDD, pp. 426–434 (2008)
5. Koren, Y., Bell, R.M., Volinsky, C.: Matrix factorization techniques for recommender systems. IEEE Computer 42(8), 30–37 (2009)
6. Lemire, D., Maclachlan, A.: Slope one predictors for online rating-based collaborative filtering. In: Proceedings of SIAM Data Mining (SDM 2005) (2005)
7. Ma, H., King, I., Lyu, M.R.: Learning to recommend with social trust ensemble. In: SIGIR, pp. 203–210 (2009)
8. Ma, H., Yang, H., Lyu, M.R., King, I.: Sorec: social recommendation using probabilistic matrix factorization. In: CIKM, pp. 931–940 (2008)
9. Ma, H., Zhou, D., Liu, C., Lyu, M.R., King, I.: Recommender systems with social regularization. In: WSDM, pp. 287–296 (2011)
10. Salakhutdinov, R., Mnih, A.: Probabilistic matrix factorization. In: Advances in Neural Information Processing Systems, vol. 20 (2008)
11. Su, X., Khoshgoftaar, T.M.: A survey of collaborative filtering techniques. Adv. in Artif. Intell., 4:2 (January 2009)
12. Tang, J., Zhang, D., Yao, L.: Social network extraction of academic researchers. In: ICDM, pp. 292–301 (2007)
13. Yang, X., Steck, H., Liu, Y.: Circle-based recommendation in online social networks. In: KDD, pp. 1267–1275 (2012)

Authenticating Users of Recommender Systems Using Naive Bayes

Zhengang Wu, Liangwen Yu, Huiping Sun*, Zhi Guan, and Zhong Chen

Institute of Software, EECS, Peking University, Beijing, China
MoE Key Lab of High Confidence Software Technologies (PKU)
MoE Key Lab of Network and Software Security Assurance (PKU)
{wuzg,yulw,sunhp,guanzhi,chen}@infosec.pku.edu.cn

Abstract. Knowledge Based Authentication (KBA) verifies the credibility of claimed identities by matching various user-related data. Popular recommender systems hold abundant personalized data that are valuable for KBA. This paper studies how to authenticate users with abundant rating data in recommender systems. For this, we propose a measurable user authentication scheme for recommender systems with secure personalized data under the Naive Bayes model. Next, we analyze its usability and security for knowledge sources under possible guessing strategies and experimentally evaluate its performance in real datasets. And the proposed scheme is practical in recommender systems.

Keywords: Knowledge Based Authentication, Bayesian Decision, Naive Bayes, entropy, recommender systems.

1 Introduction

With the rapid development of web applications, real-world recommender systems manage massive users and recommendation data. Knowledge Based Authentication (KBA) verifies real identities by assessing personal information from claimants. Some large-scale e-commerce sites such as eBay and Amazon check their users using KBA.

Existing KBA applications mainly focus on common personal information such as Home Address and preset questions and therefore probably share same secret information with risky ones. To reduce the vulnerability on secret leakage, we pay much attention to private application-dependent data for KBA in the setting that recommender systems can hold some rating data securely.

Contribution. To the best of our knowledge, this paper first identify the practical problem that how to authenticate users with abundant user-related application data of recommender systems. The contribution is two-fold as follow.

(1)For this problem, we propose a measurable KBA-based approach to analyze personalized data in recommender systems, using Naive Bayes. This paper describes the solution in detail and analyze its usability and security briefly.

---

* Corresponding author.

X. Lin et al. (Eds.): WISE 2013, Part I, LNCS 8180, pp. 199–208, 2013.

(2)We implement the solution and experimentally evaluate its performance to demonstrate its practicability, using real datasets from a movie rating site.

The rest of this paper is organized as follows. Section 2 introduces related works. Section 3 describes the solution in detail. Section 4 analyzes its usability and security. Section 5 evaluates the solution using real datasets, followed by a conclusion in Section 6.

2   Related Works

Several recent works measure the guess ability on password[1]. Unfortunately, the entropy of these secrets is not high enough, especially in the background knowledge accessible to adversaries. In fact, it is very difficult that finding a secret which is easy to remember and of high entropy in massive users[1,2].

The Bayesian decision theory[3] is commonly used in information security areas such as privacy protection[4] and spam filtering[5]. Chen et al. propose BN-KBA[6,7] that is the first KBA scheme based on the Bayesian method. By comparison, password-based authentication[8] simply checks the consistency of passwords. BN-KBA assesses traditional personal information data such as Name and Credit Card which are independent for specific applications. The proposed approach focuses on application-dependent user data and so has different details, adopting the assumptions of BN-KBA and the Bayesian method.

3   KBA Based on Recommendation Data

3.1   Models

System Model. The KBA system in Fig.1 makes an authentication decision (true or false) according to knowledge sources (private data from recommender systems) and evidences from the claimant. The evidence is some pieces of knowledge or information for verifying the claimed identity from a claimant to the KBA system. The KBA server and the recommender system server are located on the trusted network. Threats to the trusted network belong to the focus of intrusion detection or network management and we think that the private data are secure enough. The recommender system server publishes public data for

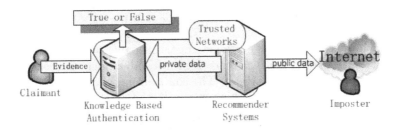

Fig. 1. System Model

providing online services. Public data reveal only the statistical information of private data. We think that these mechanisms are effective. For example, in a movie rating site such as MovieLens, every user can read the average rating of a movie and does not access exact ratings of other users.

Attack Model. We assume that all parties (users, the KBA server and the recommender system server) follow the semi-honest model. That is, an adversary (an imposter) outside the trusted network can acquire the statistical public data but not capture private personalized data. An imposter impersonates a real user by correctly guessing some private records of the user and the similar adversary is common in research works on password-based authentication and KBA.

Data Model. The private data in Fig. 1 involves m users and n knowledge sources. Using the Naive Bayes model[3], we adopt the independence assumption that knowledge sources of KBA are independent. We store and normalize the private data available for KBA and users, using the knowledge matrix R. Every row of R represents private records of a user i. The ith row vector r_i is denoted by $r_i = (r_{i1}, \cdots, r_{ij}, \cdots, r_{in})$. Every column of R represents records from a knowledge source j that is regarded as a discrete random variable x_j. The possible values of x_j is a finite set denoted by $V(x_j)$.

Definition 1 (the knowledge matrix,R). *the knowledge matrix $R = [r_{ij}]_{m \times n}$, where r_{ij} is the record that the user i knows about the knowledge source j.*

The procedure from private data to the matrix R is the original data pre-processing. In fact, we have original data from a movie rating site, a movie's rating is an integer from 1 to 5 which represents the level of favorite. Ratings of a movie from all users create a knowledge source on the movie. r_{ij} in R is the value which the user i has rated for the movie j.

3.2 Bayesian Decision

A claimant i^* claims that he is the identity i and provides an evidence vector $E = \{e_1, \cdots, e_n\}$ to KBA. An element e_j of E is an unverified record on the knowledge source x_j.

KBA asserts an authentication result A by matching E and R. A_{true} represents that KBA accepts the claimed identity and A_{false} means that KBA rejects one. KBA outputs A_{true} if $Pr(A_{true}|E) > Pr(A_{false}|E)$, according to the Bayesian decision rule for minimum error[3].

The posterior probability $Pr(A_{true}|E) = \frac{Pr(E|A_{true})Pr(A_{true})}{Pr(E)}$, according to Bayes' Rule[3]. Similarly, $Pr(A_{false}|E) = \frac{Pr(E|A_{false})Pr(A_{false})}{Pr(E)}$. Therefore, we can assert A_{true} by calculating the Likelihood Ratio (LR) as follow:

$$LR(E) \stackrel{\text{def}}{=} \frac{Pr(E|A_{true})}{Pr(E|A_{false})} > \frac{1 - Pr(A_{true})}{Pr(A_{true})} \stackrel{\text{def}}{=} threshold \tag{1}$$

A specific system gives $Pr(A_{true})$ and the threshold is a constant for a given $Pr(A_{true})$. If $LR(E) > threshold$, the class variable A equals A_{true}, which represents that KBA accepts the claimed identity. We can set a higher threshold for a higher assurance.

3.3   Estimation of Prior Probabilities

To calculate the likelihood ratio $LR(E)$ and the threshold in Equation 1 , we estimate three prior probabilities, $Pr(E|A_{true})$, $Pr(E|A_{false})$ and $Pr(A_{true})$.

1) $Pr(E|A_{true})$ is the memory availability probability that the real user i recalls or provides his knowledge correctly. If $e_j = r_{ij}$, $f(e_j, r_{ij}) = 1$ and otherwise $f(e_j, r_{ij}) = 0$. The random function $f(e_j, r_{ij})$ accords with the Bernoulli distribution $B(1, p_j)$ because the range of $f(e_j, r_{ij})$ is $\{0, 1\}$.

We can calculate $Pr(E|A_{true}) = \prod_{j=1}^{n} Pr(e_j|A_{true})$ with the time complexity of $O(m)$, under the independence assumption.

$$Pr(e_j|A_{true}) = Pr(f(e_j, r_{ij})) = \begin{cases} p_j & if \ e_j = r_{ij} \\ 1 - p_j & otherwise \end{cases} \qquad (2)$$

2) $Pr(E|A_{false})$ is the guessing probability that an adversary guesses the ith row vector, $\boldsymbol{r}_i = (r_{i1}, \dots, r_{ij}, \dots, r_{in})$, for impersonating the targeted identity i. The jth column vector $\boldsymbol{r_j}$ of R is regarded as the distribution of x_j. Because the possible values of the random variable x_j belong to the set $V(x_j)$, x_j is subject to the multinomial distribution $multinomial(a_j(t_1), \dots, a_j(t_{|V(x_j)|}))$, where $a_j(t)$ is the probability that a value $t \in V(x_j)$ appears in the jth column vector $\boldsymbol{r_j}$. We can calculate $a_j(t)$ using Equation 3 where $count(t)$ means the number of elements which equal t in the jth column vector $\boldsymbol{r_j}$.

$$a_j(t) = Pr(r_{ij} = t|j) = \frac{count(t)}{m} = \frac{|\{r_{ij}|r_{ij} = t\}|}{m}, t \in V(x_j) \qquad (3)$$

An imposter responds with the guessing random variable x_j^* which is subject to a multinomial distribution $multinomial(a_j^*(t_1), \dots, a_j^*(t_{|V(x_j)|}))$.

The knowledge source x_j's guessing ability $g(x_j)$ can be defined as the mathematical expectation of the probabilities that an imposter correctly guesses possible values of x_j according to [7]. Namely, $g(x_j) \overset{def}{=} E(a_j^*(t)) = \sum_{t \in V(x_j)} a_j(t)a_j^*(t)$. Intuitively, $g(x_j)$ means the average success rate of guessing.

Under the independence assumption, $Pr(E|A_{false}) = \prod_{j=1}^{n} Pr(e_j|A_{false})$ where we can estimate $Pr(e_j|A_{false})$ using $g(x_j)$ as follow.

$$Pr(e_j|A_{false}) = \begin{cases} g(x_j) & if \ e_j = r_{ij} \\ 1 - g(x_j) & otherwise \end{cases}, where \ g(x_j) = \sum_{t \in V(x_j)} a_j(t)a_j^*(t) \quad (4)$$

We can estimate $Pr(E|A_{false})$ under the Independent and Identically Distributed (IID) assumption [3] and the Rational Attacker (RA) assumption[6] respectively.

The IID Assumption. The evidences from the prospective imposters are still subject to the past distribution of x_j under this assumption. Namely in every column of R, $a_j^*(t) = a_j(t)$. Therefore we can calculate $g(x_j) = \sum_{t \in V(x_j)} a_j^2(t)$ on n knowledge sources using Algorithm 1 with the time complexity of $O(mn)$.

Algorithm 1. GAIID: Guessing Abilities under the IID Assumption

---

    Data: The knowledge matrix, $R = [r_{ij}]_{m \times n}$.
    Result: The guessing abilities: $g(x_1), \ldots, g(x_j), \ldots, g(x_n)$.
1 **for** $j \leftarrow 1$ **to** m **do**
2      $g(x_j) \leftarrow 0$;
3   **for** $t \in |V(x_j)|$ **do**
4          $count \leftarrow 0$;
5   **for** $i \leftarrow 1$ **to** n **do**
6   **if** t *equals* r_{ij} **then**
7                  $count \leftarrow count + 1$;
8          $a(t) \leftarrow count/n$;
9          $g(x_j) \leftarrow g(x_j) + a^2(t)$;
10 **return** $g(x_1), \ldots, g(x_j), \ldots, g(x_n)$;

The Rational Attacker Assumption. We can obtain the lower bound of $LR(\boldsymbol{E})$ by substituting Equation 2 and 6 into $LR(\boldsymbol{E}) = \frac{\prod_{j=1}^n Pr(e_j|A_{true})}{\prod_{j=1}^n Pr(e_j|A_{false})}$. According to Equation 5 that is a proved conclusion in [6] under the RA assumption, Equation 6 can be proofed as follow.

$$\frac{1}{|V(x_j)|} \le a_j^*(t) \le \max_{t \in V(x_j)} a(t) \tag{5}$$

$$Pr(e_j|A_{false}) \le \begin{cases} \max_{t \in V(x_j)} p(t) & if \ e_j = r_{ij} \\ 1 - \frac{1}{|V(x_j)|} & otherwise \end{cases} \tag{6}$$

Proof (Equation6). Note:(1)$g(x_j) = \sum_{t \in V(x_j)} a_j(t) a_j^*(t)$;(2)$\sum_{t \in V(x_j)} a_j(t) = 1$.

If $e_j = r_{ij}, Pr(e_j|A_{false}) = g(x_j) \le \sum_{t \in V(x_j)} a_j(t) \max_{t \in V(x_j)} a(t) = \max_{t \in V(x_j)} a(t)$.

If $e_j \ne r_{ij}, Pr(e_j|A_{false}) = 1 - g(x_j) \le 1 - \sum_{t \in V(x_j)} a_j(t) \frac{1}{|V(x_j)|} = 1 - \frac{1}{|V(x_j)|}$.

    3) $Pr(A_{true})$ is the probability that the real user initiates the authentication request. The maximum likelihood estimation of $Pr(A_{true})$ is a/b . The total number of authentication request is b and the total number of authentication success is a. We estimate values of a and b according to authentication statistics or security situation reports. According to equation 1, $Pr(A_{true}) = 1/(1 + threshold)$. Namely, there is one real identity in $1 + threshold$ claimants.

Knowledge Source Selection. Knowledge sources have a large uncertainty to provide enough information for distinguishing imposters and real identities from claimants. We can choose n entities of available knowledge sources denoted by the set S to achieve a larger entropy based on less ones. All possible choices create a n-dimensional vector space S^n. According to the maximum entropy principle, knowledge source selection is an optimization problem $x = \arg\max_{x \in S^n} H(x)$.

4   Analysis

Security. KBA needs enough data to distinguish real identities from claimants and is unable to prevent imposters who know shared secrets fully. KBA servers choose valid secret information and restrict the request rate. In Fig. 1, imposters can employ three attack strategies mentioned in [7]. The x_j's guessing variable x_j^* is subject to the distribution $multinomial(a^*)$ where $a^*(t)$ denotes the probability that the imposter chooses a value t from $V(x_j)$ by an strategy.

1) Using the sophisticated guessing, an imposter chooses t from $V(x_j)$ with the past probability distribution. In fact, this guessing strategy is the IID assumption as follow.

$$a^*(t) = a(t), \text{where } \forall t \in V(x_j) \tag{7}$$

2) Using the uniform guessing, an imposter chooses t from $V(x_j)$ with the same likelihood as follow. This uniform guessing implies the lower bound of $a_j^*(t)$ in Equation 5.

$$a^*(t) = 1/|V(x_j)|, \text{where } \forall t \in V(x_j) \tag{8}$$

3) Using the deterministic guessing, an imposter chooses a value of x_j^* whose probability is maximum in the distribution $multinomial(a)$ on $V(x_j)$ as follow. This strategy is optimal in guessing and indicates the upper bound of $a_j^*(t)$ in Equation 5.

$$x_j^* = \arg\max_{t \in V(x_j)} a(t) \tag{9}$$

Shannon entropy[9](Equation 11) is the metric of the uncertainty in information and the generalization of Shannon entropy is Renyi entropy[10](Equation 10) where the argument n is a nonnegative integer such as 0,1,2 and ∞.

$$H_n(x_j) = \frac{1}{1-n} \log(\|x_j\|_n) = \frac{1}{1-n} \log\left(\sum_{t \in V(x_j)} a_j^n(t) \right) \tag{10}$$

$$H_1(x_j) = \lim_{n \to 1} H_n(x_j) = - \sum_{t \in V(x_j)} a_j(t) \log(a_j(t)) \tag{11}$$

Three variants of Renyi entropy measure the three attack strategies. The collision entropy $H_2(x_j) = - \log(\sum_{t \in V(x_j)} a_j^2(t))$ measures the ability that the random variable x_j resists imposters with the sophisticated guessing. The Hartley entropy $H_0(x_j) = \log(|V(x_j)|)$ measures the ability that x_j resists the sophisticated guessing. The min-entropy $H_\infty(x_j) = - \log(\max_{t \in V(x_j)} a_j(t))$ measures the resistance to the deterministic guessing. The inequality, $H_0(x_j) \leq H_1(x_j) \leq H_2(x_j) \leq H_\infty(x_j)$, is proofed in [11].

Usability. KBA can verify claimed identities for real users correctly. The user provides correct authentication data via a challenge-response process which some human factors may affect. Therefore, the memory availability is extremely important. p_j in Equation 2 measures the memory availability of x_j for average users. The authentication system can improve the memory availability by carefully choosing reliable memory.

An available method is assistant applications which manage the authentication data to help users. e.g. When users rate the objects in recommender sites, the assistant software can record partial or whole user recommendation data in a trusted local device. In this way, users almost provide correct memory into KBA perfectly but rely on additional software.

Another method depends on the human memory unfaithful. We can help people recall the authentication information by designing a humanized user-interface. People have long-term memory to audiovisual narratives such as movies and DVDs[12], although they are not good at mechanical memory. Movie rating data associate long-term memory. e.g. We can recall easily whether a watched movie is favorite.

5   Experiments

This solution is implemented using JAVA and the program is executed in a laptop that has a 2GHz Core i7 CPU and 4GB RAM. We evaluate it with an anonymized corpus from a movie rating site, MovieLens. Every user has at least 20 ratings in the corpus. Two datasets from the sample are summarized as follow.

dataset	the number of users	the number of movies	the number of ratings
1	500	3,172	73,871
2	6,040	3,883	1,000,209

Performance in Different Datasets. We compared the results on two datasets as shown in Figure 2. The whole authentication procedure includes database operations. Figure 2(a) illustrates that the time cost for authenticating one user is proportional to the size of the dataset and the number of knowledge sources. For a given knowledge matrix $R = [r_{ij}]_{m \times n}$, the complexity of the whole process is $O(mn)$. The complexity of computing the memory abiltiy is $O(n)$ because we set all probabilities of memory ability as a constant. The complexity of the algorithm 1 is $O(mn)$.

For evaluating the classification performance we measure two main metrics, False Rejection Rate (FRR) and False Acceptance Rate (FAR). In general, FAR and FAR are as low as possible (lower is better) because it means that KBA is more accurate. Several important parameters affect this performance.

Figure 2(b) shows that the shapes of two FAR curves are similar on two different datasets, and so does FRR. Namely, the larger dataset does not improve classification performance significantly compared with the rapidly increasing time cost, although Dataset 2 is over ten times bigger than Dataset 1. Therefore,

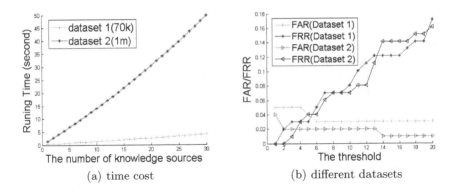

Fig. 2. Performance in Different Datasets

in practical KBA system, we can partition several smaller subdatabases from massive data to achieve sound time cost.

Effect of Threshold and Knowledge Sources. We adjusted the threshold in an appropriate range in the experiments. Figure 3(a) illustrates that FAR decreases slowly but FRR increases rapidly, with the increasing threshold. The sound range of the threshold is narrow. The optimal threshold on 10 knowledge sources is about 3 at Equal Error Rate (EER) and the threshold on 20 knowledge sources is about 5. However, the assigned threshold relies on the outside security situation. We employ other security mechanisms to ensure that there is at least one real identity in $1 + threshold$ claimants.

We can choose more knowledge sources to improve both FRR and FAR, comparing two subplots in Figure 3(a). However, more knowledge sources may increase user participation and thus weaken the whole system's usability. We trade off the number of knowledge sources and user-related usability.

Figure 3(b) illustrates that the memory availability is an extremely important parameter. This figure intuitively shows that the memory availability of the practical knowledge sources is larger than 0.8 and a knowledge source whose memory availability is less than 0.5 has no authentication ability. Some additional mechanisms can improve the memory availability.

Effect of Different Assumptions and Guessing Strategies. Figure 4(a) illustrates that the approach on the IID assumption is more accurate on verifying real identities according the FRR curves and on the contrary the approach on the Rational Attacker (RA) assumption is more accurate on distinguishing the active imposters according to the FAR curves.

Figure 4(b) illustrates that the approach on the RA assumption has extreme accuracy for distinguishing the imposter with the uniform guessing and the deterministic guessing. However, the approach on the RA assumption is slightly weak in the sophisticated guessing.

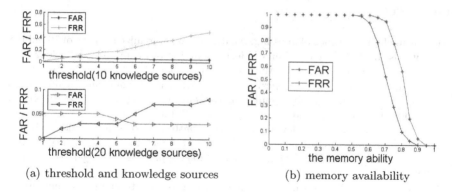

(a) threshold and knowledge sources   (b) memory availability

Fig. 3. Effect of Threshold, Knowledge Sources and Memory Availability

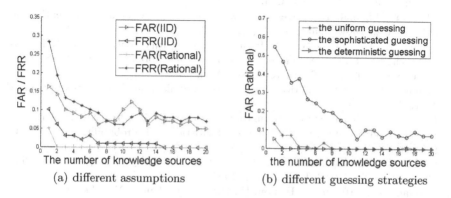

(a) different assumptions   (b) different guessing strategies

Fig. 4. Effect of Different Assumptions and Guessing Strategies

The result is consistent with experiences. Intuitively, the real users are more possible to be in line with the IID assumption and the active imposters are inclined to employ the deterministic guessing strategy.

6 Conclusion

We explore that how to authenticate users with abundant user-related rating data in recommender systems. The proposed scheme analyzes recommendation data using Naive Bayes. This paper makes a brief discussion on possible attack strategies and evaluates this scheme experimentally in the real-world datasets.

Acknowledgment. This work is partially supported by the HGJ National Significant Science and Technology Projects under Grant No. 2012ZX01039-004-009, Key Lab of Information Network Security, Ministry of Public Security under Grant No.C11606, the National Natural Science Foundation of China under Grant No. 61170263. We thank Shyong Lam and Jon Herlocker for their generating and releasing the original MovieLens dataset.

References

1. Bonneau, J.: The science of guessing: Analyzing an anonymized corpus of 70 million passwords. In: IEEE Symposium on Security and Privacy, pp. 538–552 (2012)
2. Kelley, P.G., Komanduri, S., Mazurek, M.L., Shay, R., Vidas, T., Bauer, L., Christin, N., Cranor, L.F., Lopez, J.: Guess again (and again and again): Measuring password strength by simulating password-cracking algorithms. In: IEEE Symposium on Security and Privacy, pp. 523–537 (2012)
3. Webb, A., Copsey, K., Cawley, G.: Statistical pattern recognition. Wiley (2011)
4. Dwork, C.: Differential privacy. In: Bugliesi, M., Preneel, B., Sassone, V., Wegener, I. (eds.) ICALP 2006, part II. LNCS, vol. 4052, pp. 1–12. Springer, Heidelberg (2006)
5. Caruana, G., Li, M.: A survey of emerging approaches to spam filtering. ACM Comput. Surv. 44(2), 9:1–9:27 (2008)
6. Chen, Y.: A bayesian network model of knowledge-based authentication. PhD thesis, Madison, WI, USA, AAI3261360 (2007)
7. Chen, Y., Liginlal, D.: Bayesian networks for knowledge-based authentication. IEEE Trans. Knowl. Data Eng. 19(5), 695–710 (2007)
8. Karaca, K., Levi, A.: Towards a framework for security analysis of multiple password schemes. In: Proceedings of the 1st European Workshop on System Security, EUROSEC 2008, pp. 16–21. ACM, New York (2008)
9. Shannon, C.: A mathematical theory of communication. Bell System Technical Journal (1948)
10. Renyi, A.: On measures of information and entropy. In: Proceedings of the 4th Berkeley Symposium on Mathematics, Statistics and Probability, pp. 547–561 (1960)
11. Cachin, C.: Entropy measures and unconditional security in cryptography. Ph.D Dissertation No. 12187. ETH Zürich (1997)
12. Furman, O., Dorfman, N., Hasson, U., Davachi, L., Dudai, Y.: They saw a movie: Long-term memory for an extended audiovisual narrative. Learning & Memory 14, 457–467 (2007)

Taxonomy Based Personalized News Recommendation: Novelty and Diversity*

Junyang Rao, Aixia Jia, Yansong Feng, and Dongyan Zhao

ICST, Peking University, China
{raojunyang,jiaaixia,fengyansong,zhaodongyan}@pku.edu.cn

Abstract. Recommender systems are designed to help users quickly access large volumes of information according to their profiles. Most previous works in recommender systems have put their emphasis on the accuracy of finding the most similar items according to a user's profile, while often ignoring other aspects that may affect users' experiences in practice, e.g., the novelty and diversity issues within a recommendation list. In this paper, we focus on utilizing taxonomic knowledge extracted from an online encyclopedia to boost a content-based personalized news recommender system without much human involvement. Given a recommendation list, we improve a user's satisfaction by introducing the taxonomy based novelty and diversity metrics to include novel, but potentially related items into the list, and filter out redundant ones. The experimental results show that the coarse grained knowledge resources can help a content-based news recommender system provides accurate as well as user-oriented recommendations.

Keywords: Personalized Recommender System, Novelty and Diversity, Taxonomy, Online Encyclopedia.

1 Introduction

Recommender systems, nowadays, have been recognized as one of the essential components in not only traditional e-commerce websites, e.g., books or goods recommendations in Amazon, but also various neat applications that facilitate people with a better access to the '*Big Data*', e.g., news articles, tweets, movies, music, games, research papers, friends, or even which jobs you should think over. The mystery about these recommender systems can be roughly categorized into the following streams: *collaborative filtering* and *content-based filtering*. There are also many attempts in practice to combine the two paradigms together in order to benefit from both sides.

Take personalized news recommendation as an example, the dominant paradigm is the *content-based* framework which can easily make predictions based on users' reading histories. Usually, a vector space model (i.e., the bag-of-words format) with the TF-IDF weighting [1] is utilized to represent a news article, and all articles read by a user are accumulated to model the user's profile. Taking background knowledge into

---

* This work is partially supported by the 863 Program (No. 2012AA011101) and the Natural Science Foundation of China (No. 61272344 and 61202233). Any question please refer to fengyansong@pku.edu.cn

X. Lin et al. (Eds.): WISE 2013, Part I, LNCS 8180, pp. 209–218, 2013.

consideration has been shown a good way to incorporate semantic analysis during recommendation so as to deal with *the mismatch of vocabulary* issue naturally associated with the bag-of-words format [2]. The potential knowledge resources can be in the form of finely crafted knowledge bases for a specific domain that would provide more accurate relation estimations but are costly to build. For example, [3, 4] use finely crafted ontologies to model user profiles and estimate the similarity between two words. On the other hand, [5] uses coarse grained taxonomies extracted from encyclopedia websites to render the similarity between words. These automatically built taxonomies may contain noises but can be easily obtained from the web without much human involvement.

In addition, most existing content-based models optimize their systems with regard to the item-user similarity solely and the quality of recommendations is only evaluated in terms of accuracy, while ignoring that there are still several issues that may affect user experiences in practice. In a real news recommender system, a user may not be happy to read a whole list of news articles focusing on only one single aspect about his/her interest, though this list could produce the highest similarity scores according to his/her profile. Previous studies [6–8] have argued that a more diverse recommendation list will increase the probability of items being chosen by the user. Those approaches address the diversity of a recommendation list for a given user through the dissimilarity between items, meantime, maintaining a relatively high level of similarities according to the user profile. *Novelty*, as an important quality dimension of a recommendation list, has received relative less attention, and mainly discussed in a bag-of-words format. All these issues discussed above demand more intensive investigations with regard to the *novelty* and *diversity* issues, especially on a semantic rich platform, which has already shown advantages in terms of capturing relevance [5].

In this paper, we will concentrate on utilizing coarse grained knowledge resources to improve personalized news recommender systems, in terms of not only the accuracy, but also real user experiences. We first harvest taxonomy knowledge resources from free online encyclopedia web sites without much human involvement and exploit these resources into a content-based news recommendation platform. We argue that the real user experience is as important as the recommendation accuracy, we thus explore the feasibility of making a recommendation output more user-oriented by taking its novelty and diversity into account on our coarse grained knowledge resources. Different from previous work, we model the two aspects as comparisons between news articles and a user profile on our coarse grained knowledge resources, where a recommendation list with a broad coverage of existing user interest points as well as new but potentially related points are encouraged in order to improve the user experiences.

2 Related Work

Novelty. Novelty is being identified as a fundamental quality of recommendation effectiveness and added-value in recent years, several approaches to assess the novelty dimension have been proposed which can be divided into two main categories: *Popularity-based Novelty* and *Distance-based Novelty* [9–11]. In general, *Popularity-based Novelty* is defined as a negative correlate to the click rate of the item. In *Distance-based Novelty*, the novelty of an item is modeled with respect to a set of items on a

Euclidean view, and it is defined as the *average* or *minimum distance* between the item at hand and the items in the set .

Zhang et. al. define another *Distance-based Novelty* in [7, 10, 11]. They definite the novelty that an item $i \in L$ brings to a set L as follows:

$$novelty_L^i = \frac{1}{p-1} \sum_{j \in L} d(i,j) \tag{1}$$

Assuming L is the set of items that a user likes, then the novel items by this definition correspond to the more unusual tastes of the user.

Diversity. Diversity is another important quality of recommender systems, especially in content-based recommendation techniques. Almost all of the techniques for diversity are proposed as the intra-distance of the set of items.

For example, in [6, 7, 12], diversity is modeled as the aggregate or equivalently average dissimilarity of all pairs of items in the set. Cai-Nicolas Ziegler et. al. introduce the *intra-list similarity metric* to assess the topical diversity of recommendation lists [8]. The *intra-list similarity* is defined as follows:

$$ILS(R) = \frac{\sum_{i \in R} \sum_{j \in R, i \neq j} c_o(i,j)}{2} \tag{2}$$

where R is the set to be recommended, and c_o, an arbitrary function measuring the similarity between items i and j, defined on a taxonomy in [13]. Hence, higher scores of *intra-list similarity* denote lower diversity.

3   User Profile Construction

3.1   Building a Taxonomy from Online Encyclopedia

It is known that there are many collaboratively edited, free encyclopedia sites available on the Web, which have been widely used as background knowledge resources in various research. Examples of general purposes include Wikipedia in English, and Hudong encyclopedia, Baidu encyclopedia in Chinese[1]. Here, in order to provide deep semantic analysis for news recommendation, we build a taxonomy with wide coverage automatically from Hudong encyclopedia, which hosts over 5 million concepts.

In the Hudong encyclopedia, each concept appears as a web page containing hyperlinks to other concepts; concepts of the same category are stored in the same directory, and all directories form a hierarchy. We thus construct the taxonomy as follows: 1) build the skeleton of the taxonomy by scanning the hierarchy of directories; 2) extract instances from every concept page and append them as leaf nodes of the taxonomy.

Compared to other manually built taxonomies, our taxonomy is admittedly noisy in nature, but saves time and human labors, and more importantly, it is encoded with general knowledge about the relationship between concept pairs with a wider coverage. A snapshot of the taxonomy is shown in Figure 1.

---

[1] http://www.wikipedia.org/, http://www.hudong.com/, http://baike.baidu.com/

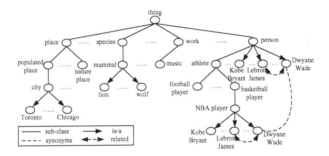

Fig. 1. A snapshot of our Hudong taxonomy: we translate all concept names into English for the ease of exposure, and use *related* to represent the actual relation names for brevity

3.2   User Profiling

With the background knowledge in hand, we are now ready to first model the news articles and user profiles.Formally, we assume this taxonomy contains a set of n concepts with their relations denoted as $Taxonomy^W = \{c_1^w, c_2^w, ..., c_n^w\}$.

For a given article, we only consider its concept words appearing in the taxonomy concept list(black circles in Figure 2), represented as $News = \{< c_1^n, w_1^n >, ..., < c_p^n, w_p^n >\}$, where $c_i^n \in Taxonomy^W$, w_i^n is the TF-IDF weighting of concept $c_i^n (1 \leq i \leq p)$, and p the number of concepts found in the article. For a given user, we construct the user profile by accumulating all concepts found in the articles that the user has read before(grey points in Figure 2), denoted as $User = \{< c_1^u, w_1^u >, ..., < c_q^u, w_q^u >\}$, where q is the total number of concepts found in the user's reading history, w_j^u is the average weighting of concept $c_j^u (1 \leq j \leq q)$ in the articles containing this concept and read by this user.

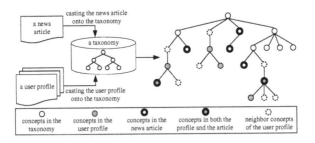

Fig. 2. The framework of User Profile Construction

4   Taxonomy-Based Novelty

Most existing recommender systems optimize their systems with regard to the item-user relevance, while ignoring the real user experiences. Now we will move a step further to improve the users' satisfaction by investigating the novelty and diversity issues involved in a given recommendation list. Here, we assume that an item is novel to a user if the two constraints are met: **1)** the user is not familiar with the item (it is not in the user's profile); **2)** the user is interested in the item (it is related to the user's profile).

As we can see in Figure 2, the dotted nodes around the user profile nodes (grey) on the taxonomy can be used to model the article's novelty to the user.

Concept-Concept Novelty. Given an taxonomy, we first define the novelty between two concepts as :

$$
nov(c_1, c_2) = \begin{cases} 0, & d = 0 \ or \ isSynonyms(c_1, c_2) \\ \dfrac{e^\delta}{e^\delta + 1} \cdot \dfrac{e^{-\lambda}\lambda^d}{d!}, & otherwise \end{cases} \tag{3}
$$

where, c_1 and c_2 are two concepts, d is the shortest distance from c_1 to c_2 on the taxonomy, while δ is the shortest distance from their lowest common ancestor to the root node of the taxonomy, λ is Poisson parameter. Basically, we design the novelty as a *Poisson function*: when two concepts are too close, the novelty are small; when their distance increases, the novelty will increase accordingly. But when they are too far to each other, the novelty should not increase.

In Formula 3, $\frac{e^\delta}{e^\delta + 1}$ is designed as a weight and will prefer more concrete concept pairs at a lower level, based on the assumption that two adjacent concrete concepts at the bottom of the taxonomy tend to maintain a high relevance level while being more easily accepted by users than those at a higher level.

News-User Novelty. Then the novelty of an article to a profile can be modeled as:

$$
nov(News, User) = \frac{1}{p} \sum_{i=1}^{p} \max_{1 \leq j \leq q} \{nov(c_i^n, c_j^u) \times w_{i,j}\} \tag{4}
$$

where $w_{i,j}$ is taken as a confidence of $nov(c_i^n, c_j^u)$,

$$
w_{i,j} = \frac{2}{1 + e^{k\tau}}, \ with \ \tau = \frac{abs(w_i^n - w_j^u)}{\max(w_i^n, w_j^u)}. \tag{5}
$$

We can see that when w_i^n and w_j^u are about equal, i.e., c_i^n and c_j^u have similar importance to their corresponding concept sets, the concept novelty $nov(c_i^n, c_j^u)$ will have a higher confidence ($w_{i,j}$). The smoothing factor k in Formula 5 is used to control the sensitivity of the confidence factor $w_{i,j}$. Larger k will lead to a more sensitive confidence function, which in turn penalize the concept pairs that have distinct importance in their own sets.

By substituting Formula 4 with Formula 3, we are able to compute the novelty of an article given a user profile.

5   Taxonomy-Based Diversity

A number of definitions of diversity have been proposed in the literature, which mainly model the diversity as the aggregated or average (equivalently) dissimilarity, and ignore the semantic relevance between two items. In this section, we introduce a taxonomy-based diversity between two items.

Given a recommendation list, we spread all concepts of these articles, as well as the user profile, onto the taxonomy and record their occurrences. We can see some seriously overlapped groups, such as the dotted oval in Figure 3; the more and bigger seriously overlaps we can find, the less diversity the candidate list has. Therefore, in order to allow diversity in the recommendation list, i.e., *diversifying* the list, we should reduce these seriously overlapped concepts.

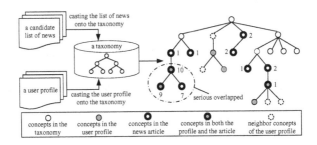

Fig. 3. The graphical illustration of the taxonomy based diversity

Concept-Concept Diversity We first define the diversity between two concepts as:

$$div(c_1, c_2) = \begin{cases} 0, & d = 0 \ or \ isSynonyms(c_1, c_2) \\ \frac{1}{e^{\delta}+1} \cdot (1 + \log_{2H} \frac{d}{2H}), otherwise \end{cases} \tag{6}$$

where H is the height of the taxonomy, $(1+\log_{2H} \frac{d}{2H})$ indicates that the further the two concepts are, the more diversity they allow to each other. δ is the same as in Formula (3): when the depth of their lowest common ancestor increases, the diversity between the two concepts gets smaller.

News-News Diversity. Then the diversity of two articles can be defined as the aggregated maximum diversity of the concepts in the two articles:

$$div(News, News') = \frac{1}{p} \sum_{i=1}^{p} \max_{1 \leq j \leq p'} \{div(c_i^{News}, c_j^{News'}) \times w_{i,j}\} \tag{7}$$

where p' is the number of concepts in $News'$, $w_{i,j}$ is the same as in Formula (4).

Finally, the diversity of a set of articles is defined as the equivalently average diversity of all pairs of articles in this set.

$$div(S) = \frac{2}{m(m-1)} \sum_{i \in S} \sum_{j \neq i \in S} div(i, j), \quad m = |S|. \tag{8}$$

The diversity of article i regarding to article list S is defined as:

$$div_S^i = div(S \cup \{i\}) - div(S) \tag{9}$$

6   Recommendation Generation

Given a ranked article list S (can be produced by any relevance model), and a given size K, we can improve the list in terms of novelty(diversity) by finding a subset with K items from S, which is optimized by balancing between relevance and novelty(diversity), shown in Algorithm 1 (replace val in Algorithm 1 with nov or div_S^i to improve the novelty or diversity, respectively).

Algorithm 1. RecommendationImpr

Require: Input: candidate set S, K items remained, threshold th.
           Output: improved result R for input set S.
1: sort items in S on $relevance$;
2: select top-K items from S and insert into R;
3: $pos \leftarrow K + 1$;
4: select the item i with minimal val in R;
5: **while** $i.relevance - S[pos].relevance \le th$ **do**
6:    **if** $i.val < S[pos].val$ **then**
7:       remove i from R;
8:       insert $S[pos]$ into R;
9:    $pos + +$;
10:   **if** $pos > |S|$ **then**
11:      break;
12:   select the item i with minimal val in R;
13: **return** R;

The main idea is to select the K articles with highest relevance scores from S into the recommendation list R, then swap the item with the lowest novelty(diversity) score in R with the article with next highest relevance score from the remaining items of S. To maintain a high level relevance for R, a pre-defined threshold th can be used to stop the swapping when the highest relevance score of the remaining items in S is no longer high enough.

7   Evaluation

We build a news recommendation platform (NRS) to conduct our experiments, which have about 1,080,000 news articles from Sina News (http://news.sina.com.cn) and 581 users, about 400 have read more than 10 articles. Our taxonomy is constructed with 5 million entries extracted from Hudong encyclopedia.

Novelty. We evaluate the proposed novelty computation following [14]'s method: 1) For each user, we randomly select 10 topics from the news pool, and for each topic, the NRS randomly selects 10 articles as a group. Within each group, we ask every user (20 users in total, all native Chinese speakers) to read the 10 articles sequentially, marks whether the current one is redundant or not compared to other articles he/she previously read in this group, and if yes, record the redundant pairs. 2) For each group, the

annotated data is randomly split into two sets: 40% for profiling , 60% for recommendation. 3) Aggregate the profile/test sets from all groups, respectively, to form the finally
profiling set and testing set. 4) For each user, a relevance model generates a recommendation list R, then we will check each article in the list whether it is redundant (either
compared to this user's profiling set, or compared to other articles in the list R). The
percentage of redundant news articles in the recommendation list R is used as the metric of *redundancy*. And calculate the *score* of the list R with Formula 10. 5) Repeat
Step 2 – Step 4 100 times for each user.

We optimize the parameter λ in Formula 3 and th in Algorithm 1 to maximize *score*,
the linear combination of relevance and redundancy:

$$score(R) = \alpha * relevance(R) + (1 - \alpha) * (1 - redundancy(R)) \quad (10)$$

where α is treated as a human-oriented factor and can be edited by users, we set $\alpha = 0.7$
in our experiments.

We search the parameters (λ, th) using grid search with cross-validation, where $\lambda \in
[0, 15]$, and $th \in [0.0, 1.0]$. When $\alpha = 0.7$, the value of *score* with respect to different
λ and th is shown in Figure 4, and the optimized parameters (λ, th) are set to 2 and 0.3,
respectively.

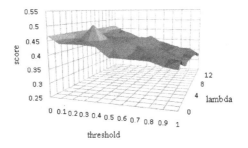

Fig. 4. The values of *score* with respect to different λ and th when $\alpha = 0.7$

To evaluate the proposed the novelty based improvement, we compare two types of
recommendation lists produced by a taxonomy based relevance recommendation model
[5] and its improved version modified by our proposed novelty computation. We generated 436 recommendation lists in total. The redundancy rates of the two recommendation lists generated by the relevance model (denoted as *relevance*) and the novelty
improved lists (denoted as *relevance + novelty*) are compared in Figure 5. Most recommendation lists, generated by the relevance model solely, have redundant rates larger
than 0.6. When the recommendation lists are improved with the taxonomy-based novelty, more than half of these recommendation lists have a redundant rate below 0.3. The
result shows that, when the relevance of a recommendation list is ensured (the average
difference between two types of lists is 0.03) , the proposed method can reduce the
redundancy effectively.

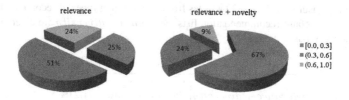

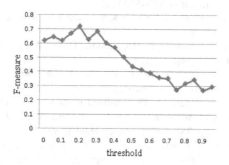

Fig. 5. The distribution of redundancy rates of recommendation lists generated by the default relevance model and its improved version with our proposed taxonomy-based novelty

Diversity. In our human experiment, we have 20 participants in total, all native Chinese speakers; each reader has to read 200 news articles randomly provided by NRS and indicates each story whether it is interesting or not. The obtained data are divided into two parts: one for parameter tuning and the other for recommendation. The latter is randomly split into two sets: 40% for profiling , 60% for testing. For each participant, the profiling set is used to construct the user's reading profile. For each recommendation list, we compute the precision and recall for the top 20 items as well as their F-measure, and repeat the process 100 times by randomly splitting the profile/testing sets and calculate the average performance. We tune the parameter th^2 on a range of $[0, 1]$. As we

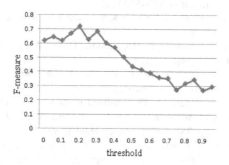

Fig. 6. The values of $F - measure$ with respect to different ths

can see in Figure 6, the results of F-measure indicate that when $th = 0$, the step 5 in Algorithm 1 would never be executed, the recommendation list R is the default recommendation without diversification. Higher th makes the list R more diverse while keeping a higher F-measure. However, when the value of th is larger than 0.35, more unrelated news articles are inserted into R to increase the diversity, but, at the cost of F-measure decreasing by a large margin.

We then set the parameter $th = 0.2$ when used in Algorithm 1 for diversifying a list. The recommendation performance with/without diversity is shown in Table 1. We can see that the proposed diversifying process has significantly improved the default recommendation list.

---

[2] Note: this threshold is different from the one used in the novelty experiments.

Table 1. The F-measures of the top 20 articles from two different types of recommendation lists, *relevance* are the default recommendation lists, *relevance* + *diversity* are diversified by our proposed taxonomy-based method

	Precision	Recall	F-measure
relevance	65.67%	58.01%	61.60%
relevance + *diversity*	78.00%	65.50%	71.21%

8   Conclusion

In this paper, we exploit taxonomic structures encoded with background knowledge to boost the personalized news recommendation. Importantly, our taxonomy is automatically constructed from free online encyclopedia websites without intensive human involvement. Our model considers the issues of novelty and diversity in a taxonomy environment. The experimental results show that the taxonomic knowledge resources can help a content-based news recommender system provide accurate as well as user-oriented recommendations.

References

1. Salton, G., Buckley, C.: Term-weighting approaches in automatic text retrieval. Information Processing & Management 24(5), 513–523 (1988)
2. Salton, G., Wong, A., Yang, C.S.: A vector space model for automatic indexing. Commun. ACM 18(11), 613–620 (1975)
3. IJntema, W., Goossen, F., Frasincar, F., Hogenboom, F.: Ontology-based news recommendation. In: EDBT/ICDT Workshops (2010)
4. Hliaoutakis, A., Varelas, G., Voutsakis, E., Petrakis, E.G.M., Milios, E.E.: Information retrieval by semantic similarity. Int. J. Semantic Web Inf. Syst. 2(3), 55–73 (2006)
5. Rao, J., Jia, A., Feng, Y., Zhao, D.: Personalized news recommendation using ontologies harvested from the web. In: Wang, J., Xiong, H., Ishikawa, Y., Xu, J., Zhou, J. (eds.) WAIM 2013. LNCS, vol. 7923, pp. 781–787. Springer, Heidelberg (2013)
6. Smyth, B., McClave, P.: Similarity vs. diversity. In: Aha, D.W., Watson, I. (eds.) ICCBR 2001. LNCS (LNAI), vol. 2080, pp. 347–361. Springer, Heidelberg (2001)
7. Zhang, M., Hurley, N.: Avoiding monotony: improving the diversity of recommendation lists. In: RecSys, pp. 123–130 (2008)
8. Ziegler, C.N., McNee, S.M., Konstan, J.A., Lausen, G.: Improving recommendation lists through topic diversification. In: WWW, pp. 22–32 (2005)
9. Vargas, S., Castells, P.: Rank and relevance in novelty and diversity metrics for recommender systems. In: RecSys, pp. 109–116 (2011)
10. Hurley, N., Zhang, M.: Analysis of methods for novel case selection. In: ICTAI (2), pp. 217–224 (2008)
11. Zhang, M., Hurley, N.: Novel item recommendation by user profile partitioning. In: Web Intelligence, pp. 508–515 (2009)
12. Zhang, M.: Enhancing diversity in top-n recommendation. In: RecSys, pp. 397–400 (2009)
13. Ziegler, C.N., Lausen, G., Schmidt-Thieme, L.: Taxonomy-driven computation of product recommendations. In: CIKM, pp. 406–415 (2004)
14. Zhang, Y., Callan, J., Minka, T.: Novelty and redundancy detection in adaptive filtering. In: Proceedings of the 25th Annual International ACM SIGIR Conference on Research and Development in Information Retrieval, SIGIR 2002, pp. 81–88. ACM, New York (2002)

Distinguishing Social Ties in Recommender Systems by Graph-Based Algorithms

Xiaochi Wei, Heyan Huang, Xin Xin, and Xianxiang Yang

School of Computer Science and Technology
Beijing Institute of Technology, Beijing, China
{wxchi,hhy63,xxin,yangxianxiang}@bit.edu.cn

Abstract. Incorporating the social network information into recommender systems has been demonstrated as an effective approach in improving the recommendation performance. When predicting ratings for an active user, his/her taste is influenced by the ones of his/her friends. Intuitively, different friends have different influential power to the active user. Most existing social recommendation algorithms, however, fail to consider such differences, and unfairly treat them equally. The problem is that the friends with less influential power might mislead the rating predictions, and finally impair the recommendation performance. Some previous work has tried to differentiate the influential power by local similarity calculations, but it has not provided a systematic solution and it has ignored the propagation of the influence among the social network. To solve the above limitations, in this paper, we investigate the issue of distinguishing different users' influence power in recommendation systematically. We propose to employ three graph-based algorithms (including PageRank, HITS, and heat diffusion) to distinguish and propagate the influence among the friends of an active user, and then integrate them into the factorization-based social recommendation framework. Through experimental verification in the Epinions dataset, we demonstrate that the proposed approaches consistently outperform previous social recommendation algorithms significantly.

Keywords: Recommender Systems, Social Network, Collaborative Filtering, Graph-based Algorithms.

1 Introduction

As an indispensable technique to overcome the information overload problem on the Web, the research of recommender systems has been investigated deeply over the decades. Collaborative filtering plays the key role in recommender systems. It predicts the rating for each user-item pair by mining common behavior patterns from the historical log. The technique of recommender systems has successfully enforced the development of the Web applications, such as Amazon, Netflix, Youtube, and etc.

Recently, incorporating the social network information has been demonstrated effective in improving the recommendation performance. The underlining assumption is that the taste of an active user is not only determined by himself/herself, but also by his/her trusted friends. Figure 1 shows an example of the social network of an active user u. In the figure, if two users are friends, there will be a link between them, which is

X. Lin et al. (Eds.): WISE 2013, Part I, LNCS 8180, pp. 219–228, 2013.
© Springer-Verlag Berlin Heidelberg 2013

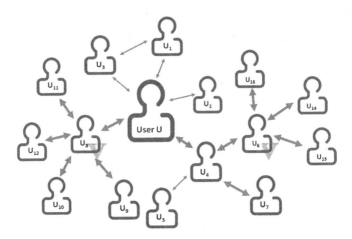

Fig. 1. An Example of Social Graph in Recommendation

also called a "social tie".It can be observed that u_1, u_2, u_3, u_4 and u_8 are his/her trusted friends. In factorization-based social recommendation algorithms, latent features of the active user will be correlated with the ones of these friends, which achieves significant improvements for the recommendation performance [8].

Originally, most social recommendation algorithms unfairly treat each user's friends equally. In these algorithms, u_1, u_2, u_3, u_4 and u_8 have the same influence power to the active user in Fig. 1. Nevertheless, this might not be accurate in reality. For instance, the active user might be more likely to trust the opinions of an expert rather than the ones of others. In Fig. 1, u_8 is an expert user (denoted by "V"). Thus its impact should be larger than the ones of u_1, u_2, u_3 and u_4. If recommendation algorithms do not distinguish the strength of the influential power, the impacts from u_1, u_2, u_3 and u_4 might mislead the estimation of the active user's latent feature vector, and consequently, mislead the performance of rating predictions.

Some previous work has tried to distinguish the influential power by simply calculating the local similarity between the user and his/her friends, such as Pearson correlation coefficient (PCC) [10], and etc. The problem is that the influence propagation among the social network is ignored in these algorithms. Nevertheless, in estimating the influence among the social network, the propagation is indeed an important factor. For example, in Fig. 1, if the propagation is ignored, the active user will be independent with u_6; and if only the local similarity is calculated, the influence power of u_2 and u_4 might be similar. But in reality, u_6 may also influence the active user indirectly by the propagation through u_4; and therefore u_4 is much likely to have more influence power to the active user than u_2, as it absorbs u_6's expertise. Consequently, the local similarity alone is not enough for accurately estimating the social influence among the users, and the propagation in the network could not have been ignored.

The above limitations naturallymotivate the investigation of graph-based algorithms for distinguishing the influential power of different users in social recommendations. Graph-based algorithms differentiate each user based on the information propagation in the network. Thus the above limitations could be solved. Particularly, we choose three graph-based algorithms, including PageRank, HITS and heat diffusion. We utilize the probabilisticmatrix factorization (PMF) [9] as our recommendation framework, and incorporate graph-based algorithms to distinguish the social ties between different users. Through experimental verification, it is demonstrated that our methods are very effective in improving the recommendation performance.

2   Related Work

2.1   Collaborative Filtering

Collaborative filtering is a kind of significant approaches in recommender systems. It mainly contains two types, the neighbor-based method and the model-based method. The neighbor-based method utilizes the similar user or item to predict the ratings [6,5]. It is deeply investigated in research communities, and is also widely utilized in industry [5], such as Amazon and Netflix. The model-based method relies on machine learning techniques, which trains models by utilizing the observed user-item rating matrix and then makes predictions via the trained models instead of directly calculating the ratings [14,2].

2.2   Collaborative Filtering with Social Information

Social-based recommender systems [11,8] are proposed to overcome the defect of the assumption that users are independent and the neglect of the social relationship among users. It utilizes the information of trusted friends to represent the active user. Xin et al. proposed a social recommendation model to overcome the sparseness of data [11]. In recent years, Ma et al. [10] proposed two social recommendation methods with social regularization terms to improve the accuracy of recommender systems.

Unfortunately, most previous work only treats the social ties among users as equal binary values [9], which could not describe the social ties accurately and might make the friends with less influential power impair the recommendation performance in some cases. Although some work simply employs local similarities, such as PCC, to distinguish the social ties [10], it has omitted the propagation of the influential power in the social graph, which cannot describe the social ties accurately either.

2.3   Graph-Based Algorithms

HITS [4] and PageRank[1] are two prominent graph-based algorithms to search authority nodes. They all have been deeply investigated in many areas [3,12]. Chakrabarti et al. [3] have utilized HITS for automatically compiling resource lists for general topics. Xing et al. [12] have proposed weighted PageRank algorithm, which takes into account the importance of both in-links and out-links and distributes rank scores based on the popularity of the pages. Different from HITS and PageRank, heat diffusion calculates the correlation among different nodes in the graph by utilizing the physical heat diffusion process. Recently, heat diffusion based approaches have been widely used in various aspects [13,7]. Yang el al. [13] has proposed a ranking algorithm with heat diffusion process. Ma et al. [7] has proposed heat diffusion models on social network to describe the diffusion of the influence among people.

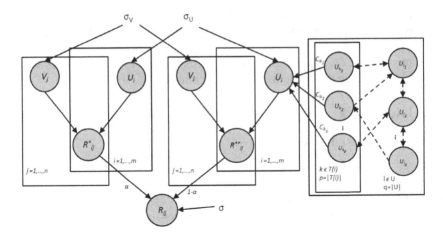

Fig. 2. Recommendations with Social Trust Ensemble

3   Probabilistic Matrix Factorization Framework

3.1   Probabilistic Matrix Factorization with Social Information

PMF is an effective approach to recommender systems. It factorizes the user-item rating matrix R into two low-dimensional matrices, user latent matrix U and item latent matrix V.

Social recommendation approaches are proposed to describe the social information in recommender systems. They fuse the information of the friends of the users into the models. So both the facts of the users and the friends should be reflected in the rating $R_{i,j}$. Therefore the objective function $L(R, S, U, V)$ is modeled as

$$
\begin{aligned}
L(R, S, U, V) &= \\
&= \frac{1}{2} \sum_{i=1}^{m} \sum_{j=1}^{n} I_{i,j}^{R} (R_{i,j} - g(\alpha U_i^T V_j + (1 - \alpha) \sum_{k \in T(i)} S_{i,k} U_k^T V_j))^2 \\
&+ \frac{\lambda_U}{2} \|U\|_F^2 + \frac{\lambda_V}{2} \|V\|_F^2,
\end{aligned}
\tag{1}
$$

where $S = \{S_{i,j}\}$ is a matrix representing the social trust between users, and T_i represents the set of friends of user u_i. The objective function could be solved by gradient descent algorithms.

3.2   Probabilistic Matrix Factorization with Distinguished Social Information

PMF with distinguished social information is the method we proposed to solve the defect of previous social recommendation methods, which only consider the social information as binary values. Though some previous research distinguishes the friends by simply using local similarity such as PCC, they can hardly consider the influence of the whole social graph and the propagation of the influential power. In our approaches, graph-based algorithms are employed to determine the trust value, and we could get a trust rank from them. Thus, the objective function $L(R, S, U, V, G)$ is defined as

$$
\begin{aligned}
L(R, S, U, V, G) = \\
= \frac{1}{2} \sum_{i=1}^{m} \sum_{j=1}^{n} I_{i,j}^{R} (R_{i,j} - g(\alpha U_i^T V_j + (1-\alpha) \sum_{k \in T(i)} C_{i,k}(S, G) U_k^T V_j))^2 \\
+ \frac{\lambda_U}{2} \|U\|_F^2 + \frac{\lambda_V}{2} \|V\|_F^2,
\end{aligned}
\tag{2}
$$

where $C_{i,j}(S, G)$ is the trust value we utilized to describe the influential power of u_j to u_i, as it is shown in Fig. 2. Considering that in real world the influence to an active user is dominated by only a small percent of his/her total friends, we employ the power-low distribution $C_{i,j}(S, G) = e^{-a \log r_{i,j}}$ ($r_{i,j}$ is the rank of the user u_j in the friend list of user u_i which could be obtained from the graph-based algorithms) to describe the trust value.

4 Graph-Based Algorithms

4.1 PageRank

PageRank assumes that the influence of the social information is randomly transferred in the social graph based on the theory of Markov chain. In a directed graph $G(V, E)$, V is the set of nodes, and E represents the set of links. Each link from v_i to v_j is denoted by (v_i, v_j). So the Markov chain contains $|V|$ states, with an $|V| \times |V|$ transition probability matrix $\mathbf{P}$. So we could get the next state $\mathbf{X}_{k+1} = \mathbf{P} \mathbf{X}_k$ from the current state $\mathbf{X}_k$, where

$$
P_{i,j} = \begin{cases} 1/O_i, & (v_i, v_j) \in E, \\ 0, & otherwise. \end{cases}
\tag{3}
$$

Based on the ergodic theorem, a unique steady probability distribution is obtained after several steps. And the trust rank is obtained by the PageRank values of each user.

4.2 HITS

Different from PageRank, HITS gives two different types of relevance, Authorities and Hubs. A great hub node has many out-links to authority nodes and a great authority node is linked to by many great hub nodes. Similar to PageRank, a directed graph $G(V, E)$ is employed, Each node is given an authority score $a(v_i)$ and a hub score $h(v_i)$, and they are respectively represented as

$$
a(v_i) = \sum_{j:(v_j, v_i) \in E} h(v_j), \quad h(v_i) = \sum_{j:(v_i, v_j) \in E} a(v_j).
\tag{4}
$$

We express it into a matrix form, and after substitution we could solve it with the similar method with PageRank

$$\mathbf{a}_{k+1} = \mathbf{L}^T \mathbf{L} \mathbf{a}_k, \quad \mathbf{h}_{k+1} = \mathbf{L} \mathbf{L}^T \mathbf{h}_k, \tag{5}$$

where

$$L_{i,j} = \begin{cases} 1, & (v_i, v_j) \in E, \\ 0, & otherwise. \end{cases} \tag{6}$$

Similar to PageRank, the rank list is also obtained based on the authority score.

4.3 Heat Diffusion

We utilize heat diffusion to describe correlations among different users, which is inspired by the physical phenomenon that the heat always flows from a high temperature position to a low temperature position. We describe the process in a directed graph $G(V, E)$, V is the set of all users in the social graph, and E represents the set of pipes. Each pipe from v_i to v_j is denoted by (v_i, v_j), meaning that v_j trusts v_i and the heat flows only from v_i to v_j. By using diffusion process, the node v_i receives $\sum_{j:(v_j, v_i) \in E} \alpha f_j(t) \Delta t / d_j$ amount of heat from all its neighbors. α is the heat diffusion coefficient; d_j is the out degree of node v_j; and $f_j(t)$ is the heat at node v_j at time t. At the same time, v_i will also diffuse $\alpha f_i(t) \Delta t / d_i$ amount of heat to each of its subsequent nodes. To sum up, the value of the heat node v_i changed from time t to time $t + \Delta t$ is formulated as

$$\frac{f_i(t + \Delta t) - f_i(t)}{\Delta t} = \alpha(-\tau_i f_i(t) + \sum_{j:(v_j, v_i) \in E} \frac{f_j(t)}{d_j}), \tag{7}$$

where τ_i is the indicator function to identify whether node v_i has any out-links or not. We express it into a matrix form and in the limit $\Delta t \to 0$, it becomes

$$\frac{d}{dt} \mathbf{f}(t) = \alpha \mathbf{H} \mathbf{f}(t), \tag{8}$$

where

$$H_{i,j} = \begin{cases} 1/d_j, & (v_j, v_i) \in E, \\ -\tau_i, & i = j, \\ 0, & otherwise. \end{cases} \tag{9}$$

This differential equation could be solved easily. In the consideration of the high time complexity, a discrete approximation is employed to calculate the formula [7]

$$\mathbf{f}(t) = (\mathbf{I} + \frac{\alpha t}{P} \mathbf{H})^P \mathbf{f}(0), \tag{10}$$

where P is a positive integer. By using heat diffusion, the value of the diffused heat indicates the trust value among users. In order to get the rank of the trust value, for each user v_i, we set $\mathbf{f}(0) = \mathbf{0}$, and $f_i(0) = h_0 > 0$, then execute the heat diffusion process, and finally obtain $\mathbf{f}(t)$. In $\mathbf{f}(t)$, $f_j(t)$ with a higher score means v_i is much easier to influence v_j. So we get the rank $\mathbf{r}(i)$, which indicates who is more likely to be influenced by v_i. We utilize $r_j(i)$ to approximate the rank of v_i in the trust list of v_j.

Table 1. Dataset Statistics

Statistics	User	Item	Trust	Be Trusted
Min. Num.	8	2	1	1
Max. Num.	442	910	1,060	1,135
Avg. Num.	34.70	34.53	35.79	35.79

Table 2. Overall Performance Comparisons

Training Data	Metrics	D=5						
		PMF	Social	PCC	PageRank	HITS	HD	PCC-HD
90%	MAE	1.1000	1.0413	1.0395	**1.0325**	1.0344	1.0383	1.0354
	RMSE	1.3375	1.2953	1.2950	**1.2894**	1.2905	1.2938	1.2895
80%	MAE	1.1292	1.0410	1.0394	1.0378	1.0350	1.0361	**1.0342**
	RMSE	1.3498	1.2975	1.2963	1.2925	1.2912	1.2900	**1.2853**
Training Data	Metrics	D=10						
		PMF	Social	PCC	PageRank	HITS	HD	PCC-HD
90%	MAE	1.0978	1.0565	1.0532	1.0406	1.0409	1.0401	**1.0384**
	RMSE	1.3590	1.3182	1.3238	1.3111	1.3103	1.2809	**1.2798**
80%	MAE	1.0986	1.0505	1.0491	1.0483	1.0477	1.0469	**1.0450**
	RMSE	1.3751	1.3500	1.3247	1.3186	1.3210	1.3175	**1.3170**

5   Experiments

5.1   Dataset

We used Epinions[1] as the dataset of our experiments. Users could share their opinions on different products and rate them from 1 star to 5 stars. The ratings could be reviewed and influence others' opinion on the product. On Epinions, each user has a trust list, which is a set of directed links from the user to the trusted ones.

The dataset we used in our experiments consists of 49,289 users who rated a total of 139,738 different items at least once. The number of rating is 664,824. The density of the user-item matrix is less than 0.01%. The dataset also contains 487,183 issued trust statement. In order to make the experiment more efficient, we chose 5,095 active users and 5,206 hot items. The statistics of the data source is shown in Table 1.

5.2   Baselines

In our experiments, in order to show the performance of our approaches, we made comparisons with PMF, social-based PMF, social-based PMF with PCC and our graph-based PMF .

1. PMF is the most commonly used approach, which only factorize the User-Item matrix into User latent matrix and Item latent matrix to make the prediction.

2. Social-based PMF is a traditional recommender system fusing social information, it treats the friends of a user equally.

---

[1] www.epinions.com

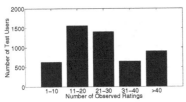

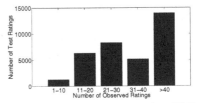

(a) Distribution of Test Users(80% as Train-   (b) Distribution of Test Ratings(80% as
ing Data)   Training Data)

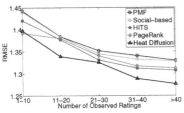

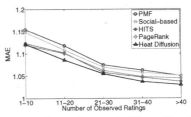

(c) RMSE Comparison on Different User   (d) MAE Comparison on Different User Rating
Rating Scales (80% as Training Data)   Scales (80% as Training Data)

Fig. 3. Performance Comparison on Different Users

3. Social-based PMF with PCC is another social-based method, it utilizes PCC to evaluate the weight of trust value $S_{i,j}$.

4. Graph-based PMF is our approach, 3 graph algorithms are used in our experiments, PageRank, HITS and Heat Diffusion.

5. Additionally, in order to illustrate the contribution of distinguishing the social ties in the process of influence flowing, we utilized PCC to distinguish the heat diffusion coefficient α in the heat diffusion process and named it PCC-HD.

We randomly selected 90% and 80% of all the ratings as the training set, and the remaining as the test set. In the experiment we used 5 dimensions and 10 dimensions respectively to describe latent factors. The parameter α was set to 0.8, and $\lambda_U = \lambda_V = 0.001$. The Mean Absolute Error (MAE) and the Root Mean Square Error (RMSE) were used to measure the effectiveness of our recommendation methods.

5.3   Overall Performance

Experimental results are shown in Table 2. It can be observed that, among the methods, social recommendation methods generally perform better than the method without social information (PMF), which is consistent with previous work. Our social recommendation methods with distinguished social ties outperform the ones without distinguished social ties. Within the our four recommendation algorithms, PCC-HD performs better than other graph-based methods, especially the heat diffusion without distinguishing social ties. So it is effective in distinguishing the social ties in influence propagation process. In order to illustrate our approaches perform well on different rating number users, we divided the users into five groups based on the number of ratings in the

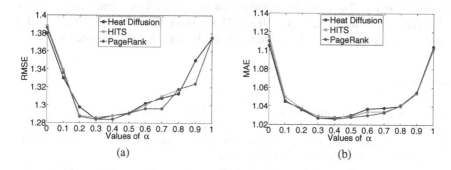

Fig. 4. Impact of Parameter α (Dimensionality=10)

training set, and predicted the ratings on each group. Fig. 3 shows the results of the experiment. The number of ratings is respectively $1 - 10, 11 - 20, 21 - 30, 31 - 40$ and > 40. The user distribution is shown in Fig. 3(a) and the rating distribution is shown in Fig. 3(b). Fig. 3(c) shows the RMSE of each group(80% as training data) with five different methods, and Fig. 3(d) shows the MAE of each group(80% as training data) with five different methods. It is clear that our graph-based PMF models perform well on different groups, especially when there are little observed ratings of each user.

5.4 Parameter Analysis

In our graph-based method, the parameter α is used to balance the properties of the user himself and the properties of his friends. In this experiment, we set α from 0 to 1 with the step of 0.1 to observe the performance change with the change of α. In our approaches, $\alpha = 1$ means the model only contains the user-item matrix itself without any social information, and $\alpha = 0$ means the model only contains social information. In the experiment we used 80% percent training data. The results are shown in Fig. 4(a) and Fig. 4(b) which indicates that the value of α could change the performance of our models by balancing the percentage of the user's own property and the social information. The optimal value of the parameter α is around 0.3.

6 Conclusion

In this paper, we have proposed a graph-based model for recommender systems. The purpose of our work is to distinguish the social ties of each user to make the social graph fused in recommender systems more accurately, rather than to unfairly treat each social tie equally. We have utilized PageRank, HITS and Heat Diffusion algorithms in our methods. Experiments conducted have illustrated that the proposed approaches are effective on different users with different observed number of ratings. Compared with PMF and previous social recommendation methods, our approaches have achieved a significant improvement under different configurations.

Acknowledgments. The work described in this paper was fully supported by the National Basic Research Program of China (973 Program, Grant No. 2013CB329605). The authors also would like to thank the reviewers for their helpful comments.

References

1. Brin, S., Page, L.: The Anatomy of a Large-Scale Hypertextual Web Search Engine. Computer Networks and ISDN Systems J. 30, 107–117 (1998)
2. Canny, J.: Collaborative Filtering with Privacy via Factor Analysis. In: Proceedings of SIGIR 2002, pp. 238–245. ACM Press, New York (2002)
3. Chakrabarti, S., Dom, B., Gibson, D., Kleinberg, J., Raghavan, P., Rajagopalan, S.: Automatic Resource Compilation by Analyzing Hyperlink Structure and Associated Text. Computer Networks and ISDN Systems 30(1), 65–74 (1998)
4. Kleinberg, J.: Authoritative sources in a hyperlinked environment. In: 9th ACM-SIAM Symposium on Discrete Algorithms, pp. 604–632. ACM Press, New York (2008)
5. Linden, G., Smith, B., York, J.: Amazon.com Recommendations: Item-to-item Collaborative Filtering. IEEE Internet Computing J. 7(1), 76–80 (2003)
6. Ma, H., King, I., Lyu, M.R.: Effective Missing Data Prediction for Collaborative Filtering. In: Proceedings of SIGIR 2007, pp. 39–46. ACM Press, New York (2007)
7. Ma, H., Yang, H.X., Lyu, M.R., King, I.: Mining Social Networks Using Heat Diffusion Processes for Marketing Candidates Selection. In: Proceedings of CIKM 2008, pp. 233–242. ACM Press, Napa Valley (2008)
8. Ma, H., Yang, H., Lyu, M.R., King, I.: SoRec: Social Recommendation Using Probabilistic Matrix Factorization. In: Proceedings of CIKM 2008, pp. 931–940. ACM Press, New York (2008)
9. Ma, H., King, I., Lyu, M.R.: Learning to Recommend with Social Trust Ensemble. In: Proceedings of SIGIR 2009, pp. 203–210. ACM Press, Boston (2009)
10. Ma, H., Zhou, D.Y., Liu, C.: Recommender System with Social Regularization. In: Proceedings of WSDM 2011, pp. 287–296. ACM Press, Hong Kong (2011)
11. Xin, X., King, I., Deng, H., Lyu, M.R.: A social recommendation framework based on multiscale continuous conditional random fields. In: Proceedings of CIKM 2009, pp. 1247–1256. ACM Press, Hong Kong (2009)
12. Xing, W., Ghorbani, A.: Weighted Pagerank Algorithm. In: 2nd Annual Conference on Communication Networks and Services Research, pp. 305–314. IEEE Press, Fredericton (2004)
13. Yang, H., King, I., Lyu, M.R.: Diffusionrank: a possible penicillin for web spamming. In: Proceedings of SIGIR 2007, pp. 431–438. ACM Press, Amsterdam (2007)
14. Zhang, Y., Koren, J.: Efficient Bayesian Hierarchial User Modeling for Recommendation System. In: Proceedings of SIGIR 2002, pp. 47–54. ACM Press, New York (2007)

Personalized Location-Aware QoS Prediction for Web Services Using Probabilistic Matrix Factorization

Yueshen Xu, Jianwei Yin, Wei Lo, and Zhaohui Wu

School of Computer Science and Technology, Zhejiang University, China
{xyshzjucs,zjuyjw,spencer_w_lo,wzh}@zju.edu.cn

Abstract. QoS prediction is critical to Web service selection and recommendation, with the extensive adoption of Web services. But as one of the important factors influencing QoS values, the geographical information of users has been ignored before by most works. In this paper, we first explicate how Probabilistic Matrix Factorization (PMF) model can be employed to learn the predicted QoS values. Then, by identifying user neighbors on the basis of geographical location, we take the effect of neighbors' experience of Web service invocation into consideration. Specifically, we propose two models based on PMF, i.e. *L-PMF* and *WL-PMF*, which integrate the feature vectors of neighbors into the learning process of latent user feature vectors. Finally, extensive experiments conducted in the real-world dataset demonstrate that our models outperform other well-known approaches consistently.

Keywords: QoS Prediction, Web Service, Probabilistic Matrix Factorization, Data Sparsity.

1 Introduction

A Web service is a self-describing programmable application used to achieve interoperability and accessibility over a network, which is implemented in a standard language and published through a specific protocol [1]. Based on Service-Oriented Architecture (SOA), Web service plays an important role in developing inner- and inter-enterprise information systems. Meanwhile, since Web service technologies are widely utilized in cloud computing, especially in Software-as-a-Service (SaaS) platform, the number of Web services is exploding.

Apart from the functionality of Web services, the non-functional properties termed as Quality of Service (QoS) are also identified as distinguishing characteristics of Web services, in which price, response time and some other properties are included. QoS plays a critical role in Web service selection, discovery and recommendation, especially in the case that the candidates faced by users have similar functions [2]. But in most cases, users can only access a small number of QoS values due to the following reasons: 1) Since the number of Web services is extremely large, it is unrealistic to invoke all of them. 2) QoS values of many

X. Lin et al. (Eds.): WISE 2013, Part I, LNCS 8180, pp. 229–242, 2013.

Web services are inconstant and sensitive to the alteration of their infrastructure, for instance, network bandwidth. In real world applications, those missing values need to be predicted for Web service selection and recommendation.

As one of the most widely adopted approaches to predicting missing ratings in recommender systems, Collaborative Filtering (CF) has been used to solve the problem of QoS prediction by Web service community in recent years [3–5]. Two kinds of memory-based CF algorithms have been mainly employed, i.e. user-based and item-based model, both of which utilize similar users or items sharing similar historical records [9]. The core process of CF is the similarity calculation between two users or items, which usually employs Pearson Correlation Coefficient (PCC) as the similarity measurement. But such kinds of approaches suffer from the following drawbacks due to the utilization of PCC: 1) Techniques based on PCC fail to solve the problem so-called 'cold-start', which means that a user has never invoked any Web services or a Web service has never been invoked by any users. 2) PCC can only take the subjective preferences of users into consideration, but ignores the fact that QoS is mainly determined by the physical environment [6]. Therefore, these drawbacks degrade the prediction accuracy of CF techniques, and limit the usability.

Recently, a few works have noticed the effect of users' location on QoS values, which are based on the observation that users located in different areas can undergo different experience in the invocation process of the same Web service, due to the diverse physical infrastructure [4, 6]. A concrete example is given in the following. According to the report of Akamai in the second quarter of the year 2012, South Korea and Japan are the top two countries in the average connection speed, while China has relatively low connection speed [8]. So in such a case that users in Seoul, Tokyo and Beijing invoke the same Web service for weather forecast, users in Seoul and Tokyo will experience shorter response time than users in Beijing do. Meanwhile, the response time may be further shorter for users in Seoul than in Tokyo. Figure 1(a) shows the scenario of location-aware Web service invocation.

Besides, another intractable issue in QoS prediction is data sparsity, the reasons of which have been presented in the second paragraph of this section. High data sparsity means that most entries in the user-service invocation matrix (shown in Fig. 1(b)) are empty, which leads to the failure of techniques based on PCC. Figure 1(b) shows the task to fulfill missing values in the matrix. As the same situation with QoS prediction, data density is usually extremely sparse in recommender systems [9]. Recently, Probabilistic Matrix Factorization (PMF) has been proposed to predict ratings, especially in the case of extreme data sparsity and massive data volume, and has been proved more effective than other approaches [10, 12, 13]. But QoS values are principally affected by objective operating environment of Web services, which are different from user ratings mainly affected by users' subjective preferences. So approaches based on PMF need much modification to model QoS prediction effectively.

To address these problems, in this paper, we propose a method of predicting QoS values based on the combination of PMF model and geographical information.

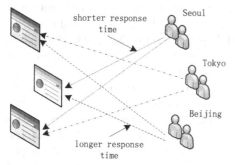

	service1	service2	service3	service4
user1	q_{11}			
user2		q_{22}		q_{24}
user3				
user4	q_{41}			q_{44}
user5			q_{53}	

(a) Location-aware Web Service Invocation

(b) User-Service Invocation Matrix

Fig. 1. Web Service Invocation Scenario in Real World

First, we gain preliminary predicted values by learning latent user and service feature vectors with the basic PMF model. Then, after identifying the k nearest neighbors of a user, the basic PMF model is extended to be an enhanced location-aware model through fusing the QoS values of the user and neighbors together. Further, the predication accuracy of the model is promoted by weighting the contribution of each neighbor, and we get the final weighted location-aware PMF model. In the end, extensive experiments are conducted, demonstrating that our models outperform other well-known algorithms consistently.

In summary, the contributions of this paper can be concluded as follows:

1. We explicate and verify the effectiveness of PMF model in QoS prediction for Web services.
2. We expound and demonstrate the importance of the utilization of geographical information for Web service QoS prediction.
3. We propose two ways to utilize the geographical information by identifying users' neighbors, and further gain two corresponding location-aware PMF models to improve the predication accuracy.
4. We conduct comprehensive experiments with the real-world dataset, demonstrating the effectiveness of our methods.

The rest of this paper is organized as follows: Section 2 summarizes related works in QoS prediction for Web services and PMF model. Section 3 first presents how PMF model can be used to predict QoS values, and then gives a detailed explanation of the two models based on the fusion of PMF model and the geographical information. Section 4 presents the experimental results in performance, and investigates the effects of the parameters. Section 5 concludes the paper and discusses the future work.

2   Related Work

QoS prediction is critical to many key problems in Web service domain, such as Web service selection [14], discovery [15], composition [16] and recommendation

[5]. Currently, Collaborative Filtering is the most widely adopted algorithm to predict QoS values in Web service community due to the simplicity and maturity.

Shao et al. [3] utilized the user-based CF algorithm for QoS prediction, which was modified with the combination of the positive and negative correlation. Zheng et al. [5] proposed a hybrid CF model fusing the user-based and item-based CF algorithms together, in which *confidence weights* were used to balance the respective weights of the two models. Though the CF-based methods are easy to implement and relatively effective, they suffer from sharp accuracy declination in the case that QoS values are sparse, and can hardly integrate any other factors into the model, for instance, the geographical information utilized in this paper.

The contribution of geographical information to the improvement of QoS prediction accuracy has been studied recently. Chen et al. [4] developed a hierarchical clustering algorithm to identify users' neighbors with similar historical Web service invocation experience, and then, these people were supposed to be in the same region. This approach is unreasonable since for example, though users in Seoul and in Tokyo may have similar QoS values in a certain period, changes of infrastructure in Seoul cannot make any difference on users' experience of Web service invocation in Tokyo. Lo et al. [6] took the influence of users' neighbors into consideration from the real geographical sense, which appended a third regularization term at the end of the objective function of SVD-like Matrix Factorization [11]. Since the main purpose of such a kind of usage is to forbid overfitting in the learning process, it is hard to give a persuasive interpretation from the perspective of neighbors' contributions to QoS values. Besides, SVD-like Matrix Factorization has been proved a special case of PMF model, in the sense that the distribution of QoS values is assumed to be the Gaussian distribution [10]. So our models can be viewed as frameworks since any other distributions can be integrated into them.

PMF model has been used for rating prediction in recommender systems recently. Salakhutdinov et al. [10] built the basic PMF model and demonstrated its effectiveness in the well-known dataset of Netflix [17]. Ma et al. [12, 13] proposed two different ways of integrating the contributions of users' trusted friends to the final ratings into PMF model. In one work, a rating consisted of two parts, in which one part was learned from the user's own preference and the other was learned from the preferences of friends. In this paper, the predicted QoS value is also decomposed into two parts and takes advantage of the whole invocation experience of the user and his or her neighbors.

3 QoS Prediction with Probabilistic Matrix Factorization

In this section, we first build the basic PMF model to learn latent user features and service features. Then, in Sect. 3.2, we give an explication of the two location-aware PMF models. In the end, the complexity of our algorithms is analyzed in detail.

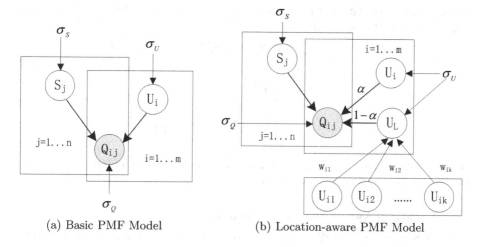

(a) Basic PMF Model   (b) Location-aware PMF Model

Fig. 2. Graphical Models for User and Service Features Learning

3.1   User and Service Features Learning

As a probabilistic factor analysis model, PMF can factorize the high rank user-service invocation feature space into a joint low rank feature space, in which users and Web services are both represented by low dimensional feature vectors. Then the predicted QoS value is gained as the inner product of two corresponding feature vectors. The basic insight of PMF is that the number of factors dominating the quality of Web service invocation is limited, and the final QoS value is mainly determined by how those factors act upon users and Web services. So the key task is to learn the latent feature matrices of both sides.

Let $Q = \{q_{ij}\}$ represents the $m \times n$ user-service invocation matrix depicted in Fig. 1(b). Note that $q_{ij} \in [0, \infty)$, in which 0 means that a user has never invoked the Web service before or the QoS value has been missing. Let $U \in R^{d \times m}$ and $S \in R^{d \times n}$ represent latent user and service feature matrices respectively, in which the column vector U_i and S_j represent the latent user and service feature vectors respectively, and the dimensionality of them is both d. The conditional distribution over the observed QoS values is defined as

$$P(Q|U, S, \sigma_Q^2) = \prod_{i=1}^{m} \prod_{j=1}^{n} \left[\mathcal{N}(q_{ij}|U_i^T S_j, \sigma_Q^2) \right]^{I_{ij}}, \tag{1}$$

where $\mathcal{N}(q_{ij}|U_i^T S_j, \sigma_Q^2)$ denotes the probabilistic density function of the Gaussian distribution with the mean $U_i^T S_j$ and standard deviation σ_Q [10]. I_{ij} is the indicator whose value equals to 1 if user i invoked Web service j before and equals to 0 otherwise. Meanwhile, we set zero-mean spherical Gaussian prior on user and service feature vectors as

$$P(U|\sigma_U^2) = \prod_{i=1}^{m} \mathcal{N}(U_i|0, \sigma_U^2 I), \quad P(S|\sigma_S^2) = \prod_{j=1}^{n} \mathcal{N}(S_j|0, \sigma_S^2 I), \tag{2}$$

where I is the identity matrix. Then, the posterior probability of (U, S) is obtained through Bayesian inference as

$$P(U, S|Q, \sigma_Q^2, \sigma_U^2, \sigma_S^2) \propto P(Q|U, S, \sigma_Q^2) \times P(U|\sigma_U^2) \times P(S|\sigma_S^2)$$

$$= \prod_{i=1}^{m} \prod_{j=1}^{n} \left[\mathcal{N}(q_{ij}|U_i^T S_j, \sigma_Q^2) \right]^{I_{ij}} \tag{3}$$

$$\times \prod_{i=1}^{m} \mathcal{N}(U_i|0, \sigma_U^2 I) \times \prod_{j=1}^{n} \mathcal{N}(S_j|0, \sigma_S^2 I).$$

We denominate this algorithm as *basic PMF model*, the probabilistic graphical model of which is depicted as Fig. 2(a).

3.2   Location-Aware Features Learning

Since QoS values of the Web service invoked by users in the same area tend to be similar, we hold the view that it is not enough to predict QoS values just employing the user's own latent features. So in the following two new models, the predicted QoS value is divided into two parts, one is learned from the user's own latent features, and the other is learned from the neighbors' latent features.

Integrating K Nearest Neighbors' Features. First, the neighbors L of user u are identified by the distance d_{ul} between user l and user u, which can be calculated in advance via Euclid Distance of user's longitude and latitude. Then, the average predicted QoS value of a Web service invoked by the neighbors is learned as

$$\hat{q}_L = \frac{1}{|L|} \sum_{l=1}^{k} U_l^T S_j,$$

where L represents those users who are in the k nearest neighbors set of user u, so $|L|$ exactly equals to k. Further, the final QoS value can be learned as

$$q_{ij} \approx \hat{q}_{ij} = \alpha U_i^T S_j + \frac{(1-\alpha)}{|L|} \sum_{l=1}^{k} U_l^T S_j,$$

where α is a regulatory factor to adjust the weight of the two parts. Therefore, with the same Bayesian inference procedure in the basic PMF model, we can gain the *location-aware PMF model (L-PMF)* as

$$P(U, S|Q, \sigma_Q^2, \sigma_U^2, \sigma_S^2)$$

$$\propto \prod_{i=1}^{m} \prod_{j=1}^{n} \left[\mathcal{N}(q_{ij}|(\alpha U_i^T S_j + \frac{(1-\alpha)}{|L|} \sum_{l=1}^{k} U_l^T S_j), \sigma_Q^2) \right]^{I_{ij}} \tag{4}$$

$$\times \prod_{i=1}^{m} \mathcal{N}(U_i|0, \sigma_U^2 I) \times \prod_{j=1}^{n} \mathcal{N}(S_j|0, \sigma_S^2 I).$$

The probabilistic graphical model of location-aware PMF model is depicted as Fig. 2(b), with $w_{ij} = 1$ $(j = 1...k)$ in the graph consistently.

Distinguishing Neighbors' Significance with Weight. In location-aware PMF model, the contributions of all neighbors are treated equally, which are not so corresponding with the real scenario of Web service invocation. In reality, the infrastructure of neighbors who are nearer is more similar with that of the user, so QoS values of the same Web service invoked by them are more close. Taking this fact into account, we define the similarity between two users as

$$Sim_{il} = exp(-d_{il}),$$

where d_{il} is the distance between user i and user l, and $Sim_{il} \in (0,1]$. $Sim_{il} = 1$ means that d_{il} equals to 0, namely that user i and user l live in the exactly same place, and $Sim_{il} \rightarrow 0$ means that the two users live extremely far, for instance, user i live in Tokyo while user l live in Kabul. Note that the form of the similarity calculation formula is not exclusive, since any similarity formula satisfying the properties that $exp(-d_{il})$ owns can be a candidate. Further, the normalized similarity as the measurement of the individual importance of each neighbor is calculated as

$$w_{il} = \frac{Sim_{il}}{\sum_{g \in L} Sim_{ig}},$$

Finally, the final QoS value of Web service j invoked by user i is learned more properly as

$$q_{ij} \approx \hat{q}_{ij} = \alpha U_i^T S_j + (1 - \alpha) \sum_{l=1}^{k} w_{il} U_l^T S_j,$$

Therefore, we get the *weighted location-aware PMF model (WL-PMF)* as

$$
\begin{aligned}
&P(U, S | Q, \sigma_Q^2, \sigma_U^2, \sigma_S^2) \\
&\propto \prod_{i=1}^{m} \prod_{j=1}^{n} \left[\mathcal{N}(q_{ij} | (\alpha U_i^T S_j + (1 - \alpha) \sum_{l=1}^{k} w_{il} U_l^T S_j), \sigma_Q^2) \right]^{I_{ij}} \\
&\times \prod_{i=1}^{m} \mathcal{N}(U_i | 0, \sigma_U^2 I) \times \prod_{j=1}^{n} \mathcal{N}(S_j | 0, \sigma_S^2 I),
\end{aligned}
\tag{5}
$$

Further, the logarithmic form of the posterior probability is expressed as

$$
\begin{aligned}
&\ln P(U, S | Q, \sigma_Q^2, \sigma_U^2, \sigma_S^2) \\
&= -\frac{1}{2\sigma_Q^2} \sum_{i=1}^{m} \sum_{j=1}^{n} I_{ij} (q_{ij} - (\alpha U_i^T S_j + (1 - \alpha) \sum_{l=1}^{k} w_{il} U_l^T S_j))^2 \\
&\quad -\frac{1}{2\sigma_U^2} \sum_{i=1}^{m} U_i^T U_i - \frac{1}{2\sigma_S^2} \sum_{j=1}^{n} S_j^T S_j \\
&\quad -\frac{1}{2} ((\sum_{i=1}^{m} \sum_{j=1}^{n} I_{ij}) \ln \sigma_Q^2 + MD \ln \sigma_U^2 + ND \ln \sigma_S^2) + C,
\end{aligned}
\tag{6}
$$

where C is a constant independent of those variables. Maximizing the above log of the posterior distribution over latent feature matrices with fixed hyper-parameters (i.e. the standard deviation of the prior probability) is equivalent to gaining the minimum sum of squared errors formalized as the objective function with quadratic regularization terms as follows:

$$
\begin{aligned}
E = & \frac{1}{2} \sum_{i=1}^{m} \sum_{j=1}^{n} I_{ij} (q_{ij} - (\alpha U_i^T S_j + (1-\alpha) \sum_{l=1}^{k} w_{il} U_l^T S_j))^2 \\
& + \frac{\lambda_U}{2} \|U\|_F^2 + \frac{\lambda_S}{2} \|S\|_F^2,
\end{aligned}
\tag{7}
$$

where $\lambda_U = \sigma_Q^2/\sigma_U^2$ and $\lambda_S = \sigma_Q^2/\sigma_S^2$, and $\| \cdot \|_F^2$ denotes the *Frobenius norm*. Gradient descent can be used over matrix U and S to find a local minimum of the objective function in Eq. (7). So for the 'cold-start' user or service, even if the result learned from the own corresponding feature vectors is not so satisfying, the accuracy of the predicted value can be further improved by the rectification of learning results of neighbors' feature vectors. The partial derivatives of the objective function over latent user and service feature vectors are as follows:

$$
\begin{aligned}
\frac{\partial E}{\partial U_i} = & \alpha \sum_{j=1}^{n} I_{ij} S_j (\alpha U_i^T S_j + (1-\alpha) \sum_{l=1}^{k} w_{il} U_l^T S_j - q_{ij}) \\
& + (1-\alpha) \sum_{g \in G(i)} \sum_{j=1}^{n} I_{gj} w_{gi} S_j ((\alpha U_g^T S_j + (1-\alpha) \sum_{l=1}^{k} w_{gl} U_l^T S_j) \\
& - q_{gj}) + \lambda_U U_i, \\
\frac{\partial E}{\partial S_j} = & \sum_{i=1}^{m} I_{ij} ((\alpha U_i^T S_j + (1-\alpha) \sum_{l=1}^{k} w_{il} U_l^T S_j - q_{ij}) \\
& \times (\alpha U_i + (1-\alpha) \sum_{l=1}^{k} w_{il} U_l)) + \lambda_S S_j,
\end{aligned}
\tag{8}
$$

where $G(i)$ contains all the users whose neighbors include user i. The form of the partial derivatives over U_i and S_j in *L-PMF* is almost the same with those in Eq. (8), except that w_{il} equals to 1 and $(1-\alpha)$ is substituted by $(1-\alpha)/|L|$. In the following experiments, λ_U and λ_S are set to the same value for convenience.

3.3   Complexity Analysis

The main computational cost of the two models arises from the procedure of gradient descent based on Eq. (8), the number of iterations of which is an absolutely small constant. So we only need to analyze the computational complexity of $\partial E/\partial U_i$ and $\partial E/\partial S_j$.

The computational complexity of $\partial E/\partial U_i$ and $\partial E/\partial S_j$ in a single iteration is $O(\rho_Q kd + \rho_Q \bar{g} kd)$ and $O(\rho_Q kd + \rho_Q k)$ respectively, where ρ_Q denotes the non-empty values in the user-service invocation matrix and d is the dimensionality of

Table 1. Statistics on users' neighbors

k	$\bar{g}$	Matrix Density=5%		Matrix Density=10%	
		k/ρ_Q	$\bar{g}/\rho_Q$	k/ρ_Q	$\bar{g}/\rho_Q$
40	41.035	0.04%	0.042%	0.02%	0.021%
100	100.593	0.1%	0.1%	0.05%	0.051%
160	160.873	0.16%	0.16%	0.08%	0.082%

latent feature vectors, which also is a small constant. And $\bar{g}$ denotes the average number of the users whose k nearest neighbors include user i. In most cases, k and $\bar{g}$ are far less than ρ_Q, which are verified by the statistical results calculated on the dataset described in Sect. 4 and shown in Table 1. Therefore, the computational complexity above can be simplified into $O(\rho_Q \bar{g} kd)$ and $O(\rho_Q kd)$, and finally approximately combined into $O(\rho_Q \bar{g} kd)$ together, which indicates that the computational complexity is linearly scalable to the size of datasets, so the two models can be employed on very large datasets.

4 Experiments and Evaluation

In this section, extensive experiments are conducted by comparison to other well-known approaches to answer the following questions: 1) How does the performance of our two models compare with the state-of-the-art algorithms in QoS prediction for Web services? 2) What is the best k to incorporate the effect of neighbors into this problem? 3) What is the most proper α to regulate the influence of the user and neighbors in the learning process of QoS values? 4) How many latent features should be employed to learn the final QoS value?

4.1 Dataset Description

The dataset offered by Zheng et al. [18], which contains 339 users and 5825 Web services, is used to conduct our experiments. Geographical information and response time are gained from the file *userlist.txt* and *rtmatrix.txt* respectively.

4.2 Metrics

As one of the most commonly used evaluation metrics in recommender systems [9], *Root Mean Squared Error (RMSE)* is employed here to measure the prediction accuracy of all approaches, which is defined as follows:

$$RMSE = \sqrt{\frac{1}{N} \sum_{i,j} (q_{ij} - p_{ij})^2},$$

where q_{ij} and p_{ij} are the true and predicted values respectively, and N is the number of values in the testing dataset. Meanwhile, *Mean Absolute Error (MAE)* is gained with RMSE synchronously after the exactly same learning process, which is defined as:

$$MAE = \frac{1}{N} \sum_{i,j} |q_{ij} - p_{ij}|.$$

4.3   Comparison and Performance

Several other state-of-the-art approaches are chosen to compare with our two models, including:

1. UPCC: This approach is much similar with user-based CF, which first calculates the similarity between users based on PCC and then gains the predicted value as the weighted average of the known values of the similar users [3].
2. IPCC: This approach is similar with UPCC, except that the key procedure is the similarity calculation between items [7].
3. UIPCC: This approach combines the advantage of UPCC and IPCC by balancing the proportions of them in the final result [5].
4. RegionKNN: This approach classifies services and users into different regions, and modifies UPCC by the similarity computation between regions and the identification of similar services and users in the same regions [4].
5. Basic-PMF: This approach is proposed by [10], and has been verified to be effective in recommender systems. A detailed explanation is given in this paper to show how basic PMF model can be used for QoS prediction.
6. LBR2: This approach first calculates the difference of latent feature vectors between the user and the neighbors, and then appends the *Frobenius norm* of the difference to the objective function of Matrix Factorization model[6].

The whole dataset is divided into training data and testing data by randomly removing a large number of QoS values in user-service invocation matrix. For instance, 95% values are removed randomly as testing data, and the 5% left in the matrix is trained to predict those removed ones. More specifically, four types of data sparsity are used to conduct the experiments, which are 95%, 90%, 85% and 80% respectively. In our experiments, the default parameter settings are $k = 40, \alpha = 0.6$ and $d = 10$, in which d represents the dimensionality of the latent feature vector. Moreover, λ_U and λ_S are set to 0.001 equally in all of the following experiments.

Table 2 shows that *Basic-PMF* gets better prediction accuracy than *RegionKNN* and *UIPCC*, which verifies its effectiveness. Meanwhile, our two models achieve smaller RMSE and MAE under all situations of data sparsity. Besides, *WL-PMF* gets better performance than *L-PMF* in most cases, which is consistent with the fact that the neighbors who are nearer share more similar invocation experience with the user. Further, *WL-PMF* gains 5.68% and 3.80% performance improvement in RMSE, as well as 10.27% and 6.61% improvement in MAE on average, in comparison with *Basic-PMF* and *LBR2* respectively. Moreover, prediction accuracy raises with the increasing of data density, which is a natural phenomenon showing that more historical records can depict user features more accurately.

Table 2. Accuracy Comparison(A Smaller RMSE or MAE value means better performance)

Approach	Matrix Density (MD)							
	MD=5%		MD=10%		MD=15%		MD=20%	
	RMSE	MAE	RMSE	MAE	RMSE	MAE	RMSE	MAE
UPCC	1.6670	0.7839	1.6012	0.7445	1.4745	0.6824	1.4179	0.6418
IPCC	1.5231	0.7838	1.4585	0.7296	1.4184	0.6839	1.3430	0.6111
UIPCC	1.5059	0.7639	1.4349	0.6862	1.4065	0.6698	1.3341	0.5919
RegionKNN	1.4932	0.7620	1.4047	0.6659	1.3564	0.6483	1.3134	0.5911
Basic-PMF	1.4995	0.7450	1.3790	0.6183	1.3326	0.6020	1.2730	0.5649
LBR2	1.4671	0.7132	1.3286	0.6130	1.3196	0.5726	1.2608	0.5332
L-PMF	**1.4347**	**0.6635**	**1.2955**	**0.5667**	**1.2411**	**0.5292**	**1.2106**	**0.5095**
WL-PMF	**1.4345**	**0.6702**	**1.2908**	**0.5656**	**1.2389**	**0.5275**	**1.2099**	**0.5077**

In the following parts, we conduct experiments on the impact of the parameters α, k, d only with *WL-PMF* for space limitation.

4.4 Impact of α

The parameter α regulates the respective proportions of latent feature vectors of the user and his or her neighbors in the learning process of missing QoS values. We investigate the impact of α from the range of 0.1 to 0.9 in the experiment settings of $k = 40$ and $d = 10$, under the matrix density equals to 10% and 15% respectively.

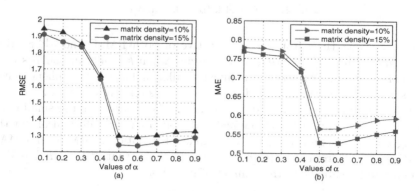

Fig. 3. Impact of α ($k = 40$, $d = 10$)

As shown in Fig. 3, RMSE and MAE both reach the minima in the range of 0.5 to 0.7, and get acceptable and relatively small values when α continues increasing. But they suffer from drastic fluctuation and get large values below 0.5. This changing trend indicates that though the contributions of the feature vectors of neighbors take importance roles, but the significance of the user's own feature vectors are always dominant. Finally, we can draw such a conclusion that if neighbors feature vectors account for a too large proportion, it will lead to the deviation of the predicted value from the real QoS value, so α must be not less than 0.5. Besides, if α equals to 1, *WL-PMF* model is degenerated into the basic PMF model, the performance of which is comparatively lower.

4.5   Impact of k

The parameter k determines the number of neighbors whose latent feature vectors are integrated into the targeted QoS value. In ideal conditions, only those neighbors who are near enough to ensure that they really share similar infrastructure with the user should be involved in. We investigate the impact of k in the experiment settings of $\alpha = 0.6$ and $d = 10$.

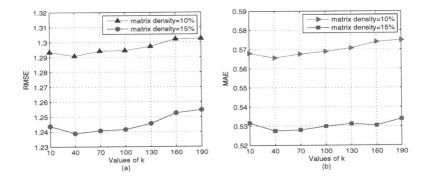

Fig. 4. Impact of k ($\alpha = 0.6$, $d = 10$)

Figure 4 shows that, from the perspective of both RMSE and MAE, *WL-PMF* achieves the highest prediction accuracy when $k = 40$, and suffers from performance degradation when k is much smaller or bigger. This phenomenon demonstrates that on the one hand, those neighbors relatively farther away may introduce noise into the predicted value, and on the other hand, too small number of neighbors cannot provide enough valuable auxiliary information for the learning process. Meanwhile, since the extent of variation of RMSE and MAE is relatively small from $k = 10$ to $k = 190$, it shows that *WL-PMF* has high scalability and flexibility to the number of neighbors.

4.6   Impact of d

The parameter d controls the number of latent features of a user, which cannot be observed directly. We investigate the impact of d in the value range of 5 to 30 in the experiment settings of $k = 40$ and $\alpha = 0.6$. As shown in Fig. 5, as d increases, both RMSE and MAE first decrease, and reach the minima at $d = 10$, and then begin to increase again.

Figure 5 also shows that *WL-PMF* gains satisfactory accuracy among values from 5 to 20, but performs not so well from the value 25, which indicates that the actual number of features dominating the experience of a user in Web service invocation is limited. Meanwhile, since the threshold with the smallest RMSE is 10, it is inferred that the factors influencing the process of Web service invocation is relatively various in the real world. But this result does not mean that in every practical scenario, the optimal value of d is always 10 due to the variety and complexity of factors influencing QoS values.

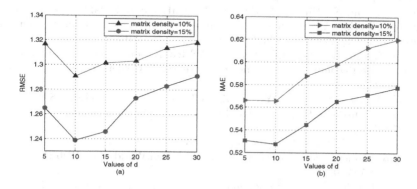

Fig. 5. Impact of d ($k = 40$, $\alpha = 0.6$)

5 Conclusions and Future Work

In this paper, we first expound the significance of involving geographical information in the problem of QoS prediction for Web services. Then an interpretation describing how PMF model can be used to solve this problem is given in detail. Afterwards, two models based on PMF that combine with the effects of neighbors are proposed to gain higher prediction accuracy. In the two models, we explain how to incorporate the invocation experience of neighbors into the learning process of latent user feature vectors. Finally, our approaches are verified to be effective through sufficient experiments.

In the future, for one thing, we will explore the way of utilizing the interaction between QoS properties, since some QoS members are interrelated, for instance, availability and throughput. For another, we will try to utilize the geographical information of Web services to further improve the prediction accuracy.

Acknowledgement. The research in this paper is supported by National Science and Technology Supporting Program of China (No.2012BAH06F02), National Natural Science Foundation of China under Grant (No.61272129), Research Foundation for the Doctoral Program by Ministry of Education of China (No.20110101110066), New-Century Excellent Talents Program by Ministry of Education of China (No.NCET-12-0491), Zhejiang Science Fund for Distinguished Young Scholars (R13F020004).

References

1. Zhang, L., Zhang, J., Cai, H.: Services Computing. Springer, Tsinghua University Press (2007)
2. Papazoglou, M.P.: Service-Oriented Computing: Concepts, Characteristics and Directions. In: International Conference on Web Information System Engineering, pp. 3–12 (2003)
3. Shao, L., Zhang, J., Wei, Y., Zhao, J., Xie, B., Mei, H.: Personalized QoS Prediction for Web Services via Collaborative Filtering. In: International Conference on Web Services, pp. 439–446 (2007)
4. Chen, X., Liu, X., Huang, Z., Sun, H.: RegionKNN: A Scalable Hybrid Collaborative Filtering Algorithm for Personalized Web Service Recommendation. In: International Conference on Web Services, pp. 9–16 (2010)
5. Zheng, Z., Ma, H., Lyu, M.R., King, I.: WSRec: A Collaborative Filtering Based Web Service Recommender System. In: International Conference on Web Services, pp. 437–444 (2009)
6. Lo, W., Yin, J., Deng, S., Li, Y., Wu, Z.: Collaborative Web Service QoS Prediction with Location-Based Regularization. In: International Conference on Web Services, pp. 464–471 (2012)
7. Sarwar, B.M., Karypis, G., Konstan, J.A., Riedl, J.: Item-based collaborative filtering recommendation algorithms. In: International World Wide Web Conference, pp. 285–295 (2001)
8. Akamai. The State of the Internet, http://www.akamai.com/stateoftheinternet/
9. Su, X., Khoshgoftaar, T.M.: A Survey of Collaborative Filtering Techniques. In: Advances in Artificial Intellegence 2009, pp. 1–20 (2009)
10. Salakhutdinov, R., Mnih, A.: Probabilistic Matrix Factorization. In: Advances in Neural Information Processing Systems, vol. 20 (2008)
11. Koren, Y., Bell, R.M., Volinsky, C.: Matrix Factorization Techniques for Recommender Systems. IEEE Computer, 30–37 (2009)
12. Ma, H., Yang, H., Lyu, M.R., King, I.: SoRec: social recommendation using probabilistic matrix factorization. In: ACM International Conference on Information and Knowledge Management, pp. 931–940 (2008)
13. Ma, H., King, I., Lyu, M.R.: Learning to recommend with social trust ensemble. In: ACM SIGIR, pp. 203–210 (2009)
14. Sreenath, R.M., Singh, M.P.: Agent-based service selection. Journal of Web Semantics, 261–279 (2004)
15. Ran, S.: A model for web services discovery with QoS. ACM SIGecom Exchanges 4(1), 1–10 (2003)
16. Jaeger, M.C., Rojec-Goldmann, G., Mühl, G.: QoS Aggregation for Web Service Composition using Workflow Patterns. In: Proc. of the Enterprise Distributed Object Computing Conference, pp. 149–159 (2004)
17. Netflix. Netflix Prize Competition (2006), http://www.netflixprize.com/
18. Zheng, Z., Zhang, Y., Lyu, M.R.: Distributed QoS Evaluation for Real-World Web Services. In: International Conference on Web Services, pp. 83–90 (2010)

Verifying Transactional Requirements of Web Service Compositions Using Temporal Logic Templates

Scott Bourne, Claudia Szabo, and Quan Z. Sheng

School of Computer Science
The University of Adelaide, SA 5005, Australia
{scott.bourne,claudia.szabo,michael.sheng}@adelaide.edu.au

Abstract. Ensuring reliability in Web service compositions is of crucial interest as services are composed and executed in long-running, distributed mediums that cannot guarantee reliable communications. Towards this, transactional behavior has been proposed to handle and undo the effects of faults of individual components. Despite significant research interest, challenges remain in providing an easy-to-use, formal approach to verify transactional behavior of Web service compositions before costly development. In this paper, we propose the use of temporal logic templates to specify component-level and composition-level transactional requirements over a Web service composition. These templates are specified using a simple format, configured according to scope and cardinality, and automatically translated into temporal logic. To verify design conformance to a set of implemented templates, we employ model checking. We propose an algorithm to address state space explosion by reducing the models into semantically equivalent Kripke structures. Our approach facilitates the implementation of expressive transactional behavior onto existing complex services, as demonstrated in our experimental study.

1 Introduction

Service-oriented architectures and Web services have been the focus of active research in the past decade [1–3]. Despite significant interest in techniques for designing, deploying, and ensuring reliability of Web services, many existing services experience severe issues such as timeout, dependability and unexpected behavior [4]. Market pressures that require ad-hoc deployment without proper quality assurance contribute to this issue. An important challenge remains verifying the correctness of a Web service composition with respect to fault-handling logic at *design-time*. This will permit developers to identify design flaws before costly development and improve the quality of the service composition [2, 3, 5].

Transactional behavior can be used to contain, handle, and undo the effects of faults in the execution of a Web service composition [3, 5, 6]. Requirements for transactional behavior need to be drawn from application-specific business logic that dictate which faults are acceptable, retriable, or recoverable. For example,

X. Lin et al. (Eds.): WISE 2013, Part I, LNCS 8180, pp. 243–256, 2013.

a service to retrieve customer details can be retried safely, but services to commit payment or place orders may require recovery or replacement upon failure. Business logic can also dictate requirements at the composition-level, such as compensatory processes to rollback all execution effects [7].

A formal yet practical process of specifying transactional behavior and requirements, followed by the verification of the Web service composition design can significantly reduce development and maintenance cost, and also increase credibility and reliability in the deployed service. Previous approaches have proposed the detailed specification of failure models [5], composition risk levels [7], or the definition of vital components for the successful completion of Web services [3]. These approaches offer significant improvements towards the verification of transactional requirements, but tend to lead to state space explosion when verified [8] or restrict the transactional requirements that can be verified. A method to specify transactional requirements that, to the best of our knowledge, is yet to be explored, is the adaptation of temporal logic patterns [9–11]. These are frequently used structures of temporal logic properties [12], that can be implemented to specify sophisticated requirements. This allows users to take advantage of the expressive power of temporal logic, while reducing the effort and error-prone nature of pure manual specification.

In this paper, we propose a modeling approach for the design of transactional Web service compositions that allows users to specify transactional requirements formally using temporal logic templates and verify conformance at design time. Our earlier work has proposed a novel model that separates the service behavior into *operational* and *control* behaviors, allowing for flexible design, development, and verification of complex Web services [2]. The operational behavior contains the underlying business logic of the system, while the control behavior maintains the transactional state and guides the execution of the service. A set of pre-defined messages enable conversations between operational and control behavior that trigger the execution of components, indicate faults, specify recovery operations, and signal completion, among other operations [13]. We extend our previous work by enabling the control and operational behaviors to be verified against transactional requirements drawn from the business logic specified by the user with temporal logic templates. The main contributions of our work are:

- A set of temporal logic templates to formally specify component-level and composition-level transactional requirements derived from business logic to facilitate the verification of composite Web services.
- A service verification approach based on model checking that utilizes temporal logic properties obtained from implemented templates to verify transactional requirements, while addressing state space explosion with model reduction measures.
- A prototype implementation that facilitates our verification approach over Web service composition designs as control and operational behavior models.

The remainder of the paper is organized as follows. Section 2 presents an overview of the control and operational behavior approach for Web service modeling. Section 3 outlines our approach for capturing transactional requirements with temporal logic templates. Section 4 describes how a design can be verified against these requirements in our approach. Section 5 reports the prototype implementation and experimental study. Finally, Section 6 contrasts our work with related work and Section 7 concludes this paper and discusses future directions.

2   Background

To specify transactional Web service compositions at design-time, we adapt the modeling method proposed in our previous work [2], based on the separation of Web service behavior into *control* and *operational* behaviors. The control behavior is an application-independent model that maintains the transactional state of the composition, while the operational behavior contains the application-dependent flow of business tasks. The execution and recovery operations of the service are directed by the control behavior, according to events reported from the operational behavior. Using these models, this design provides a detailed view of the functional and transactional behavior of a composition, moreover, each perspective to be designed and modified independently.

The control and operational behavior models are expressed as a 5-tuple $\mathcal{B} = \langle \mathcal{S}, \mathcal{L}, \mathcal{T}, s_0, \mathcal{F} \rangle$ where $\mathcal{S}$ is a finite set of state names, s_0 is the initial state, $\mathcal{F} \subseteq \mathcal{S}$ is a set of final states, and $\mathcal{L}$ is a set of event-condition-action labels. $\mathcal{T} \subseteq \mathcal{S} \times \mathcal{L} \times \mathcal{S}$ is the transition relation where each $t \in \mathcal{T}$ consists of a source and target state, and a transition label. We can express these models using statecharts, as shown in the online payment composition in Figure 1. The control behavior model contains the transactional states of the composition, while the operational behavior contains the flow of business tasks of the process.

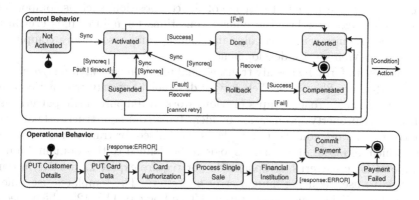

Fig. 1. The control and operational behaviors of a basic online payment composition

To enable communication between the behavior models, we use a set of *inter-behavior messages*. These allow the control behavior to direct execution, and the operational behavior to report events and status. The messages are classified as *initiation messages* that are sent from the control behavior, and *outcome messages* that are responses from the operational behavior. The initiation messages include `Sync` to initiate or resume execution of the service, `Recover` to trigger recovery operations, `Delay` to force a response from an operational behavior state, and `Ping` to test the liveness of a state. The outcome messages include `Success` to indicate the successful commit of the service, `Fail` to signal an abort, `Fault` to indicate the presence of a fault that requires recovery, `Ack` to report the liveness of a state, and `Syncreq` to request a `Sync` message to retry a component or resume execution following the recovery. These inter-behavior messages enable us to specify transactional behavior over the design. For example, in the online payment example in Figure 1, a `Sync` message could be used to trigger the process from *PUT Customer Data*, and a `Success` or `Fail` message could be sent from *Commit Payment* and *Payment Failed* respectively.

Our previous work [13] proposes a method to ensure well-formed inter-behavior conversations that avoid deadlock, incomplete execution, and prevent inconsistency between the behavior models. However, these properties are insufficient for the designer to ensure that the transactional behavior of the model conforms to their expectations, as the design cannot be verified against application-dependent requirements, such as the success of critical components prior to commit, or acceptable alternative operations given a failed component.

3   Temporal Logic Templates

A developer must have confidence that a potentially long-running composition of distributed and heterogeneous Web services will conform to a set of transactional requirements for containing and handling faults. These requirements could apply to failures of individual components, e.g., required recovery operations to undo the component's effect, or to the scope of the composition, such as components critical for success, or relaxed atomicity conditions for failure. It is crucial to specify these requirements formally, and identify and resolve any compliance issues prior to Web service development.

We propose to formally specify transactional requirements with *temporal logic templates* that are filled by a Web service designer to specify transactional requirements. This approach adapts previous work in temporal logic patterns [9], which simplify property specification by identifying common structural patterns. In contrast, our work constructs templates specialized for transactional requirements, which allow detailed business logic to be defined to this domain. Similar to temporal logic patterns, templates do not require expert knowledge in temporal logic to use, reducing human error and effort. The templates use Linear Temporal Logic (LTL) and Computational Tree Logic (CTL) [12], which specify properties of a system over two different timeline representations. We employ both languages since the properties they can specify are not equivalent [14].

Each template is defined with fields for *description, design prerequisites, required variables, scope, cardinality, temporal logic* and an *example use*. The *Scope* field allows the user to limit when the transactional requirement should be applied, while the *Cardinality* field reduces template ambiguity by allowing users to customize the relationship between variables. Since transactional requirements can apply to the behavior of the whole composition, or to specific components, we group our templates into two categories. *Component-level* templates specify requirements specific to components, and *composition-level* templates apply to the transactional behavior of the entire composition. The separation of control and operational behaviors in our design model means the components of interest to component-level templates are the operational behavior states. In contrast, the composition-level templates utilize control behavior states, as they present a transactional view of the composition. The descriptions of the component-level and composition-level templates are shown in Table 1 and 3 respectively. The full specifications of all our templates are omitted for space, but considerators and readers are referred to the author's website for the full list[1]. A formal proof of completeness for our template set is difficult to obtain, but in this paper we provide a foundation of examples that can be extended easily.

Table 1. Names and descriptions of the component-level temporal logic templates

Template Signature	Description
CompensateFailure <Component,Recovery, Card,Scope>	Specifies a component and a condition. The failure of the component requires the condition to be satisfied in the future, to recover from the failure.
CompensateSuccess <Component,Recovery, Card,Scope>	Specifies a component and a condition. When the composition must be undone, the condition must be satisfied to undo the effect of the component.
Alternative <Component,Recovery, Card,Scope>	Following the failure of a component, one or several alternative operations, expressed as a condition, are considered acceptable replacements.
NonRetriable <Component,Scope>	Following failure of a component, retrial is either not possible, or the user is not interested in it.
RetriablePivot <Component,Scope>	A component that may be retried, but not undone. Following its success, the service must commit.
NonRetriablePivot <Component,Scope>	A component that may not be retried or undone, and leads to commit or abort depending on success.

3.1   Component-Level Templates

Component-level templates specify transactional requirements for handling failures of individual components in a Web service composition. For example, a component-level transactional requirement of the online payment composition could specify that Card Authorization cannot be retried without first reattempting PUT Card Data. Other common transactional requirements applied to components include *compensatable, retriable, pivot, replaceable,* or similar labels [3, 5–7]. However, using templates to specify these requirements, instead of applying labels to components, is a more expressive method, as it allows scope and

---

[1] www.adelaide.edu.au/directory/scott.bourne?dsn=directory.file;field= data;id=24812;m=view

other options to be adjusted, and enables complex requirements to be specified, such as satisfactory recovery conditions for specific failures.

We propose six component-level templates, as shown in Table 1. Compensate Failure, CompensateSuccess, and Alternative, require both the component and the compensatory or alternative operations as variables. The Card attribute specifies the cardinality relationship between these variables. Templates Non Retriable, RetriablePivot and NonRetriablePivot contain only a single component and specify how that component may be treated following failure or success. All templates contain a variable to define the Scope of the requirement.

Table 2 contains a complete specification of the CompensateFailure template. The template is implemented by specifying an operational behavior state as the *component*, and a boolean *Recovery* condition that when satisfied, reflects satisfactory failure recovery. The cardinality field allows the user to specify whether *Recovery* only applies to a single failure, or to several failures of the same component. The Scope can be global (G), or a function over a condition P. Cardinality and scope determine the LTL property to be used. As shown in the example, the LTL property of cardinality $1:1$ and global scope can be informally translated as: it is always the case that if there is a *FAULT* message sent from the component, the component will not be executed until its *Recovery* operation is performed; nevertheless, a fault of the component will always be followed by *Recovery*. This template can be implemented as shown by the example row, which specifies the required fault handling at Card Authorization in the online payment design.

3.2   Composition-Level Templates

Composition-level templates differ from component-level templates by specifying requirements over the entire composition, such as preconditions, triggers, or reachability conditions for entering control behavior states. For example, in the online payment design, the success of Card Authorization and Commit Payment could be preconditions for commit. Our templates can capture these requirements by using control behavior states.

The proposed set of composition-level temporal logic templates is shown in Table 3. ControlStateCritical, ControlStateTrigger, ControlStateReachable, and ControlStateUnreachable allow users to specify pre-conditions, triggers and reachability conditions for entering control behavior states. Compensation and ConditionalCompensation allow users to verify that the compensation actions of the composition meet requirements. The underlying properties of these compensation templates partially overlap some component-level templates, but the Web service designer can determine which template type is appropriate. For example, if one operation undoes the effect of several components, it would be simpler to implement a single composition-level template. Conversely, if several components each have a corresponding rollback operation, verifying each relationship individually through component-level templates is preferable.

Table 4 shows the full template specification of ControlStateCritical, which specifies a precondition for entering a control behavior state. To express this property in temporal logic, LTL with past-time operators is applied [12]. The O

Table 2. Template specification for `CompensateFailure`

Name	CompensateFailure <Component,Recovery,Card,Scope>		
Type	Component-level		
Variables	*Component*	An operational behavior state that requires recovery upon failure.	
	Recovery	A condition that undoes the effect of the failure. This can be a single component or a set of components structured with $\wedge$ and $\vee$ operators.	
	Card	One of the cardinality options below.	
	Scope	One of the scope options below.	
Description	The failure of *Component* leaves an impact an effect, which must be compensated by *Recovery* becoming true in the future.		
Prerequisite	A `Fault` message originating from *Component* in the operational behavior is necessary for this requirement to be verified.		
Cardinality	1:1	*Recovery* undoes one failure of *Component*.	
	Many:1	*Recovery* can undo many failures of *Component*.	
Scope	G	The template applies in all executions.	
	P	Applies during the satisfaction of a condition P.	
	$\neg P$	Applies during the negation of a condition P.	
	Before P	*Recover* must precede the satisfaction of P.	
LTL	1:1	G	$G(Component.FAULT \rightarrow ((\neg(Activated \wedge Component) \cup Recovery) \wedge F(Recovery))$
		P	$F(P) \rightarrow G(Component.FAULT \rightarrow ((\neg(Activated \wedge Component) \cup Recovery) \wedge F(Recovery))$
		$\neg P$	$G(\neg P) \rightarrow G(Component.FAULT \rightarrow ((\neg(Activated \wedge Component) \cup Recovery) \wedge F(Recovery))$
		Before P	$G(Component.FAULT \rightarrow (((\neg(Activated \wedge Component) \wedge \neg P) \cup Recovery) \wedge F(Recovery))$
	Many:1	G	$G(Component.FAULT \rightarrow F(Recovery))$
		P	$F(P) \rightarrow G(Component.FAULT \rightarrow F(Recovery))$
		$\neg P$	$G(\neg P) \rightarrow G(Component.FAULT \rightarrow F(Recovery))$
		Before P	$G(Component.FAULT \rightarrow ((\neg P \cup Recovery) \wedge F(Recovery)))$
Example	CompensateFailure <Card Authorization,PUT Card Data,1:1,G>		

operator is used to specify a property that must have occurred previously, while the H operator defines a property that must hold in *all* previous states. The example specifies the critical condition for entering the `Done` state.

Table 3. Names and descriptions of the composition-level temporal logic templates

Template Signature	Description
`ControlStateCritical` `<ControlState,Condition,Scope>`	A condition required for entering a control behavior state.
`ControlStateTrigger` `<ControlState,Condition,Scope>`	A condition that must trigger a control behavior state in the future.
`ControlStateReachable` `<ControlState,Condition,Scope>`	A condition that indicates a control behavior is reachable.
`ControlStateUnreachable` `<ControlState,Condition,Scope>`	A condition that indicates a control state should not be reachable in the future.
`Compensation` `<CompCondition>`	Specifies a condition that must be met during any compensation process.
`ConditionalCompensation` `<ExecCondition,CompCondition>`	Given a condition that can be satisfied during successful execution, specifies a second condition for compensation.

4   Proposed Verification Approach

We apply model checking [14] to verify that a Web service composition designed using our control and operational behavior model conforms to a set of transactional requirements specified with temporal logic templates. Model checking is a method to formally verify a system against a set of properties by exhaustively exploring the system state space. If a contradiction is found, a stack trace demonstrating the violation is produced. We employ NuSMV [15] for model checking, since it provides support for properties specified in LTL and CTL.

To address the state explosion problem inherent in model checking [14], we automatically reduce the state space of the control and operational behavior model as much as possible, by removing states and messages not relevant to the requirements being verified. To this end, we generate a Kripke structure [16] based on the temporal relations between the variables specified in the templates. Template variables can be defined as $\mathcal{V} \subseteq \mathcal{S}_{co} \cup \mathcal{S}_{op} \cup \mathcal{M}$ where $\mathcal{S}_{co}$ and $\mathcal{S}_{op}$ are control and operational behavior states and $\mathcal{M}$ is the set of inter-behavior messages. The Kripke structure will capture all instances of elements from $\mathcal{V}$ in the design, and the transition relation between those instances.

A Kripke structure is a finite-state system model defined as $\mathcal{K} = \langle \mathcal{S}_k, \mathcal{I}, \mathcal{T}_k, \mathcal{L} \rangle$, where $\mathcal{S}_k$ is a finite set of states, $\mathcal{I} \subseteq \mathcal{S}_k$ is the set of initial states, $\mathcal{T}_k \subseteq \mathcal{S}_k \times \mathcal{S}_k$ is the transition function, and $\mathcal{L}$ is the labelling function that assigns *atomic propositions* to each state. The atomic propositions are the unique set of properties that hold at a given state. We identify a set of three atomic propositions for each state in the Kripke structure; *i)* the control behavior state; *ii)* the operational behavior state; and *iii)* the most recent inter-behavior message, represented as a pair (s_{op}, m) of related operational behavior state and message type. Since we aim to reduce the model to the properties we wish to verify, the Kripke structure only contains the states with atomic propositions with elements from $\mathcal{V}$.

Table 4. Template specification for `ControlStateCritical`

Name	`ControlStateCritical<ControlState,Condition,Scope>`	
Type	Composition-level	
Variables	*ControlState*	The control behavior state this critical condition applies to.
	Condition	The precondition for entering this control behavior state. This can be a single component or a set structured with $\wedge$ and $\vee$ operators.
	Scope	One of the scope options below.
Description	*Condition* denotes the precondition for entering *ControlState*. When *ControlState* is entered, *Condition* must have been met previously on the execution path.	
Scope	G	The template applies in all executions.
	P	Applies during the satisfaction of a condition P.
	$\neg P$	Applies during the negation of a condition P.
	Before P	*ControlState* is entered before P is met.
LTL	G	$G(\textit{ControlState} \rightarrow O(\textit{Condition}))$
	P	$F(P) \rightarrow G(\textit{ControlState} \rightarrow O(\textit{Condition}))$
	$\neg P$	$G(\neg P) \rightarrow G(\textit{ControlState} \rightarrow O(\textit{Condition}))$
	Before P	$G(\textit{ControlState} \rightarrow (O(\textit{Condition}) \wedge H(\neg P)))$
Example	`ControlStateCritical` `<Done,Card Authorization,Commit Payment,G>`	

To build the Kripke structure, the control and operational behavior model must be exhaustively traversed, so the Kripke states can be created and linked as elements from $\mathcal{V}$ are encountered. Due to space constraints, we omit the detailed algorithm, but provide a description of the verification process. The control and operational behavior are explored with a depth-first traversal, starting from the control behavior state `Not Activated` and the operational behavior yet to be initialized. From this state, the control behavior activates and explores each possible way to trigger operations in the operational behavior through inter-behavior messages. The traversal explores every possible execution path within the operational behavior, and every reachable inter-behavior message. When an element from $\mathcal{V}$ is encountered, a Kripke state with the atomic properties currently true in the traversal is either created, or a transition to an existing Kripke state containing those properties is added. The traversal handles cycles in the model by backtracking when an exiting Kripke state is linked to, a set of atomic properties are revisited since the last addition to the Kripke structure, or the control behavior terminates through `Done`, `Abort` or `Compensated`. The Kripke structure is complete once the traversal returns to `Not Activated`.

Figure 2 shows a Kripke structure of the design in Section 2 and the example template inputs of Tables 2 and 4, such that $\mathcal{V} = \{$`Card Authorization`, `PUT Card Data`, `Commit Payment`, `Done`$\}$. Each state is labeled with three atomic propositions as described above. While a Kripke structure of all reachable atomic

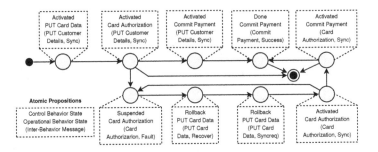

Fig. 2. An example Kripke structure generated from the online payment composition

propositions of the model of Figure 1 would contain 31 states, the reduction measures minimize this structure to 11, creating less model checking overhead.

5   System Implementation and Experiments

We have implemented our verification approach in a prototype tool, with an interface for specifying control and operational behaviors as shown in Figure 3. The prototype reduces the model to the Kripke structure, and writes it to an SMV (Symbolic Model Verifier) file with the temporal logic representations of the implemented templates. NuSMV uses this file to exhaustively verify the model against the set of temporal properties, and lists the properties found to be true, plus any violating state sequences. To further help the designer, our future work will interpret these sequences to diagnose design flaws within the control and operational behavior.

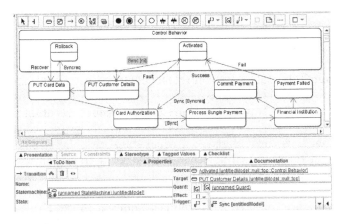

Fig. 3. Specifying a Web service composition as control and operational behaviors

We demonstrate our proposed approach with an extension to the online payment example of Section 2. This design uses the PayLane Web service API[2], and

---
[2] http://devzone.paylane.com

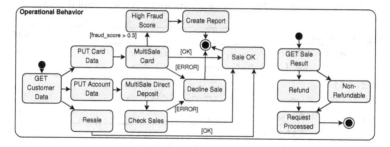

Fig. 4. The operational behavior model of the complex online payment composition

extends the online payment example from Section 2 by incorporating multiple payment options, namely, `card charge` and `direct debit`. While the control behavior model remains the same as the earlier example, the operational behavior and inter-behavior messages of the complex online payment design are shown in Figure 4 and Table 5 respectively. The composition enables users to pay by credit card or direct deposit, either by entering new data or retrieving details from a previous transaction. When a card payment is processed, a `fraud score` is returned to indicate the likeliness of fraud being committed. In cases where this score is above a given threshold, a report of the transaction is made. When a direct deposit is made, the result of the transaction is not immediately available and must be checked. The design also contains a compensatory process that determines whether to process a refund based on retrieved sales details.

Table 5. Inter-behavior messages specified over the online payment design

Message	Source	Target	Guard
Sync	Activated	GET Customer Data	[no message]
Sync	Activated	PUT Card Data	[Resale.SYNCREQ]
Sync	Activated	PUT Account Data	[Resale.SYNCREQ]
Syncreq	Resale	Activated	–
Delay	Activated	MultiSale Card	–
Fault	MultiSale Card	Activated	–
Fault	Check Sales	Activated	–
Success	Sale OK	Activated	–
Success	Request Processed	Rollback	–
Fail	Decline Sale	Rollback	–
Fail	Create Report	Rollback	–
Recover	GET Sale Result	Rollback	[Sale OK.SUCCESS]
Recover	High Fault Score	Rollback	[MultiSale Card.FAULT]
Recover	Decline Sale	Rollback	[FAULT]

We identify seven critical transactional requirements to be verified over this design, as shown in Table 6. Requirement TR1 specifies that when a committed sale needs to be undone, a refund should be processed, excepting non-refundable

Table 6. Requirements implemented with temporal logic templates and verified

ID	Requirement	Result
TR1	CompensateSuccess `<Sale OK, Refund	Non-Refundable, 1:1, G >`	Passed
TR2	Alternative `<Resale, PUT Account Data	PUT Card Data, 1:1, G>`	Passed
TR3	RetriablePivot `<Check Sales, G>`	Passed
TR4	ControlStateCritical `<Aborted, Create Report, High Fraud Score>`	Passed
TR5	ControlStateTrigger `<Done, Sale OK & (MultiSale Card |` `MultiSale Direct Deposit | Resale), G >`	Passed
TR6	ControlStateReachable `<Done, Resale.SYNCREQ, G>`	Passed
TR7	ControlStateReachable `<Done, MultiSale.DELAY, G>`	Failed

sales. TR2 requires that the failure of historical transaction data sales should lead to card or account data being updated before the sale is retried. Requirement TR3 asserts that the result of direct deposit sales should be requested until one is obtained, which then leads to commit or abort. In the event that a high `fraud score` is detected, requirement TR4 specifies that a report must be generated before the composition aborts. TR5 requires that the composition should always commit following the successful processing of card or direct deposit payment. Requirement TR6 specifies that commit should be reachable following the failure of resale, while TR7 requires that commit should be reachable even if the response of `MultiSale Card` is delayed. NuSMV took 0.03 seconds to verify these requirements and determined that the design violates TR7, since `Done` is unreachable following a `Delay` message to `MultiSale Card`. To satisfy TR7, the design must be refined to enable `MultiSale Card` to be retried or replaced following a delay. The remaining requirements were satisfied for all states in the design.

6   Related Work

Ensuring and verifying transactional requirements in Web service composition has been an area of increasing research interest in recent years. Bhiri et al. [5] and Montagut et al. [6] present Web service composition methods to ensure failure atomicity properties defined as an Accepted Termination States (ATS) [17] model. While an ATS model allows detailed specifications of failure atomicity, it is an exhaustive method that increases exponentially in size as the composition grows, whereas our template set allows requirements to be flexibly defined only to components the designer wishes to verify. Montagut et al. [8] limit the ATS model to zones of the composition where transactional requirements are considered critical. Despite this, all valid termination state combinations must be exhaustively defined, while in our approach the designer can determine the level of requirement detail. The FACTS framework [3] enables users to verify that the transactional behavior and exception handling of a design model supports components specified critical for success. Our composition-level templates enables more detailed composition-level requirements with the definition of critical preconditions, triggers, and reachability conditions for commit, abort, and other transactional states. Another approach proposed by El Haddad et al. [7] uses *risk levels* provided by the user to specify whether the resulting composition should be compensatable, but reduces user control over the transactional behavior of their composition. In contrast, our approach allows compensatory activities to be explicitly specified and verified against business requirements.

Our use of templates draws inspiration from existing work in business rule compliance with temporal logic patterns. Dwyer et al. [9] defined a set of patterns in LTL and CTL with various scope options, based on commonly recurring temporal logic property structures found across surveyed specifications. This initial pattern set has been adapted and expanded in subsequent research. Smith et al. [10] extended this set into templates, allowing them to be customized according to cardinality and other fine-grained options. Elgammal et al. [11] produced an expanded pattern set with a framework that enables atomic patterns to be composed together, and includes a set of basic composite patterns. Since our focus is on transactional requirements instead of general compliance, we produce a set of temporal logic templates highly specialized and more appropriate towards that use. Furthermore, our template set applies to a specific design model with control and operational models, which allows our templates to specify requirements at the component-level and composition-level, increasing granularity and expressiveness at no additional computational cost. Finally, Yu et al. [18] apply temporal logic patterns towards verifying Web service compositions. However, their work analyzes WS-BPEL schemas for conformance to general functional properties, while our approach can be applied prior to any development.

7 Conclusion

The verification of transactional requirements of Web service compositions using information derived from business logic remains an important challenge despite increasing research interest in recent years. Identifying and resolving compliance issues to transactional requirements early in development is desirable. Furthermore, the formal specification of transactional requirements is error-prone and an approach to hide the specification complexities from the Web service designer will increase the reliability of the design. In this paper, we propose a set of temporal logic templates to formally verify component-level and composition-level transactional requirements of Web service compositions at design time. Each template is defined by the Web service designer according to a specification that contains a description, variables, and scope and cardinality variants. The implemented templates are automatically translated into temporal logic properties that are verified using a model checker. The proposed approach has been successfully implemented and verified with several example scenarios. Our future work includes supporting parallel workflow patterns, enabling more sophisticated guard conditions, and scalability tests with increasingly complex scenarios.

References

1. Papazoglou, M.P., Traverso, P., Dustdar, S., Leymann, F.: Service-Oriented Computing: State of the Art and Research Challenges. IEEE Computer 40(11) (2007)
2. Sheng, Q., Maamar, Z., Yahyaoui, H., Bentahar, J., Boukadi, K.: Separating Operational and Control Behaviors: A New Approach to Web Services Modeling. IEEE Internet Computing (3), 68–76 (2010)
3. Liu, A., Li, Q., Huang, L., Xiao, M.: FACTS: A Framework for Fault-tolerant Composition of Transactional Web Services. IEEE Transactions on Services Computing 3(1), 46–59 (2010)
4. Domingue, J., Fensel, D.: Toward a Service Web: Integrating the Semantic Web and Service Orientation. IEEE Intelligent Systems 23(1), 86–88 (2009)
5. Bhiri, S., Perrin, O., Godart, C.: Ensuring Required Failure Atomicity of Composite Web Services. In: Proceedings of the 14th International Conference on World Wide Web, pp. 138–147. ACM (2005)
6. Montagut, F., Molva, R., Golega, S.: Automating the Composition of Transactional Web Services. International Journal of Web Services Research (IJWSR) 5(1) (2008)
7. El Hadad, J., Manouvrier, M., Rukoz, M.: TQoS: Transactional and QoS-Aware Selection Algorithm for Automatic Web Service Composition. IEEE Transactions on Services Computing 3(1), 73–85 (2010)
8. Montagut, F., Molva, R., Tecumseh Golega, S.: The Pervasive Workflow: A Decentralized Workflow System Supporting Long-Running Transactions. IEEE Transactions on Systems, Man, and Cybernetics, Part C: Applications and Reviews (2008)
9. Dwyer, M.B., Avrunin, G.S., Corbett, J.C.: Patterns in Property Specifications for Finite-State Verification. In: Proceedings of the 1999 International Conference on Software Engineering, pp. 411–420. IEEE (1999)
10. Smith, R.L., Avrunin, G.S., Clarke, L.A., Osterweil, L.J.: PROPEL: an Approach Supporting Property Elucidation. In: Proceedings of the 24th International Conference on Software Engineering, pp. 11–21. ACM (2002)
11. Elgammal, A., Turetken, O., van den Heuvel, W.-J., Papazoglou, M.: Root-Cause Analysis of Design-Time Compliance Violations on the Basis of Property Patterns. In: Maglio, P.P., Weske, M., Yang, J., Fantinato, M. (eds.) ICSOC 2010. LNCS, vol. 6470, pp. 17–31. Springer, Heidelberg (2010)
12. Emerson, E.: Temporal and Modal Logic. In: Handbook of Theoretical Computer Science, vol. 2, pp. 995–1072 (1990)
13. Bourne, S., Szabo, C., Sheng, Q.Z.: Ensuring Well-Formed Conversations Between Control and Operational Behaviors of Web Services. In: Liu, C., Ludwig, H., Toumani, F., Yu, Q. (eds.) ICSOC 2012. LNCS, vol. 7636, pp. 507–515. Springer, Heidelberg (2012)
14. Baier, C., Katoen, J.P., et al.: Principles of Model Checking. MIT Press (2008)
15. Cimatti, A., Clarke, E., Giunchiglia, E., Giunchiglia, F., Pistore, M., Roveri, M., Sebastiani, R., Tacchella, A.: NuSMV 2: An Opensource Tool for Symbolic Model Checking. In: Brinksma, E., Larsen, K.G. (eds.) CAV 2002. LNCS, vol. 2404, pp. 359–364. Springer, Heidelberg (2002)
16. Kripke, S.: Semantical Considerations on Modal Logic. Acta Philosophica Fennica 16, 83–94 (1963)
17. Kim, W.: Modern Database Systems: The Object Model, Interoperability, and Beyond. ACM Press/Addison-Wesley Publishing Co. (1995)
18. Yu, J., Manh, T.P., Han, J., Jin, Y., Han, Y., Wang, J.: Pattern Based Property Specification and Verification for Service Composition. In: Aberer, K., Peng, Z., Rundensteiner, E.A., Zhang, Y., Li, X. (eds.) WISE 2006. LNCS, vol. 4255, pp. 156–168. Springer, Heidelberg (2006)

Transitional Resource Meta-model: Generating Restful Service to Implement Complex Activity

Lu Fang, Hongming Cai, Cheng Xie, and Lihong Jiang

School of Software, Shanghai Jiao Tong University, Shanghai, China
{illusion,hmcai,chengxie}@sjtu.edu.cn, jiang-lh@cs.sjtu.edu.cn

Abstract. Service-oriented architecture is widely adopted in the development of web information system. The construction of web service, however, is challenging, since it should not only meet up with intensively changing requirements but also be capable of handling complex activity where multiple entities are involved. The traditional RESTful service is restricted in implementing complex activity since it is single-resource-oriented. Using SOAP services to implement is possible but time-consuming, thus not adaptive to the sensitive requirements change. Therefore, a Transitional Resource Meta-model (TRM) is proposed in this paper to generate the RESTful service with the capability of executing complex activity in a flexible and fast way. Our proposed model functions on describing the complex activity by using a state transfer sequence for multiple entities as well as generating the service interface and controlling the execution of the service. A case study is given to represent the construction process and generation results of TRM, and a comparison with REST-based architecture and SOAP-based architecture is provided at the end to show the advantages of TRM approach.

Keywords: RESTful service, service generation, complex activity, state transfer, multiple entities, web information system.

1 Introduction

In today's construction of web information systems, there is an increasing demand to implement the complex and intensively changing business requirements [1]. The web service is a popular choice for enterprises to implement such an adaptive system [2]. Correspondingly, the construction of the web service is faced with two main challenges: (1) to proceed in a fast mean (since the intensive change of requirements makes the modification or creation of service a frequent matter); (2) and to support complex business activities (a typical example is the merge payment for multiple orders, where conditional judgments, iterations can be found and more than one type of resource is involved in). Therefore, finding a fast approach to construct web services that can implement complex activities is the key problem to solve.

REST style architecture [3] is proposed to develop light-weighted services. It is advantageous in providing scalable components using generic uniform interface and independent deployment by regarding every involved entity as a resource. Under the

X. Lin et al. (Eds.): WISE 2013, Part I, LNCS 8180, pp. 257–266, 2013.

support of many brilliant frameworks (JAX-RS [4] or ruby on rails [5]), the development and deployment of RESTful service is simple and convenient. The functionality of RESTful service is, however, limited since it can hardly handle the complex activities where the state transfers of more than one type of resource entity are executed. Therefore, pure REST style architecture is only suitable for the representation of single entities, but not suitable for the further complex demands.

A great many developers attempt to integrate REST with SOAP to absorb the advantages of both [6]. With the high flexibility, the RPC-style service owns the capacity to implement complicated logics in the way of traditional programming. It does not provide high usability for the need of rapid development. This style is also criticized to be a bad mean which is actually not RESTful [7], bringing inconvenience for service consumers.

Considering the advantages and the shortages of the current research situations, we proposed a Transitional Resource Meta-model (TRM) to facilitate the rapid construction of RESTful services supporting implement complex activities. The main contribution of our paper lies in three aspects: (1) Proposes the concept of entity loaders to support activity with multiple entities involved; (2) Proposes the state transfer sequence to support complex logic; (3) Implements a platform to realize the automatic generation and execution of RESTful services based on TRM, saving a great deal of development cost.

The rest of the paper is organized as follows. In Section 2, we explain the modeling of TRM in details. In Section 3, we expand on how service is generated based on TRM. In Section 4, a case study from a typical B2C scenario is presented. In Section 5, we make conclusions on our studies and present our directions of future works.

2   Overall Framework

To provide the capability of handling complex activity for the web service in a resource-oriented architecture, we propose TRM based on REST architecture.

TRM is the meta-model of "transitional resource". Different from entity resources in traditional REST architecture, transitional resource is a representation of the execution process of complex activities and related event logs. It consists of four partitions: the event logs to record the log information for each activity; the inputs and outputs of the service; the involved entities during the process; and finally the execution sequence of the entities.

In Figure 1, we show how TRM is constructed, and how it controls the automatic execution of corresponding web service.

The construction of TRM consists of four procedures, the definition of service log, the definition of input and output for the service interface, the abstraction of involved entities to form the entity loader model and finally the abstraction of transition sequence to form the transition sequence model. A TRM formatted in XML will be generated based on the above procedures. Then a corresponding service interface exposing to the service consumer is auto-generated, where request from the client is received, and corresponding response is sent.

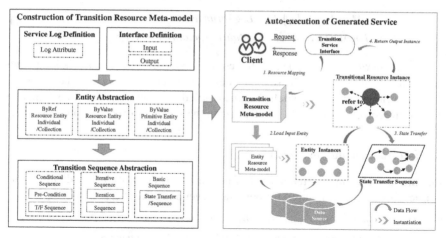

Fig. 1. Construction and Execution Process of Web Service based on TRM

In the first step, the request and URI pattern from the client is resolved to map corresponding TRM in the server.

In the second step, the defined Entity Loaders firstly pre-loads the required entity resource meta-models, which would play a role in fetching the instance data from the distributed data sources. Then the returned instance data would be assembled into representable entity resource instances (encapsulated in XML or JSON format). Primitive entities are also initialized in this step.

In the third step, the State Transfer Sequence Model controls the execution of process by scheduling on how and in what order the entity resources would be manipulated. Through interpreting, the sequence is executed and corresponding resource instances would be state-transferred. A transitional resource instance representing the transition process itself would also be generated and could be a member that involved in the sequence. Meanwhile, the updated resource instances would be persisted and written back to the data sources.

In the last step, the TRM then determines what instances would be returned to the client according to its Output Definition and Service Log Model. So the selected instances would be finally encapsulated and sent back to the client as responses.

Above procedures described how the "POST" interface is provided to the client. Moreover, the "GET" and "DELETE" interface is also supported to fetch or delete the history log through the traditional implementation approach of REST architecture.

3   Transitional Resource Meta-model

3.1   Log Attributes Model

The Log Attributes Model defines a set of attributes that record the log information generated in the execution process. It provides a representation of attribute format as well as enough information to access the data source. For instance, the *"timestamp"* records the time when the service is executed and the *"status"* records whether the whole service is executed successfully or not.

$$LM ::= \{la_1, ..., la_n\}, la_i \in LogAttribute \tag{1}$$
$$LogAttribute ::=< attributeName, type, attributeDesc >$$

The *attributeName* is equivalent to the column name of a table in a relational database, and the *attributeDesc* is the semantic description for the attributes. The data of all these log attributes could be fetched and modified by accessing the transitional resource instance itself using a *"this"* reference in the sequence execution process.

3.2   Interface Definition Model

In the Interface Definition Model, a set of parameters is defined to provide the format and structure of input and output for both the server and client:

$$Input ::=< type, name, isCollection?, itemname? >$$
$$Output ::=< type, name, source, isCollection?, itemname? > \tag{2}$$

The range of *type* in the parameter not only includes the basic data type, but also a "resource" type representing the entity resources, which are represented in XML or JSON data format. The *name* attribute of input and output defines the variable names appeared in the HTTP entity body.

An input/output with a collection of data is defined by the *isCollection* attribute with a value of true or false. The *source* attribute defines the origins of the output instance data, through searching the *source* variable name in the processed entities.

3.3   Entity Loading Model

An Entity Loading Model is a significant sub-model that pre-loads the actual resource entities or primitive entities as the inputs of the transition sequence.

$$EntityLoader ::=< type, source, isCollection?, \tag{3}$$
$$variableName, link|iterativelink|value >$$

The Entity Loaders are components that define how the involved entities are initialized before the process execution. They play the role to assemble an entity from the data source or initialize a new entity based on a specified type. Every entity owns a unique key defined by the required attribute called *"variableName"*.

The instance data of an entity may be initialized or loaded through two approaches: one is by using the data contained in the request input directly, the other one is by referencing to the resource entity in the server using a link URI with an id or scoping information. These two approaches are classified by using the attribute called *"source"*, the former one is defined as *"byValue"*, while the latter one is defined as *"byRef"*. Correspondingly, a *link/iterativelink/value* pattern is defined to fetch the data.

A "byRef" entity requires a link to reference to the required resource entity. In the link, there are some actual data that should come from the input of the client or pre-defined entities, we use a bracket to define these uncertain variables.

A *"byValue"* entity contains a value. In the value pattern, both a single variable and a calculation expression based on the pre-defined variables are allowed. It could also be defined as a collection one if it is a collection type. Different from the "byRef" entity, it initialized the variable directly without request to the other resources.

3.4   Transition Sequence Model

The Transition Sequence Model defines the execution process of the service, consisting of a series of sequences with a tree structure. The non-leaf nodes are all sequences, while the leaf nodes are all transfers.

A Sequence is an ordered set of sub-sequence or state transfers that manipulate over the involved entities. There are three types of independent sequence to allow three types of logic control:

1. Basic Sequence

Inherited from sequence, it has no limitations, with an ordered execution process.

$$Sequence_{basic} ::= < Sequence^*, Transfer^* > \tag{4}$$

2. Conditional Sequence

$$Sequence_{conditional} ::= < pre-condition, tSequence, fSequence? > \tag{5}$$

Conditional sequence allows a selective branch to execute different sequence. *tSequence* and *fSequence* are dependent sequences, representing the sequences when the pre-condition is obeyed or not. In a *pre-condition*, there exists a logic expression consisting of the declared entity variables in the Entity Loaders and some basic arithmetic operations as well as logic operations; it must return a Boolean value.

3. Iterative Sequence

$$Sequence_{iterative} ::= < iterator, \; iSequence > \tag{6}$$

$$Iterator ::= < collection, \; itemName > \tag{7}$$

Iterative Sequence is a sequence that executes a sequence repeatedly by passing iterable items of a collection. *iSequence* is the iterative one that would be repeatedly executed. An iterator owns a collection variable and an itemName to represent the item.

The atomic unit under each sequence is defined as Transfer, which implements the state transfer on each entity:

$$Transfer ::= < target, \; method, \; valueSetter+ > \tag{8}$$

$$valueSetter ::= < value, target? > \tag{9}$$

The method defines the transfer method on the resource, a *target* references to the objective of the transfer, i.e. the resource entity or primitive entity to be transferred. For each transfer, there is a list of valueSetters to set the attributes to required values or set the primitive entity directly to the wanted value. The value could be an expression but must return a legal type consisted with the type defined in the entity meta-model or in the Entity Loader Model. Once the transfer is executed, corresponding method will operate on the target, and the data of the resource would also be persisted back into the data source.

To execute the sequence, an interpretation mechanism is required, the corresponding interpretation algorithm is shown as following:

```
program Trans{TranSequence}
  foreach (s  in TranSequence)
    if(s •Transfer)
     setState(s.target, s.valueSetters)
      transfer(s.target, s.method)
    else if(s•Sequence_conditonal) conditionalTrans(s)
    else if(s•Sequence_iterative) iterativeTrans(s)
    else Trans(s)
  return true
program conditionalTrans(TranSequence)
  if TranSequence.pre-conditon is true
    Trans(TranSequence.tSequence)
  else Trans(TranSequence.fSequence)
program iterativeTrans(TranSequence)
  foreach(item in TranSequence.Iterator.collection)
  Trans(TranSequence.iSequence)
```

4   Case Study and Analysis

This section will present a case study, showing the construction and generation process of the web service based on a common B2C Scenario.

4.1   Case Study

In our scenario, a customer needs to pay for a list of orders together with a coupon code he has owned. The coupon code contains a discount rate if it is available. The service should calculate the total price of the unpaid orders with the discount rate recorded in the coupon, and then transfer the money from the buyer account to the seller account if the balance of the buyer account is enough. Since this is a B2C website, the seller account is a constant one. And in the final, a bill is generated if orders are paid successfully.

This whole activity could be implemented through providing a "Merge Payment" service to the client. To construct the TRM for such a scenario, we follow a series of following steps to abstract pre-requisite meta-data for the TRM construction:

Step 1: Definitions of the Input / Output and the log attributes.

According to the scenario, the input and output should include a coupon, a buyer account and a list of orders, the output should include a bill for this payment. The service log should include not only status and timestamp but also a remark.

Step 2: Abstraction of the entities.

In this step, we abstract the entities that are involved in the execution of a payment. According to the scenario, involved entities should include five resource entities and one primitive variable entity to sum up prices, as shown in Table 1.

Table 1. Entity Abstraction

Entities Name	Type	Mapped Resource	Individual /Collection	ByRef /ByValue	Initial Link/Value
Orders	Resource	Order	Collection	ByValue	Orders
Buyer Account	Resource	Account	Individual	ByRef	/example/Account/{buyer AccountId}
Seller Account	Resource	Account	Individual	ByRef	/example/Account/seller
Bill	Resource	Bill	Individual	ByRef	/example/Bill/_new
Coupon	Resource	Coupon	Individual	ByRef	/example/Coupon/{coupo nId}
Total Price	Double	Null	Individual	ByValue	0.0

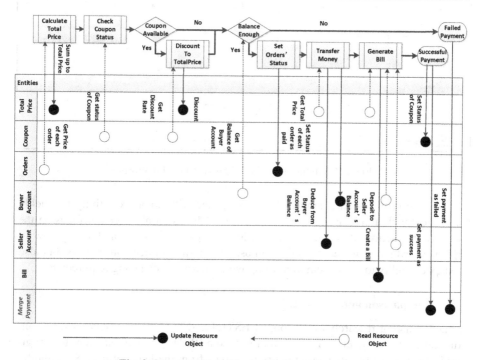

Fig. 2. Sequence Flow Chart with Status Transfers

Step 3: Abstraction of the sequence and the state transfers.

In the third step, the inner execution process of a payment could be abstracted as the flow chart in Figure 2 showing how the process is executed and how involved entities are state-transferred. The flow chart on the horizontal axis shows in what sequence the state transfers are organized, modeling by the three types sequence. While the actions over the entities on the vertical axis shows how each state transfer is conducted, i.e. the method and the target state.

Step 4: Generate TRM and execute the service over the implemented framework.

In the last step, we construct TRM through a GUI configuration platform as shown in Figure 3. Except for the configuration of four sub models, a *DBConfig* is also configured for the data source to support distributed data access and a URI is configured to provide service interface to the client. The platform supports the generation of an xml-formatted TRM. Based on the constructed TRM, our implemented framework would provide a service interface and generate corresponding WADL files for the call of the service. User can invoke the service through basic HTTP requests.

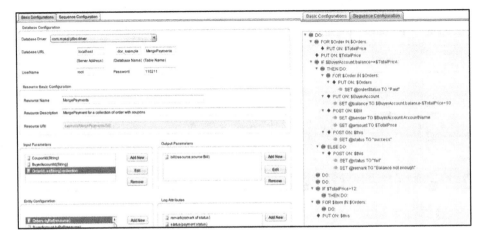

Fig. 3. Configuration of TRM based on GUI platform

This case provides an example of how a service with complex activity is quickly constructed with almost no programming efforts. The user only needs to focus on the business abstraction of the scenario to a logic sequence instead of thinking how to implement, which provides not only high usability to the service constructors, but also good extensibility to the system to meet up with intensively changing requirements.

4.2   Comparison and Analysis

In many existing research works, both RESTful and SOAP are widely used in constructing the legal systems. Yan Liu represented a REST-based architecture to reengineer legal systems by abstracting entity resources from the legal system [8]. Timothy K. Shih applied SOAP and WSDL techniques to construct a learning management system [9]. Both of them have advantages over traditional non-SOA architecture but still remain some problems and shortages.

Therefore, based on these two cases, we make a comparison of our TRM-based architecture with RESTful and SOAP to show how TRM-based win over either of two traditional services from four aspects of characteristics, functionality, adaptability, usability and scalability according to the functionality and basic quality attributes measurements of software architecture proposed in [10]. A detailed comparison is shown in table 2.

Table 2. Comparison of TRM-based with REST-based and Pure RESTful

Architectural Index	TRM-based Architecture	REST-based Architecture	SOAP-based Architecture
Functionality			
Complex logic	Strong	Weak	Strong
Manipulation over multiple Entity	Strong	Weak	Medium
Operational Persistence	Strong	Weak	Weak
Status Monitoring	Strong	Weak	Medium
Adaptability			
Modification	Easy	Medium	Hard
Adding new function	Easy	Hard	Hard
Uniform Interface	High	High	Low
Client Diversity	High	High	High
Loose Coupling with client	High	High	Low
Usability			
Implementation Efforts	Low	Medium	High
Configurability	High	Medium	Low
Learning Cost	Low	Medium	High
Scalability			
Support Distributed Data Source	High	Low	Low
Support Distributed Calculation	High	Low	Low

According to the comparison, we can conclude that TRM-based architecture is more suitable on implementing services that involve manipulation for multiple entities and complicated logics over pure RESTful services from the view of functionality. Generated service would support the complex activity well.

On the other hand, our case study shows that the usability is quite high particularly for non-programmers, which provides great convenience on modifying or creating new services when requirements changes are coming. Therefore, with high usability and adaptability that requires less developing cost and adjustment cost, it is also more suitable for the scenarios with increasing and flexible requirements compared with the REST architecture and particularly SOAP-based architecture.

5   Conclusion And Future Work

In this paper, we proposed the TRM to solve the problem of generating web service that supports complex activity. It provides a convenient approach to define and automatically execute the complex activity by regarding the activity itself as a resource. By proposing transition sequence, where the state transfers of multiple resources are modeled, the construction and configuration of execution process become not only possible but also flexible to meet up with the sensitive business requirements.

Our future work will focus on the automatic composition of the transitional RESTful service based on TRM to support the process with larger granularity. The ontology will be involved as a significant data layer and corresponding semantic descriptions will be added to the TRM.

Acknowledgment. This research is supported by the National Natural Science Foundation of China under Grant No.71171132, the National High Technology Research and Development Program of China ("863" Program) under No.2008AA04Z126.

References

1. Sawyer, P., Bencomo, N., Whittle, J., Letier, E., Finkelstein, A.: Requirements-aware systems: A research agenda for Re for self-adaptive systems. In: 2010 18th IEEE International Requirements Engineering Conference (RE), pp. 95–103. IEEE (2010)
2. Bieberstein, N., Bose, S., Walker, L., Lynch, A.: Impact of service-oriented architecture on enterprise systems, organizational structures, and individuals. IBM Systems Journal 44(4), 691–708 (2005)
3. Fielding, R.T.: Architectural styles and the design of network-based software architectures (Doctoral dissertation, University of California) (2000)
4. Burke, B.: RESTful Java with Jax-RS. O'Reilly Media (2010)
5. Dix, P.: Service-oriented design with ruby and rails. Addison-Wesley Professional (2010)
6. Richardson, L., Ruby, S.: RESTful web services. O'Reilly Media (2008)
7. Renzel, D., Schlebusch, P., Klamma, R.: Today's Top "RESTful" Services and Why They Are Not RESTful. In: Wang, X.S., Cruz, I., Delis, A., Huang, G. (eds.) WISE 2012. LNCS, vol. 7651, pp. 354–367. Springer, Heidelberg (2012)
8. Liu, Y., Wang, Q., Zhuang, M., Zhu, Y.: Reengineering legacy systems with RESTful web service. In: 32nd Annual IEEE International Computer Software and Applications, COMPSAC 2008, pp. 785–790. IEEE (July 2008)
9. Shih, T.K., Chang, W.C., Lin, N.H., Lin, L.H., Hsu, H.H., Hsieh, C.T.: Using SOAP and .NET Web Service to build SCORM RTE and LMS. In: 17th International Conference on Advanced Information Networking and Applications, AINA 2003, pp. 408–413. IEEE (March 2003)
10. Bass, L., Clements, P., Kazman, R.: Software Architecture in Practice, 2nd edn. Pearson Education, India (1998)

UCOSAIS: A Framework for User-Centered Online Service Advertising Information Search

Hai Dong[1], Farookh Khadeer Hussain[2], and Elizabeth Chang[1]

[1] School of Information Systems, Curtin University of Technology, Australia
[2] School of Software, University of Technology, Sydney, Australia

Abstract. The emergence of Internet advertising brings about an economic and efficient marketing means for small and medium enterprises in service industries. Every day, massive service advertising information is published over the Internet. Nevertheless, on the other side, service consumers find it difficult to quickly and precisely retrieve their desired services. This problem is partly caused by the ubiquitous, heterogeneous, and ambiguous nature of online service advertising information. In this paper, we propose a systematic framework – UCOSAIS – for online service advertising information search. Inspired by the philosophy of user-centered design, this framework comprises an ontology-learning-based focused crawler for service information discovery and classification, a faceted semantic search component for service concept selection, and a user-click-based similarity computing component for service concept ranking adjustment.

Keywords: online service advertising information, service discovery, service search, user-centered design.

1 Introduction

It is well recognized that the information technology has a profound effect on the conduct of the business, and the Internet has become the largest marketplace in the world. It is estimated that, by June 2012, over one third of worlds population (over 2.4 billion) uses Internet, with an estimated annual growth of nearly 18%, in contrast to 360 million users in 2000[1]. Innovative business professionals have realized the commercial applications of the Internet for their customers and strategic partners. They therefore turn the Internet into an enormous shopping mall and a huge catalogue. Consumers are able to browse varieties of product or service advertisements from the Internet, and directly buy those goods through online transaction systems [1]. In the service industry, Internet advertising is also popular among small and medium enterprises, due to the advantages of low cost, high flexibility, and ease of publishing. Nevertheless, many service consumers find it difficult to quickly and precisely retrieve their desired service advertising information from the Internet, not only owing to the lack of specialized service

---

[1] http://www.internetworldstats.com/

X. Lin et al. (Eds.): WISE 2013, Part I, LNCS 8180, pp. 267–276, 2013.

information registration and retrieval platforms, but also because of the following features of online service advertising information.

Ubiquity. Service advertising information can be registered in various business information registries and stored in geographically dispersed servers in the global network.

Heterogeneity. Given the diversity of services in the real world, many schemes have been proposed to classify services from various perspectives. Nevertheless, there is not a publicly agreed scheme available for classifying service advertising information over the Internet. Furthermore, many commercial registries often mix up product and service advertising information [2].

Ambiguity. Most of the online service advertising information does not retain a consistent format or standard. They are described by natural languages and embedded in vast Web information, the content of which is sometimes ambiguous for service consumers to understand [2]; on the other hand, if service consumers do not have clear knowledge regarding their service queries, the queried results may be ambiguous and inaccurate.

Service search is not a new topic in academia. Substantial works has been published in the areas of Web services and other digital services. Nevertheless, few research efforts have been put into the area of online service advertising information search, by taking into account the above features of the advertising information. Although our previous research has focused on this issue by designing a novel framework [3,4], the overall performance of the prototype is not sufficiently convincing, due to a lack of understanding users perceptions and behaviors in the search process.

User-centered design (UCD) is a broad term to describe how system design interacts with end-users, and is a term widely used in the field of user interface design [5]. In this paper, we propose a User-Centered Online Service Advertising Information Search (UCOSAIS) framework, inspired by the philosophy of UCD. This framework comprises an ontology-learning-based focused crawler for service information discovery and classification, a faceted semantic search component for service concept selection, and a user-click-based similarity computing component for service concept ranking adjustment.

The rest of this paper is organized as follows: in Section 2, we review the previous research in service advertising information search and the theory of UCD; in Section 3, we introduce the technical details of the proposed UCOSAIS framework; in Section 4, we carry out a series of experiments to evaluate the framework; conclusion is drawn in the final section.

2   Related Works

As described previously, very few studies have been carried out on online service advertising information search.

Dong et al. [4,10] proposed an online transport service advertising information search approach. Within this framework, an OWL-annotated transport service ontology was designed to represent transport service domain knowledge and to classify standardized and semanticized transport service information, namely transport service metadata. The principle of this approach is 1) matching service requesters queries and concepts from the transport service ontology by means of an Extended Case-Based Reasoning (ECBR) model and 2) retrieving associated transport service metadata from matched concepts.

Dong et al. [3] also introduced a similar service search approach for the health service domain. The major characteristics of their work are as follows: 1) an RDF/RDFS-annotated health service ontology and an RDF/RDFS-annotated health service metadata schema, 2) a more efficient Index-based ECBR (IECBR) model for request-concept match, and 3) the UMLS SPECIALIST Lexicon[2], which is an online medical thesaurus, for query expansion.

The limitations of the two methodologies above is that performance, especially the efficiency of the previous generic service search models is not very convincing.

3 System Architecture

By taking into account the three features of online service advertising information and the limitations of the existing approaches in this area, we propose a UCOSAIS framework in this paper that assimilates the philosophy of UCD. The primary functionality of the framework is to enable service users to search required online service advertising information based on their own service queries. The framework follows a keyword-based search style, due to the fact that most of the popular goods or service search engines, such as eBay, Amazon.com, etc., are employing this style, and a study [6] shows that users are inclined to enter short queries (2.4 terms/query) and less likely (less than 5%) to use complex query functions. This framework comprises three components as follows:

- An ontology-learning-based focused crawler which is able to utilize service domain ontologies to automatically discover, extract, annotate, and classify domain-relevant service advertising information from the Internet as well as to employ the crawled information to evolve the ontologies, taking into account the ubiquitous, heterogeneous, and ambiguous feature of the information. The technical details of the crawler can be referenced from [2], which is not discussed in this paper.
- A Hybrid Service Concept Selection (HSCS) Module which enables a service user to select an appropriate service concept to clearly denote his/her service request to encounter the issue of heterogeneity and ambiguity.
- A Click-Enhanced Service Concept Ranking (CESCR) Module which provides a complementary function to enhance the ranking of relevant service concepts in the HSCS module based on users perceptions towards the results of the HSCS module.

Two bases are also planned in the framework for knowledge or data storage, which are:

---

[2] http://www.nlm.nih.gov/research/umls/

- A Service Knowledge Base, including a Service Ontology Base that stores service domain ontologies for service advertising information and user query disambiguation and denotation and a Service Metadata Base that stores standardized online service advertising information, i.e., service metadata, in which the metadata are denoted by relevant ontological concepts.
- A Query-Click Database, which stores historical user-click data for specific queries.

In the rest of this section, the technical details of the modules and the bases are introduced.

3.1   Hybrid Service Concept Selection Module

The HSCS Module designates an appropriate concept from a service ontology to a service users service query. This is a semi-automatic process comprising two steps: 1) automatically recommending a list of semantically relevant concepts to the user; and 2) allowing the service user to manually select a concept to denote his/her query intention, guided by the philosophy of UCD. A hybrid match model is the key component to realize the function of automatic recommendation, which is designed to choose semantically relevant concepts for a query. This model combines a pseudo-logic-based match model and a semantic-based match model. The pseudo-logic-based match model is to match the extent of subsumption between a query and a concept. The reason why we employ the pseudo-logic-based match model other than the logic-based match is that the latter needs the higher computing cost, which is not practical in the real environment. The output of this hybrid match model is a combined view of ranked concepts. The hybrid match model is mathematically presented as follows:

Let Q be a service query, $S(Q)$ be the synset of Q, C be a concept of a service ontology, and C has a group of concept descriptions CD, five match levels are defined to determine the relatedness between Q and C as follows:

Exact match. $(Q \equiv C \Leftrightarrow (\forall CD \in C) \sqcap [(Q = \exists CD) \sqcup (\exists S(Q) = \exists CD)])$
Plug-in match. $(Q \subseteq C \Leftrightarrow (\forall CD \in C) \sqcap [(Q \subseteq \exists CD) \sqcup (\exists S(Q) \subseteq \exists CD)])$
Subsume match. $(Q \equiv C \Leftrightarrow (\forall CD \in C) \sqcap [(Q \supseteq \exists CD) \sqcup (\exists S(Q) \supseteq \exists CD)])$
Intersection match. $((Q \sqcap C \neq \phi \Leftrightarrow (Q \neq C) \sqcap (Q \not\subseteq C) \sqcap (Q \not\supseteq C) \sqcap [((Q \sqcap \exists CD \neq \phi) \sqcup (\exists S(Q) \sqcap \exists CD \neq \phi)])$
Fail match. $((Q \sqcap C = \phi \Leftrightarrow (Q \neq C) \sqcap (Q \not\subseteq C) \sqcap (Q \not\supseteq C) \sqcap [\neg(Q \sqcap \exists CD \neq \phi)])$

where the pseudo-logic-based match is applied for the exact, plug-in, and subsume match and the semantic-based match model is used for the intersection and fail match.

The semantic-based match model computes the extent of intersection between a query Q and a concept C when Q and C do not fit the exact, plug-in, and subsume match. The semantic-based match model consists of a pre-processing process and a real-time match process. The primary task of the pre-processing process is to obtain synonyms and to calculate weights for terms in concept descriptions CD, in which the weights refer to the particularness values of the terms. Here we utilized the inverse document frequency (IDF) model to measure the particularness values. The procedure of the pre-processing process is shown in Fig. 1.

Require: C is a concept of a service ontology O, C has a group of concept descriptions
CD[j], and each concept description CD[j] has a group of terms T[j][h].
Ensure: root and synonyms of T[j][h], and weight of T[j][h] – W[j][h].
1: **for** each concept description CD[j] **do**
2:     **for** each term T[j][h] in concept description CD[j] **do**
3:         Find synonyms of T[j][h] from WordNet
4:         T[j][h]←T[j][h]⊔ synonyms of T[j][h]
5:         $W[j][h] \leftarrow \log \frac{\{|C_\alpha||\forall C_\alpha \in O\}}{\{|C_\beta||(\forall C_\beta \in O)\sqcap(\forall CD \in C_\beta)\sqcap(T[j][h]\in \exists CD)\}}$
6:     **end for**
7: **end for**

Fig. 1. Procedure of the pre-processing process

The real-time match process calculates the semantic similarity value between a query and a concept by combining Dong et al.'s [7] semantic similarity model and Plebani et al.'s [8] bipartite graph model.

In WordNet, terms/concepts are linked by the hypernym/hyponym relationship, and thus terms can be viewed as having a hierarchical structure. Dong et al.'s [7] semantic similarity model calculates the semantic relatedness between two terms in WordNet by taking into account the position of the lowest common hyponym of the terms in the hierarchy. The mathematical expression of the model is as follows:

$$sim_{Dong}(C_1, C_2) = \begin{cases} \frac{max_{C \in S(C_1, C_2)}\{-\log[P(C)]\}}{max_{C \in \Theta}\{-\log[P(C)]\}} & if\ C_1 \neq C_2; \\ 1 & if\ (C_1 = C_2) \sqcup (C_1 \in \delta(C_2)) \sqcup \\ & (C_2 \in \delta(C_1)) \end{cases}$$

(1)

where C_1 and C_2 are two concepts in WordNet, $S(C_1, C_2)$ is the set of concepts that subsume both C_1 and C_2, $P(C)$ is the probability of encountering a subconcept of C, $\delta(C_1)$ is the synset of C_1, and $\delta(C_2)$ is the synset of C_2.

Plebani et al.'s [8] bipartite graph model chooses the optimal match between the group of terms in a query and the group of terms in a concept description, given all the semantic similarity values between the two groups of terms. The mathematical expression of the model is as follows:

Given a graph $G = (V, E)$, where V is a group of vertices and E is a group of edges that link between V. A matching in G is defined as $M \subseteq E$ so that no two edges in E share a common end vertex. An assignment in G is a matching M so that each vertex in G has an edge in M. Let us suppose that the set of vertices is partitioned into two sets S (namely the terms $S[i]$ in the query Q) and T (namely the terms $T[j]$ in the concept description CD), and each edge in this graph has an associated value v, i.e., the semantic similarity value between each pair of terms in the query and in the concept description within the interval $[0, 1]$ given by Equation (1)). A function f returns the maximum weighted assignment, i.e., an assignment for which the average weight of the edge is maximum. Eventually the semantic similarity value Sim between a query Q and a concept description CD is obtained by using the weights of query terms $S[i] - SW[i]$ (introduced below) – to compute the weighted mean of the assignment [8]. The assignment in bipartite graphs can be expressed in a linear programming model, which is

$$Sim_S(Q, CD) = maxSim_S(v, SW, S, T) = \frac{\sum_{i \in I}^{j \in J} f(S[i], T[j]) \cdot SW[i]}{\sum_{i \in I} SW[i]} \quad (2)$$

subject to

$$f(S[i], T[j]) = \max(v) = \max(\sum_{i \in I}^{j \in J} sim_{Dong}(S[i], T[j])), \forall i \in I, \forall j \in J, \quad (3)$$
$$I = [1...|S|], J = [1...|T|]$$

The weights of terms in the query – $SW[i]$ – are obtained by a query term processing method which searches for the weights of the counterparts from the processed terms in the ontology, instead of calculating them, with the purpose of saving computing cost in the real-time match process. If a term or the synonyms of the term cannot be found from the ontology, the term will be assigned with the maximum particularness value in terms of the IDF model. The procedure of this process is described in Fig. 2.

Require: Q is a service query, and Q contains a group of terms S[i]. O is a service ontology, and O contains a group of terms T, each term T is associated with a weight W.
Ensure: Roots of S[i] and weight of S[i] – SW[i].
1: **for** each term S[i] in P **do**
2:     **for** each term T in O **do**
3:         **if** T == S[i] **then**
4:             SW[i]← W
5:         **end if**
6:     **end for**
7:     **if** SW[i]==NULL **then**
8:         SW[i]← log $\{|C| |\forall C \in O\}$
9:     **end if**
10: **end for**

Fig. 2. Procedure of the query term processing process

When the combined concept ranking view is generated by the hybrid match model, a Manual Selection process is offered to allow a service user to manually denote his/her service request, following the philosophy of UCD. In this process, the service user is required to manually select a concept from the list of concepts to denote his/her query intention. If the selected concept is a parent concept in the service ontology, the service user may choose to browse its associated service metadata or its more specific children concepts. Simply clicking the concept will cause all the children concepts to be unfolded and ranked in the same pattern. This is a recursive process until the service user eventually chooses the appropriate concept.

3.2   Click-Enahnced Service Concept Ranking Module

The limitation of the hybrid match model in the HSCS Module is that this process does not take service users perceptions into account in the query-concept match process, which may have the consequence that the match results cannot fulfill service users real needs. To remedy this defect, we propose a CESCR Module in this section to improve the rank of service concepts that have an intersection match with a service query. The principle of this model is to track past service users click behaviors for the same query and recalculate the semantic similarity values of the concepts within the intersection match level. A Query-Click Monitor is designed to track service users query and click behaviors and to store the record in a Query-Click Database. Every time the semantic-based match model starts to calculate the semantic similarity values, the CESCR Module is invoked to retrieve the relevant record from the database. It then employs a click-enhanced similarity adjustment (mathematical) model to fine-tune these intersection match values.

In a search process, a service user may click many concepts in an attempt to identify which concept best represents his/her service request. To realize the CESCR Module, it is therefore necessary to identify the positive clicks from service users click records. Consequently, we make the following assumption regarding service users click behaviors to enable the identification of positive clicks:

After the semantic-based concept rank view is displayed to the service user according to a query, the service user may click as many concepts as required, and will eventually click the concept that will best represent his/her service request. This is the last clicked concept in this session for this query; the service user will also have browsed the detailed information about its associated metadata.

Once the positive clicks for a query are identified from a users click record, the click-enhanced similarity adjustment model is employed to adjust the semantic similarity values between the query and all the concepts in the intersection match. The click-enhanced similarity adjustment model is mathematically presented as follows:

Let C be the concepts of a service ontology O, Q be a service query, and $Click$ be the function that retrieves the positive click number of a concept for the query. The click-enhanced similarity adjustment model contains the following three steps.

In the first step, we need to calculate the relocated similarity value for each concept, i.e. $ReSim$, which is obtained by relocating the total semantic similarity value of the concepts that have an intersection match with the query, based on the number of positive clicks on each concept. The relocated similarity value between a query Q and a concept $C[i]$ is mathematically expressed as follows:

$$ReSim(Q, C[i]) = \frac{\sum Sim(Q, C[j]) \cdot Click(Q, C[i])_{\forall C[j] \in C[\theta] | sim_{\forall C[\theta] \in O}(Q, C[\theta] > 0)}}{\sum Click(Q, C[j])_{\forall C[j] \in C[\theta] | sim_{\forall C[\theta] \in O}(Q, C[\theta] > 0)}}$$

(4)

where $Click(Q, C[i])$ is the number of positive clicks on $C[i]$ for Q.

The defect of the first step is that the relocated similarity value could be larger than 1 when many concepts have high semantic similarity values. To solve this defect, we use a pure click-based similarity value ($ClickSim$) algorithm in the second step to deal with this exception, which is mathematically presented as follows:

$$ClickSim(Q, C[i]) = \begin{cases} Resim(Q, C[i]) \; if \; Max[ReSim(Q, C[i])] \leq 1 \\ \frac{Click(Q, C[i])}{MaxClick(Q)} \quad if \; Max[ReSim(Q, C[i])] > 1 \end{cases}$$

(5)

with

$$MaxClick(Q) = max(Click(Q, C[j]))_{\forall C[j] \in C[\theta] | sim_{\forall C[\theta] \in O}(Q, C[\theta] > 0)}$$

(6)

The defects of the second step are: 1) Click(Q, C[i]) must be large enough to be accurate; and 2) the click-based similarity value relies totally on the number of positive clicks and ignores the original semantic similarity value. To counter these two defects, in the third step, we aggregate the click-based similarity values obtained in Equation (5) and semantic similarity values obtained in Equation (2) by a weighted mean to calculate the adjusted similarity value – $AdjSim(Q, C[i])$, which is

$$AdjSim(Q, C[i]) = \alpha \cdot ClickSim(Q, C[i]) + (1 - \alpha) \cdot Sim(Q, C[i])$$

(7)

where α is higher, $AdjSim(Q, C[i])$ is more sensitive to the click-based similarity value.

It should be noted that the prerequisite of this model is that the majority of service users need to select the correct concepts to represent their service requests. To fulfill this requirement, the number of positive clicks must be large enough to ensure accuracy.

4 Evaluation

In this section, we carry out a series of experiments to evaluate. First, we implemented a prototype of the UCOSAIS framework – a Customized Semantic Service Search Engine (CSSSE) by means of Java Server Pages (JSP), JavaScript, Java Servlet, Asynchronous JavaScript and XML (AJAX), Protege API, WordNet API, and MySQL. To evaluate the HSCS Module and the CESCR Module involved in the UCOSAIS framework, we employed the evaluation approach from the field of Information Retrieval (IR), which compares the performance of the hybrid match model enhanced by the CESCR Module (abbreviated as click-enhanced hybrid match model) and the existing service search models, i.e.,

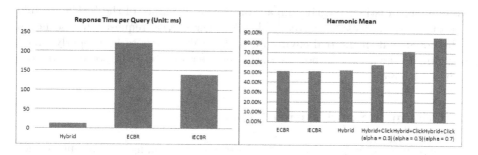

Fig. 3. (a) Comparison of the hybrid, ECBR and IECBR models on response time and (b) Comparison of the ECBR, IECBR, and click-enhanced hybrid models on harmonic mean

ECBR [4] and IECBR [3], based on the performance indicators of harmonic mean and response time. The test data set is obtained from our previous research [4], which includes a transport service ontology comprising 304 concepts and 100 transport-service-related queries.

In the experiment, WordNet is used in all the candidate models, in which the hybrid match model uses it for the synonym search of query terms and semantic similarity calculation, and the other models use it for the synonym search of query terms.

The performance comparison of the hybrid match model and the existing models on response time is shown in Fig. 3(a). It can be seen that the response time of the hybrid match model was 90% less than the other models, resulting from that the pseudo-logic-based match model simplifies the match process and thus saves the response time. This could make the hybrid match model more applicable than the other models when being applied in the real environment, as response time plays a key role in users satisfaction level [9].

To test the positive impact of the CESCR Module on the hybrid match model, we set up an experiment by recruiting 10 volunteers and allowing each volunteer to try the 100 queries (once per query). Prior to typing the queries, we showed the volunteers the peer-reviewed results. The volunteers then entered the queries into the search engine and selected appropriate concepts to represent their service requests. Meanwhile, their click record was stored in the Query-Click Database by the Query-Click monitor. The CESCR Module was subsequently invoked to adjust the semantic similarity values of concepts at the intersection match level. We eventually tested the performance of the click-enhanced hybrid match model at $\alpha = 0$ (hybrid match model only), 0.3, 0.5, and 0.7. The performance comparison between the existing service search models and the proposed model on harmonic mean is illustrated in Fig. 3(b). It is found that the click-enhanced hybrid match model has an outstanding improvement on harmonic mean compared with the existing models and the pure hybrid match model, and this improvement has become more distinct along with the increase of , since most of the persons in this experiment selected the correct concepts to represent their service requests as a result of our hint. Hence, it can be deduced that the CESCR

Module has a distinct positive impact on the performance of the HSCS Module if most service requesters correctly denote their service requests. It also needs to be noted that in the real environment the value needs to be properly tuned according to the actual proportion of the service requesters who correctly denote their service requests.

5   Conclusion and Future Work

In this paper, we proposed a UCOSAIS framework aiming at enabling service consumers to quickly and precisely retrieve and select desired online service advertising information. The philosophy of UCD is harnessed in this research, which is reflected in two aspects: the involvement of service users in ontology-based service request denotation, and the reference of users click behaviors for service concept recommendation.

The limitation of the currently research is that this framework can only be used for retrieving and selecting single services. Therefore, for the future research, we will study how composite service requests are decomposed and matched with service advertising information according to the contextual information of service users and information, guided by the philosophy of user-centered design.

References

1. Wang, H., Lee, M.K.O., Wang, C.: Consumer privacy concerns about Internet marketing. Commun. ACM 41, 63–70 (1998)
2. Dong, H., Hussain, F.K., Chang, E.: Ontology-learning-based focused crawling for online service advertising information discovery and classification. In: Liu, C., Ludwig, H., Toumani, F., Yu, Q. (eds.) ICSOC 2012. LNCS, vol. 7636, pp. 591–598. Springer, Heidelberg (2012)
3. Dong, H., Hussain, F.K.: Semantic service matchmaking for digital health ecosystems. Knowl.-Based Syst. 24, 761–774 (2011)
4. Dong, H., Hussain, F.K., Chang, E.: A service search engine for the industrial digital ecosystems. IEEE Tran. Ind. Electron. 58, 2183–2196 (2011)
5. Abras, C., Maloney-Krichmar, D., Preece, J.: User-Centered Design. In: Bainbridge, W. (ed.) Berkshire Encyclopedia of Human-Computer Interaction, vol. 2, pp. 763–768. Berkshire Publishing Group, Great Barrington (2004)
6. Spink, A., Wolfram, D., Jansen, M.B.J., Saracevic, T.: Searching the web: The public and their queries. J. Am. Soc. Inf. Sci. Tec. 52, 226–234 (2001)
7. Dong, H., Hussain, F.K., Chang, E.: A context-aware semantic similarity model for ontology environments. Concurr. Comp. - Pract. E. 23, 505–524 (2011)
8. Plebani, P., Pernici, B.: URBE: Web service retrieval based on similarity evaluation. IEEE Trans. Knowl. Data Eng. 21, 1629–1642 (2009)
9. Shneiderman, B.: Designing the User Interface: Strategies for Effective Human-Computer Interaction. Addison-Wesley, Reading (1998)
10. Dong, H., Hussain, F.K., Chang, E.: A human-centered semantic service platform for the digital ecosystems environment. World Wide Web 13, 75–103 (2010)

Near Duplicate Text Detection
Using Frequency-Biased Signatures

Yifang Sun, Jianbin Qin, and Wei Wang*

The University of New South Wales, Australia
{yifangs,jqin,weiw}@cse.unsw.edu.au

Abstract. As the use of electronic documents are becoming more popular, people want to find documents completely or partially duplicate. In this paper, we propose a near duplicate text detection framework using signatures to save space and query time. We also propose a novel signature selection algorithm which uses collection frequency of q-grams. We compare our algorithm with Winnowing, which is one of the state-of-the-art signature selection algorithms. We show that our algorithm acquires much better accuracy with less time and space cost. We perform extensive experiments to verify our conclusion.

Keywords: near duplicate text detection, Winnowing, k-stability, collection frequency.

1 Introduction

It is now a common practice to use electronic documents in business communications (e.g., Web pages, word documents) and personal life (e.g., emails), as these digital documents are easy and cost-effective to store, retrieve and share. Given a collection of such documents, it is often needed to find documents that are nearly duplicate from a given query document either *completely* or *partially*. We call this the *near duplicate text detection* problem, and it has wide applications such as copyright enforcement, plagiarism detection, and version control.

To scale to large collection of document, the prevalent method for near duplicate text detection is based on *signatures*: a set of signatures are extracted and indexed for the documents at indexing time, and at query time, the query document's signatures are produced in the same manner; this generates a set of candidate texts which will be finally compared with the query document. However, many of the existing methods, such as (mod p) = 0 scheme [13], local maximum [2], spotSig [19] and I-match [7], are based on heuristics that cannot even guarantee 100% detection of exact copies.

In this paper, we first propose a general framework for the problem based on the Winnowing-family algorithms [17], which have the *locality* property that *exact* copies exceeding a certain length are guaranteed to be detected. In order to quantify the ability to detect *near* duplicate copies (i.e., copying with a small

---

* This work is partially supported by ARC DP Projects DP130103401 and DP130103405.

X. Lin et al. (Eds.): WISE 2013, Part I, LNCS 8180, pp. 277–291, 2013.

amount of errors). we propose a novel and useful concept, k-stability. We compute the k-stability for all Winnowing-family algorithms and the result reveals that the original Winnowing algorithm trades quality (i.e., recall) for better efficiency. We then proposed a simple yet effective variation of the Winnowing algorithm, named frequency biased Winnowing, which achieves both good efficiency and high quality. We also consider candidate text generation methods as well as optimizations to further reduce the number of similarity computations. We experimentally evaluated our method in a plagarism detection benchmark, and our method is shown to achieve higher recall with the superior time and indexing space efficiency than the method based on the original Winnowing.

The rest of the paper is organized as follows: Section 2 gives the problem definition and notations. Section 3 introduces our proposed framework. Section 4 analyzes the k-stability of Winnowing-family algorithms and proposes an improved signature selection method based on collection frequencies. Section 5 introduces the candidate text generation method and Section 6 gives an effective improvement by eliminating unnecessary computations. Section 7 shows experimental results. Related works are introduced in Section 8 and Section 9 concludes the paper.

2   Problem Definition and Notations

We first give the formal definition of near duplicate text detection problem.

Definition 1 (Near Duplicate Text Detection). *Given a collection $\mathcal{C}$ of documents and a query document Q, a near duplicate text detection algorithm will return the best near duplicate text of Q in $\mathcal{C}$, indicated by $d \in \mathcal{C}$ and the start and end positions of the text in d (denoted by pos_{start} and pos_{end}), respectively.*

Although the precise definition of near duplicate is application-dependent, in most cases they are evaluated by a similarity function, which returns high scores when two text strings share a large portion of *identical* or *highly similar* substrings [18].

Note that the above problem definition is general enough to support several important applications. For example, the *near duplicate document detection* problem [17] can be deemed as a special case where the starting and ending positions are always the beginning and the end of the documents, respectively. For another example, the *text reuse problem* [18,23] can be solved by issuing multiple near duplicate text detection queries, each with a sentence as the query document.

Notations. All array indexes start from 1. Given a string T, $len(T)$ denotes its length. T has $len(T) - q + 1$ q-grams, which forms its *q-gram set* and is denoted as $grams_T$. The cardinality of a (multi-)set S is denoted by $|S|$. Given a q-gram g in a document, pos_g denotes the offset of its first character in the document.

3   A Framework of Near Duplicate Text Deteciton

Obviously, the naïve algorithm which performs character-to-character comparison between Q and every document $d \in \mathcal{C}$ is too costly and does not scale well with the size of the document collection. Existing works are mainly based on selecting a small set of candidates $C' \subseteq C$ by extracting and matching document *signatures* [18]. The most prevalent form of signatures are q-grams, which are substring of q characters. A document of length l will generate $l - q + 1$ overlapping q-grams, and usually only a subset of them will be selected as the signatures of the document by a *signature selection* process.

We capture such approaches in a general framework as follows:

- In the *indexing phase*, for each document d in C, a set of signatures is selected from its q-grams. The signature selection method (denoted as SelectSigs in Algorithm 1 and discussed in Section 4) could be any algorithm that will be introduced in Section 8, including *Winnowing* [17] and our *frequency biased Winnowing*. An inverted index, I, is then built that maps a signature to each of its occurrences (identified by document ID and the position within the document).

- In the *query processing phase* (See Algorithm 1), a set of signatures S_q is selected using the same signature selection method SelectSigs(Line 3). All the occurrences of each signature are collected via probing the index I, and then grouped by document (Lines 4–6). For each document returned, several candidate texts will be generated by the GenCandTexts function (to be discussed in Section 5) and stored in $CAND$ (Lines 7–8). Similarities between the query and each candidate text will be calculated (Lines 9–10) and the one achieving the maximum similarity will be returned (Line 11).

Algorithm 1. Query(Q)

---

1 $CAND \leftarrow \emptyset$;
2 Initialize sim and G to be empty hashtables;
   /* select Q's signatures   */
3 $S_q \leftarrow$ SelectSigs(Q);
   /* find and group all occurrences of Q's signatures by document   */
4 for each *signature* $s \in S_q$ do
5 | for each *pair* $(d_i, pos_j) \in I[s]$ do
6 | ⌊ $G[d_i] \leftarrow G[d_i] \cup \{pos_j\}$;

   /* generate candidate texts for each candidate document   */
7 for each $d_i \in G$ do
8   ⌊ $CAND \leftarrow CAND \cup$ GenCandTexts($G[d_i]$);

   /* find the best candidate text   */
9 for each *candidate* $c_i \in CAND$ do
10 ⌊ $sim[c_i] \leftarrow$ CalcSim(Q, c$_i$);
11 **return** $\arg\max_{c_i \in CAND} sim[c_i]$;

The function CalcSim computes the similarity of a candidate text c_i against the query Q. In this paper, we consider one-sided Jaccard of q-gram multisets of c_i and Q, or

$$sim(c_i, Q) = \frac{|grams_{c_i} \cap grams_Q|}{|grams_Q|} \qquad (1)$$

We break ties by favoring the shortest text.

4   Signature Selection Algorithms

While it is possible to select all the q-grams of a document as its signatures, this usually results in too many comparisons in practice due to the existence of some frequently occurring q-grams. On the other hand, selecting very few q-grams tends to miss many of the query results, or limit the flexibility of the algorithm (e.g., can only detect near duplicate sentences [23] or documents [15]). Hence, the signature selection process is a trade-off between efficiency, space and effectiveness (specifically recall). While many heuristic selection methods exists (such as [19,18,23]), we consider the Winnowing-family algorithms [17], as it has the guarantee that exact copy of substrings exceeding a certain length will always be detected.

In this section, we first briefly introduce and analyze Winnowing-family algorithms, including the original Winnowing method, and then identify a novel concept of k-stability, which is essential to quantifies the probability that a near duplicate text will be detected under the Winnowing-family algorithms. We then point out the limitation of the original Winnowing algorithm due to a dilemma between high stability and low efficiency. Finally, we propose a simple yet effective alternative Winnowing-based algorithm, named Frequency Biased Winnowing, that achieves a better trade-off than the original Winnowing method.

4.1   Winnowing-Family Algorithms

A Winnowing-family algorithm firstly calculates $f(g_i)$ for all the q-grams g_i in the document using an *injective function* $f(x)$. It then uses a sliding window of size w to select signatures. Within each window, it selects the q-gram g_{min}, such that $f(g_{min})$ is the smallest in the window, as the signature. If there is a tie, then the *rightmost* occurrence will be selected.

Example 1. Let $q = 3$ and $w = 4$. Consider the document "abcdedcba", whose q-grams are: {abc, bcd, cde, ded, edc, dcb, cba}. Assume $f(x) = (c_1 \cdot 7^2 + c_2 \cdot 7 + c_3) \bmod 23$, where c_i indicates the ASCII code of the i-th character of the q-gram. Then the corresponding values are $\{1, 14, 4, 15, 20, 7, 17\}$.

Given $w = 4$, we have four windows: (__1__, 14, 4, 15), (14, __4__, 15, 20), (__4__, 15, 20, 7), and (15, 20, __7__, 17). q-grams corresponding to underlined bold numbers are signatures selected in each window. Thus, abc, cde and dcb with values 1, 4 and 7 will be selected as the signatures of the document.

The original Winnowing algorithm [17] belongs to this Winnowing-family by using a random hash function with a sufficiently large codomain as $f(x)$. Later in Section 4.3 we will propose another frequency biased instance of the Winnowing-family algorithms.

Winnowing-family algorithms hold an important property named *locality*. An algorithm is *l-local* if, for any two identical strings with length at least *l*, they will always have at least one identical signature and thus will be guaranteed to be detected by the algorithm. This property is essential to detect *exact* copying. Consider two identical strings of length $l = w + q - 1$ where w is the window size. It is obvious that a Winnowing-family algorithm will always select the same minimum-valued q-gram in the windows as signatures. Therefore, all the Winnowing-family algorithms are $(w + q - 1)$-local.

4.2 k-Stability of Winnowing-Family Algorithms

While the locality property of Winnowing-family algorithms is essential for *exact* duplicate text detection, it does not help to analyze the performance of the algorithm for *near* duplicate text detection, which is arguably the more common and difficult case. To this end, we propose a novel concept named *k-stability*, which capture the ability for a Winnowing-family algorithm to detect text with small (or k) errors.

Definition 2 (k-stability of Winnowing-family algorithms). *Given a Winnowing-family algorithm M, consider randomly and independently changing k q-grams in a window W, which results in W'. The k-stability is the expected probability of that the signatures of W and W' are the same under the algorithm M.*

Obviously, the k-stability depends on content of the window W. In order to get a general, closed-formula characterization for an algorithm, in the following, we compute the k-stability for a window where its constituent q-grams are randomly and independently selected from the entire document collection (e.g., the distribution q-grams in the window are the same as those in the collection).

First, we establish the following Lemma.

Lemma 1. *For any discrete random variable X with possible values $\{x_1, x_2, \ldots, x_n\}$, the following equation holds for sufficiently large $t > 0$:*

$$\sum_{i=1}^{n} (p(x_i) \cdot (F(x_i) - \frac{p(x_i)}{2})^t) \approx \frac{1}{t+1}$$

where $p(x_i)$ is the probability mass function and $F(x_i)$ is the cumulative distribution function, i.e., $F(x_i) = \sum_{j=1}^{i} p(x_j)$.

Proof. Without loss of generality, we assume $x_i < x_{i+1}$. We also additionally define x_0, such that $x_0 < x_1$, and $F(x_0) = p(x_0) = 0$.

Consider a function

$$F_c(y) = \begin{cases} 0 & \text{, when } y < x_0 \\ \frac{a-b}{x_i - x_{i-1}} y + \frac{b \cdot x_i - a \cdot x_{i-1}}{x_i - x_{i-1}} & \text{, when } x_{i-1} \leq y < x_i, i \in [1, n] \\ 1 & \text{, when } y \geq x_n \end{cases}$$

where a = $\sqrt[t+1]{(t+1)\cdot(F(x_i)-\frac{p(x_i)}{2})^t\cdot F(x_i)}$,   and b =
$\sqrt[t+1]{(t+1)\cdot(F(x_i)-\frac{p(x_i)}{2})^t\cdot F(x_{i-1})}$.

Since $F_c(y)$ is monotonous, bounded and right continuous, there must exist a random variable Y such that $F_c(y)$ is the cumulative distribution function of Y.
Then for any $x_1 \le x_i \le x_n$, we have:

$$\int_{x_{i-1}}^{x_i}(p_c(y)\cdot F_c^t(y))dy = \lim_{x\to x_i^-}\frac{1}{t+1}F_c^{t+1}(y)\Big|_{x_{i-1}}^{x} = p(x_i)\cdot(F(x_i)-\frac{p(x_i)}{2})^t$$

where $p_c(y)$ is the probability density function of Y. Then we have:

$$\sum_{i=1}^{n}(p(x_i)\cdot(F(x_i)-\frac{p(x_i)}{2})^t) = \sum_{i=1}^{n}\int_{x_{i-1}}^{x_i}(p_c(y)\cdot F_c^t(y))dy$$

$$=\frac{1}{t+1}F_c^{t+1}(x_n)-\frac{1}{t+1}F_c^{t+1}(x_0)-o(\frac{1}{t})\approx\frac{1}{t+1}\qquad\qquad\square$$

Theorem 1. *The k-stability of a Winnowing-family algorithm with window size w is approximately $\frac{w-k}{w+k}$.*

Proof. Assume we randomly and independently pick w q-grams from the collection to form a window W, and another k q-grams to form a set S_{new}. We will then randomly and independently pick k q-grams from W and replace them with q-grams in S_{new}. We name these k q-grams as S_{old} and the rest q-grams as S_{rest}.

Apparently, the signature of W will not change after we subsitute k q-grams, only when the signature sig is in S_{rest}. In Winnowing-family algorithms, this indicates that $f(sig)$ is the rightmost samllest value among all the $w+k$ picked q-grams. The probability of this event can be estimated using Lemma 1:

$$Pr = \sum_{i=1}^{n}\left(p(x_i)\cdot(\sum_{j=1}^{i-1}p(x_j)+\frac{p(x_i)}{2})^{w+k-1}\right)$$

$$=\sum_{i=1}^{n}\left(p(x_i)\cdot(F(x_i)-\frac{p(x_i)}{2})^{w+k-1}\right)\approx\frac{1}{w+k}$$

There are $w-k$ q-grams in S_{rest} and we need to consider each of them. Thus for the event "signature in W is not changed" will happen with probability

$$\binom{w-k}{1}\cdot\frac{1}{w+k}=\frac{w-k}{w+k}\qquad\qquad\square$$

Note if we change one character, it will affect at most q q-grams. This observation straightly leads us to the following corollary.

Corollary 1. *Assume a Winnowing-family algorithm with gram length q and window size w. if we change m characters in a window, the signatures of it will remain the same, in the worst case, with probability $p = \max(0, \frac{w-mq}{w+mq})$.*

Table 1. Stability of Winnowing-family algorithms

Setting		m			
		1	2	3	4
$q = 50$	Worst Case	33.33%	0%	0%	0%
$w = 100$	Average	52.25%	25.70%	12.51%	6.11%
$q = 4$	Worst Case	94.67%	89.61%	84.81%	80.25%
$w = 146$	Average	94.77%	89.82%	85.12%	80.67%

Remark 1. *By letting $m = 0$, the Winnowing-family algorithms have 0-stability of 100%, which agrees with the locality property. So in this sense, we can deem k-stability as an extension of the locality property.*

Stability Analysis for the Original Winnowing Algorithm. Table 1 shows the probabilities of the signature in a window remaining the same after changing m characters, in worst case (i.e., in Corollary 1) as well as on average. With the typical setting of Winnowing from [17], where $q = 50$ and $w = 100$, changing even few characters will bring a significant decreasing to its stability, as well as the robustness of a near duplicate text detection method based on Winnowing.

However, from Corollary 1, as as showing in Table 1, we know that with the same locality of $(w + q - 1)$, a smaller q is much more preferable (e.g., $q = 4$) with respect to stability. Unfortunately, Winnowing cannot benefit from such q's. When q is smaller, the average occurence of q-grams is higher due to the reduction of distinct q-grams in the corpus. Then a random hash function $f(x)$ will have more chance to select a frequently-occurring q-gram as signature, which will affect the number of candidates as well as the query time for a Winnowing based method. This motives us to propose following Winnowing-family algorithm to fight against these problems.

4.3   Collection Frequency Biased Winnowing

We propose *Frequency Biased Winnowing*, which achieves a better stability by using small q's, yet it stilll achieves good efficiency for query processing.

Collection frequency, defined as the number of times that a term appears in the collection, is a statistical measurement to evaluate the importance of a term (or q-gram) in a collection. This leads us to use frequency of q-grams when selecting signatures. Rare q-grams are more preferable because they are more representative and able to make the length of posting lists shorter.

Our proposed signature selection algorithm, **frequency biased Winnowing**, is a Winnowing-family algorithm which takes collection frequency for each q-gram as their hash values. Since the frequency of two different q-grams might be the same, the *alphabet order* of the q-grams will be used to break such tie.

Example 2. Consider the same setting and document as in Example 1, where $q = 3$ and $w = 4$, q-grams of the document are: $\{abc, bcd, cde, ded, edc, dcb, cba\}$. Assume their corresponding collection frequencies are $\{18, 62, 50, 43, 30, 79, 30\}$.

Then underlined numbers (also in bold face) can be selected from 4 windows:
*(**18**, 62, 50, 43), (62, 50, 43, **30**), (50, 43, **30**, 79), and (43, 30, 79, **30**). In the*
last window, the last 30 is selected because of the alphabet order *of* cba *and* edc.

Obviously our proposed frequency biased Winnowing is a Winnowing-family algorithm, therefore it holds the *locality* property. And its k-*stability* is $\frac{w-k}{w+k}$.

Comparison with Winnowing. Frequency biased Winnowing prefers small q. Because when q becomes larger (e.g., $q > 10$), the number of possible distinct q-grams tends to be extremly large (i.e., $|\Sigma|^q$, where Σ is the alphabet), and most of them will have the frequency of 0 or 1. Then the algorithm selects signatures almost only based on their alphabet orders and has no benifit from collection frequencies. According to the experiments, q between 3 and 5 is the best setting for our method.

According to Corollary 1, a smaller q is more preferable for Winnowing-family algorithms with respect to its stability. On the other hand, since we always choose the q-gram with smallest collection frequency in the window, the posting list in our algorithm will not be very long. Therefore, our method will improve the effectiveness compared with Winnowing with large q's and also improve the efficiency compared with Winnowing with small q's. Our experiments have verified our analysis.

5   Generate Candidate Texts

In this section, we introduce the methodology of generating candidate texts in our near duplicate text detection method (i.e., the GenCandTexts function, Line 8 of Algorithm 1).

For each candidate document, we now have a sorted list that contains positions of matching signatures in the document. We do not use order among matching signatures, as it is quite common to have near duplicate text with reordered subparts. Instead, candidate texts are generated by applying the following heuristic rules of *merge signatures* and *determine boundaries*.

– **Merge Signatures.** We first combine *continuous* signatures together. Two signatures are continuous if they *may* be derived from two overlapping or adjacent windows. We can either store the window positions where the signature are generated. Otherwise, given the positions of two signatures s_i and s_j (assuming $pos_{s_i} < pos_{s_j}$), they are considered to be continuous if $pos_{s_j} - pos_{s_i} \leq 2w + q - 2$, as this is the worst case where the two signatures are the first and last signature in two adjacent windows, respectively.
– **Determine Boundaries.** Given an ordered merged list of sigantures $\{s_1, s_2, \ldots, s_m\}$ of document d, we generate the candidate text that has the *longest possible length*: we take the substring between positions pos_{start} and pos_{end}, where $pos_{start} = \max(1, pos_{s_1} - w + 1)$ and $pos_{end} = \min(pos_{s_m} + w + q - 1, len(d))$.

6 Heap Based Optimization

Lines 9-11 of Algorithm 1 compute the similarity for every candidate text and returns the largest one. This is not efficient if there is a large number of candidate texts which are long, or not similar to Q. In this section, we propose a heap-based optimization to reduce the number of similarity computations.

We observe that an upper bound of the similarity between two strings can be easily computed based on their lengths. In Equation (1), $|grams_Q|$ is fixed for a given query, thus the similarity between c and Q is only affected by $|grams_c \cap grams_Q|$. Since $|grams_c \cap grams_Q| \le \min(|grams_c|, |grams_Q|)$, we can easily work out an upperbound of $sim(c, Q)$ as follows:

$$sim(c, Q) \le sim_{ub}(c, Q) = \frac{\min(|grams_c|, |grams_Q|)}{|grams_Q|} = \min(\frac{|grams_c|}{|grams_Q|}, 1)$$

In addition, $sim_{ub}(c, Q)$ increases monotonically with $|grams_c| = len(c) - q + 1$.

The optimized query algorithm is shown in Algorithm 2, which should replace Lines 7–11 of Algorithm 1. The major modifications are:

- We use a max-heap H to organize candidate texts, based on their upper bound similarity ub_score.
- We maintain the current maximum score in max_score, and we terminate the loop only when the head of the heap H's upper bound score is no more than max_score.
- We use a similarity computation function CalcSim2 which can stop earlier during the similarity computation (See Algorithm 3). Note that we convert a multiset of q-grams to a set of q-grams by annotating a q-gram g as g_i if it is the i-th occurrence of q-gram. We perform the same transformation for Q and index it so that the set membership query (Line 4) can be performed efficiently.

Our experiments show that this optimization can save up to 99% number of similarity computations.

Algorithm 2. OptimizedQuery

```
   /* generate candidate texts for each candidate document        */
1  for each d_i ∈ G do
2      for each candidate text c_i ∈ GenCandTexts(G[d_i]) do
3          ub_score ← SimUB(c_i);
4          H.enqueue(c_i, ub_score);

   /* find the best candidate text                                */
5  max_sim ← 0;
6  while H.head.ub_score > max_sim do
7      c ← H.dequeue();
8      max_sim ← max(max_sim, CalcSim2(Q, c, max_sim));
9  return c;
```

Algorithm 3. CalcSim2(Q, c, max_sim)

---

1 $max\_err \leftarrow |grams_c| \cdot (1 - max\_sim)$;
2 $err \leftarrow 0$;
3 for each q-gram $g \in grams_c$ **do**
4   **if** $g \notin Q$ **then**
5 $err \leftarrow err + 1$;
6   **if** $err \geq max\_err$ **then**
7   **return** 0;

8 return $(|grams_c - err|)/|grams_Q|$;

---

7   Experimental Results

In this section, we report our experiment results with two different implementations of our near duplicate text detection method, based on Winnowing [17] and frequency biased Winnowing respectively. We compare the performance of our proposed algorithm against Winnowing. We also show the improvments of heap based optimization introduced in Section 6.

7.1   Experiments Setup

Our near duplicate text detection system is implemented in Java and compiled using JDK 1.6.0. We use the Lucene library (Version 3.3.0)[1] to help build and retrieve the indexes. All experiments are carried out on a PC with a Quad-Core AMD Opteron 8378@2.4GHz Processor and 96GB RAM, and running Ubuntu 4.4.3.

Dataset. We use **PAN-PC-10**[2], a publicly available real dataset, to test our method. The **PAN-PC-10** dataset is published and used in Plagiarism Detection Task of PAN Workshop and Competition of Year 2010, which contains $11,148$ source documents and $68,558$ plagiarism cases. For each plagiarism case, the corresponding source sections are provided in the annotation of the dataset.

We remove those non-English documents from dataset, as our method is not designed for cross-lingual plagiarism. For both source documents and plagiarism cases, we converte all the non alphanumeric characters to '_' and all the uppercase characters to lowercase. Thus we finally have $10,482$ documents with average length of $149,354$ in the dataset. The total size of the dataset is 1.57 GB and the alphabet size is 37 (i.e., [a-z0-9_]).

Parameter Setting. We implemented our method on both Winnowing and frequency biased Winnowing under variant settings of q and w. We keep $q+w = 150$, such that the same locality guarantee will hold.

---

[1] http://lucene.apache.org/
[2] http://www.webis.de/research/events/pan-10

As suggested in [17], We set $q = 50$ and $w = 100$ for Winnowing. We also try other possible q's from 3 to 60 and Winnowing achieves its best performance considering both efficiency and effectiveness on PAN-PC-10 dataset when $q = 10$ and $w = 140$. Therefore, we report the experiment results for Winnowing on these two settings.

For frequency biased Winnowing, we also try different q's from 3 to 5, and results for q equals to 4 and 5 are reported.

Queries and Measurements. There are two main different types of near-duplications in the **PAN-PC-10** dataset [16], which are *artificial* (automatic) plagiarism cases and *simulated* (manual) plagiarism cases. In artificial plagiarism cases, there are three different obfuscation levels (i.e., *none, low* and *high*). For each of the four types above, we randomly select 100 plagiarism cases as queries and use the facts in the annotation as the ground truth for evaluation.

We focus on the following 5 measurements (all measurements are averaged over all queries):

- **Index Size**, which is the space needed to store the index.
- **Index Time**, which is the total time needed to index the whole collection.
- **Accuracy.** We use recall, precision and F_1 score to measure the accuracy of our method. Given query Q, its recall and precision are defined as $|\Omega| / |S|$ and $|\Omega| / |Q|$ respectively. Where Ω represents the detected plagiarized paragraph, and S indicates the real plagiarized paragraph.
- **Query Time**, which is the total time to process a query.
- **Calculated Candidate Texts**, which is the number of candidate texts whose similarity to Q is calculated. We report the number before and after applying optimization. We also report the total length of calculated candidate texts, as they are approxmately proportional to the query time, and the query time before applying optimization is extremly long thus we do not report it.

7.2   Indexing Time and Size

We plot the index size and indexing time for both algorithms with different parameters in Figure 1(a). The spots and line show indexing time. It is clearly that Winnowing takes much more time on indexing than frequency biased Winnowing. This is mainly due to the following two reasons. Firstly, the number of distinct signatures in Winnowing is much more than those in frequency biased Winnowing, (E.g., $30,982,703$ vs. $486,248$). Secondly, the time cost for calculating hash values in Winnowing is much longer than looking up the frequency table in frequency biased Winnowing.

The bars show index size of two different algorithms. Apparently Winnowing also has a larger index size, especially when q is large. This is because of the different number of distinct signatures two algorithms, also the length of signatures in Winnowing is longer. It usually takes more space to store a String (e.g., signatures) than integers (e.g., positions), thus Winnowing requires more space.

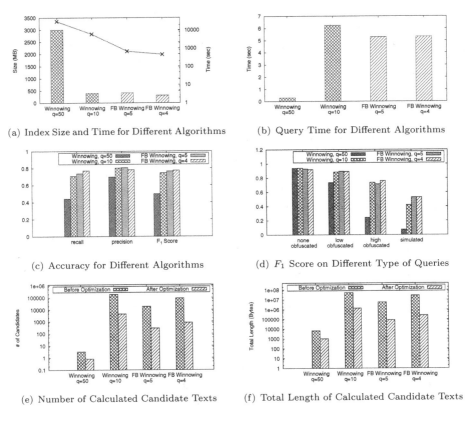

(a) Index Size and Time for Different Algorithms

(b) Query Time for Different Algorithms

(c) Accuracy for Different Algorithms

(d) F_1 Score on Different Type of Queries

(e) Number of Calculated Candidate Texts

(f) Total Length of Calculated Candidate Texts

Fig. 1. Experimental Results

7.3   Accuracy

We plot the average recall, precision and F_1 score over 400 queries on both algorithms with different parameters in Figure 1(c). Clearly, frequency biased Winnowing has a much better accuracy than Winnowing.

More specifically, frequency biased Winnowing with $q = 4$ achieves F_1 score of 0.775 while Winnowing achieves 0.496 with $q = 50$ and 0.745 with $q = 10$. Considering that both algorithms achieve similar precisions, this gap is mainly due to the low recall of Winnowing algorithm. As we stated in Section 4.2, the stability affects recall for Winnowing-family algorithm. Winnowing with larger q has lower stability thus lower recall than frequency biased Winnowing (i.e., $44.00\% - 70.66\%$ vs. $76.56\% - 77.15\%$).

We also plot F_1 score for both algorithms on *different types* of queries in Figure 1(d). Both algorithms perform well on none obfuscated plagiarism cases. Winnowing starts to fail on low obfuscated artificial plagiarism cases, especially with $q = 50$. And frequency biased Winnowing completely beats Winnowing on hard queries (i.e., high obfuscated artificial and simulated plagiarism cases). This is also due to Winnowing's low recall, especially for hard queries, where the loss of stability will bring non-negligible impact on its accuracy.

It is worth mentioning the results of PAN-10 competition. The first place achieves average recall of 69.17% and F_1 score of 0.797, while the second place achieves 62.99% and 0.709. Although our current method cannot support cross-lingual plagiarism cases, and our queries are generated based on the ground truth, it is still justified to say that our near duplicate text detection method is competitive against the top works in the area.

7.4   Query Time

We plot the average query time for both algorithms with different q's in Figure 1(b). We observe that the query time of frequency biased Winnowing is smaller than Winnowing with $q = 10$, but much larger than Winnowing with $q = 50$. Winnowing generates very few or even no candidate for hard queries when $q = 50$. But when $q = 10$, it generates more candidates than frequency biased Winnowing, which leads to more similarity computations thus more time cost.

7.5   Calculated Candidate Texts

In order to verify our analysis of query time as well as show the improvement of heap based optimization, we plot the number of calculated candidate texts and the total length of them, before and after using heap based optimization, in Figure 1(e) and Figure 1(f) respectively. Our optimization brings significant improvements. Up to 99% of candidate texts are skipped, so does the total length of them. Our optimization also saves approximately 99% of query time, as most time is spent on similarity computations. This is due to that most queries have a high similarity answer, once we find it, we can almost ignore the rest candidates.

8   Related Work

Winnowing is considered very important in various areas and used in a number of works. For example, [22] uses it to partition the files and further detect the redundancy in the file. [10] uses it to generate variable sized blocks in order to perform accelerating multi-pattern matching. It also used to quickly find the possible plagiarism parts [5], but only "copy and paste" plagiarism cases are explored. [11] uses it to shorten the size of input data on its secure file scanning system on enterprise networks. However, seems no one focuses on improving Winnowing.

There are many works focusing on near duplicate text detection by using different signature selecting methods. [3] selects every l-th q-grams, which is susceptible to positional changes such as insertion or deletion. The (mod p) = 0 scheme [13] selects q-grams whose hash values can be divided by p, but it is possible to select nothing from a document. Very similar to Winnowing, [2] selects the q-gram whose hash value is smaller than its previous and next h q-grams. It holds a weaker locality which only guarantees to return same or no signature for identical substrings. Spotsigs [19] takes a chain of words that follows a stopword as signatures to find near duplicate Web documents. [1] takes the idea of Spotsigs but also considers the standard tf-idf weighting. It uses sampling to detect duplicate news stories and achieves a good performance. All metioned methods, including other methods like [9,15,18,21], either offer no locality guarantees or suffer from large number of false positives.

There are previous works exploiting collection statistics (mainly inverse document frequency). [8] uses words with the first 30 to 60 highest idf, [7] selects terms with high idf, and its extention [12] uses external collection statistics.

Another category of approaches to detect near duplicate document is to find documents that are highly similar to the query document as a whole. Representative approaches include those emplying Jaccard similarities based on tokens or word n-grams, and those employing Hamming distance based on a binary feature vectors constructed from the documents. There are efficient exact computation algorithms [20,14,24] as well as approximate algorithms based on locality sensitive hashing [4,1,6].

9   Conclusion

In this paper, we propose a new near duplicate text detection framework using signatures selected by Winnowing-family algorithms. We raise a new concept named k-stability with theoretical analysis to measure the stability of Winnowing-family algorithms when small errors happening, and propose a new frequency biased Winnowing algorithm.We also propose candidate text generation methods and optimization to improve the performance of our framework. Our experimental result shows the significant improvement of our proposed algorithm and the good performance on a plagarism detection benchmark.

References

1. Alonso, O., Fetterly, D., Manasse, M.: Duplicate news story detection revisited. Tech. Rep. 60, Microsoft Research (2013)
2. Bjørner, N., Blass, A., Gurevich, Y.: Content-dependent chunking for differential compression, the local maximum approach. J. Comput. Syst. Sci. 76(3-4), 154–203 (2010)
3. Brin, S., Davis, J., Garcia-Molina, H.: Copy detection mechanisms for digital documents. In: SIGMOD Conference, pp. 398–409 (1995)
4. Broder, A.Z., Charikar, M., Frieze, A.M., Mitzenmacher, M.: Min-wise independent permutations (extended abstract). In: STOC, pp. 327–336 (1998)
5. Butakov, S., Scherbinin, V.: On the number of search queries required for internet plagiarism detection. In: ICALT, pp. 482–483 (2009)
6. Charikar, M.: Similarity estimation techniques from rounding algorithms. In: STOC, pp. 380–388 (2002)
7. Chowdhury, A., Frieder, O., Grossman, D.A., McCabe, M.C.: Collection statistics for fast duplicate document detection. ACM Trans. Inf. Syst. 20(2), 171–191 (2002)

8. Conrad, J.G., Guo, X.S., Schriber, C.P.: Online duplicate document detection: signature reliability in a dynamic retrieval environment. In: CIKM, pp. 443–452 (2003)
9. Hamid, O.A., Behzadi, B., Christoph, S., Henzinger, M.R.: Detecting the origin of text segments efficiently. In: WWW, pp. 61–70 (2009)
10. Hua, N., Song, H., Lakshman, T.V.: Variable-stride multi-pattern matching for scalable deep packet inspection. In: INFOCOM, pp. 415–423 (2009)
11. Jiang, J., Tang, Y., Liu, B., Xu, Y., Wang, X.: Skip finite automaton: A content scanning engine to secure enterprise networks. In: GLOBECOM, pp. 1–5 (2010)
12. Kolcz, A., Chowdhury, A., Alspector, J.: Improved robustness of signature-based near-replica detection via lexicon randomization. In: KDD, pp. 605–610 (2004)
13. Manber, U.: Finding similar files in a large file system. In: USENIX Winter, pp. 1–10 (1994)
14. Manku, G.S., Jain, A., Sarma, A.D.: Detecting near-duplicates for web crawling. In: WWW, pp. 141–150 (2007)
15. Mittelbach, A., Lehmann, L., Rensing, C., Steinmetz, R.: Automatic detection of local reuse. In: Wolpers, M., Kirschner, P.A., Scheffel, M., Lindstaedt, S., Dimitrova, V. (eds.) EC-TEL 2010. LNCS, vol. 6383, pp. 229–244. Springer, Heidelberg (2010)
16. Potthast, M., Barrón-Cedeño, A., Eiselt, A., Stein, B., Rosso, P.: Overview of the 2nd international competition on plagiarism detection. In: CLEF (Notebook Papers/LABs/Workshops) (2010)
17. Schleimer, S., Wilkerson, D.S., Aiken, A.: Winnowing: Local algorithms for document fingerprinting. In: SIGMOD Conference, pp. 76–85 (2003)
18. Seo, J., Croft, W.B.: Local text reuse detection. In: SIGIR, pp. 571–578 (2008)
19. Theobald, M., Siddharth, J., Paepcke, A.: Spotsigs: robust and efficient near duplicate detection in large web collections. In: SIGIR, pp. 563–570 (2008)
20. Xiao, C., Wang, W., Lin, X., Yu, J.X., Wang, G.: Efficient similarity joins for near-duplicate detection. ACM Trans. Database Syst. 36(3), 15 (2011)
21. Yang, H., Callan, J.P.: Near-duplicate detection by instance-level constrained clustering. In: SIGIR, pp. 421–428 (2006)
22. Zhang, J., Suel, T.: Efficient search in large textual collections with redundancy. In: WWW, pp. 411–420 (2007)
23. Zhang, Q., Wu, Y., Ding, Z., Huang, X.: Learning hash codes for efficient content reuse detection. In: SIGIR, pp. 405–414 (2012)
24. Zhang, X., Qin, J., Wang, W., Sun, Y., Lu, J.: Hmsearch: An efficient hamming distance query processing algorithm. In: SSDBM (2013)

Reconciling Folksonomic Tagging with Taxa
for Bioacoustic Annotations

Anthony Truskinger, Ian Newmarch, Mark Cottman-Fields, Jason Wimmer,
Michael Towsey, Jinglan Zhang, and Paul Roe

Queensland University of Technology
Brisbane, Queensland Australia
{anthony.truskinger,ian.newmarch,
m.cottman-fields}@student.qut.edu.au,
{j.wimmer,m.towsey,jinglan.zhang,p.roe}@qut.edu.au

Abstract. Acoustic sensors are increasingly used to monitor biodiversity. They can remain deployed in the environment for extended periods to passively and objectively record the sounds of the environment. The collected acoustic data must be analyzed to identify the presence of the sounds made by fauna in order to understand biodiversity. Citizen scientists play an important role in analyzing this data by annotating calls and identifying species.

This paper presents our research into bioacoustic annotation techniques. It describes our work in defining a process for managing, creating, and using tags that are applied to our annotations. This paper includes a detailed description of our methodology for correcting and then linking our folksonomic tags to taxonomic data sources.

Providing tools and processes for maintaining species naming consistency is critical to the success of a project designed to generate scientific data. We demonstrate that cleaning the folksonomic data and providing links to external taxonomic authorities enhances the scientific utility of the tagging efforts of citizen scientists.

Keywords: tagging, citizen science, folksonomy, taxonomy, linking, annotation.

1 Introduction

Acoustic sensors are an effective method for monitoring biodiversity over large spatial and temporal scales. Sensors generate large volumes of data, continuously and objectively, without the need for a field worker to be constantly present.

The data collected must be analyzed to identify the individual species that vocalize in the recordings in order to understand the biodiversity of a region. Identification allows for estimates of species richness and the monitoring of changes in the ecosystem over time.

We have developed a system for annotating spectrogram images of audio recordings, to identify individual species within recordings. When a target acoustic event is identified, it is tagged. The tags that are applied are short, textual, freeform labels.

X. Lin et al. (Eds.): WISE 2013, Part I, LNCS 8180, pp. 292–305, 2013.

A full analysis of 480 hours of acoustic sensor data was performed and each species' vocalization was identified in each one minute segment.

When applying the tags to acoustic events, we chose to use an open taxonomy (a folksonomy) to allow citizen scientists to perform tagging. The flexibility of a folksonomy allowed for rich descriptions of the vocalizations for species. A folksonomy also allows for more information to be gathered than what would have been possible had the analysis been restricted to a fixed ontology that permitted only known species names to be used as tags. When utilizing a folksonomy it is expected that some problems in the data will occur due to user error. Accordingly, after the data was analyzed, we discovered inconsistencies in the *common name* tags that were used to identify the species in an acoustic event.

Our system included a restriction of one tag per annotation of an acoustic event. Many of the citizen scientists wanted to contribute more information to each annotation. They achieved this by encoding multiple pieces of information into the one allowed tag. This violated the common practice standard that defines each tag as containing only a single piece of information [1]. To complicate matters, there were several instances of *common name* tags that were ambiguous or incomplete, resulting in an inaccurate species identification.

Using this data for scientific work like determining species richness, calling rates, and species abundance required extensive data manipulation. The manipulations included: data normalization to ensure consistent capitalization, spacing and hyphenation; the correction of inconsistent spelling; splitting apart combined tags into separate tags; and mapping different *common name* tags to their correct taxon.

This paper presents contributions in three ways. First, through experience we demonstrate a single tagging methodology is not ideal when combined with a folksonomy. Second, we argue that when an ontology is available it should be used to enforce consistency in what could be called a hybrid taxonomy. Third, we demonstrate a practical method for repairing a damaged folksonomy similar to our dataset.

2   Project Description

Our project uses sensors to record audio. The majority of the sensors are Song Meter SM2+ made by Wildlife Acoustics. They run on batteries (for up to a week of continuous recording) and record data onto multiple SD memory cards. They are deployed in the field and either the data or the unit is collected after an allotted time. When the SD cards are retrieved, their data is uploaded to servers for storage. Once integrated with our system, processing and analysis of the data is then possible.

Traditionally we have approached the analysis problem with a variety of automated methods. Some of these methods have successfully detected certain species with some reliability [2, 3]. However, automated methods are complex, difficult to train using real-world sensor data, time-consuming, expensive, and generally require a dedicated analysis algorithm per species.

As an alternative, we created a semi-automated system for processing our data. We provided online tools for volunteer participants to analyze the recordings from the sensors [4]. We asked our participants to listen through sections of audio and to identify vocalizations of fauna they heard. A spectrogram, an image representing the amplitude over time at each frequency (see Fig. 1), was shown along with the audio.

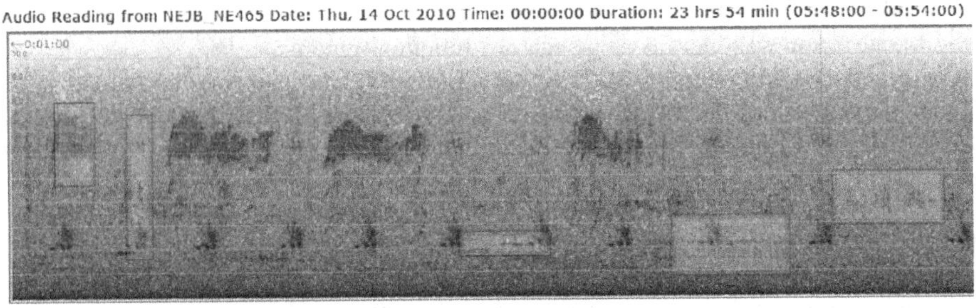

Audio Reading from NEJB_NE465 Date: Thu, 14 Oct 2010 Time: 00:00:00 Duration: 23 hrs 54 min (05:48:00 – 05:54:00)

Fig. 1. An example spectrogram with participant annotations shown

The annotation user interface has interactive drawing areas that allow a user to marquee an acoustic event of interest and associate a tag with it. We define an annotation as the combination of a marquee of an acoustic event with one or more tags.

When a vocalization was found, the participants drew a rectangle around the event on the spectrogram to specify the time and frequency bounds of the event. Each bounding rectangle was labelled with a tag representing the common name of the species that they believed had generated the vocalization.

The marquee data, associated tags, time of day, location, and the species that called, are the core pieces of data generated by our research for use by ecologists.

3   Tagging

Tags are simply textual labels that can be associated with a particular piece of data. Tags are frequently used in situations that require additional meaning to be associated with a resource.

As a result of the Web 2.0 phenomenon, tagging has become an increasingly popular Internet principal for classifying the data of many websites and their contents [5].

Tags allow content to be easily labeled by users – be it by the authors of the content or by others within a community. Each tag is ideally just a single keyword descriptor (sometimes a short phrase) associated with a piece of content. Most tagging systems allow the association of more than one tag to a piece of content. Tagging systems have become popular because they have a low barrier of entry for users [6]. This is a central reason for the proliferation of tagging throughout many popular websites, including Del.ico.us, Flicker, Twitter, Gmail, Facebook, YouTube, SoundCloud, and Tumblr.

Tags are useful because of their size and form. They are generally short and usually adjectives, thus they do not contain most of the unnecessary language that would usually exist in an equivalent descriptive sentence [1, 7]. This lack of function words required for grammatical correctness results in tags being linguistically simple enough to be easily used for comparison, categorization, and classification purposes with relative ease by both humans and algorithms

A tag is intended to describe one unique concept; tags should not combine atomic pieces of information [1]. If a tag encodes multiple pieces of information, it loses its ability to describe uniquely a single concept, thereby resulting in less effective categorizations of content.

Tagging systems are also interesting because they are extensible – a fixed taxonomy can be enforced in a tagging system but usually is not. Some tagging systems even allow for the definition of hierarchical tag structures that allow for the casual indirect association of data with different classifications. For example, Gmail can create nested labels to organize emails into a folder like hierarchy.

4   Tagging in a Bioacoustic Application

4.1   Reasons for Using Tags

We chose to use a tagging system for labeling faunal bioacoustic events. A tagging system was the easiest way to associate common names of fauna with the bioacoustic events. Tags enabled our users – many of whom are birdwatchers – to describe the events they saw in familiar terminology. The language that is produced from this freeform tagging is a known advantage to using tagging systems and is often referred to as a folksonomy [6, 7].

Initially, we considered only allowing tags to be used from a fixed set of tags – essentially using a general taxonomy. This however proved problematic for several reasons.

Firstly, the vocalizations made by species of interest do not have a fixed, formal, taxonomy available for describing how they sound. For example, when describing the sound of a vocalization tags can take forms similar to 'screech', 'ch-wik ch-wik', 'click', and 'laughing'. Often we are not even sure how many different vocalizations a target species can produce and we were interested in allowing our participants to describe these vocalizations.

Secondly, a folksonomy allows for creativity. The main reason we use a semi-automated process for analysis of audio data is to take advantage of the superior classification ability of human participants. Part of what makes our participants better, is their ability to creatively describe differences between acoustic events. This creativity extends to the freedom to choose the text they tag an acoustic event with.

Lastly, existing fixed taxonomies of species names proved problematic to use. Species' scientific names and common names do exist as taxonomies; however, they are prone to changes, new spellings, or reclassification (e.g. changing the family name of a species) [8]. We decided that allowing our participants to tag without a fixed list of taxa would allow them to work without considering evolving taxa.

A. Problems

Despite the above, we still encountered some problems in practice.
1)   The one tag policy

Initially we chose to support only one tag per acoustic event [9]. The single tag we applied to these events was a common name, e.g. 'Eastern Yellow Robin'. We wanted to collect multiple, descriptive tags for each acoustic event but we found the participants spent too much time tagging each event. We made the decision to allow only a single tag to minimize the time a user would spend annotating. After a large amount of analysis had been done, we realized the one-tag policy was a mistake. Our participants wanted to include more information than just the common name. This forced them to combine information into one tag that which should have been in separate tags. As a result, this had the effect of polluting our tag database with unreliable and inconsistent tags.

2)   Inherent problems with folksonomies

The disadvantages of a folksonomy became apparent when our participants started annotating data. The issue with giving users free form control over tag composition is that users unlike computers make mistakes.

Textual errors are common but are relatively easy to correct. Common human textual mistakes included misspellings, inconsistent or unnecessary punctuation, inconsistent pluralization, and grammatical errors.

Other errors we found in the tag data were semantic: tags can be dependent in context, have ambiguous meaning (e.g. abbreviations / shorthand), or can just be applied incorrectly. The semantic errors we see in tags are similar in part to the problems of polysemy (when one word has multiple meanings), synonymy (when many words share a meaning), abstraction level (how specific is the tag), and detail, as described by [7].

5   Tag Correction

Our participants have generated over 90 000 annotations over three years. All of these annotations were created under a one-tag policy.

As described previously, there are errors in this dataset that need to be fixed to ensure the data is rigorous enough for scientific study. Additionally, we wanted the ability to link our folksonomy to external taxonomic data sources to allow the retrieval of extra data. To achieve both of these goals, the dataset must be relatively consistent.

Apart from ensuring each tag represents a single concept, the text itself is difficult to correct or standardize, especially when it was generated as a set of folksonomic tags. This is due to the semantics of a tag. An algorithm can compare the characters of a string, but is currently incapable of comprehending the full linguistic meaning and context of the text and thus cannot know what a participant has actually meant when using that tag. Therefore, just like the analysis of audio, the correction of tags requires at least some human involvement.

Apart from the use-case of ensuring scientifically rigorous data, correcting the tags in the datasets and ensuring consistency represented an opportunity. We speculated the rules and algorithms needed to clean the annotation dataset would be useful after the initial clean, in the form of preventative heuristics that detect when a user is about to make a mistake.

As another by product of cleaning our dataset, we reasoned that both the new consistency and a deep understanding of the data would help improve some of our automated analysis that rely on tagged acoustic events. Automated algorithms that are currently benefitting include a suggestion algorithm (for suggesting what acoustic event a user might be looking at based on previous examples found in the annotation dataset) and generalized acoustic event recognizers that use the annotation data as training data sets [10].

Lastly, one of our largest motivating reasons for correcting the data with an algorithm stems from the behaviors of those consume it. Providing our annotation data to ecologists is our project's main research goal. Previously, when providing our data to ecologists, typical usage patterns include inserting the data into spreadsheet software or statistical packages. No software package that we are aware of is capable of correcting, or even detecting, erroneous tags without prohibitive effort.

Despite the difficulty, there have been cases of end-users trying to manually correct tags anyway. Their effort often requires several days of work, is usually partially incorrect, and wasted because the result is not shared. If we can ensure the dataset is consistent before being sent to ecologists, more than one party will benefit.

As our database of annotations increases in size, it becomes increasingly impractical to correct the tags manually.

5.1   Tag Correction Implementation

The correction of our 90 000 tags consists of two stages. The first stage was a preliminary normalization of the dataset that corrects simple textual inconsistencies. The second stage involves matching the tags against a taxonomy and providing spell-checker style suggestions for amendments. The first stage was automated and the second stage was partially automated.

5.2   Basic Normalization

The first stage consists of cleaning the data by applying simple text normalization transformations. These transformations include applying a capitalization convention and removing, adjusting, or adding certain characters.

It was decided that a capitalization convention should be applied to the dataset for readability. The rest of algorithm uses case-insensitive string comparisons. The first letter of every tag is capitalized and all other letters are set to lowercase – essentially coercing the tags to the Pascal Case standard.

The next step in cleaning the data entails removing unwanted tags and unwanted characters from tags.

Table 1. Example Corrections (white space marked with • (U+00B7))

Original Text	Cleaned Text	Additional Tags	Actions(S)
/Eastern•Yellow•Robin•	Eastern•Yellow•Robin		- Trailing white space - Special character gets stripped
Pied••Butcherbird2?	Pied•Butcherbird	*Requires Verification,* 2	- Duplicate white space - Question mark → Requires Verification - Numerical suffix separated
Torresian••Crow1	Torresian•Crow	1	- Numerical suffix separated - Duplicate white space
INTENTIONALLY LEFT EMPTY	Unknown		- Transformed to Unknown

White Space. Extra white space causes problems in string comparison and must be removed. Redundant whitespace occurred in 202 annotations in our dataset. Redundant white space is problematic because just a few tags with redundant whitespace introduces a relatively large number of false classes in the set of unique tags. Redundant whitespace was most common at the end of a tag where it is difficult for a human to spot unnecessary white space (see Table 1. for examples). The cleaning algorithm removes all leading and trailing white space from tags, as well as any repeated white space characters.

Characters with Special Semantics. Discussed previously, was the single tag policy that this dataset was generated under. Enthusiastic users that wanted to add more information did so by adding in characters or abbreviations with special meaning. This in turn corrupted the core concept of a tag, making differentiating between unwanted characters and characters with additional meaning difficult. Since purging unwanted characters outright was not possible without losing the valuable additional information associated with them, a set of heuristics are defined for the processing of different characters. Special characters that had no consistent clear meaning were stripped.

1. **Numbers:** These numeral suffixes were attached to tags to indicate what type of vocalization the fauna had made. The numeral suffixes were independent for each type of *common name* tag. We found no cases where a numeral suffix did not act as a distinguisher for the vocalization type of a particular species. The algorithm separates these values out to another tag.
2. **Question marks (?):** Participants created a convention where the uncertainty of a classification was represented with a question mark appended to a tag. The algorithm separates these question marks out and instead associates an additional 'Requires Verification' tag.
3. **Special suffix 'sp':** The 'sp' suffix is another participant convention that emerged when participants could not determine the species of an acoustic event. The 'sp' suffix is an abbreviation for the term 'species' and in this case meant that only an identification to the genus level of a taxonomy was possible. The 'sp' tag is a special type of suffix that describes the abstraction level of the other words it is associated with. This suffix was stripped and not preserved since the linking algorithm (described further on) is capable of operating without this extra information.
4. **Special Characters:** These characters provide no further meaning to the tag. They are removed from the tag. Examples include characters such as *, &, or @.
5. **Unknown:** 0.74% of tags were found to have tags that meant 'unknown' in some way. Examples of this include 'blank', '' (actually blank), 'unknown', '??tt38?', and 'something'. These tags are converted into one standard unknown tag: 'Unknown'.

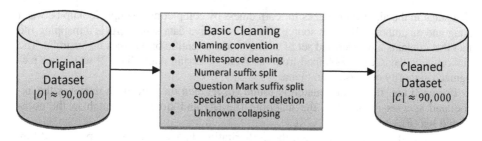

Fig. 2. The flowchart describing the data normalization step

5.3   Problem Detection and Resolution

The problem detection part of the algorithm deals with subtler problems in the tags like spelling, grammar, or other mistakes that are caused by user error.

The first stage of problem detection was implementing an algorithm often used by spell-checkers. We use the spell-checker algorithm by referencing a species taxonomy as its dictionary. Automatic text correction with spell-checkers have a number of inherent and challenging problems. The primary of which is the Cupertino effect – the phenomenon where spell checkers change words from a correct value to an incorrect or less accurate value [11]. Thus, this part of the algorithm does not change values automatically, since correcting values automatically is analogous to inferring a tag's semantics – previously established as bad practice. Instead, a process of flagging errors is done so users, who are capable of understanding the context in which a tag was applied, can verify the tags manually. The problem detection algorithm was implemented in two phases:

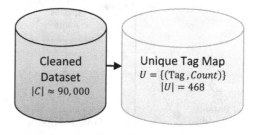

Fig. 3. The flowchart demonstrating the collapse from a full data set to a unique tag map

Preliminary Scale Reduction. See Fig. 3. There are over 90 000 annotations in the full cleaned dataset (**C**) but only 468 unique tags. Thus, a map (i.e. an associative array) is created using the tag text as a key and grouped tag counts as the values (**U**). This reduced the amount of processing power needed for the rest of the algorithm by two orders of magnitude. Examples of the result of this process are listed in Table 2.

Simple Taxa Comparison. See Fig. 4. If a tag already has the correct formatting and spelling then there is no need for further correction. The

Table 2. Unique Tag Count map

Note: Tags are all in lower case for case insensitive matching

Common Name	Count
crow	245
sacred kingfisher	349
scared kingfisher	210

first stage in the algorithm checks for correctness by completing a simple lookup between a tag and an authoritative data source. The authoritative data source (**A**) is a mapping file that that contains a pre-verified set of common and scientific bird names in Australia. The authoritative dataset was obtained from the Atlas of Living Australia [12] which is a recognized authority on Australian fauna.

For each tag, if an exact match is found (in **A**) then the algorithm retrieves the associated species name from the mapping file (**A**) and stores the match in the completed list (**F**).

Tags that do not exactly match the authoritative data source (A) are processed further with a step that attempts to match a tag via an approximate match. An approximate match is determined by calculating the edit distance between a unique tag (from U) and each of the entries in the authoritative data source (A).

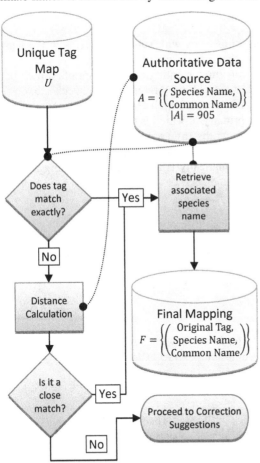

The distance between two strings is determined using the Damereaus-Leveshtein distance algorithm [13]. This is an algorithm traditionally used as a spell-checking algorithm. This metric is defined as the number of operations required to transform one string into another. Recognized operations are switching the positions of two characters, changing one character to another, deleting a character, and inserting a character into a string. For example: the distance between 'apple' and 'azple' is one, for only one operation is required to match the strings (change the 'z' to 'a').

For each tag, if an approximate match is found (in A) then the algorithm retrieves the associated species name from the mapping file (A) and stores the match in the completed list (F). An approximate match occurs when the edit distance is less than two (2) operations.

The approximate match threshold is adjustable. An increase in the threshold results in more matches correlated with a decrease in precision. A decrease in the

Fig. 4. The flowchart demonstrating the simple taxa comparison algorithm

threshold results in fewer matches (sometimes none at all) but often has higher precision.

If the distance between two items being compared is below the threshold, the suggested output is then added to the suggested correction queue.

Correction Suggestions. See Fig. 5. If a tag did not match or approximately match any taxa from the authoritative data source (**A**), the tag is then sent to the suggestion part of the algorithm. In this stage, the algorithm attempts to find an appropriate correction for the tag by using only the data from the unique tag map (**U**) itself. This stage only suggests – no automatic action is taken without human intervention.

First, the algorithm calculates the Dameraus-Leveshtein distance between a tag itself and each other tag in the unique tag map (**U**). If the distance between the tag itself and the items from the unique tag map (**U**) is less than or equal to the edit operations threshold, then all matching tags are grouped together. For each group the matches' tags are sorted by their unique count (their frequency within the full dataset **C**) in descending order. As the grouping process continues, the tag with the highest frequency is updated as the best match. It then becomes the suggested correction for all the other tags in its group.

Our assumption here is that the most popular variant of a tag is the correct one. In our experience, this has always been the case. However, given a situation where this is not the case human input can correct this assumption.

Finally, when all items from the unique tag map (**U**) have been processed the suggested corrections are exported to a CSV file, to be reviewed by an appropriate participant, usually a citizen-expert.

Participants. We define a participant as any user participating in the analysis of data. The participants that annotated most of this data are experts in recreational ornithology.

The role of a participant is to review the suggested corrections CSV file and mark if the corrections are valid or invalid. Once the file has been reviewed, it is read back

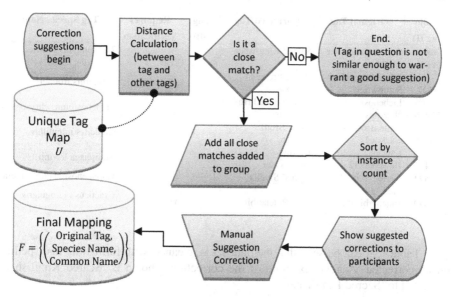

Fig. 5. The flowchart demonstrating the frequency based suggestion corrections and the manual process involved with accepting changes

Table 3. Part of a Suggested Corrections File

Problem	Replacement	Reason	Problem	Replacement	Reason
willie-wagtail	willie wagtail	Dict	lewins honeyeater	lewin's honeyeater	Dict
lorikeet	lorikeets	Tag Count	yellow faced honey eater	yellow-faced honey-eater	Dict
lewin's honey eater	lewin's honeyeater	Dict	yellow faced honeyeater	yellow-faced honey-eater	Dict
straited pardalote	striated pardalote	Dict	white-throated treecreepe	white-throated tree-creeper	Dict
rufus whistler	rufous whistler	Dict	white throated honeyeater	white-throated honey-eater	Dict

into the correction program so the reviewed corrections can be applied. The manual corrections are persisted across program runs and thus can be consulted when verifying any new tags. Included in Table 3 is an example of what the suggested correction file looks like when it shown for review by a participant.

5.4   Results

Having completed the processing of the data, the differences between the old and new data can be seen in Table 4. The table shows a small sample of the annotation dataset with original tags with their final corrections. The previously mentioned issues of inconsistent capitalization, spelling errors, and tag suffixes are highlighted for clarity.

Table 4. The final result of cleaning

Audio Tag ID	Original Tag	Corrected Tag	Tag Suffix	Requires Verification	Tag Species Name
4001	Scared Kingfisher	Sacred Kingfisher			Todiramphus sanctus
60638	??js84	Js	84	Yes	CORRECTLY BLANK
3627	White-browed Scrubwren?	White-browed Scrubwren		Yes	Sericornis frontalis
92266	Lichenostomus chrysops•	Yellow-faced Honeyeater			Lichenostomus chrysops
10188	Little Bronze-Cuckoo1	Little Bronze-cuckoo	1		Chalcites minutillus
68437	Lewins Honeyeater2	Lewin's Honeyeater	2		Meliphaga lewinii
37893	Varied Sitella1	Varied Sittella	1		Daphoenositta chrysoptera
91089	Pied Butcherbird4?	Pied Butcherbird	4	Yes	Cracticus nigrogularis

In Table 4, it can be seen that the data set is, cleaner, with formatting and spelling errors corrected. A good example of the corrections shown is 'Scared Kingfisher', corrected to 'Sacred Kingfisher'.

Statistics. Fig. 6 shows relevant correction statistics that were made by the algorithm. Interestingly, out of the 90 225 tags that were processed only 1.12% (1 011) of the tags were flagged as incorrect after pre-processing. This was at first surprising, giving cause for concern for the algorithms' correctness. Post-analysis showed the actual reason for a small amount of the tags being flagged as incorrect is due to the pre-processing (the data normalisation) that was completed on the dataset. If no pre-processing was done then 87% (78 553) of the tags would of been flagged as incorrect. This demonstrates how important basic normalization of the data set is. Of the 1 011 tags that were flagged as incorrect, 876 suggested corrections were considered a valid correction when reviewed. This means the suggested correction algorithm has an accuracy of 82%.

Furthermore, of the 90 255 tags processed, 85 957 (95%) of the tags were populated with a species name. This is an ideal result because it means that 95% of the data set can be considered valid after cleaning and can be potentially linked to external data sources. There are some errors in the final dataset and future work is targeted at improving accuracy.

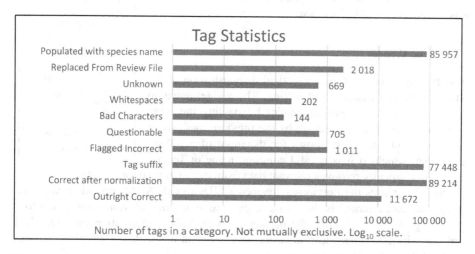

Fig. 6. A bar graph the shows various statistics about the tags in our dataset. Note the x-axis is in log scale.

6   Tag Linking

With the tags normalized, cleaned, and mapped to species names, the data is ready to be mapped to a formal taxonomic structure.

The Atlas of Living Australia (ALA) is an online resource that provides information on faunal species in Australia. This information can be accessed through their provided web API (Application Programming Interface). We choose to link to the ALA because the majority of our data involves birds (for which the ALA has a wealth of information) and their API was accessible and well documented.

As a species name is usually less ambiguous than a common name, the species name was chosen as the link between the local dataset and ALA's records. The goal is to use the *species name* tag on each annotation to search for the associated LSIDs. A LSID is a Life Science Identifier; it is a convention for uniquely identifying a Life

Sciences resource on the web using a uniform resource name (URN). LSIDs are the unique identifier ALA uses to identify species.

To link the LSID's, a similar map to the unique tag count map is created but instead of holding a tuple of (Tag, Count) it holds a tuple of (SpeciesName, LSID) – where the LSID part of the tuple is empty to begin with.

Each item in the species name map is searched for using the ALA API. The query is returned containing a list of search results. For each search result, an attribute named "score" is used to score of the relevancy of the search result. The higher the value of score the more accurate the result is. The best search result is taken and the LSID associated with the tag that was searched for.

Recall that of the 90 225 tags, 85 957 successfully had a species name associated with them. Of that 85 957, a LSID was linked to 78 478 (87%) as well.

Fig. 7. A screenshot of the web widget

6.1   The Widget

A web widget was built as a practical demonstration of the utility of the tag cleaning and linking. The widget reacts to input (either typed or programmatic) by passing the input through the tag mappings set up by the previous analysis of our annotation dataset. It allows for the input of a folksonomic tag, which, if it is a valid common name or species name, automatically retrieves relevant data from the ALA taxonomic data source.

The widget uses the previously mapped LSIDs to call ALA's provided API. The taxonomic data returned is then used to provide additional relevant information to a user. Its goal was to provide the participant with further information on a searched bird that will make classification easier, in a way that is simple and efficient. An example of the widget is shown in Fig. 7.

7   Conclusion

Our annotation dataset represents a large and valuable investment. For our area of research, the dataset is unique resource that was created for the scientific study of the environment. To use this data for further research, we must ensure the data rigorous.

In this paper, we have detailed the reasons for choosing a folksonomy and the subsequent problems that ensued. We saw consumers of our dataset waste many hours of effort cleaning the data by hand and we saw an opportunity to not only clean our dataset but also to learn from it.

Our dataset was cleaned so that we were able to ensure that 95% of our annotations could be associated with a species name – a large improvement over the 13% before cleaning. After the mapping, it was then possible to link to an external data source to provide additional structured information to 87% of our tags when classifying through the web widget.

As future work: the simple heuristics we used for cleaning have already been used in part in the construction of the web widget. We are currently working on incorporating these rules into our online analysis tool to proactively limit future errors that may occur in our data source.

We have also decided to implement a hybrid folksonomy and taxonomy system. The new analysis system we are designing is a multi-tagging folksonomy, with a seeded taxonomy that makes it easier for participants to choose the correct taxon for a classification. Additionally, we are strongly considering separate tag fields that will group tags based on their concept. For example: there may be a 'Common Name' field, a 'Species Name' field, a 'Looks Like' field, a 'Sounds Like' field, and an 'Other' field – all designed to keep the flexibility of a folksonomy but with the added advantage of enforcing some structure. In another vein of research, we will formalize our method for applying contextual tags, which detail the methods of analysis used or describe the known / unknown / unheard status for a section of audio.

References

1. Xu, Z., Fu, Y., Mao, J., Su, D.: Towards the semantic web: Collaborative tag suggestions. In: Collaborative Web Tagging Workshop at WWW 2006, Edinburgh, Scotland (2006)
2. Planitz, B., Roe, P., Sumitomo, J., Towsey, M.W., Williamson, I., Wimmer, J., Zhang, J.: Listening to nature: acoustic monitoring of the environment (2009)
3. Towsey, M., Planitz, B., Nantes, A., Wimmer, J., Roe, P.: A toolbox for animal call recognition. Bioacoustics 21, 107–125 (2012)
4. Wimmer, J., Towsey, M., Planitz, B., Williamson, I., Roe, P.: Analysing environmental acoustic data through collaboration and automation. Future Generation Computer Systems (2012)
5. Cuff, D., Hansen, M., Kang, J.: Urban Sensing: Out of the Woods. Communication of the ACM 51, 24–33 (2008)
6. Gasc, A., Sueur, J., Jiguet, F., Devictor, V., Grandcolas, P., Burrow, C., Depraetere, M., Pavoine, S.: Assessing biodiversity with sound: Do acoustic diversity indices reflect phylogenetic and functional diversities of bird communities? Ecological Indicators 25, 279–287 (2013)
7. Marlow, C., Naaman, M., Boyd, D., Davis, M.: HT06, tagging paper, taxonomy, Flickr, academic article, to read. In: Proceedings of the Seventeenth Conference on Hypertext and Hypermedia, pp. 31–40. ACM, Odense (2006)
8. Ames, M., Naaman, M.: Why we tag: motivations for annotation in mobile and online media. In: Proceedings of the SIGCHI Conference on Human Factors in Computing Systems, pp. 971–980. ACM, San Jose (2007)
9. Mason, R., Roe, P., Towsey, M., Jinglan, Z., Gibson, J., Gage, S.: Towards an Acoustic Environmental Observatory. In: IEEE Fourth International Conference on eScience, pp. 135–142 (2008)
10. Truskinger, A., Yang, H.F., Wimmer, J., Zhang, J., Williamson, I., Roe, P.: Large Scale Participatory Acoustic Sensor Data Analysis: Tools and Reputation Models to Enhance Effectiveness. In: 2011 IEEE 7th International Conference on E-Science (e-Science), pp. 150–157. IEEE (2011)
11. Zimmer, B.: The Cupertino Effect. Language Log, March 9 (2006)
12. Atlas of Living Australia (2013), http://www.ala.org.au/
13. Oommen, B., Loke, R.: Pattern recognition of strings with substitutions, insertions, deletions and generalized transpositions. Pattern Recognition 30, 789–800 (1997)

Multi-relational PageRank
for Tree Structure Sense Ranking

Roberto Interdonato and Andrea Tagarelli

DIMES - University of Calabria, Italy
{rinterdonato,tagarelli}@dimes.unical.it

Abstract. In this paper, we study the problem of structural sense ranking for tree data using a multi-relational PageRank approach. By considering multiple types of structural relations, the original tree structural context is better leveraged and used to improve the ranking of the senses associated to the tree elements. Upon this intuition, we advance research on the application of PageRank-style methods to semantic graphs inferred from semistructured/plain text data by developing the first PageRank-based formulations that exploit heterogeneity of links to address the problem of structural sense ranking in tree data. Experiments on a large real-world benchmark have confirmed the performance improvement hypothesis of our proposed multi-relational approach.

1 Introduction

Tree-shaped data are pervasively used to model real-life objects and their structural relationships. Since the advent of XML, semantic-rich information with an inherent (hierarchical) logical organization has found a convenient way to be managed and exchanged. However, the ambiguity embedded in the flexibility in using (meta)languages for coding information in tree data makes it such data heterogeneous by nature. Disclosing the semantics underlying the structural constituents of tree data is essential to enable a number of applications, ranging from the mapping and integration of conceptually related information in tree-structured schemas, to the semantics-aware similarity search in heterogeneous Web data, from the organization of semantically related documents, to the definition of summaries for different semantic views over data collections.

The presence of varying degrees of structuredness that are used to explain the logical organization of the information in tree data makes the coupling of (tree structural) syntactic information with the appropriate semantics a different problem to be solved than lexical ambiguity related ones in plain text, particularly word sense disambiguation. Moreover, it is quite common that multiple fine-grained senses may be correct (at different confidence levels) for a given term; consequently, it might be more useful for retrieval and data management purposes to produce a contextual ranking of the senses rather than to decide exactly for a single sense and regard it as the only appropriate one.

Structural sense ranking is recognized as challenging in database and information retrieval research, whereby solutions to other semantics-aware problems

X. Lin et al. (Eds.): WISE 2013, Part I, LNCS 8180, pp. 306–319, 2013.

(e.g., schema matching, keyword searching, feature extraction, document clas-
sification/clustering) can be complemented or supported. Note that the studies
in [17,18] have demonstrated that tree data does require a sense ranking method
that takes into account the structural relations in the data.

The network of meanings underlying the structural constituents of tree data
can be conveniently represented as a labeled (weighted) graph, therefore graph-
based ranking methods are natural candidates to solve the structural sense ranking problem. In particular, eigenvector-centrality methods, like PageRank, have
been already used in semantic networks inferred from plain text (e.g., [13,1], but
the list would be clearly longer if other natural language processing tasks were
included) and they have also shown to generally outperform non-PageRank yet
graph-based approaches.

Surely a question remains: if the subtlety and multiplicity of tree structural
relations that hold among the underlying concepts in a tree data would not be
disregarded, then a ranking algorithm would propagate the importance scores
through different multi-typed relations modeled as a *heterogeneous information
network* (HIN). Our intuition is that a multi-relational ranking method should
in principle be able to better leverage the semantics of annotations in tree data
(i.e., markup tag names) that are structurally related at different levels.

Contributions. While existing research has already filled a lack of knowledge
on the suitability of PageRank-style methods to semantic networks for the structural sense ranking problem [17,18], no investigation on the presumed benefits
deriving from a HIN representation of the structural semantics in tree data has
been made so far. In this work we are hence interested in exploring the structural sense ranking problem in semantic networks inferred from tree data, when
multiple types of tree structural relations are taken into account. We believe this
joins an important issue due to the ever increasing demand for knowledge-driven
applications to manage tree data through the emerging paradigm of dealing with
mixed type information in graph models. With the purpose of pushing towards
the study of multi-relational PageRank-style methods in multi-typed semantic
networks, we propose a novel PageRank-based framework for structural sense
ranking, for which different approaches are developed to deal with multiple types
of tree structural relations. More specifically, we define an approach that consists
of a weighted PageRank model for a tree-structure-aware semantic multidigraph.
We also present two alternative formulations of the PageRank-based structural
sense ranking problem, the first essentially leading to a decomposition into multiple independent PageRanks for single-type tree structural relations, and the
second based on the assumption of biasing the PageRank by means of multi-
typed structural relations. Our extensive experimentation on a large real-world
benchmark of XML data has assessed the significance of a multi-relational approach to the structural sense ranking problem, and finally demonstrated that
better ranking solutions are obtained when multiple types of tree structural relations are taken into account.

The rest of the paper is organized as follows. Section 2 provides background
notions on semantic relatedness measures and also briefly mentions PageRank for

word sense disambiguation and ranking methods in HINs. Section 3 describes our proposed structural sense ranking framework, and provides formal details about the construction of context graphs and the ranking methods. Section 4 presents experimental methodology and results. Section 5 concludes the paper.

2   Background and Related Work

Semantic Relatedness and PageRank for Word Sense Disambiguation. Knowledge-based approaches to sense ranking and disambiguation assume the availability of a knowledge base as a source of information about the word meanings. In this respect, *WordNet*[1] is widely used in knowledge-based data management tasks, while semantic relatedness measures are the essential tools to choose the most plausible sense to assign to each word in an input text, or in general to determine the ranking of its senses. We now very briefly recall the definition of classic measures as they will be used in our experimental evaluation, while the interested reader can refer to [4] for further details. Focusing on the content affinity of the descriptions (glosses) associated with any two concepts c_1, c_2, an effective *gloss-overlap-based* measure is $go\text{-}rel(c_1, c_2) = \sum_{go \in GO(c_1, c_2)} |go|^2$, where $GO(c_1, c_2)$ denotes the set of disjoint, maximal word-sequences shared between the c_1's gloss and c_2's gloss (overlaps), and $|go|$ indicates the number of words in the overlap go. *Path-based* measures are instead defined as functions of the location (*depth*) of concept-nodes in the lexical ontology. Concept specificity and commonality properties are well-encompassed by the Wu & Palmer measure: $p\text{-}rel(c_1, c_2) = \frac{2 \times depth(lcs(c_1, c_2))}{depth(c_1) + depth(c_2)}$, where $lcs(\cdot, \cdot)$ computes the least common subsumer for any two concepts. The above measure has also an *information-content-based* counterpart, known as Lin measure and hereinafter denoted as *ic-rel*, whereby the notion of concept-node depth is replaced with the amount of information a concept provides, i.e., $IC(c) = -\log \Pr(c)$, where $\Pr(c)$ is typically estimated by the relative frequency of usage of concept c in a corpus—note that lexical ontologies like WordNet embed statistics about the usage of concepts.

Concerning the application of PageRank to semantic networks inferred from natural language texts specifically for word sense disambiguation problems, we acknowledge the existence of important studies (e.g., [13,1]). The basic idea common to all those approaches is to represent a lexical ontology like WordNet as a graph whose vertices are concepts (synsets) and edges correspond to lexical/conceptual relations, and then to apply over it a (possibly weighted or biased) PageRank method. Due to space limits, we cannot discuss the aforementioned works, however here we remark that they were already the focus of a comparative evaluation in our previous works [18,17], which generally showed a poor effectiveness of word sense disambiguation methods conceived for plain text when applied to tree-structured text.

Ranking in Heterogeneous Information Networks. The advent of multityped interconnected social media and bibliographic networks, scientific (e.g.,

---

[1] http://wordnet.princeton.edu/

medical) information systems, and next-generation e-commerce systems has posed new challenges in managing large-scale HINs. Ranking models are central to address such challenges, and in fact they have been developed for a variety of tasks such as keyword search in databases (e.g., ObjectRank [2]), Web object ranking (e.g., PopRank [14]), expert search in digital libraries (e.g., [20,6]), link prediction (e.g., [5]), recommender systems and Web personalization (e.g., [12,11]). Moreover, there has been an increasing interest in integrating ranking with mining tasks, like the case of ranking-based clustering addressed by RankClus [15] and NetClus [16] methods.

Besides the novelty of the application domain in which ranking in HINs is addressed in this work, the HIN in our framework differs from others in that while vertices are all of the same type, multiple structural relations induce multi-typed edges that can also be drawn between the same pair of vertices; by contrast, parallel edges are not handled in most existing HINs. We take into account the weighting of edge types by unsupervised learning schemes, which do not require neither any training set based on a domain-expert-provided ranking [14] nor ad-hoc specified criteria [2]. Moreover, our HIN does not need to follow a particular topology like a bipartite graph, as in [15,12], or star network schema, as in [16].

3   Structural Sense Ranking Framework

Let $\mathcal{D}$ denote a labeled tree data instance rooted in a node with label t_0, and let $T(\mathcal{D}) = \{t_0, t_1, \ldots, t_n\}$ be the set of tree element labels in $\mathcal{D}$. We will refer to $T(\mathcal{D})$ as T, if the input tree data is clear from the context, and to the elements in $T(\mathcal{D})$ as *tags*. For each tag $t \in T$, the set of concepts or *senses* of t available in the reference *lexical ontology* is denoted as $\mathcal{C}(t)$. Our general goal for structural sense ranking in tree data is as follows:

> Given a labeled tree data instance $\mathcal{D}$ and assuming the availability of a lexical ontology, a semantic network is built over the tag concepts and such that it is aware of the multiple structural relations underlying the tags in $\mathcal{D}$. A ranking of all concepts associated with each tag-label in $\mathcal{D}$ is to be computed using a PageRank-style method applied on the constructed semantic network.

We present next our solutions to accomplish this goal, which adopt different approaches to handle multi-typed tree structural relations.

3.1   The Multi-structure Semantic PageRank approach

Tree-Structure-Aware Semantic Multidigraph. We build the ranking context graph upon the following methodology. We consider all concepts of the tags in a tree data instance as vertices of the context graph. Edges are drawn between two tags' concepts if a selected *structural relation* holds in the tree instance for any two nodes that are respectively labeled with the two tags. Concepts of the same tag should not be connected to each other in order to avoid undesired mutual reinforcement effects in the concept ranking; as an exception, since the same concept can in principle belong to different tags, self-loops might be drawn if the concept is shared by two structurally connected tag nodes. Edge weights are computed to express the strength of association between any two connected concepts: this should rely primarily on the semantic relatedness between the concepts but should also consider the impact of the repetition of substructures across the input tree instance. Formally, we define the ranking context graph as a directed multigraph (multidigraph) of the form $\mathcal{G} = \langle \mathcal{V}, \mathcal{T}, \mathcal{E}, w \rangle$ such that:

- $\mathcal{V} = \{c \mid c \in \mathcal{C}(t), t \in T\}$.
- $\mathcal{T} \subseteq \mathcal{T}_0$, where $\mathcal{T}_0$ denotes the domain of *structural relations* for the tag nodes in $\mathcal{D}$. Hence, $\mathcal{T}$ is regarded as the selected set of structural relations that corresponds to the set of edge-types in $\mathcal{G}$.
- $\widetilde{\mathcal{E}} = \bigcup_{\tau \in \mathcal{T}} \widetilde{\mathcal{E}}(\tau)$, such that $\widetilde{\mathcal{E}}(\tau) = \{(c_i, c_j, \tau) \mid c_i \in \mathcal{C}(t), c_j \in \mathcal{C}(t'), t, t' \in T \wedge t' \xrightarrow{\tau} t\}$. Function $t' \xrightarrow{\tau} t$ applies to a pair of tags t, t' and returns a boolean value depending on whether the structural relation $\tau \in \mathcal{T}$ holds in $\mathcal{D}$ between two nodes labeled with t' and t, respectively.
- $w : \widetilde{\mathcal{E}} \to \Re^*$ is an edge weighting function defined, for each $(c_i, c_j) \in \widetilde{\mathcal{E}},$[2] as:

$$w(c_i, c_j) = semrel(c_i, c_j) \times sf(c_i, c_j) \tag{1}$$

In (1), *semrel* is a non-negative real-valued function that corresponds to a selected measure of word semantic relatedness. Function sf calculates the frequency of occurrence of a direct structural relation underlying the associated tag nodes relating to two concepts, and is defined as:

$$sf(c_i, c_j) = 1 + \log_{fo(\mathcal{D})} \left(\prod_{t,t'} (1 + freqPC(t, t')) \right) \tag{2}$$

where t, t' are such that $c_i \in \mathcal{C}(t), c_j \in \mathcal{C}(t')$, $fo(\mathcal{D})$ is the average fan-out of $\mathcal{D}$, and $freqPC(t, t')$ is the number of times that t' is a child node of t in $\mathcal{D}$. Function sf acts as an augmenting factor for those concept edges whose associated tag nodes are more frequently linked in the tree instance.
- $\mathcal{E} \subseteq \widetilde{\mathcal{E}}$ such that $\mathcal{E} = \{e = (c_i, c_j) \mid e \in \widetilde{\mathcal{E}} \wedge w(e) > 0\}$. Note that condition $w(e) = 0$ holds only if $semrel(e) = 0$, for any edge e.

The above definition is general as it does not impose any particular (set of) structural relations (for drawing the edges) and semantic relatedness measures (for weighting the edges). To provide a complete specification of the ranking context graph, here we define the domain of structural relations ($\mathcal{T}_0$) by focusing on binary functions that capture the relative position of nodes in a subtree:

- $\tau = childOf$: $t' \xrightarrow{\tau} t$ holds if t' is child of t;
- $\tau = descOf$: $t' \xrightarrow{\tau} t$ holds if t' is descendant of t;

---

[2] Edge notation is simplified (i.e., pair of vertices) when there is no dependency on a particular structural relation type, as for the edge weighting function.

- $\tau = child|siblchildOf$: $t' \xrightarrow{\tau} t$ holds if t' is child of t or child of a t's sibling;
- $\tau = desc|sibldescOf$: $t' \xrightarrow{\tau} t$ holds if t' is descendant of t or descendant of a t's sibling.

Moreover, we instantiate function *semrel* as one of the standard dictionary-based semantic relatedness measures previously discussed in Section 2; of course, other measures could be used in alternative, including those recently developed that utilize Wikipedia or Web-based knowledge sources (e.g., [17,19,9,8]), however evaluating their impact on the structural sense ranking performance is out of the scope of this work.

Structural Relation Weighting Schemes.   To deal with multiple structural relations, we define *weighting schemes* (alternative to uniformly weighting) which, assuming the unavailability of user-specified requirements or prior knowledge, are based on characteristics of the tree data instance. One approach would rely on the assumption that the most frequent instances of a structural relation are the most important ones; this obviously implies that more complex (i.e., indirect) structural relations would be assigned with higher weights, since the frequency of occurrence is a non-decreasing function for increasing complexities. However, this approach might also have the shortcoming of further penalizing the score propagation through graph edges that belong to simpler yet direct relations (e.g., *childOf*), which already have a lower support in the tree data instance. The opposite approach would hence assign higher weights to less frequent relations, thus aiming to balance the properties of rarity (low support) and locality that a structural relation has in the tree when propagating the ranking score in the context graph. We hereinafter refer to the two weighting approaches as *support-aware* and *locality-aware* weighting schemes, respectively.

Given a structural relation $\tau \in \mathcal{T}$, if we denote with $n(\mathcal{D}, \tau)$ the number of edges in $\mathcal{D}$ of type τ, the support-aware weight of τ is defined as:

$$\omega_\tau{}^{(s)} = \frac{n(\mathcal{D}, \tau)}{\sum_{\tau' \in \mathcal{T}} n(\mathcal{D}, \tau')} \tag{3}$$

whereas the locality-aware weight of τ is defined as:

$$\omega_\tau{}^{(l)} = \frac{\sum_{\tau' \in \mathcal{T}, \, \tau' \neq \tau} n(\mathcal{D}, \tau')}{(|\mathcal{T}| - 1) \sum_{\tau' \in \mathcal{T}} n(\mathcal{D}, \tau')} \tag{4}$$

Note that both the above definitions are such that $\sum_{\tau \in \mathcal{T}} \omega_\tau = 1$, which is a requirement in the application of the weighting scheme to the ranking models that will be presented next.

Multi-structure Semantic PageRank.   Our proposed ranking method, named *multi-structure semantic PageRank* (MSSPR), adapts a weighted Page-Rank formulation to deal with a multi-relational, edge-typed graph. Essentially, the underlying random-walk model is expressed by as many transition probability matrices as the different edge types. Given the ranking context graph $\mathcal{G}$ with structural relation set $\mathcal{T}$ and corresponding $|\mathcal{T}|$ weights ω_τ, the ranking score of any concept c_i is computed as:

$$r_i = \alpha \left(\sum_{\tau \in \mathcal{T}} \omega_\tau \sum_{j \in B_\tau(i)} \frac{w(j,i)}{out_\tau(j)} r_j \right) + \frac{1-\alpha}{|\mathcal{V}|} \tag{5}$$

where $B_\tau(i)$ is the set of concepts that are linked to c_i through τ, $out_\tau(j)$ is the sum of weights on outgoing edges of type τ for c_j, and α is a damping factor ($\alpha \in [0,1]$, commonly set to 0.85). Equivalently, the matrix form of MSSPR is:

$$\mathbf{r} = \alpha \left(\sum_{\tau \in \mathcal{T}} \omega_\tau \mathbf{S}_\tau \mathbf{r} \right) + (1-\alpha)\mathbf{v} \tag{6}$$

where $\mathbf{v} = \frac{1}{|\mathcal{V}|}\mathbf{1}$ is the teleportation vector, and $\mathbf{S}_\tau$ denotes the column-stochastic transition probability matrix associated to the structural relation τ, i.e., only edges of type τ are considered in $\mathbf{S}_\tau$. Note that (6) can also be written as $\mathbf{r} = \alpha \mathbf{S}_{\mathcal{T}}\mathbf{r} + (1-\alpha)\mathbf{v}$, where $\mathbf{S}_{\mathcal{T}}$ is a convex combination of all the $\mathbf{S}_\tau$ matrices weighted by the corresponding ω_τ.

Upon the above MSSPR formulation, we introduce a variant into the definition of $\mathbf{v}$ to bias MSSPR according to the usage frequency of the concepts in $\mathcal{V}$. The rationale here is that a-priori importance of the concepts can be estimated based on their linguistic popularity as known from annotated text corpora, and hence the probability of moving to a concept-vertex c_i might be defined as proportional to its usage frequency. Formally, the ith element of the teleportation vector, for each $c_i \in \mathcal{V}$, is computed as: $v_i = (usage\_freq(c_i)+1)(\sum_{c \in \mathcal{V}} usage\_freq(c) + |\mathcal{V}|)$, where $usage\_freq(c)$ is the c's frequency of usage as stored in the reference lexical ontology (cf. Section 2), and the Laplace smoothing is introduced to handle unavailability of information about a concept's usage count. We will refer to the biased version of MSSPR as MSSPR-uf.

3.2   Alternative Multi-relational Methods

We devise two alternative approaches to structural sense ranking in tree data, whose common characteristic is a relaxation of the assumption of multidigraph definition of the context graph while maintaining the information on all selected types of tree structural relations. In particular, we raised two generic questions: *(Q-1) What if multiple instances of a basic PageRank model are separately built and performed over all structural relation types? (Q-2) What if information on all structural relation types is used only to bias a single instance of a basic PageRank model?* In the following we elaborate on each of the above points.

Weighted Combination of PageRank Vectors. To answer question (Q-1), we perform MSSPR for each of the structural relation types in $\mathcal{T}$, and the final ranking is obtained as a weighted linear combination of the multiple PageRank stationary vectors π_τ produced by the $|\mathcal{T}|$ runs of MSSPR:

$$\pi = \sum_{\tau \in \mathcal{T}} \omega_\tau \pi_\tau \tag{7}$$

We will refer to this approach as the pSSPR method.

Multi-structure Aware Personalized PageRank. To answer question (Q-2) we develop an adaptation of personalized PageRank, named mS-PPR, in which the bias in the ranking model relies on the tree structural relations of various types.

Let $\mathcal{G}_p = \langle \mathcal{V}_p, \mathcal{T}, \mathcal{E}_p, w_p \rangle$ be the ranking context graph with vertex set $\mathcal{V}_p$ coinciding with $\mathcal{V}$ of MSSPR, and edge set $\mathcal{E}_p = \{(c_i, c_j) \mid c_i \in \mathcal{C}(t), c_j \in \mathcal{C}(t'), t, t' \in T \wedge t' \to t\}$, where $t' \to t$ means that t' is child of t. For each $(c_i, c_j) \in \mathcal{E}_p$, a weight $w_p(i, j)$ is computed to express the probability that any tag associated to c_i implies any tag associated to c_j through a direct structural relation in the tree; formally, $w_p(i, j) = \mathrm{avg}_{t, t' \in T} \Pr(t, t') = \Pr(t \cap t') / \Pr(t) = freqPC(t, t') / freq(t)$ such that $c_i \in \mathcal{C}(t), c_j \in \mathcal{C}(t')$, where $freq(t)$ is the total number of occurrences of tag t in $\mathcal{D}$ (and $freqPC(t, t')$ is defined as for MSSPR). If we denote with $out_p(j)$ the sum of weights w_p on out-going edges of c_j, and with $R_\tau(i)$ the set of concept vertices that are pointed by c_i through edges of type τ, the mS-PPR score of any c_i is computed as:

$$r_i = \alpha \sum_{j \in B(i)} \frac{w_p(j, i)}{out_p(j)} r_j + (1 - \alpha) \left(1 - \frac{\sum_{\tau \in \mathcal{T}} |R_\tau(i)|}{\sum_{h \in \mathcal{V}} \sum_{\tau \in \mathcal{T}} |R_\tau(h)|} \right) v_i \qquad (8)$$

with $v_i = 1/(|\mathcal{V}| - 1)$ if $R_\tau(i) \neq \emptyset$, otherwise $v_i = 0$. The teleportation factor in (8) is defined to ensure that the teleportation matrix is stochastic, and that the probability of teleportation increases with smaller τ-specific out-neighbor sets.

4 Experimental Evaluation

4.1 Data and Assessment Methodology

The official INEX 2009 collection[3] is a corpus of semantically annotated XML documents representing Wikipedia articles, which perfectly fits our evaluation needs due to its semantic and structural heterogeneity. Annotations consist in assigning each tag with two attributes: *wordnetid*, whose value corresponds to a unique sense id in WordNet 3.0, and *confidence*, whose value (typically within 0.6 and 1) expresses the confidence the annotator originally had in assigning that wordnetid to the tag. From this benchmark dataset, we extracted a very large set consisting of 1,289,309 XML documents (4 GB size), whose main characteristics are summarized in Table 1. We processed the articles to keep only the structure information, so to obtain trees of tags, rooted in `article`. We finally treated the document trees either separately or conveniently merged into a single huge tree (rooted in a fictitious tag node `articles`): the two choices, henceforth referred to as *homogeneous evaluation* and *heterogeneous evaluation* cases, respectively, actually correspond to two different realistic scenarios. A reason that should make this twofold evaluation worthy of investigation is that we expect that a

---

[3] http://www.mpi-inf.mpg.de/departments/d5/software/inex/

Table 1. Evaluation dataset: structural and semantic characteristics

# tags	# distinct tags	min depth	avg depth	max depth	avg fanout*	max fanout*	avg polysemy	max polysemy	# monose- mous tags	avg polysemy**
159,094,497	5,203	3	9.01	74	1.36	4,643	2.45	33	2,390	4.01

*Fanout values refer to sublevels of the `article`'s level.   **Monosemous tags not considered.

relatively conceptual homogeneity of the tags in a tree would justify the use of structural contexts that rely on more complex relations; conversely, for a tree covering a larger variety of topics (i.e., tag labels), building the context graph over (directly) related tags would reduce the disambiguation "noise" which might be produced by more complex structural contexts.

Following the methodology in [17], we generated a gold standard, or *reference ranking*, for the evaluation dataset. Concisely, for each tag a probability distribution over its senses is computed by taking into account the multiple occurrences as well as the differently assigned wordnetid and confidence values the tag can have in the collection. The interested reader is referred to [17] for details.

To assess the effectiveness of the proposed methods, we used criteria that are standard in ranking tasks: *normalized discounted cumulative gain (nDCG)*, *Binary preference function (Bpref)*, and *Fagin's intersection metric (F)*; for each of them, the higher the score the better the ranking evaluation. However, such criteria needed to be adapted to our setting, as described next.

Let $\mathcal{L}^*$ and $\mathcal{L}$ denote the reference ranking and the ranking produced by an algorithm, respectively. *Normalized discounted cumulative gain* (nDCG) [10] measures the usefulness (gain) of an item based on its relevance and position in a list. Formally, nDCG is the ratio between the discounted cumulative gain to its ideal (reference) counterpart taking into account the top-k-ranked items in two lists. Contextualized to each tag t, nDCG is defined as: $nDCG_t^{(k)} = \frac{DCG_t^{(k)}}{IDCG_t^{(k)}}$. Discounted cumulative gain is based on the assumption that highly relevant items appearing in lower positions in a list should be more penalized as the graded relevance value is reduced logarithmically proportional to the position of the result. For a ranking $\mathcal{L}$, we use symbol $\mathcal{L}_t(i)$ to denote the ranking value associated to the sense of tag t that is ranked in position i, and symbol $\mathcal{L}_t[i]$ to denote the ranking value associated to the ith sense of tag t. For a tag t, the DCG is computed as: $DCG_t^{(k)} = \mathcal{L}_t^*[\arg \mathcal{L}_t(1)] + \sum_{i=2}^{k} \frac{\mathcal{L}_t^*[\arg \mathcal{L}_t(i)]}{\log_2(i+1)}$, where symbol $\arg \mathcal{L}_t(i)$ is used to denote the sense number of the sense ranked at position i in the algorithm's ranking, and hence $\mathcal{L}_t^*[\arg \mathcal{L}_t(i)]$ is the reference ranking value for that sense. Term $IDCG_t^{(k)}$ is calculated w.r.t. the reference ranking values for the senses of t: $IDCG_t^{(k)} = \mathcal{L}_t^*(1) + \sum_{i=2}^{k} \frac{\mathcal{L}_t^*(i)}{\log_2(i+1)}$. Hence we obtain the final $nDCG$ as the average of the $nDCG_t^{(k)}$ computed over all tags t.

We also compared an algorithm's ranking with the reference ranking without averaging over the tag-specific distributions. This leads to a problem of comparing *partial rankings*, since elements in one list may not be present in the other list. The *Fagin's intersection metric* [7] is commonly used to solve the

Table 2. Performance of MSSPR methods

T	ω_T	heterogeneous evaluation								homogeneous evaluation							
		MSSPR				MSSPR-uf				MSSPR				MSSPR-uf			
		nDCG	Bpref	F_1	F_2	nDCG	Bpref	F_1	F_2	nDCG	Bpref	F_1	F_2	nDCG	Bpref	F_1	F_2
{c}	–	0.929	0.240	0.401	0.200	0.936	0.263	0.440	0.232	0.699	0.560	0.442	0.318	0.899	0.604	0.534	0.369
{d}	–	0.931	0.253	0.403	0.202	0.938	0.275	0.441	0.233	0.663	0.501	0.428	0.294	0.894	0.532	0.530	0.349
{c,d}	u	0.936	0.247	0.406	0.203	0.938	0.271	0.442	0.236	0.726	0.559	0.436	0.316	0.899	0.608	0.539	0.372
	l	0.934	0.248	0.407	0.205	**0.939**	0.272	**0.443**	**0.237**	0.725	0.564	0.439	0.318	**0.902**	0.607	**0.541**	**0.373**
	s	0.822	0.247	0.405	0.202	0.886	0.271	0.442	0.235	0.727	0.561	0.437	0.315	0.899	**0.609**	0.540	0.370
{sc}	–	0.930	0.262	0.394	0.201	0.936	0.284	0.426	0.232	0.694	0.534	0.441	0.320	0.895	0.562	0.533	0.363
{c,sc}	u	0.931	0.253	0.396	0.202	0.937	0.274	0.431	0.233	0.708	0.551	0.440	0.319	0.897	0.593	0.536	0.367
	l	0.932	0.255	0.398	0.203	0.938	0.276	0.431	0.234	0.706	0.562	0.443	0.323	0.901	0.598	0.537	0.369
	s	0.803	0.263	0.395	0.201	0.872	0.285	0.427	0.232	0.705	0.542	0.439	0.318	0.896	0.583	0.534	0.363
{c,d,sc}	u	0.935	0.254	0.401	0.202	0.938	0.276	0.436	0.234	0.734	0.553	0.438	0.318	0.898	0.601	0.540	0.372
	l	0.934	0.256	0.403	0.202	**0.939**	0.277	0.435	0.236	0.732	0.560	0.440	0.321	0.901	0.603	0.540	**0.373**
	s	0.812	0.262	0.396	0.201	0.876	0.284	0.428	0.231	0.727	0.542	0.437	0.316	0.897	0.589	0.536	0.366
{sd}	–	0.932	0.263	0.396	0.202	0.937	0.289	0.427	0.232	0.627	0.480	0.427	0.296	0.893	0.502	0.526	0.347
{sc,sd}	u	0.933	0.264	0.397	0.204	0.938	**0.292**	0.426	0.233	0.716	0.532	0.438	0.317	0.898	0.569	0.536	0.368
	l	0.933	0.265	0.398	0.203	0.938	0.289	0.424	0.234	0.718	0.529	0.433	0.313	0.897	0.567	0.536	0.365
	s	0.811	0.264	0.396	0.202	0.872	0.290	0.425	0.232	0.725	0.534	0.436	0.317	0.896	0.573	0.538	0.367

Results correspond to average performance over the various semantic relatedness measures. Bold values refer to the best scores per evaluation case and assessment criterion.

above problem and applies to any two top-k lists: $F(\mathcal{L}^*, \mathcal{L}, k) = \frac{1}{k} \sum_{i=1}^{k} \frac{|\mathcal{L}^*_{:i} \cap \mathcal{L}_{:i}|}{i}$, where $\mathcal{L}^*_{:i}$, $\mathcal{L}_{:i}$ denote the sets of senses from the 1st to the ith position in the respective rankings. Therefore, F is the average over the sum of the weighted overlaps based on the first k senses in both rankings. We defined two variants of F, henceforth denoted as F_1 and F_2, which are based on different setups of $\mathcal{L}_{:i}$ and $\mathcal{L}_{:i}$. In F_1, the actual reference ranking is obtained by simply sorting all scores in the original reference ranking, whereas the algorithm's ranking scores are first normalized by tag (to resemble the tag-specific probability distributions in the original reference ranking), and then sorted. In F_2, for both the reference and algorithm's rankings, each concept's score is multiplied by the logarithm of the number of senses of the unique tag associated to the concept (recall that a concept is treated as a pair tag-IDsense, i.e., a synset in WordNet).

Bpref [3] computes a preference relation of whether judged relevant candidates R of a list $\mathcal{L}'$ are retrieved (in a list $\mathcal{L}''$), ahead of judged irrelevant candidates N. It is formulated as $Bpref(R, N) = (1/|R|) \sum_r (1 - (\#\text{of } n \text{ ranked higher than} r)/|R|)$, where r is a relevant retrieved candidate, and n is a member of the first $|R|$ irrelevant retrieved candidates. As queries, we used the root-to-leaf tag-paths in $\mathcal{D}$, judging the top-1 ranked senses of each tag in the path as relevant candidates, and all the other senses of these tags as not relevant. The overall $Bpref$ score was obtained as a weighted average over the tag-path $Bpref$ scores weighted by the number of occurrences of a particular path.

4.2 Results

We discuss our experimental evaluation in terms of effectiveness and efficiency.[4] To avoid cluttering the presentation in the result tables, we will use the

---

[4] Experiments ran on an Intel Core i7-3960X CPU @ 3.30GHz, 64GB RAM machine.

Table 3. Performance of pSSPR methods

$\mathcal{T}$	ω_τ	heterogeneous evaluation												homogeneous evaluation							
		pSSPR				pSSPR-uf				pSSPR				pSSPR-uf			
		$nDCG$	$Bpref$	F_1	F_2	$nDCG$	$Bpref$	F_1	F_2	$nDCG$	$Bpref$	F_1	F_2	$nDCG$	$Bpref$	F_1	F_2
{c,d}	u	0.821	0.245	0.402	0.200	0.882	0.270	0.443	0.234	0.726	0.547	0.437	0.308	0.900	0.595	0.541	0.365
	l	0.820	0.247	0.404	0.201	**0.886**	0.271	**0.445**	**0.236**	0.724	0.550	0.439	0.311	**0.901**	0.592	**0.544**	0.366
	s	0.818	0.247	0.400	0.198	0.879	0.269	0.442	0.235	0.703	0.538	0.438	0.315	0.896	0.582	0.536	0.362
{c,sc}	u	0.805	0.250	0.398	0.201	0.877	0.270	0.436	0.233	0.704	0.550	0.441	0.317	0.897	0.594	0.539	0.366
	l	0.807	0.252	0.400	0.201	0.878	0.272	0.435	0.233	0.703	0.559	0.442	0.320	0.900	**0.596**	0.538	**0.368**
	s	0.802	0.263	0.396	0.199	0.872	0.285	0.427	0.232	0.627	0.480	0.427	0.296	0.893	0.502	0.526	0.347
{c,d,sc}	u	0.818	0.252	0.400	0.197	0.884	0.273	0.441	0.232	0.731	0.545	0.438	0.313	0.900	0.592	0.544	0.366
	l	0.817	0.254	0.401	0.199	0.885	0.275	0.440	0.234	0.729	0.549	0.439	0.315	**0.901**	0.591	0.545	**0.368**
	s	0.812	0.262	0.395	0.200	0.875	0.284	0.428	0.231	0.723	0.525	0.436	0.310	0.896	0.561	0.541	0.361
{sc,sd}	u	0.814	0.261	0.395	0.202	0.874	**0.289**	0.438	0.230	0.716	0.526	0.439	0.310	0.900	0.559	0.540	0.362
	l	0.812	0.263	0.396	0.200	0.874	0.288	0.426	0.232	0.714	0.519	0.434	0.306	0.898	0.556	0.541	0.359
	s	0.812	0.262	0.394	0.198	0.873	0.288	0.427	0.229	0.663	0.501	0.428	0.294	0.894	0.532	0.530	0.349

Results correspond to average performance over the various semantic relatedness measures. Bold values refer to the best scores per evaluation case and assessment criterion. Rows corresponding to singleton sets $\mathcal{T}$ are the same as in Table 2, hence are not reported.

following abbreviated notations for the selected structural relations: c for *childOf*, d for *descOf*, sc for *child|siblchildOf*, and sd for *desc|sibldescOf*; for the structural relation weighting schemes, we will use notations u, s, l for the uniform, support- and locality-aware scheme, respectively. It should be noted that, since the singleton sets (i.e., {c}, {d}, {sc}, and {sd}) correspond to PageRank methods each based on a single-type structural semantic graph, those methods actually play the role of competitors against our proposed multi-relational setting.

Effectiveness. Tables 2–4 report on performance results according to all evaluation criteria, by varying structural context and weighting scheme, and also distinguishing between the heterogeneous and homogeneous case (cf. Section 4.1); in the latter case, results are averages over the individual trees. Also, we chose not to include monosemous tags in the ranking evaluation in order to avoid a bias in the result presentation. Reported results correspond to a setup of the $nDCG$'s parameter k to 3 (which is close to the average polysemy, cf. Table 1) and of the F's parameter k to 5000; the latter setting was chosen to take into account a reasonably large portion of the global rankings produced by the methods (about 10% of the size of the vertex set in the ranking context graph).

Looking at Table 2, MSSPR-uf outperformed MSSPR in terms of all criteria, in both heterogeneous and homogeneous evaluation cases. This supports our expectation that exploiting information on the concepts' usage frequency is beneficial to the ranking performance. Results were generally higher in the homogeneous case, where the advantage taken by MSSPR-uf w.r.t. MSSPR is also more evident, on all criteria. More importantly to the purpose of our study was to find out that the best results per evaluation case and criterion indeed were obtained for multi-typed structural contexts, particularly on the combinations {c,d} and {c,d,sc}. This aspect was emphasized in the homogeneous case, for which the higher cohesiveness of the tags enables a multi-typed structural context to significantly improve upon each of its corresponding subsets.

The weighting schemes impacted differently over the various criteria in the heterogeneous case, with s performing worse than the other schemes for $nDCG$,

Table 4. Performance of mS-PPR

$\mathcal{T}$	heterogeneous evaluation				homogeneous evaluation			
	nDCG	Bpref	F_1	F_2	nDCG	Bpref	F_1	F_2
{c}	0.634	0.238	0.390	0.188	0.443	0.562	0.387	0.285
{d}	**0.638**	0.231	0.377	**0.194**	0.449	0.542	0.392	0.285
{c,d}	0.637	**0.239**	**0.394**	**0.194**	**0.451**	**0.566**	**0.406**	**0.291**
{sc}	0.635	0.232	0.391	0.190	0.447	0.560	0.402	0.279
{c,sc}	0.636	0.235	**0.394**	0.192	0.448	0.563	0.404	0.280
{c,d,sc}	0.637	0.237	0.392	0.193	0.450	0.564	0.405	0.281
{sd}	0.635	0.231	0.377	0.193	0.450	0.540	0.404	0.288
{sc,sd}	**0.638**	0.232	0.393	0.192	0.449	0.561	0.401	0.283

Bold values refer to the best scores per evaluation case and assessment criterion.

but comparably or slightly better in terms of the other criteria. In the homogeneous case, relative differences among the weighting schemes were more consistent over the criteria; moreover, as we expected, scheme l led to better performance than s and u for contexts that involve the c relation. As concerns the impact of the semantic relatedness measures (results not shown), better performance over the sets $\mathcal{T}$, regardless of the weighting scheme, was generally obtained by using p-rel (in terms of $nDCG$ and $Bpref$) and go-rel (in terms of Fs), although relative differences were scarcely significant (e.g., 2.0E-4 $nDCG$). We also evaluated the ranking performance of the MSSPR methods when only the *semrel* term would be considered in the edge-weighting function, i.e., $sf = 1$ for all edges; in that case, we observed a general decrease in the performance, with order of 1.0E-3 for each evaluation criterion, which would indicate that the sf term in the edge-weighting function does serve for the purpose of weighing the impact of the repetition of substructures across the input tree instance.

The pSSPR approach (Table 3) also performed better on multi-typed structural contexts, although its overall performance did not improve upon MSSPR (except for F_1, but with average gap of just 0.003); interestingly, the impact due to the weighting scheme tended to be irrelevant in the heterogeneous case, while a predominance over s was observed in the homogeneous case. Concerning mS-PPR (Table 4), there is a less clear evidence of the benefits that can derive from using multi-typed structural contexts, however in any case it was generally outperformed by MSSPR methods (even by the non-personalized MSSPR in most cases) for all criteria, with average gaps up to 0.449 $nDCG$, 0.045 $Bpref$, 0.136 F_1 and 0.082 F_2.

Note also that an evaluation of the methods' best performances led to findings similar to those observed for the average performances; particularly, MSSPR outperformed pSSPR especially in terms of $nDCG$ and $Bpref$, while mS-PPR performance was significantly lower than both MSSPR and pSSPR (e.g., $nDCG$ gaps from MSSPR up to 0.472 for the homogeneous case).

Efficiency. Table 5 shows the times that were required for building the ranking context graphs and for performing the ranking by MSSPR and pSSPR; results were averaged over the semantic relatedness measures and corresponded to the heterogeneous evaluation case. Considering MSSPR, the graph building times increased for increasing structural complexity of the corresponding sets $\mathcal{T}$.

Table 5. Time performances (milliseconds): MSSPR vs. pSSPR

$\mathcal{T}$	MSSPR		pSSPR	
	graph	*ranking*	*graph(max)*	*ranking(max)*
{c}	1.066E+06	1.862E+04	1.066E+06	1.862E+04
{d}	1.299E+06	2.428E+04	1.299E+06	2.428E+04
{c,d}	1.531E+06	4.484E+04	1.299E+06	2.428E+04
{sc}	1.150E+06	3.671E+04	1.150E+06	3.671E+04
{c,sc}	1.249E+06	6.011E+04	1.150E+06	3.671E+04
{c,d,sc}	1.634E+06	8.707E+04	1.299E+06	3.671E+04
{sd}	1.812E+06	4.405E+04	1.812E+06	4.405E+04
{sc,sd}	2.488E+06	8.279E+04	1.812E+06	4.405E+04

The ranking times mainly depended on the graph size, which was inferior of one order of magnitude when *go-rel* was used (this might be explained by a lower access rate to WordNet w.r.t. *p-rel* and *ic-rel*). The ranking times clearly depended on the different rate of convergence as well, which on average was around 25 iterations for *p-rel* and *ic-rel* and 60 iterations for *go-rel*.

We also compared MSSPR with pSSPR under a "parallel" runtime configuration, i.e., in which the maximum runtimes per structural context were taken. The comparison was clearly in favor of pSSPR, albeit both methods' runtimes were of the same order of magnitude. Moreover, as concerns mS-PPR (results not shown), the graph building time was comparable to the MSSPR graph building time with context *childOf* (in fact, it is independent on the choice of $\mathcal{T}$), while the ranking time was always slower than the MSSPR one on the corresponding set $\mathcal{T}$ (usually about one order of magnitude).

5   Conclusion

We addressed the problem of structural sense ranking in tree data by proposing a multi-relational PageRank framework over a structure-aware semantic network. We developed different formulations of the problem, mainly focusing on the modeling of a semantic multidigraph as ranking context graph and on PageRank methods that differently handle multi-typed structural relations in tree data. Results have demonstrated that dealing with multi-typed structural relations in tree data indeed leads to the expected improvements in performance w.r.t. the case of single-type structural contexts.

While we have taken XML document trees as a case in point for the experimental evaluation, our approach to structural sense ranking can be readily applied to any kind of domain where semantic-rich data attributes have an inherent hierarchical organization over possibly multiple types of structural relations.

References

1. Agirre, E., Soroa, A.: Personalizing PageRank for Word Sense Disambiguation. In: Proc. 12th Conf. of the European Chapter of the Association for Computational Linguistics (EACL), pp. 33–41 (2009)
2. Balmin, A., Hristidis, V., Papakonstantinou, Y.: ObjectRank: Authority-Based Keyword Search in Databases. In: Proc. Int. Conf. on Very Large Data Bases (VLDB), pp. 564–575 (2004)

3. Buckley, C., Voorhees, E.M.: Retrieval evaluation with incomplete information. In: Proc. ACM SIGIR Conf. on Research and Development in Information Retrieval (SIGIR), pp. 25–32 (2004)
4. Budanitsky, A., Hirst, G.: Evaluating WordNet-based Measures of Lexical Semantic Relatedness. Comput. Ling. 32(1), 13–47 (2006)
5. Davis, D.A., Lichtenwalter, R., Chawla, N.V.: Multi-relational Link Prediction in Heterogeneous Information Networks. In: Proc. Int. Conf. on Advances in Social Networks Analysis and Mining (ASONAM), pp. 281–288 (2011)
6. Deng, H., Han, J., Lyu, M.R., King, I.: Modeling and exploiting heterogeneous bibliographic networks for expertise ranking. In: Proc. Int. Joint Conf. on Digital Libraries (JCDL), pp. 71–80 (2012)
7. Fagin, R., Kumar, R., Sivakumar, D.: Comparing Top k Lists. SIAM Journal on Discrete Mathematics 17(1), 134–160 (2003)
8. Gabrilovich, E., Markovitch, S.: Computing Semantic Relatedness Using Wikipedia-based Explicit Semantic Analysis. In: Proc. Int. Joint Conf. on Artificial Intelligence (IJCAI), pp. 1606–1611 (2007)
9. Gracia, J.L., Mena, E.: Web-Based Measure of Semantic Relatedness. In: Bailey, J., Maier, D., Schewe, K.-D., Thalheim, B., Wang, X.S. (eds.) WISE 2008. LNCS, vol. 5175, pp. 136–150. Springer, Heidelberg (2008)
10. Järvelin, K., Kekäläinen, J.: Cumulated gain-based evaluation of IR techniques. ACM Trans. Information Systems 20(4), 422–446 (2002)
11. Kashyap, A., Amini, R., Hristidis, V.: SonetRank: leveraging social networks to personalize search. In: Proc. ACM Conf. on Information and Knowledge Management (CIKM), pp. 2045–2049 (2012)
12. Lee, S., Song, S., Kahng, M., Lee, D., Lee, S.: Random walk based entity ranking on graph for multidimensional recommendation. In: Proc. ACM Conf. on Recommender Systems (RecSys), pp. 93–100 (2011)
13. Mihalcea, R., Tarau, P., Figa, E.: PageRank on Semantic Networks, with Application to Word Sense Disambiguation. In: Proc. 20th Int. Conf. on Computational Linguistics (COLING) (2004)
14. Nie, Z., Zhang, Y., Wen, J.-R., Ma, W.-Y.: Object-level ranking: bringing order to Web objects. In: Proc. ACM Conf. on World Wide Web (WWW), pp. 567–574 (2005)
15. Sun, Y., Han, J., Zhao, P., Yin, Z., Cheng, H., Wu, T.: RankClus: integrating clustering with ranking for heterogeneous information network analysis. In: Proc. Int. Conf. on Extending Database Technology (EDBT), pp. 565–576 (2009)
16. Sun, Y., Yu, Y., Han, J.: Ranking-based clustering of heterogeneous information networks with star network schema. In: Proc. ACM SIGKDD Int. Conf. on Knowledge Discovery and Data Mining (KDD), pp. 797–806 (2009)
17. Tagarelli, A.: Exploring Dictionary-based Semantic Relatedness in Labeled Tree Data. Information Sciences 220, 244–268 (2013)
18. Tagarelli, A., Gullo, F.: Evaluating PageRank methods for structural sense ranking in labeled tree data. In: Proc. 2nd Int. Conf. on Web Intelligence, Mining and Semantics (WIMS), 36 (2012)
19. Tsatsaronis, G., Varlamis, I., Nørvrag, K.: SemanticRank: Ranking Keywords and Sentences Using Semantic Graphs. In: Proc. Int. Conf. on Computational Linguistics (COLING), pp. 1074–1082 (2010)
20. Zhang, M., Feng, S., Tang, J., Ojokoh, B., Liu, G.: Co-ranking multiple entities in a heterogeneous network: Integrating temporal factor and users' bookmarks. In: Airong, J. (ed.) ICADL 2011. LNCS, vol. 7008, pp. 202–211. Springer, Heidelberg (2011)

Towards Content-Aware SPARQL Query Caching for Semantic Web Applications

Yanfeng Shu, Michael Compton, Heiko Müller, and Kerry Taylor

Computational Informatics, CSIRO, Australia
{yanfeng.shu,michael.compton,heiko.mueller,kerry.taylor}@csiro.au

Abstract. Applications are increasingly using triple stores as persistence backends, and accessing large amounts of data through SPARQL endpoints. To improve query performance, this paper presents an approach that reuses results of cached queries in a *content-aware* way for answering subsequent queries. With a focus on a common class of conjunctive SPARQL queries with filter conditions, not only does the paper provide an efficient method for testing whether a query can be evaluated on the result of a cached query, but it also shows how to evaluate the query. Experimental results show the effectiveness of the approach.

1 Introduction

With the popularity of Semantic Web technologies, applications are increasingly using triple stores as persistence backends, accessing large amounts of data through SPARQL endpoints. As the number of queries increases, and data grow in size, scalability becomes an issue. To address this, much work has been done to improve the performance of triple stores through better storage, indexing and query optimisation. However, little has been done so far to take advantage of caching.

The work by Martin et al. [6] represents a first step towards filling the gap, where caching is performed by a proxy layer residing between an application and a SPARQL endpoint, and the proxy answers a SPARQL query without accessing the triple store if the query is identical to a cached query. Caching in [6] is basically *content-blind*, unaware of the content of cached results. In this paper, we go one step further and explore reusing cached results in a *content-aware* way, so that the proxy can not only answer a query that exactly matches a cached query, but it can also answer a query by processing the result of a cached query. Such a caching approach requires SPARQL query containment checking, i.e. checking whether the result of a query is contained in that of a cached query. Containment checking for full-SPARQL in general is undecidable [5]. Considering this, we focus on a fragment of SPARQL which is commonly used in real world queries [8], conjunctive queries with simple filter conditions (CQSFs). As our first contribution, we define SPARQL query containment based on the (set) semantics of SPARQL, and give sufficient conditions for containment checking of CQSFs.

Containment checking alone, however, is not enough. It is possible that a query is contained in a cached query but cannot be evaluated on its result. For example,

X. Lin et al. (Eds.): WISE 2013, Part I, LNCS 8180, pp. 320–329, 2013.

given query Q_1, returning the names of all students of age 20 from a university database, and query Q_2, returning the names of all students. It is easy to see that the result of Q_1 is contained in that of Q_2. However, Q_1 cannot be evaluated on the result of Q_2, since it does not contain enough information to evaluate the age constraint. Another issue with containment checking is that its cost could be considerable, which potentially compromises the benefit of caching. Our second contribution addresses these two issues. We introduce a notion of *evaluability* and define requirements for queries to be answered using cached results. We further provide an efficient method for checking whether a query can be answered using the result of a cached query (i.e. query evaluability checking), and show how query answering is done. We evaluate our approach experimentally based on the LUBM benchmark. The results show that our approach achieves much better performance than no caching and content-blind caching cases. The rest of the paper is organised as follows. Section 2 introduces the concepts. Section 3 describes our caching approach. Section 4 presents experimental results. Finally, Section 5 concludes the paper and points out future work.

2   Preliminaries

2.1   Syntax and Semantics of SPARQL

SPARQL is the official W3C recommendation for querying RDF graphs. In this paper, we focus on SELECT queries on ground RDF graphs. Let V be a set of variables disjoint from U (URIs) and L (Literals). Variables in V are prefixed by the symbol ?. We denote a SELECT query by $Q(S, P)$, where $S \subset V$ is the set of variables to be returned, P is the graph pattern to be matched. For simplicity, we restrict our discussion to conjunctive queries with simple filter conditions (CQSFs), i.e. queries composed of AND and simple FILTER operators. We refer to a filter condition as *simple*, if it involves at most one variable. Given a graph pattern P, we use $vars(P)$ to denote the set of variables in P (for a triple pattern t, we use $vars(t)$), and $ftrs(P)$ to denote the set of filter conditions in P.

We define the semantics of SPARQL by following the set semantics defined in [9,7]. A *solution mapping*[1] μ from V to $U \cup L$ is a partial function $\mu : V \rightarrow U \cup L$. The domain of μ, $dom(\mu)$, is the subset of V where μ is defined. Given a triple pattern t and a solution mapping μ such that $vars(t) \subseteq dom(\mu)$, we use $\mu(t)$ to denote the triple obtained by replacing the variables in t according to u. Two solution mappings μ_1 and μ_2 are compatible, denoted by $\mu_1 \sim \mu_2$, if for all $?X \in dom(\mu_1) \cap dom(\mu_2)$, $\mu_1(?X) = \mu_2(?X)$, i.e. if $\mu_1 \cup \mu_2$ is also a solution mapping. Let Ω, Ω_1 and Ω_2 be sets of solution mappings, R a filter condition, and $S \subset V$ a set of variables. We define algebraic operations join ($\bowtie$), projection (π), and selection (σ) over mapping sets: $\Omega_1 \bowtie \Omega_2 = \{\mu_1 \cup \mu_2 \mid \mu_1 \in \Omega_1, \mu_2 \in \Omega_2 \text{ and } \mu_1 \sim \mu_2\}$; $\pi_S(\Omega) = \{\mu_1 \mid \exists \mu_2, \mu_1 \cup \mu_2 \in \Omega \wedge dom(\mu_1) \subseteq S \wedge dom(\mu_2) \cap S = \emptyset\}$; $\sigma_R(\Omega) = \{\mu \in \Omega \mid \mu \models R, \text{ i.e. } \mu \text{ satisfies } R\}$.

---

[1] It is simply called *mapping* in [9,7].

Based on these operations, the evaluation of graph patterns and queries over an RDF graph G is defined as a function $[[.]]_G$ that takes a pattern or a query, and returns a set of solution mappings. Let G be an RDF graph, t a triple pattern, P, P_1, P_2 graph patterns, R a filter condition, $S \subset V$ a set of variables, and $Q(S, P)$ a SELECT query, we define: $[[t]]_G = \{\mu \mid dom(\mu)=vars(t) \text{ and } \mu(t) \in G\}$; $[[P_1 \text{ AND } P_2]]_G = [[P_1]]_G \bowtie [[P_2]]_G$; $[[P \text{ FILTER } R]]_G = \sigma_R([[P]]_G)$; $[[Q(S, P)]]_G = \pi_S([[P]]_G)$.

Example 1. Consider a SELECT query $Q(\{?x\}, ((?x, type, student)(?x, age, ?y)$ FILTER $(?y = 20)))$ and an RDF graph $G = \{(a, type, student), (a, age, 30), (b, type, student), (b, age, 20)\}$. When evaluating Q over G, we obtain $[[Q]]_G = \{\{?x \rightarrow b\}\}$.

2.2 Containment of SPARQL Queries

We define the containment of SPARQL queries based on the definition of subsumption of solution mappings. In [1], the authors introduce a definition of subsumption of solution mappings. Here, we extend their definition by considering different sets of variables that are possibly used in queries.

Definition 1 (Subsumption of Solution Mappings). *Let Ω_1 and Ω_2 be two sets of solution mappings. Ω_1 is subsumed by Ω_2, denoted by $\Omega_1 \sqsubseteq \Omega_2$, if there is a variable mapping ψ from the domain of Ω_1 (the union of domains of its solution mappings) to the domain of Ω_2 that for every $\mu_1 \in \Omega_1$, there exists $\mu_2 \in \Omega_2$ such that $\psi(\mu_1) \subseteq \mu_2$, where $\psi(\mu_1)$ denotes the solution mapping obtained from μ_1 by replacing every variable $?X \in dom(\mu_1)$ with $\psi(?X)$.*

Definition 2 (SPARQL Query Containment). *Let Q and Q' be two SPARQL queries. Q is contained in Q', denoted by $Q \sqsubseteq Q'$, if and only if for every RDF graph G, $[[Q]]_G \sqsubseteq [[Q']]_G$.*

3   Content-aware SPARQL Query Caching

In this section, we describe our caching approach. Similar to [6], the main functionality is performed by a proxy residing between the application(s) and a SPARQL endpoint. Given a query (CQSF), the proxy first parses the query. It then checks whether the query is the same as a cached query by comparing the query strings. If that is the case, the result is retrieved from the cache and returned to the user. If the query is not cached, the proxy checks whether the query can be evaluated on a cached result. Queries that cannot be evaluated on the cache are forwarded to the SPARQL endpoint and results are cached before returned to the user. We use LRU (Least Recently Used) as the replacement scheme for our cache.

Based on results from relational databases [10], we have the following proposition for checking the containment of CQSFs[2].

---

[2] Without loss of generality, we assume that all the filter conditions are safe, i.e. each variable in a condition appears in some triple pattern.

Proposition 1. *Let $Q(S, P)$ and $Q'(S', P')$ be two CQSFs. $Q \sqsubseteq Q'$ if*

1. *there exists a mapping τ from the variables of P' to the variables, URIs or literals of P that maps each triple pattern of P' to a triple pattern of P,*
2. *$ftrs(P)$ logically implies $\tau(ftrs(P'))$, i.e. $ftrs(P) \Rightarrow \tau(ftrs(P'))$, where $\tau(ftrs(P'))$ denotes the set of filter conditions obtained from $ftrs(P')$ by replacing each variable $?X'$ in $ftrs(P')$ with $\tau(?X')$, and*
3. *$S \subseteq \tau(S')$.*

We refer to a mapping that satisfies Proposition 1 a *containment mapping*. Given two queries Q and Q', $Q \sqsubseteq Q'$ only tells us that the result of Q is contained in that of Q'. We still have to compute the result of Q. Unfortunately, $Q \sqsubseteq Q'$ does not guarantee that Q can be evaluated on the result of Q', as pointed out in the Introduction. Here, we give a definition of what it means that a SPARQL query can be evaluated on the result of another SPARQL query. The definition is inspired by Larson and Yang's work on computing SQL queries from derived relations [4].

Definition 3 (Evaluability). *Let Q and Q' be two SPARQL queries. We say that Q can be evaluated on the result of Q', or simply, Q can be evaluated by Q', denoted by $Q \preceq Q'$, if the operations needed to compute the result of Q from the result of Q' contain no algebraic joins.*

In order for Q to be evaluated on the result of Q', the following conditions are required to hold.

Proposition 2. *Let $Q(S, P)$ and $Q'(S', P')$ be two CQSFs. $Q \preceq Q'$, if*

- *there exists a containment mapping τ from Q' to Q,*
- *for each triple pattern t of P, there exists a triple pattern t' of P' such that $\tau(t') = t$, and*
- *for each variable $?X'$ in P', if $\tau(?X')$ is an URI or a literal, or $\tau(?X')$ is a variable which is included in S or in a filter condition in P, then $?X' \in S'$.*

The first condition is easily understandable: for $Q \preceq Q'$, the result of Q must be contained in that of Q'. The second condition ensures that any triple in an RDF graph that matches a triple pattern of P also matches a triple pattern of P'. The third condition basically specifies the variables that have to be included in S' in order for Q to be evaluated by Q'. These three conditions can be used to terminate testing early if Q cannot be evaluated by Q'.

With this proposition, we can evaluate Q through three types of operations on the result of Q': $\pi_S(\Omega)$, $\sigma_R(\Omega)$, and substitution operation $\tau(\Omega)$. Ω denotes a set of intermediate solution mappings generated during the application of operations, S a set of variables to be returned in Q, and R a conjunction of filter conditions including those in Q, and those derived from τ in the form $?X' = c$ (when a variable in Q' is mapped to an URI or a literal in Q) or $?X' = ?Y'$ (when two different variables in Q' are mapped to the same variable in Q).

Example 2. Consider two queries, $Q_1(\{?X_1\}, (?X_1, type, stu)(?X_1, age, 20))$ and $Q_2\{?X_2, ?Y_2\}, (?X_2, type, stu)(?X_2, age, ?Y_3)FILTER(?Y_2 > 15))$. From $Q2$ to $Q1$, there is a containment mapping τ that satisfies Proposition 2: $\tau(?X_2) = ?X_1, \tau(?Y_2) = 20$. Let G is an RDF graph. We can evaluate Q_1 by Q_2 through $\pi_{\{?X_1\}}(\tau(\sigma_{(?Y_2=20)}([[Q_2]]_G)))$.

To test $Q \preceq Q'$, we need to check whether there is a containment mapping from Q' to Q that satisfies the conditions in Proposition 2. We can do this efficiently if Q' is acyclic, as described below. A key operation is checking whether each triple pattern of Q' can be mapped to a triple pattern of Q. This is done by comparing corresponding elements of triple patterns, i.e. subject to subject, predicate to predicate, and object to object. See Algorithm 1 for details. Let t' be a triple pattern of Q', t a triple pattern of Q, and e' and e a pair of corresponding elements of t' and t. There are four cases: (1) both e' and e are URIs or literals; (2) e' is an URI or a literal, but e is a variable; (3) e' is a variable, but e is an URI or a literal; (4) both e' and e are variables. In the first case, e' and e need to be equivalent in order for t' being able to be mapped to t. In the second case, t' cannot be mapped to t, as e' is more specific than e. In the last two cases, there is a mapping from e' to e. If all the mappings from t' to t are compatible, i.e. they agree on shared variables, then t' can be mapped to t. During the mapping, we can also check whether e' should be included in S', if it is a variable, and whether there are filter conditions involving e' or e, and whether the filter condition involving e' can be implied by the filter condition involving e. In doing so, we are checking whether t could be evaluated by t' with regard to return variables and filter conditions of Q and Q'. In our current implementation, if both filter conditions involving e' and e exist, we require that they be conjunctions of arithmetic comparisons of the form $?X\theta c$, where c is a numeric value, and $\theta \in \{=, \neq, <, \leq, >, \geq\}$.

A triple pattern of Q' can be potentially mapped to several triple patterns of Q. As such, there may be more than one mapping associated with a triple pattern of Q', all with the same domain: the set of the variables in the triple pattern. Such mappings are called *partial mappings* from Q' to Q, as their domains are a subset of the variables of Q'. Based on partial mappings, we are then able to check whether there is a containment mapping from Q' to Q such that $Q \preceq Q'$. Before we present our approach for testing $Q \preceq Q'$, we introduce the concept of query acyclicity.

A query (i.e. CQSF) is *acyclic* (*cyclic*) if its hypergraph is acyclic (cyclic). A query's *hypergraph* consists of a set of vertices and a set of hyperedges: each vertex corresponds to a variable in the query, and each hyperedge corresponds to a triple pattern and includes the variables in the triple pattern. A hypergraph is acyclic if its GYO-reduction results in an empty hypergraph; otherwise it is cyclic. The *GYO-reduction* [2] is a process that repeatedly applies the following two operations on a hypergraph: (1) delete a vertex that occurs in only one hyperedge; (2) delete a hyperedge that is contained in another hyperedge. If a query is acyclic, an elimination tree for the query can be constructed during the GYO-reduction: each node of the tree corresponds to a triple pattern in the

---

Input: two triple patterns t and t' of Q and Q' respectively,
two sets of return variables S and S' of Q and Q' respectively,
two sets of filter conditions F and F' in Q and Q' respectively

Output: *null*, or a mapping $\tau_{t' \to t}$ from t' to t if $t \preceq t'$

$\tau_{t' \to t} \leftarrow null$, $E \leftarrow$ the triple elements of t, $E' \leftarrow$ triple elements of t';
for $i \leftarrow 1$ **to** 3 **do**

$\quad e \leftarrow$ the ith element of E, $f \leftarrow$ the filter condition involving e in F;
$\quad e' \leftarrow$ the ith element of E', $f' \leftarrow$ filter condition involving e' in F';
$\quad$ **if** *both e and e' are URIs or literals, and $e \neq e'$* **then**   *return null*;
$\quad$ **if** *e is a variable and e' is not a variable* **then** *return null*;
$\quad$ **if** *e is not a variable and e' is a variable* **then**

$\quad\quad$ **if** *$e' \notin S'$, or $e' \in S'$, $f' \neq null$ and e does not satisfy f'* **then**
$\quad\quad$ *return null*;

$\quad$ **if** *both e and e' are variables* **then**

$\quad\quad$ **if** *$((f = null)$ and $(f' = null))$* **then**
$\quad\quad\quad$ **if** *$(e \in S$ and $e' \notin S')$* **then**   *return null*;

$\quad\quad$ **if** *$((f \neq null)$ and $(f' = null))$* **then**
$\quad\quad\quad$ **if** *$(e' \notin S')$* **then**   *return null*;

$\quad\quad$ **if** *$((f = null)$ and $(f' \neq null))$* **then**   *return null*;
$\quad\quad$ **if** *$((f \neq null)$ and $(f' \neq null))$* **then**

$\quad\quad\quad$ **if** *$e' \notin S$, or f' cannot be implied by f after replacing e' in*
$\quad\quad\quad$ *f' with e* **then**   *return null*;

$\quad$ **if** *e' is a variable* **then**

$\quad\quad$ **if** *$\{e' \to e\}$ is compatible with $\tau_{t' \to t}$* **then** $\tau_{t' \to t} \leftarrow \tau \cup \{e' \to e\}$;
$\quad\quad$ **else** *return null*;

$return\ \tau$;

Algorithm 1. Checking whether $t \preceq t'$

query; if a hyperedge E is deleted by operation (2) because it is contained in some other hyperedge F, then the tree has an edge (t_E, t_F) where t_E denotes the triple pattern corresponding to E, and t_F the triple pattern corresponding to F. If there are several hyperedges containing E when E is deleted, then a random one is picked as F. For simplicity, we restrict our discussion to queries whose hypergraphs are connected. As such, an elimination tree of a query always covers all the triple patterns of the query. However, our results can be generalised to queries with disconnected hypergraphs.

An elimination tree of a query (acyclic) captures relationships of triple patterns in a deterministic way. We can take advantage of this to find containment mappings (if any) efficiently. Since a cyclic query does not have an elimination tree, we have two cases when testing $Q \preceq Q'$, i.e. when Q' is acyclic and when it

---

Input: two queries Q and Q' (Q' is acyclic)
Output: *null*, or a mapping $\tau_{Q' \to Q}$ from Q' to Q if $Q \preceq Q'$

$T \leftarrow$ *the elimination tree of* Q';
for *each triple pattern* t' *in* T **do**
    $t'.mappings \leftarrow null$;
    for *each triple pattern* t *of* Q **do**
        if $((\tau_{t' \to t} = t \preceq t') \neq null)$ **then** *insert* $\tau_{t' \to t}$ *into* $t'.mappings$;
    return null if no such t;
return null if there exists t *of* Q *that cannot be evaluated*;
for *each triple pattern* t' *in* T *(bottom-up)* **do**
    for *each child* t'_i *of* t' **do**
        $t'.mappings \leftarrow t'.mappings \ltimes t'_i.mappings$;

return null if the mappings of T*'s root are empty*;
$\Gamma \leftarrow$ *the mappings of* T*'s root*;
for *each triple pattern* t' *in* T *(top-down)* **do**
    $\Gamma \leftarrow \Gamma \bowtie t'.mappings$;
    if $dom(\Gamma) = vars(Q')$ **then for** *each mapping* $\tau_{Q' \to Q}$ *in* Γ **do**
        if *each triple pattern* t *of* Q *can be mapped to by* $\tau_{Q' \to Q}$ **then**
        return $\tau_{Q' \to Q}$;

return null;

Algorithm 2. Checking whether $Q \preceq Q'$

is cyclic (it should become clear in the following that it is unimportant whether Q is acyclic or not). For both cases, the testing consists of two major phases. The first phase is generating partial mappings for each triple pattern of Q', as described earlier. The second phase is generating containment mappings, if any, from partial mappings. If Q' is acyclic, we generate containment mappings by traversing an elimination tree of Q'. We first traverse the tree bottom-up. If a node has children in the tree, we check whether the parent's partial mappings (i.e. the partial mappings associated with the triple pattern corresponding to the parent node) are compatible with its children's by semi-joining the parent's partial mappings with the children's. This is continued until the root of the tree is reached and its partial mappings are processed. If the resulting partial mappings of the root are empty, then there are no containment mappings from Q' to Q such that Q can be evaluated by Q'. Otherwise we traverse the tree top-down to compute containment mappings by joining each node's partial mappings with its children's, until the resulting mappings cover all variables of Q'. Algorithm 2 shows the whole process when Q' is acyclic. It is adapted from the AcyclicContainment algorithm in [2].

(a) Hypergraph for Q2   (b) An elimination tree for Q2

Fig. 1. The hypergraph and elimination tree for Q_2 in Example 3

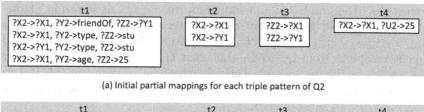

(a) Initial partial mappings for each triple pattern of Q2

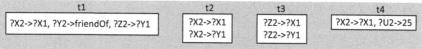

(b) Partial mappings after traversing the elimination tree of Q2 bottom-up

?X2->?X1, ?Y2->friendOf, ?Z2->?Y1,?U2->25

(c) The final containment mapping from Q2 to Q1 such that Q1 can be evaluated by Q2, obtained after traversing the elimination tree top-down

Fig. 2. Mappings generated during testing $Q_1 \preceq Q_2$ in Example 3

If Q' is cyclic, we need to check each triple pattern's partial mappings against all other triple patterns' if their domains overlap, and compute the partial mappings through semi-join operations. This is continued until there is a triple pattern whose partial mappings are empty, or there are no more changes to all triple patterns' partial mappings. If it is the former case, it means that there are no containment mappings from Q' to Q such that Q can be evaluated by Q'. In the latter case, we compute containment mappings by joining all triple patterns' partial mappings with overlapping domains one by one until the resulting mappings cover all variables of Q'. Though some heuristics can be applied to decide which triple patterns' partial mappings are processed first, the testing of $Q \preceq Q'$ when Q' is cyclic is in general inefficient in both time and space. Fortunately, most real world queries are acyclic and have few triple patterns [8].

Example 3. Let Q_1 be $(\{?X_1, ?Y_1\}, (?X_1, friendOf, ?Y_1)(?X_1, type, stu)(?Y_1, type, stu)(?Y_1, age, 25))$, Q_2 be $(\{?X_2, ?Y_2, ?Z_2\}, (?X_2, ?Y_2, ?Z_2)(?X_2, type, stu)(?Z_2, type, stu)(?X_2, age, ?U_2)FILTER(?U_2 > 20))$. To test $Q_1 \preceq Q_2$, we first

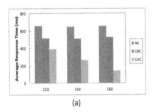
(a)

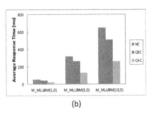
(b)

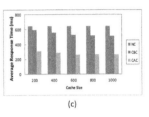
(c)

Fig. 3. Performance comparison of the three cases with respect to percentage of contained queries, dataset size, cache size

construct the hypergraph and an elimination tree for Q_2, as shown in Figure 1, where t_1, t_2, t_3, and t_4 represent triple patterns $(?X_2, ?Y_2, ?Z_2)$, $(?X_2, type, stu)$, $(?Z_2, type, stu)$, $(?X_2, age, ?U_2)$ respectively. We then generate the containment mappings from Q_2 to Q_1 such that $Q_1 \preceq Q_2$. The generation process is shown in Figure 2(a)-(c). Let τ be the final mapping in (c), G be an RDF Graph. We can evaluate Q_1 on the result of Q_2 through $\pi_{\{?X_1, ?Y_1\}}(\tau(\sigma_{(?U_2=25)(?Y_2=friendOf)}([[Q_2]]_G)))$.

4   Evaluation

We evaluated the performance of our approach through experiments. All experiments were done on a machine with the following configuration: Intel Core i5 (M540, 2.53GHz), 3.24 GB of RAM, 848GB HD, Java 1.6, 512MB of max heap size, Apache Jena 2.7.3, and TDB 0.9.3 (with the default file caching).

We generated datasets by modifying the LUBM benchmark [3]. We added a data property "value" (with the range of "xsd:double") to the benchmark ontology (univ-bench.owl) and changed the data generator so that each generated student instance would have a certain random "value". Using the modified data generator, we generated 3 datasets: MLUBM(1,0), MLUBM(5,0) and MLUBM(10,0), which contain OWL files for 1, 5, and 10 universities respectively, and then loaded each dataset into TDB after OWL inferencing. The sizes of their materialised versions are 185182, 1125339, 2279105 respectively. For query generation, we used the following template: SELECT ?X ?Y WHERE{?X rdf:type ub:Student. ?X ub:value ?Y FILTER (?Y >= %%value1%% && ?Y<= %%value2%%) ?X ub:takesCourse <http://www.Department0.University0.edu/ GraduateCourse0>}. By controlling the values which are used to replace the parameters, we generated query traces with different percentages of contained queries (i.e. the queries that are contained in other queries in the same trace). The traces we experimented with are C20, C40, and C60, with 20%, 40%, and 60% of contained queries respectively. Each trace contains 1,000 queries and has the same percentage of identical queries (20%).

Figure 3(a) shows the average response time of the three traces on M_MLUBM (10,0) in three cases, i.e. no caching (NC), content-blind caching (CBC), and content-aware caching (CAC). As expected, when the percentage of contained queries increases, the average response times with CAC decrease. This, however, has little impact on the performance of CBC, as the percentage of identical queries is the same for the three traces. We then studied the performance of CAC with respect to dataset size. For this, we ran C40 to the three datasets. As shown in Figure 3(b), CAC outperforms CBC by 44%-50%, and the performance improvement is even better for larger datasets. In the earlier experiments, we assumed unlimited cache size, i.e. there is no cache replacement. To investigate the behavior of CAC with respect to cache size, we ran C40 on M_MLUBM(10,0) at various cache sizes. From Figure 3(c), we see that even with a small cache size, CAC achieves much better performance than CBC. Also, CAC seems to be more resilient to the change of cache size.

5   Conclusions and Future Work

In this paper, we presented a caching approach for improving the performance of Semantic Web applications. Our approach is novel in that not only does it benefit a query that exactly matches a cached query, but it also benefits a query that is contained in a cached query and can be evaluated by the cached query. We tested our approach on the slightly modified version of the LUBM benchmark. Experimental results showed that our approach can achieve much better performance than no caching and content-blind caching cases. In the future, we would like to try some other cache replacement schemes, e.g. those considering access frequency or miss cost. Also, we plan to extend our approach for other fragments of SPARQL, e.g. queries with OPT and UNION operators.

References

1. Arenas, M., Perez, J.: Querying Semantic Web Data with SPARQL. In: Proceedings of PODS (2011)
2. Chekuri, C., Rajaranman, A.: Conjunctive Query Containment Revisited. Theoretical Computer Science 239(2), 211–229 (2000)
3. Guo, Y., Pan, Z., Heflin, J.: LUBM: a Benchmark for OWL Knowledge Base Systems. Journal of Web Semantics 3(2), 158–182 (2005)
4. Larson, P.A., Yang, H.Z.: Computing Queries from Derived Relations. In: Proceedings of VLDB (1985)
5. Letelier, A., Perez, J., Pichler, R., Skritek, S.: Static Analysis and Optimisation of Semantic Web Queries. In: Proceedings of PODS (2012)
6. Martin, M., Unbehauen, J., Auer, S.: Improving the Performance of Semantic Web Applications with SPARQL Query Caching. In: Aroyo, L., Antoniou, G., Hyvönen, E., ten Teije, A., Stuckenschmidt, H., Cabral, L., Tudorache, T. (eds.) ESWC 2010, Part II. LNCS, vol. 6089, pp. 304–318. Springer, Heidelberg (2010)
7. Pérez, J., Arenas, M., Gutierrez, C.: Semantics and Complexity of SPARQL. In: Cruz, I., Decker, S., Allemang, D., Preist, C., Schwabe, D., Mika, P., Uschold, M., Aroyo, L.M. (eds.) ISWC 2006. LNCS, vol. 4273, pp. 30–43. Springer, Heidelberg (2006)
8. Picalausa, F., Vansummeren, S.: What are Real SPARQL Queries Like? In: Proceedings of SWIM (2011)
9. Schmidt, M., Meier, M., Lausen, G.: Foundations of SPARQL Query Optimization. In: Proceedings of ICDT (2010)
10. Ullman, J.D.: Principles of Database Systems, 2nd edn. Computer Science Press (1982)

A Generic Tree-Like Index Framework in the Cloud

Yue Yin[1], Bin Yao[2], Yao Shen[2], Minyi Guo[2], and Changliang Xu[3]

[1,2] Shanghai Key Laboratory of Scalable Computing and Systems,
Department of Computer Science and Engineering,
Shanghai Jiao Tong University, China
[3] Alibaba Cloud Computing Company, China
t-yuyin@sjtu.edu.cn, {yaobin,yshen,guo-my}@cs.sjtu.edu.cn,
changliang.xucl@aliyun-inc.com

Abstract. In this study, we present a novel tree based index scheme for efficient indexing and serving large datasets in the cloud. It incorporates and extends the functionality of Hadoop to create a fully parallel index system. Our new scheme can be summarized as follows. First, we leverage the MapReduce framework to create an index, then publish the index meta information and write it into a meta table. Second, we use the meta information to help the system adopting an efficient method to handle a given query. Finally, we optimize the system by using cache mechanism. We conduct extensive experiments on the Hadoop cluster to demonstrate the scalability, availability and efficiency of the proposed index framework.

Keywords: distributed index, cloud computing, data warehousing.

1 Introduction

As cloud computing develops, there is an ever increasing interest in deploying storage systems that support applications requiring large datasets. Massive scale distributed data warehouses such as Facebook's Hive [20] gain more and more attentions as they allow to store data for many cloud applications. Data warehouse systems on cloud are designed to meet several key issues: scalability, availability and low latency. In Hive, datasets are automatically partitioned and replicated among computer nodes for scalability and availability. Query efficiency is achieved by both employing the MapReduce [11] framework of the underlying computing system and its built-in simple index schemes, namely the compact index and the bit map index. However, in the existing solutions, range queries, the most useful feature in data warehouse systems, are inefficient. Compact index only works properly when the data items featuring the same value are stored on the same file block. Bit map index faces the same problem as compact index, which is the limitation of usage scenarios. It can only be used on data items having few distinct values. Therefore, these index schemes used in Hive are not suited for range queries. A usual solution to this problem is to run

X. Lin et al. (Eds.): WISE 2013, Part I, LNCS 8180, pp. 330–342, 2013.

a MapReduce job that scans the entire datasets. The main drawbacks of this approach are (1) the high I/O cost by scanning the datasets and (2) the induced network transmission overhead. In practice, these drawbacks will largely reduce the system performance.

In this paper, we propose a generic tree-like secondary index framework called Cloud index (C-index) to support one-dimensional or multi-dimensional queries, which achieves low building time, low space overhead and low search latency. It is tailored for data warehouse systems. C-index is maintained in an incremental way, since in a data warehouse system, data updating is achieved by adding new data to the system. The C-index uses the traditional database index structures B-tree [9] and R-tree [14] as the base index. It leverages the innovative MapReduce framework to serve a query in parallel. C-index supports usual search operations, such as point search and range search.

C-index is designed to achieve scalability, availability and low latency. It is stored on Hadoop [1] Distributed File System (HDFS) for scalability and availability. To achieve low creation and search latencies, we partition the datasets. C-index is designed to use any user defined partitioning scheme for different given datasets. Instead of building an index for the whole datasets, C-index builds a local sub-index for each of the data partition. To route queries among the partitions in the future, all sub-indexes' meta data are recorded by a Meta Table. A query is then passed to the related sub-indexes according to the search conditions. In turn, this saves the time that would be spent searching unrelated indexes. In order to further reduce the search latency, the storage structure of C-index is optimized to minimize the I/O cost. At the same time, additional optimization on the cache is also applied. In this work, we present a distributed index system for indexing, storing and serving large datasets. Our contribution includes the following advances:

- We apply the non-distributed tree-like index to a distributed system. Tree-like indexes are very efficient on data retrieving, and distributed systems can provide powerful computing capabilities. We make C-index to leverage these advantages to perform parallel data process on large datasets.
- To reduce the I/O overhead, we change the index's storage structure, and make it suitable for distributed storage system. These changes include the node structure of the tree and the file structure of the index. This will make C-index use less I/O operations to get the data.
- We add an optimization mechanism to improve our system. The improvement is focused on the cache management.

The rest of this paper is structured as follows. Section 2 presents some related work. Section 3 describes the architecture of the index design. Section 4 presents the index operations and optimization. Section 5 focuses on the experiments and their detailed evaluation. Section 6 concludes the paper.

---

[1] http://hadoop.apache.org/

2 Related Work

Dealing with extremely large datasets in distributed environment is a challenging task. The first step is to build a scalable data storage system. Distributed file system is the most important storage system. An example of such system is Google's GFS [13] and its open-source implementation, HDFS. They are designed to store extremely large datasets that are split into equal chunks, and each chunk is replicated and randomly distributed over the cluster. Amazon's Simple Storage Service (S3) [2] is a data storage service that allows user access over the internet. Sinfonia [3] is a data sharing service that provides applications with unstructured address spaces on which to keep data. Applications are responsible for organizing and structuring their data on top of the address spaces. OceanStore [15], Farsite [1] and Ceph [22] provide a highly reliable peta-bytes level storage. Based on the above systems, some frameworks have been proposed to support various applications. Among them, MapReduce is the most efficient and the most widely used. It provides powerful functionality and simple interface for parallel data processing. On top of MapReduce, a number of systems have been proposed to process data warehousing tasks. Yahoo's Pig [3], Facebook's Hive and HadoopDB [4] are some examples. In these systems, the user writes a job in an SQL-like language and then translate it into a chain of MapReduce jobs that are executed on Hadoop. Although the underlying implementation of these systems may be different, their common goal of these systems is to provide techniques to store and retrieve huge datasets on a shared-nothing computing cluster. In this paper, we focus on providing an efficient secondary indexing solution for such kind of systems.

Index techniques in databases have been gaining more and more attentions over 20 years. Many index based methods [5,18,19,8,24,7] have been proposed to handle all kinds of queries such as point queries, range queries and nearest neighbor queries. Their common strategy consists in partitioning the space by grouping objects in a hierarchical way and then allowing access to the subsets of the space. Recently, with the increasing in the size of the dataset, distributed tree-like index algorithm [2] is proposed for indexing extremely large datasets in a cluster. The B^+-tree is distributed among the available nodes by randomly assigning each B^+-tree node to a computing node. This method exhibits two main weaknesses. First, although it uses a B^+-tree based index, the index is designed to handle dictionary lookup operations. Therefore, it is not able to effectively deal with range queries. To process a range query $[a, b]$, first, the leaf node that contains a must be located. Then, if b is not contained in the same node, it needs to retrieve the next leaf node from some other server based on the sibling pointer. Such operations continue until the whole range has been searched. Second, the whole index system induces a high maintenance cost on the server and a large memory overhead on the client side, as the client node replicates all the related internal nodes. Some different distributed solutions are

---

[2] http://aws.amazon.com/s3/

[3] http://pig.apache.org

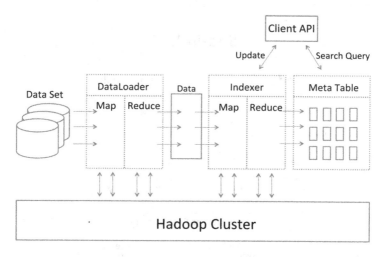

Fig. 1. System Architect

proposed to handle range queries. In [23,25,21,10], every node has a local tree index, and the whole cluster is organized as an overlay to route the query to related nodes. This strategy makes the storage system more complicated and increases the server nodes' overhead. HIndex [16] leverages HBase to store the indexes, but as a drawback the index creation remains centralized. [17] describes a built-in block-based index structure on HDFS. The aim of this solution is to perform indexing for HDFS. [6,12] make Hadoop jobs more performant through clever placement and organization of data. This work is complementary to ours, we can leverage it to accelerate job process at a lower level.

3   System Overview

Fig. 1 shows the system architecture of our index scheme. Our system is built on top of Hadoop. It consists mainly of a distributed file system called HDFS, and a MapReduce execution framework built on top of HDFS. Above the Hadoop cluster are the main components of C-index. The DataLoader is able to load any data stored using common file formats, like text file format or RCFile format. It also provides an interface to easily extend to other storage formats. When the data is loaded, it is sorted and partitioned using a search key. This stage is handled by a MapReduce Job. The Indexer is responsible for index creation. It uses another MapReduce Job to take the output of the DataLoader as input and build the indexes for all the data partitions in parallel. Finally, each index's meta information is added to the meta table for some future use. The client interacts with C-index through the Client API component. This component wraps some basic operations, such as point search and range search. Details on the data processing in C-index will be discussed throughout the next section.

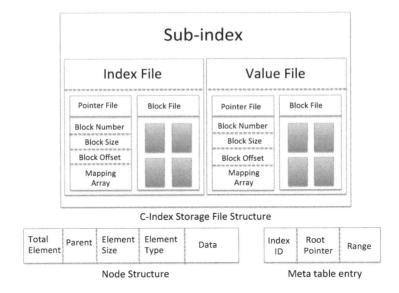

Fig. 2. C-index Storage Structure

4   The C-index

4.1   Basic Idea

The C-index scheme is based on the following observations: (1) There are usually three or more replicas stored in an HDFS to ensure the availability of data. (2) Most existing Data Warehouse systems display a lack of effective index scheme to support range queries. (3) For distributed file systems, like HDFS, random read operations are inefficient, although sometimes required. The solution adopted by C-index is to reorganize the data into a new layout to speed up both one and multi dimensional queries. C-index introduces a tree based distributed index stored in an HBase table or a structured HDFS file. It leverages the high concurrency of Hbase or Hadoop to search the related sub-indexes concurrently for a given query. C-index also caches the frequently used tree nodes in memory to improve the response time.

4.2   Index Outline

Fig. 2 shows the C-index storage outline on HDFS. A C-index is composed of a number of sub-indexes. It uses a meta table to track the sub-indexes it contains. Each entry in the meta table is mapped to a single sub-index by an unique index ID. It records the critical information of the corresponding sub-index, including a pointer to the root node and a key range for the sub-index. The meta table is stored as a structured file on HDFS. Each C-index can only have one meta table.

The B$^+$-tree based and R-tree based C-index follow the same storage outline. A sub-index is stored as two files, an index file and a value file. The index file is in charge of storing the internal node of the tree, and the value file stores the actual data or the pointers to the data. Index and value files both feature the same structure. Therefore, we only present the structure of the index file. Every index file has two sub-files, a pointer file and a block file. The pointer file stores the meta data of the index file and the block file stores the tree nodes (also called data block). The meta data in a pointer file includes the total number of data blocks, two arrays that record each data block's size and offset, and a mapping array. Since only the appending operation is available on an HDFS file, we can only append a new data block at the end of the file. To handle this, we give to each data block two numbers: a virtual number, used by the user, and a physical number, which is the actual index of the data block in the block file. We maintain an array that maps a virtual block number into a physical block number. In order to achieve an efficient operation, we store the mapping information and the real data on different files, such that we can manipulate the mapping array without impacting the real data. In summary, a sub-index is formed by 4 HDFS files. Two of them make up the pointer file, and the remaining two make up the index file. In this way, a C-index that contains n sub-indexes will have a total of $4n$ HDFS files.

The tree node structure is presented in Fig. 2. At the beginning of the node its meta data includes the total number of elements and the size and type of each element. The parent field records a pointer to its parent or -1 if it is the root node. The data field records key-pointer pairs for internal nodes, or key-value pairs for leaf nodes.

4.3   Index Creation

Since C-index is based on the underlying local B$^+$-tree or R-tree index, we start by considering how to build a tree index. One approach is to insert each record into an empty tree. However, it is quite expensive, because each entry requires us to start from the root and go down to the appropriate leaf node. For efficiency, we use the bulk loading [4] method to build a B$^+$-tree or R-tree. Here we focus on the introduction of B$^+$-tree bulk loading method. Due to the structural similarity of a B$^+$-tree and an R-tree, this method can be easily extended to the R-tree case. The method is summarized as follow:

- Sort the data entries according to a search key.
- Allocate an empty node to serve as the root, and insert a pointer to the first node that contains unindexed data entries into it.
- When the root is full, we split the root, and create a new root node.
- Keep inserting entries to the right most index node just above the leaf level, until all data entries are indexed.

---

[4] http://en.wikipedia.org/wiki/B%2B_tree

Note that, when the right-most index node above the leaf level fills up, it is split. This action may, in turn, cause a split of the upper right-most index node, and so on.

Beyond the use of the bulk loading method to speed up the index building process, we also leverage the parallel processing capabilities of the MapReduce framework on Hadoop. Algorithm 1 shows the index building process of C-index. It is divided into two phases: In the first phase, the data is partitioned and sorted then passed to the reduce function. The reduce function just writes the input data into a temporary file, and add the file name to a list. Each temporary file is mapped to a data partition. In the second phase, the map function maps each data partition to a reduced node, and the reduce function builds a local sub-index for each partition it is assigned to. Last, it writes the sub-indexes' meta information to corresponding meta table. The construction of the B^+-tree based C-index can be simply extended to the R-tree case.

4.4   Query Processing

C-index supports conventional point queries and range queries. Algorithm 2 shows a generic range query process. In line 1, we get the index meta information according to the index handler. Then we find all the related indexes that may contain data in the range r (line 3). In line 4, we decide which one of the search methods, the parallel one or the non-parallel one, is used for the given query. The search method depends on the size of the query. We can leverage the MapReduce framework to speed up the search process for some queries that need to check many sub-indexes, or just directly search the index. Starting a MapReduce job induces a large overhead. Therefore, this choice is important in terms of efficiency. When use a MapReduce job, the map function maps each sub-index to a reduce node, and the reduce function searches each sub-index it is assigned to. After

---

Algorithm 1. IndexBuilding Algorithm

Input: datasets ds
Output: index handler
 1: **Function** indexBuilding(ds)
 2: $indexHandler \leftarrow createIndexHandler()$
    phase1:
 3: $kvPair \leftarrow Map(ds)$ //for each record, output $< partitionID, dataRecord >$ pair
 4: $tempFilesList \leftarrow Reduce(kvPair)$
    phase2:
 5: $kvPair \leftarrow Map(tempFilesList)$ //output $< tempFile, tempFile >$ pair
 6: $subIndexList \leftarrow Reduce(kvPair)$ //build index for each temp file and add the new index to a list
 7: **for** each $subIndex$ in $subIndexList$ **do**
 8:     $indexHandler.addMeta(subIndex.metaInfo)$
 9: **end for**
10: return $indexHandler$

---

search all related sub-indexes, we merge the results and return them to user (line 5, 6, 7).

4.5   Index Maintenance

In this paper, we mainly focus on data warehouse systems, where the data is a bulk import to the system and is then never changed. To maximize the utilization of the system and improve the scalability, C-index provides eventual consistency. In this system, we update the index in a simple but efficient manner. An index updating operation only needs to build a new sub-index for the new dataset, and add its information to the meta table it belongs to. In some cases, it might happen that C-index contains too many small sub-indexes, which in turn will impact the overall system performance. To overcome this issue, C-index will automatically merge the small sub-indexes during the data update, if their number exceeds a certain given threshold.

4.6   Optimization

Under normal circumstance, the root node of a tree-like index always becomes the bottleneck of the system. This is reasonable as all the search operations will start from the root. In our system, since the data is partitioned into several pieces, a query may be redirected to different sub-indexes. In fact, if many queries request the same data partition at the same time, then an I/O congestion might occur. In turn, this could seriously affect the system performance.

To solve this issue, we use an in-memory buffer (also called cache in this We use an in-memory buffer (also called cache in this paper) to handle the load balance between the memory and the disk. Since the meta table is small, and every search operation on that index need to access it, it makes sense to cache it in the memory. For the sub-indexes, we adopt different strategies depending on whether we are dealing with internal or leaf nodes. As the property of tree-like index structure, the internal nodes just take a small fraction of the total space, and every tree traversing needs to access the internal nodes, so we cache as many internal nodes in the memory as possible in order to reduce the disk access time.

Algorithm 2. Range Search Algorithm

Input: range r, indexHandler idx
Output: data in the range
 1: **Function** rangeSearch(r, idx)
 2: $metaTable \leftarrow idx.getIndexMeta()$
 3: $subIndexes \leftarrow idx.getSubIndex(r)$
 4: $searchMethod \leftarrow idx.getMethod(r, metaTable)$
 5: $results \leftarrow searchMethod.doSearch(r, subIndxes)$
 6: $result \leftarrow merger(results)$
 7: return $result$

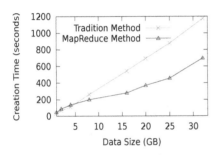

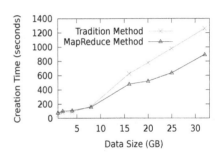

Fig. 3. Performance of CB-index Creation

Fig. 4. Performance of CR-index Creation

On the other hand, only a certain number of data nodes are cached in the memory. This is mainly because the data nodes accounts for a majority of the index structure, and for large datasets, it is unrealistic to cache all the leaf nodes in the main memory. In C-index, we use the LRU algorithm to switch in or out the data nodes.

5   Performance Evaluation

Our experimental setup includes 3 work nodes and a single master node for MapReduce. The master node is also the name node of HDFS. The worker nodes and the master node have 4 Quad-Core E5405 Intel Xeon CPUs @ 2.80GHz, 4 GB of RAM and 1 TB disk (for a total of 16 CPUs, 16GB RAM and 4T TB of disk space). We implement our system using Java 1.7.0.17. The computers run Unbuntu 12.04, kernel version 3.2.0. The version of Hadoop is the stable version 0.20.2, built from the source code to suite our setup.

The Hadoop MapReduce framework is configured to take full advantage of the available physical resources. Hadoop is given 2 GB memory in every node, and the Mapper or Reducer task is given 512 MB memory. As most of our computing task is performed in the Reducer during the building of the index, we manually set the number of Reduce task equals to the number of CPU cores. This allows us no to waste any CPU time. The number of Map tasks is automatically set by the framework.

We use real datasets (OpenStreet) from the OpenStreetMap project to evaluate our C-index scheme. The datasets represent the global road networks. Each record contains a record ID, 2-dimensional coordinate, and description. We use the real datasets to form different sizes of our test datasets, varying from 1GB to 32GB. All the test datasets are stored as files and uploaded to the HDFS. The chunk size of an HDFS file is 64MB, and all chunks have three replicas. In the rest of this section we will use CB-index and CR-index to represents B^+-tree based C-index and R-tree based C-index, respectively.

5.1   Performance of Index Creation

In this experiment, we evaluate the performance of the index building process with different data sizes for both the CB-index and CR-index methods. We compare our strategy with the traditional non-parallel one. Fig. 5 and Fig. 5 show the results of this experiment. For CB-index, when the scale of the dataset is small, the process time of a traditional method overwhelms ours. This can easily be explained as for a small dataset, the time required to start and finish a MapReduce job is longer than the building process. However, note that when increasing the size of the dataset, our method beats the non-parallel one, and the time difference becomes larger and larger as the size of the dataset increases. The performance of CR-index is similar to the one of CB-index, except that the process time takes longer. In fact, R-tree is more complicated to generate than B^+-tree. The experiment shows that the C-index building operation performs well when the size of the dataset is large, but is still tolerable when small.

5.2   Performance of Range Queries

Range queries are the most important query operations in the index. We define the selectivity as the percentage of searched data items over the whole datasets. In this experiment, we compare our index based range queries with a table scan based range queries. Fig. 5 (Fig. 6) compares the performance of a range query on a table scan and on a CB-index (CR-index) using different selectivity for a given dataset. When increasing the selectivity of a range query, the response time of our method also increased. This is due to the overhead required when reading the HDFS files and running MapReduce jobs during the index search process. Note that, the performance of a 3-dimensional query on a CR-index is similar to the one on a CB-index, but the response time is longer for both search methods. This is because the R-tree structure is more complicated than the B^+-tree structure.

Fig. 7 and Fig. 8 show the performance of a range query processed at different data sizes on a CB-index and a CR-index, respectively. When the size of the dataset is small, the table scan method is more efficient. However, when the size of the dataset increases our method becomes faster. The experiments show that the C-index performs better on large datasets, while still providing reasonable performance on small ones.

5.3   Effect of Cache

In many real systems, cache is used to solve the I/O issue. In Fig. 9 and Fig. 10, we add a cache policy to the C-index system, such that some index nodes are stored in the main memory of the computing nodes. Note that, we use the same query to test the performance on different sizes of datasets. The experiment shows that the response time of the query is reduced. Since many index nodes are already cached in the memory during the tree traversal, this reduces the number of I/O operations.

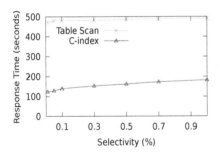

Fig. 5. Effect of Selectivity on CB-index

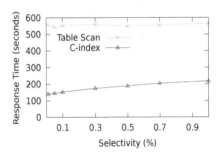

Fig. 6. Effect of Selectivity on CR-index

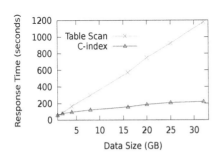

Fig. 7. Effect of Data Size on CB-index (Range Query)

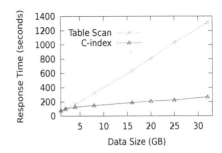

Fig. 8. Effect of Data Size on CR-index (Range Query)

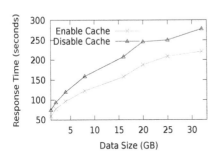

Fig. 9. Effect of Cache on CB-index

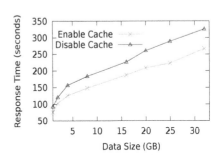

Fig. 10. Effect of Cache on CR-index

6   Conclusion

This paper presents C-index, a generic tree-like index framework for large datasets in the cloud. Our C-index model is built on the top of the B^+-tree or R-tree indexes, and provides efficient data query service for large scale cloud system. Instead of building index on the whole datasets, in C-index, we dynamically partition the datasets into pieces and build the B^+-tree or R-tree indexes for them, then publish their information in a meta table. We separate the index operations from the index storage. This renders the system easy to transplant between different storage systems. We propose algorithms to process one-dimensional or multi-dimensional range queries. To reduce the query process time, we added cache to reduce the I/O cost. Extensive experiments are conducted on the Hadoop cluster. The results confirm the effectiveness and efficiency of C-index.

Acknowledgements. This work is supported by the NSFC (No. 61100125), the 863 Program of China (No. 2011AA01A202, No. 2012AA011003), and the Program for Changjiang Scholars and Innovative Research Team in University of China (IRT1158, PCSIRT).

References

1. Adya, A., Bolosky, W.J., Castro, M., Cermak, G., Chaiken, R., Douceur, J.R., Howell, J., Lorch, J.R., Theimer, M., Wattenhofer, R.P.: Farsite: federated, available, and reliable storage for an incompletely trusted environment. In: OSDI (2002)
2. Aguilera, M.K., Golab, W., Shah, M.A.: A practical scalable distributed b-tree. Proc. VLDB Endow. 1(1), 598–609 (2008)
3. Aguilera, M.K., Merchant, A., Shah, M., Veitch, A., Karamanolis, C.: Sinfonia: a new paradigm for building scalable distributed systems. In: SIGOPS (2007)
4. Bajda-Pawlikowski, K., Abadi, D.J., Silberschatz, A., Paulson, E.: Efficient processing of data warehousing queries in a split execution environment. In: SIGMOD (2011)
5. Beckmann, N., Kriegel, H.-P., Schneider, R., Seeger, B.: The r*-tree: an efficient and robust access method for points and rectangles. In: SIGMOD (1990)
6. Borkar, V., Carey, M., Grover, R., Onose, N., Vernica, R.: Hyracks: A flexible and extensible foundation for data-intensive computing. In: 2011 IEEE 27th International Conference on Data Engineering, ICDE 2011 (2011)
7. Bozanis, P., Foteinos, P.: Wer-trees. Data Knowl. Eng. 63(2), 397–413 (2007)
8. Brakatsoulas, S., Pfoser, D., Theodoridis, Y.: Revisiting R-tree construction principles. In: Manolopoulos, Y., Návrat, P. (eds.) ADBIS 2002. LNCS, vol. 2435, pp. 149–162. Springer, Heidelberg (2002)
9. Comer, D.: Ubiquitous b-tree. ACM Comput. Surv. 11(2), 121–137 (1979)
10. Crainiceanu, A., Linga, P., Machanavajjhala, A., Gehrke, J., Shanmugasundaram, J.: P-ring: an efficient and robust p2p range index structure. In: SIGMOD (2007)
11. Dean, J., Ghemawat, S.: Mapreduce: simplified data processing on large clusters. Commun. ACM 51(1), 107–113 (2008)
12. Dittrich, J., Quiané-Ruiz, J.-A., Jindal, A., Kargin, Y., Setty, V., Schad, J.: Hadoop++: making a yellow elephant run like a cheetah (without it even noticing). Proc. VLDB Endow. 3(1-2), 515–529 (2010)
13. Ghemawat, S., Gobioff, H., Leung, S.-T.: The google file system. In: SOSP (2003)
14. Guttman, A.: R-trees: a dynamic index structure for spatial searching. SIGMOD Rec. 14(2), 47–57 (1984)
15. Kubiatowicz, J., Bindel, D., Chen, Y., Czerwinski, S., Eaton, P., Geels, D., Gummadi, R., Rhea, S., Weatherspoon, H., Wells, C., Zhao, B.: Oceanstore: an architecture for global-scale persistent storage. SIGARCH Comput. Archit. News 28(5), 190–201 (2000)

16. Li, N., Rao, J., Shekita, E., Tata, S.: Leveraging a scalable row store to build a distributed text index. In: CloudDB (2009)
17. Liao, H., Han, J., Fang, J.: Multi-dimensional index on hadoop distributed file system. In: NAS (2010)
18. Lin, K.I., Jagadish, H.V., Faloutsos, C.: The tv-tree: an index structure for high-dimensional data. The VLDB Journal 3(4), 517–542 (1994)
19. Sellis, T.K., Roussopoulos, N., Faloutsos, C.: The r+-tree: A dynamic index for multi-dimensional objects. In: VLDB (1987)
20. Thusoo, A., Sarma, J.S., Jain, N., Shao, Z., Chakka, P., Anthony, S., Liu, H., Wyckoff, P., Murthy, R.: Hive: a warehousing solution over a map-reduce framework. Proc. VLDB Endow. 2(2), 1626–1629 (2009)
21. Wang, J., Wu, S., Gao, H., Li, J., Ooi, B.C.: Indexing multi-dimensional data in a cloud system. In: SIGMOD (2010)
22. Weil, S.A., Brandt, S.A., Miller, E.L., Long, D.D.E., Maltzahn, C.: Ceph: a scalable, high-performance distributed file system. In: OSDI (2006)
23. Wu, S., Jiang, D., Ooi, B.C., Wu, K.-L.: Efficient b-tree based indexing for cloud data processing. Proc. VLDB Endow. 3(1-2), 1207–1218 (2010)
24. Xia, T., Zhang, D.: Improving the r*-tree with outlier handling techniques. In: GIS (2005)
25. Zuo, H., Jing, N., Deng, Y., Chen, L.: Can-qtree: A distributed spatial index for peer-to-peer networks. In: HPCC (2008)

Optimization of XML Queries
by Using Semantics in XML Schemas
and the Document Structure

Dung Xuan Thi Le, Moad Maghaydah, Mehmet A. Orgun, and Youliang Zhong

Department of Computing Macquarie University, Sydney, NSW 2109, Australia
{dung.le,mehmet.orgun,youliang.zhong}@mq.edu.au,
mmaghaydah@gmail.com

Abstract. This paper proposes an approach to optimizing structural-join and twig queries for XML queries by utilizing the semantics/constraints defined in XML Schemas and the existing facilities of relational database systems. In the first stage, semantic query transformations that use constraints in the schema are utilized to transform given XML queries to equivalent semantic queries. In the second stage, the structure of a given XML document is captured and stored in a table in a relational database, which is subsequently used during query translation and execution. Conducted experiments confirm the performance benefits of our approach in optimization of XML queries before and after applying semantic query transformations as well as capturing the XML document structure. We also report in detail on the results of a comparison between the performance of our approach with those of established native and modified-relational XML database systems.

Keywords: XML Data Management, XML Query Processing, Semantic Optimization.

1 Introduction

XPath queries, often expressed in twig patterns, are used to select certain nodes from a given XML document. The query condition is a Boolean expression that may involve comparisons between elements and values, and path expressions [10]. Structural constraints are represented as structural joins for a containment relationship of the XML elements (i.e., ancestor-descendant or parent-child). Reducing the amount of intermediate data created during join operations plays a crucial part in improving the performance of complex XML structural-join queries [6, 9].

In this paper, we discuss a comprehensive query optimization approach which is based on a combination of two complementary approaches: the Semantic Query Transformation approach (SQT), which utilizes the constraints in XML Schema (XDS) to rewrite XML queries into equivalent semantic queries, and the Prefix on Demand (PoD) [12,13], which uses a more compact representation of the structure of a given XML document as a set of relational tables. By integrating the two approaches, query optimization can be significantly improved on the existing database system capabilities when dealing with XPath queries with complex twig patterns. As a result, our approaches especially the latter reduce the cost of re-engineering these system kernels in particular for the relational-based solutions to be more tree-aware.

X. Lin et al. (Eds.): WISE 2013, Part I, LNCS 8180, pp. 343–353, 2013.

While the authors of [11] propose to eliminate predicates from XPath queries, we propose to eliminate the query condition component in the predicates as not all predicates can always be eliminated from a given query. In the case when a predicate in a query can be eliminated, we adopt a technique proposed earlier in [3,12,17].

We have evaluated the efficiency of our optimisation techniques using incremental XML data sets of different sizes from a well-known XML benchmark, that is, the Michigan benchmark [16]. SQT was evaluated in the relational-based solution (PoD) which also provides structural query optimisation. SQT itself is also suitable to be used on stand-alone native XML implementations. We have compared the query performance results to those experimented in eXist, a well-known native XML management system [5].

The paper is organized as follows. Section 2 overviews the related work. Section 3 illustrates the adopted approaches including semantic query transformation for XPath queries (SQT) and Prefix on Demand (PoD). Section 4 presents the framework of SQT & PoD Optimizations. In section 5, we report and discuss the results of the performance evaluation of the proposed approach and compare it with some others. Finally, we summarize our contributions and outline future work in section 6.

2   Related Work

Early studies focused on improving XML query runtime by designing new storage systems that support the hierarchical data model of the XML documents more efficiently. New operators and join algorithms have been designed from scratch in these native XML database systems to be more tree-aware [1, 2, 6].

Other studies also have developed more cost-effective solutions to harness existing relational database technology by shredding a given XML document and storing different parts of it in a proper relational schema. The document order and structure can be recovered by assigning a unique label for each node in the document [15]. Moreover, new operators and join algorithms have also been proposed [9]. However, some algorithms may require modifications to the kernel of RDBMS.

Most of the above solutions have focused on developing effective XML storage systems in the absence of a formal document schema (XML Schema XSD). Recently, the importance of capturing the document schema has been recognized since it can be used in XML query optimisation, path validations and document presentations [7, 14]. The existing works in relation to semantic query optimization [2, 18, 19] and the satisfiability problem [8] for XML queries have attracted much attention over the recent years. However, these works have focused on semantics in Document Type Definition (DTD) documents instead of actual XML Schemas (XSD). There is a big gap in semantics between DTD and XSD. The DTD lacks support of semantics of elements as it simply declares the structure of XML documents. Unlike DTD, not only can XML Schema (XSD) support what DTD already defined, it can also support the definition of a set of constraints for elements in XML documents.

Using semantics in XML Schemas for optimization purposes is still a preferred solution for query conditions especially with those that apply with both structures and values of elements. In a recent study, Le *et al* [10] proposed semantics transformation for XPath queries specified with predicates. However, their main focus was to eliminate the whole predicate from given XPath queries. As not all predicates can be eliminated from XPath queries, we consider semantic optimizations to reduce the size of the predicate instead.

Recent work on semantic query optimization proposed to construct relational tables that store semantics extracted from XML documents [20]. The semantics in a table is used to optimize a twig pattern query by avoiding patterns that make no contribution to the final result. The obvious challenge for this work is that the path referenced by IDs cannot be identified easily. Due to this problem, the work in [21] considers an extension to using ID references in DTDs to improve the processing of referenced paths. In our semantic query transformation, we use semantics in XML Schemas to provide a transformation strategy to find opportunities to optimize XPath queries. We then enhance the optimization process by applying the captured structure from an XML document to translate given queries during the processing stage.

3   Optimization Approaches

In this section, we present two approaches to XML query optimization. The first approach called Semantic Query Transformations (SQT) uses the semantics defined in the XML schema. The second approach called Prefix on Demand (PoD) is based on structural optimization. It should be noted that we are using Michigan XML Schema and queries throughout this paper. The queries we adopted for semantic query transformations focus on the hierarchies; therefore, when transforming the queries; in order to guarantee the correctness of resulting data, we made sure that the query hierarchy levels are not nested or repeated elsewhere in the data.

3.1   Semantic Query Transformation Approach

Semantic query transformations utilize semantics defined in the XML Schema to transform any given XPath query to an equivalent XPath query, which is executed more efficiently. The fragment of XPath studied in this work includes ("/", "//", "[]" or "*"). To achieve the optimization goal, we remove redundant components from the XPath query predicates before they are sent to the database engine for processing. Therefore components in the XPath query predicates are first determined for their status as explained below. To achieve the status[1] determination, we adopted the function ε defined in [10] to first determine the status of the query condition prior to making the decision to remove it. The signature of the function is given as follows:

$$\varepsilon\,(\varphi\pi) \rightarrow W$$
$$S_x$$

We now describe how the function ε works. It accepts as input an XPath query $\varphi\pi$ and uses the information extracted from a given XML Schema S_x to produce a status list $W=[w_1 \lozenge w_2 \lozenge \ldots \lozenge w_m]$. The symbol $\lozenge$ represents an optional logical operator AND or OR and $m \geq 1$, where each w_j (j=1,...,m) is the status of a query condition

---

[1] Status of query condition is either full-, partial or conflict.

that is determined by ε. π = [r₁ ◊ r₂ ◊... ◊ rₙ] is the twig-pattern of φ where π contains one or more sub twig-patterns or rather query condition(s) rᵢ , and n ≥ 1. φ is a path navigation of information filtered by π. The information extracted from Sₓ consists of a list of unique paths, namely Q and a list of constraints of elements, namely C, defined in the XML Schema S [11]. For each member q of Q, q is a full-path from the root of the schema to a schema element along the edge of the schema tree where only the parent-child relationship is used between the pair of elements. For each member c of C, c contains a sub-path of q and may carry a set of constraints and values of constraints of a target element[2] in c. The function ε first combines φ with the left part of every query condition, which is verified against each unique path q in list Q. This is to ensure that no structural conflict can be found in the path in order to avoid unnecessary processing later. Ultimately each query condition is awarded with one of the three condition statuses (full-, partial or conflict) as follows.

If a query contains no comparison value and the validation succeeds, the query is awarded the full-qualifier status 'FQ'[3]. If a series of query conditions are awarded 'PQ' where the comparison values are found to be exactly as described in the schema and they are joined by OR logical operator, the status of those 'PQ' queries is changed to 'FQ'. When the information of a query condition cannot match the information in list Q or C or both, which means a conflict has occurred, and the query condition is awarded 'CQ'.

The transformation finally removes all the conditions that are awarded with 'FQ' and the joins among them are 'OR'. The query conditions awarded with 'PQ' are also removed if they all project the same element and the joins are 'OR' among them. The query condition awarded with 'CQ' is removed only when it is joined by 'OR' with other conditions. The twig-pattern or rather the query predicate W is produced, only if it passed the determination process, in the form of determined statuses of 'PQ', 'FQ' or 'CQ' and AND/OR in between each pair of the determined statuses.

3.2   Structural Optimization Approach in a Relational Based Solution

Prefixing on Demand (PoD) [12] is a relational-based XML management approach. PoD supports query processing for data-centric XML queries in the absence of the XML document schema. PoD parses a given XML document and shreds it down to its basic components (i.e., elements, attributes...etc.) and stores them in a fixed relational schema. It maintains the document order and structure by utilizing a compressed Dewey-based labeling technique. Moreover, PoD captures the document structure summary in a small table called XML_Path. Figure 1 shows the relational schema for the PoD system. The id (binary string) is a unique Dewey label for every single node in the XML document. The id is represented as two separate Dewey values (PLabel: the parent node's label and the CLabel: the child node's label). Occurrence (integer) indicates the number of instances of the path in the actual database.

---

[2] A target element is the right most-element in an XPath query, e.g., //eNext/sixtyFour where sixtyFour is the target element.

[3] A full-qualifier (FQ) is when the values of a comparison element fully match the values of the same element in the schema.

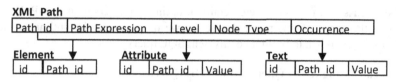

Fig. 1. Fixed Relational Schema in PoD based on Node Type

XML queries are translated into optimized SQL queries utilising features in the PoD labeling scheme, the structural information in the XML_Path table, and the nature of the relational systems. The PoD system first translates the given XML query into a query matrix (XPath matrix). The XPath matrix contains all the distinctive path expressions in the query, as shown in Table 1. The matrix is used to determine the relationships between paths.

Table 1. XPath Matrix extracted from the XML query

	P_1	...	P_n	Conditions	Result	Occurrence	Node Type
P_1	-		Relationship				E, A, or T...
...							:
P_n			-				

P_i stands for every distinguished path or node in the query; Conditions represent the type of restrictions on this path; Result indicates whether this path is part of the result set; Occurrence represents the number of instances this node in the actual database; Node Type (Element, Attribute, or Text node...etc.) identifies the right table which contains the instances of this path. The matrix will be structurally optimized and a join list table (jList) will be extracted from the final XPath matrix. jList is used to build the "WHERE" clause of the SQL query; it contains the join relationships between the different paths in the query. The join relationship can be any of the possible join values as shown in Figure 2.

Path	Path	Join Relationship	Level Path for GC
P_1	P_2	:	-
P_1	P_k	:	:
:	:	:	:
P_k	P_n	:	:
The Join Relationships			
P = P1 is Parent of P2; C = P1 is Child of P2; S = P1 is Sibling of P2; D = P1 is Descendant of P2.			
A = P1 is Ancestor of P2; CA = P1 and P2 are descendants of the same common ancestor node P3.			
VE = P1 has value equivalency condition with P2.			

Fig. 2. The optimized Join List (jList)

The fourth column in jList is used to provide the path of the common ancestor node if the join is based on a common ancestor CA relationship [12]. The jList table is ordered by the Join Relationship Priority{VE > S > (A, D) > CA}; the higher the priority the more important the join is, and possibly it is more efficient to evaluate first.

The discussion of the complete details of optimization techniques in PoD systems is beyond the scope of this paper. However, we present next the outline of the PoD translation mechanism to build an equivalent SQL query.

3.2.1 Translating XML Queries into SQL Queries

The PoD system utilises the information in the jList table and XPath_Matrix to generate equivalent SQL queries that can run directly in the relational engine. The main part of the translation process is to generate the SQL join statements. The translation algorithm, omitted due to space limitation, focuses on building the optimised structural-join part of any given XML query. For every row in the jList table, the algorithm adds references to the proper tables based on the path entry details in each row and the type of the join. The following example demonstrates the jList table in PoD and translation to SQL.

Example 1. The query //eNest[@aLevel=10]/@aUnique1 (QS3 from the Michigan benchmark [16]) is a twig query from which we can extract three path expressions:

P_1 = //eNest; P_2 = //eNest/@aLevel and aLevel = 10; P_3 = //eNest/aUnique1 (result)

Applying the PoD technique will produce the jList table given in Table 2; the PoD system removed the P_1 path since it is not required to produce correct results.

Table 2. The optimized jList for the query in Example 1

Path	Path	Join Relationship	Level Path for GC
P_2	10	VE	-
P_2	P_3	S	-

The translation algorithm would produce the following SQL query:

```
SELECT t3.value FROM Attribute t1, XML_Path t2, XML_Path t4, Attribute t3
WHERE t2.path like '%/eNest/@aLevel' and  t4.path like '%/eNest/@aUnique1'
    and t2.path_id = t1.path_id and t4.path_id = t3.path_id and t1.value = '10'
    and t3.plabel = t1.plabel
```

4   Integration of SQT and PoD

The high-level framework of XML query optimization using integrated PoD and SQT is shown in Figure 3. On start-up, the PoD module captures the document structure and the path summary when loading the initial document into the backend RDB storage. This task is shown as 1. Once the PoD completes the task, the schema pre-processing is initiated so that all constraints/semantics defined in the schemas are processed. This task is indicated with 2. The semantics are stored in the transformer component. The schema pre-processing component is terminated by now as it is no longer needed. It would be restarted if the schema is changed or modified. The transformer (indicated with 3) continuously accepts XPath queries from the user and transforms the queries where possible. If a given XPath query is valid, the transformer sends its equivalent semantic XPath query to the XML database to perform the querying task. This task is indicated with a dotted arrow line running from XML Queries Semantic Transformer to XML Database. If a conflict is detected, an informative message is returned to the user by the transformer. This task is indicated with a dotted arrow line running from XML Queries Semantic Transformer to User XML queries.

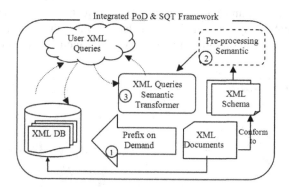

Fig. 3. Overview of Integration of PoD & SQT

5   Performance Evaluation

In this section we discuss the performance evaluation results, and highlight the improvements when one or more optimization techniques are applied to the same set of queries in different types of databases such as XML Native (eXist) and XML-enabled database system (PoD).

In our experiments, we used the PoD system in split mode (PoD-S) (i.e., two-component labels: Parent id, Child id), which has proven to be very efficient [12]. PoD was implemented in MySQL server 5.1. and Apache Java parser was used for parsing and loading XML documents. Java was also used for automating the tests. For comparison, we tested the native XML management system eXist [5] which stores XML documents in their original format. We configured eXist to use Java virtual machine of 1 GB. The sample documents were generated using the Michigan benchmark [16]. Documents of two different sizes (50MB and 500MB) were tested.

5.1   Evaluating SQT Approach in Two Different XML Storage Systems

We ran 8 queries from the Michigan benchmark, shown in Table A1 in the appendix. We first executed the original XPath queries in MySQL where PoD was applied, and once more after SQT was applied. Figure 4 shows the runtimes of the original queries and the semantically optimized ones in both PoD and eXist. The same test was repeated for a larger data set as shown in Figure 5. The results show that the query runtimes are significantly improved when either, or both, PoD and SQT applied in relational database (MySQL) and native XML database (eXist) systems. In particular, the query performance where both PoD and semantic transformations are applied, outperformed the query performance by applying only PoD. On the other hand, the original XPath queries performed in eXist indicated the worst results, however, when SQT was applied in eXist, the results outperformed the original XPath queries by 30% to 300%. The semantic transformations are able to identify the redundant components in predicates and remove them from the queries before they are sent to process on the database systems. This result demonstrates the potential of SQT as an optimization mechanism on top of a native XML management system.

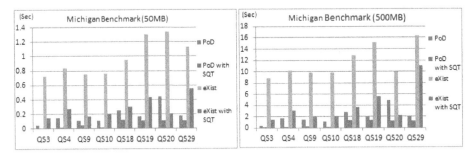

Fig. 4 & 5. Query Performance Results Using 50MB & 500MB Data Sets

5.2   Evaluating SQT Approach Integrated with PoD XML Storage System

We evaluated a scenario when XPath queries have two predicates, and based on the schema semantics either one of them but not both can be removed from the query; Table A2 in the appendix shows the details of these queries. By running these queries in PoD without SQT being applied, the queries were translated to equivalent and structurally optimised SQL statements without removing any predicates since PoD does not have a mechanism for semantic optimisation. However, applying the SQT approach to remove one redundant predicate at a time produced the same result sets but radically different run times as in Table 3.

Table 3. Query runtimes (in seconds) for the integration test

Query/Config.	Data Set of (50MB)			Data Set of (500MB)		
	PoD	PoD + SQT1	PoD + SQT2	PoD	PoD + SQT1	PoD + SQT2
AQ1	0.056	0.146	**0.036**	0.738	1.763	**0.425**
AQ2	0.034	0.1	**0.001**	0.426	1.115	**0.013**

The results show that PoD with SQT1 ran slower than PoD alone; SQT1, without inside information, removed the cheapest path to evaluate (eOccasional/@aRef), which PoD would have chosen as the first path to evaluate in relational systems based on information in the Occurrence field in the XML_Path table. However, PoD with SQT2 ran better than PoD alone because it chose to remove the other expensive path (/@aSixtyFour = 0), and ended up with the cheapest path without an overhead of extra paths. This shows the importance of integrating both frameworks to ensure better runtime in similar scenarios in which SQT could consult PoD on which option would produce better performance in the underlying storage system.

6   Conclusions

The evaluation results show that our query optimization techniques significantly improved the performance of XML twig queries by simplifying and reducing the number of required joins. Moreover, semantic optimization techniques have also improved runtime in the native XML system eXist. By comparing our complete approach to a mature XML management system, we showed that the relational systems can still be used effectively to store XML documents at an affordable cost. Future work will focus on developing further XML query rewrite rules based on integrating the semantic optimization techniques with the features in the underlying XML storage system. We also plan to conduct experiments with other native XML mamagement systems to evaluate the query performance of SQT.

References

1. Al-Khalifa, S., Jagadish, H.V., Koudas, N., Patel, J.M., Srivastava, D., Wu, Y.: Structural joins: A primitive for efficient XML Query pattern matching. In: The 18th International Conference on Data Engineering (ICDE 1902), pp. 141–152. IEEE Computer Society (1902)
2. Bao, Z., Ling, T.-W., Lu, J., Chen, B.: Semantic twig: A semantic approach to optimize XML query processing. In: Haritsa, J.R., Kotagiri, R., Pudi, V. (eds.) DASFAA 2008. LNCS, vol. 4947, pp. 282–298. Springer, Heidelberg (2008)
3. Charkravarthy, U.S., Grant, J., Minker, J.: Logic–Based Approach to Semantic Query Optimization. ACM Transactions on Database Systems 15(2), 162–207 (1990)
4. Chan, Y., Fan, W., Zeng, Y.: Taming XPath Queries by Minimizing Wildcard Steps. In: The Thirtieth International Conference on Very Large Data Bases, pp. 156–167 (2004)
5. eXist-db Open Source Native XML Database, http://exist-db.org
6. Fernández, M., Hidders, J., Michiels, P., Siméon, J., Vercammen, R.: Optimizing sorting and duplicate elimination in xQuery path expressions. In: Andersen, K.V., Debenham, J., Wagner, R. (eds.) DEXA 2005. LNCS, vol. 3588, pp. 554–563. Springer, Heidelberg (2005)
7. Goldman, R., Widom, J.: DataGuides: Enabling query formulation and optimization in semistructured databases. In: The 23rd VLDB Conference, Athens, Greece, pp. 436–445 (1997)
8. Groppe, J., Groppe, S.: Satisfiability-Test, Rewriting and Refinement of Users' XPath Queries According to XML Schema Definitions. In: Manolopoulos, Y., Pokorný, J., Sellis, T.K. (eds.) ADBIS 2006. LNCS, vol. 4152, pp. 22–38. Springer, Heidelberg (2006)

9. Grust, T., Keulen, M. V. & Teubner, J. (2003). Staircase join: Teach a relational DBMS to watch its (axis) steps. The 29th VLDB Conference. Berlin, Germany, 524-535.

10. Le, D.X.T., Bressan, S., Pardede, E., Rahayu, W., Taniar, D.: Semantic Transformation Approach with Schema Constraints for XPath Query Axes. In: Chen, L., Triantafillou, P., Suel, T. (eds.) WISE 2010. LNCS, vol. 6488, pp. 456–470. Springer, Heidelberg (2010a)

11. Thi Le, D.X., Bressan, S., Pardede, E., Taniar, D., Rahayu, W.: A utilization of schema constraints to transform predicates in xPath query. In: Bringas, P.G., Hameurlain, A., Quirchmayr, G. (eds.) DEXA 2010, Part I. LNCS, vol. 6261, pp. 331–339. Springer, Heidelberg (2010b)

12. Maghaydah, M., Orgun, M.A., Khazali, I.: Optimizing XML Twig Queries in Relational Systems. In: The 14th International Database Engineering and Applications Symposium (IDEAS), Montreal, Canada, pp. 123–129 (2010)

13. Maghaydah, M., Orgun, M.A.: Efficiently Querying Dynamic XML Documents Stored in Relational Database Systems. International Journal of Intelligent Information and Database Systems 5(4), 389–408 (2011)

14. Moro, M., Vagena, Z., Tsotras, V.: XML structural summaries. Proceedings of the VLDB Endowment 1, 1524–1525 (2008)

15. Pal, S., Cseri, I., Seeliger, O., Rys, M., Schaller, G., Yu, W., Tomic, D., Baras, A., Berg, B., Churin, D., Kogan, E.: XQuery implementaion in a relational database system. In: The 31st VLDB Conference, pp. 1175–1186. VLDB Endowment, Trondheim, Norway (2005a)

16. Runapongsa, K., Patel, J.M., Jagadish, H.V., Chen, Y., Al-Khalifa, S.: The Michigan benchmark. EECS The University of Michigan (2003), http://www.eecs.umich.edu/db/mbench/

17. Shenoy, S.T., Ozsoyoglu, Z.M.: Design and Implementation of a Semantic Query Optimizer. IEEE Transactions on Knowledge and Data Engineering 1(3), 344–361 (1997)

18. Su, H., Rundensteiner, E., Mani, M.: Semantic Query Optimization for XQuery over XML Streams. In: The 31st International Conference on Very Large Data Bases (VLDB), Trondheim, Norway, pp. 277–282 (2005)

19. Wang, G., Liu, M., Yu, J.: Effective Schema-Based XML Query Optimization Techniques. In: The Seventh International Database Engineering and Application Symposium (IDEAS), pp. 1–6 (2003)

20. Wu, H., Ling, T.W., Dobbie, G., Bao, Z., Xu, L.: Reducing Graph Matching to Tree Matching for XML Queries with ID References. In: Bringas, P.G., Hameurlain, A., Quirchmayr, G. (eds.) DEXA 2010, Part II. LNCS, vol. 6262, pp. 391–406. Springer, Heidelberg (2010)

21. Wu, H., Ling, T.W., Chen, B., Xu, L.: TwigTable: Using Semantics in XML Twig Pattern Query Processing. In: Spaccapietra, S. (ed.) Journal on Data Semantics XV. LNCS, vol. 6720, pp. 102–129. Springer, Heidelberg (2011)

Appendix

Table A1. XPath Queries and Semantic XPath Queries

Query	Original XPath Query	Semantic Query Transformation (SQT)	
		Semantic Expansion	Predicate Elimination (final optimized query)
QS3	//eNest[@aLevel=10]/@aUnique1	/eNest/eNest/eNest/eNest /eNest/eNest /eNest/eNest/eNest /eNest [@aLevel=10]/@aUnique1	/eNest/eNest/eNest/eNest /eNest/eNest /eNest/eNest/eNest /eNest/@aUnique1
QS4	//eNest[@aLevel=13]/@aUnique1	/eNest/eNest/eNest/eNest/eNest/eNest /eNest/eNest/eNest/eNest/eNest/eNest /eNest[@aLevel=13]/@aUnique1	/eNest/eNest/eNest/eNest/eNest/eNest/eNest/eNest/eNest/eNest/eNest /eNest/eNest/eNest/@aUnique1
QS9	//eNest[@aLevel=7]/eNest[position()=2]/@Unique1	/eNest/eNest/eNest/eNest/eNest/eNest /eNest[@aLevel=7]/eNest[position()=2]/@Unique1	/eNest/eNest/eNest/eNest/eNest/eNest/eNest/eNest[position()=2]/@Unique1
QS10	//eNest[@aLevel=9]/eNest[position()=2]/@Unique1	/eNest/eNest/eNest/eNest/eNest/eNest /eNest/eNest/eNest[@aLevel=7]/eNest[position()=2]/@Unique1	/eNest/eNest/eNest/eNest/eNest/eNest/eNest/eNest/eNest/eNest[position()=2]/@Unique1
QS18	//eNest[@aLevel=13][eNest[@asixteen=3]/@Unique1	/eNest/eNest/eNest/eNest/eNest/eNest /eNest/eNest/eNest/eNest/eNest/eNest /eNest[./eNest[@asixteen=3]]/@Unique1	/eNest/eNest/eNest/eNest/eNest/eNest/eNest/eNest/eNest/eNest/eNest/eNest/eNest[eNest[@asixteen=3]]/@Unique1
QS19	//eNest[@aLevel=15][eNest[@asixteen=3]/@Unique1	/ eNest/eNest/eNest/eNest/eNest/eNest/ eNest/eNest/eNest /eNest /eNest /eNest/eNest/ eNest/eNest [./eNest[@asixteen=3]]/@Unique1	/ eNest/eNest/eNest/eNest /eNest/eNest /eNest/eNest/eNest /eNest /eNest/eNest eNest/eNest[@asixteen=3]]/@Unique1
QS20	//eNest[@aLevel=11][./eNest[@aSixteen=3]]/@Unique1	/eNest/eNest/eNest/eNest/eNest/eNest /eNest/eNest/eNest/eNest/eNest[eNest /eNest[@aSixteen=3]]/@aUnique1	/ eNest/eNest/eNest /eNest/eNest /eNest/eNest/eNest /eNest /eNest/eNes[eNest/eNest/@aSixteen=3]/@aUnique1
QS29	//eNest[@aLevel=11][./eNest[@aFour=3]][./eNest[@aSixtyFour=3]]/@aUnique	/eNest/eNest/eNest/eNest/eNest/eNest /eNest/eNest/eNest/eNest/eNest[eNest[@aFour=3]][eNest[@aSixtyFour=3]]/@aUnique	/eNest/eNest/eNest/eNest/eNest/eNest/eNest/eNest/eNest/eNest/eNest[eNest/eNest[eNest[@aFour=3]][eNest[@aSixtyFour=3]]/@aUnique

Table A2. XPath Queries and Semantic XPath Queries

Query	Original XPath Query	Semantic Query Transformation	
		Equivalent Optimised Query Option 1 (SQT1)	Equivalent Optimised Query Option 2 (SQT2)
AQ1	//eNest[eOccasional/@aRef and @aSixtyFour =0]	//eNest[@aSixtyFour=0]	//eNest[eOccasional/@aRef]
AQ2	//eNest[@aLevel =12 and eOccasional/@aRef and @aSixtyFour =0]	/eNest/eNest/eNest/eNest/eNest/eNest/ eNest/eNest/eNest/eNest/eNest[@aSixtyFour =0]	/eNest/eNest/eNest/eNest/eNest/eNest/eNest/eNest/eNest/eNest/eNest/eNest[eOccasional/@aRef]

Semantic Entity Identification in Large Scale Data via Statistical Features and DT-SVM

Dingxian Wang[1], Xiao Liu[1], Hangzai Luo[2], and Jianping Fan[2]

[1] East China Normal University, Shanghai, China
[2] Northwest University of China, Xi'an, China
dingxianwang@gmail.com, xliu@sei.ecnu.edu.cn, hzluo@kunpad.com,
jfan@uncc.edu

Abstract. Semantic entities carry the most important semantics of text data. However, traditional approaches such as named entity recognition and new word identification may only detect some specific types of entities. In addition, they generally adopt sequence annotation algorithms such as Hidden Markov Model (HMM) and Conditional Random Field (CRF) which can only utilize limited context information. As a result, they are inefficient on the extraction of semantic entities that were never shown in the training data. In this paper we propose a strategy to extract unknown text semantic entities by integrating statistical features, Decision Tree (DT), and Support Vector Machine (SVM) algorithms. With the proposed statistical features and novel classification approach, our strategy can detect more semantic entities than traditional approaches such as CRF and Bootstrapping-SVM methods. It is very sensitive to new entities that just appear in fresh data. Our experimental results have shown that the precision, recall rate and F-One rate of our strategy are about 23.6%, 21.5% and 25.8% higher than that of the representative approaches on average.

Keywords: Semantic Entity Identification, New Word Identification, Decision Tree, SVM.

1 Introduction

In most multimedia applications, it is very important to understand the semantics of the input multimedia data. In most semantic models [10] of multimedia data, various semantic entities inferring the real semantics of the real world are essential to the model, because the semantics of the multimedia data can be modeled as entities and their relations in general [2]. As a result, the semantic entity extraction is the fundamental basis for multimedia semantic understanding. To resolve the semantic entity extraction problem, researchers have proposed different algorithms for different types of media. For text data, several named entity extraction algorithms [18,21] have been proposed to detect some special semantic entities, e.g. person name, location name and organization name. However, most multimedia applications need not only named entities but also other more general semantic entities (in bold and italic fonts) such as combined nouns, combined nouns and verbs and combined nouns and adjectives as shown in following examples:

X. Lin et al. (Eds.): WISE 2013, Part I, LNCS 8180, pp. 354–367, 2013.
© Springer-Verlag Berlin Heidelberg 2013

English Examples:

Our ***Palestinian brothers*** should ***declare an independent state***.

Tel Aviv will continue to ***abide by its peace treaty with Egypt*** despite the ***attack on its embassy in Cairo***.

The search for extraterrestrial life has taken another step forward - even if we are unlikely to find life as we know it any time soon, if at all. ***A team led by Swiss astronomers*** has recently discovered more than 50 exoplanets - planets orbiting stars outside the solar system.

He won three of the four ***Grand Slam titles*** this year -- at the ***Australian Open***,***Wimbledon*** and ***US Open*** -- and is talking about adding to his collection.

LAKE ARROWHEAD, Calif. (AP) - An 8-year-old boy with severe autism was found Tuesday after being lost and alone for more than 24 hours in ***the San Bernardino Mountains***.

Chinese Examples:

大概一年前, *郎咸平(Lang XianPing: person name)* 提出*中国制造业 (Chinese manufacturing industry)*危机未除, M2C才是出路。

加上*国家灾后重建配套资金(National post-disaster reconstruction funds)*250万元, 共同修*学校综合楼 (School building）*和*食堂综合楼(The canteen integrated building)*。

*美航母(U.S. aircraft carrier)*若参加*黄海军演(Yellow Sea military exercises)*, 中国有权攻击。

老人叫*张恒初(Zhang HengChu:person name)*, 今年已经*91岁(91 years old :age)*了。

*贝塔斯曼书友会(Bertelsmann book club)*撤出*中国市场(Chinese market)*。

The above examples show the semantic entities our paper intended to extract. Phrases in bold and italic fonts are the semantic entities we want to extract. More than half of the above semantic entities are not named entities such as 'declare an independent state', 'the search for extraterrestrial life' and 'Yellow Sea military exercises'. However, they carry very important semantics of the text. As a result, multimedia applications will miss important semantics if they use named entity extraction algorithms to detect the entities of the text data.

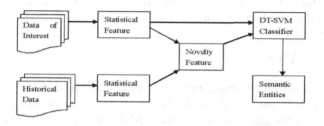

Fig. 1. Strategy Framework

In theory, the semantic entity extraction problem can be treated as a general named entity extraction problem and adopted similar algorithms such as Hidden Markov Model (HMM) [1] or Conditional Random Field (CRF) [13]. However, as the diversity of semantic entities is much higher than named entities, the extraction accuracy will be low.

In this paper, we have proposed a novel strategy to detect semantic entities. The framework of the proposed strategy is shown in Fig. 1 The basic idea of the strategy is to first extract statistical features for each potential semantic entity text string from the data of interest and then feed to a classifier to detect whether the string is a semantic entity or not. However, to achieve acceptable performance of this strategy, there are two problems to be solved first.

First, the statistical features used in the strategy must be carefully selected. Only with representative features can the classifier achieve good accuracy. In addition, the features must be extracted efficiently. Otherwise, the strategy will be running very slow because the number of potential text strings is extremely large. To resolve this problem, we have proposed a set of statistical features which are both representative and easy to be computed. Furthermore, we have proposed a set of novelty features which are sensitive to new entities occurring in the fresh data, so that the entities never shown in the training data can be detected more accurately.

Second, the dataset for the classifier is highly imbalanced. There are only around 1% to 5% semantic entities in all potential text strings. Therefore, most existing classification algorithms cannot achieve acceptable accuracy. To resolve this problem, we propose the DT-SVM algorithm that integrates DT (Decision Tree) and SVM (Support Vector Machine). The proposed algorithm in our strategy is designed to handle extremely imbalanced data.

The main contribution of this paper includes: 1)   the inner, outer and novelty statistical features are combined in our strategy and proved to be efficient and effective to help find semantic entities; 2) a two phase novel classification DT-SVM algorithm is proposed and proved to be more effective in dealing with imbalanced dataset compared with the-state-of-the-art techniques; 3) an novel and useful strategy which combines the advantage of features and DT-SVM algorithm is implemented and proved to be very effective to find semantic entities.

The remainder of this paper is organized as follows. Section 2 introduces the related works. Section 3 gives the detailed definition of the semantic entities in this study and also presents our work in detail. Section 4 shows the experimental results and analyzes the performance of our strategy. Finally, we conclude the paper and point out the future work in Section 5.

2   Related Work

Statistical based machine learning methods such as Hidden Markov Model (HMM), Decision Tree (DT) and entropy model [1, 15, 4] have been widely used in the research on English Named Entity Recognition (NER) as well as New Word Identification (NWI). The experimental results of these methods are quite good on datasets with high consistency. However, it may experience performance downgrade on datasets with high diversity, such as the web page data.

As Chinese NER is more difficult than English NER task, more advanced algorithms should be proposed to achieve better performance. Gao [9] uses statistical filtering as an important step to find the real Chinese NER. Wu [20] combines statistical model with back-off model as well as a Chinese thesaurus to help find Chinese NER. Takeuchi [17] investigates the identification and classification of technical terms in the molecular biology domain by using a combined HMM bigram model. Other methods such as CRF, class-based language model (LM), pattern-based, rule-based methods as well as hybrid methods have also been employed in Chinese NER. Chen [5] presents a Chinese NER system which incorporates basic features and additional features based on CRF and gets quite fair results on MSRA data sets. Bai [3] creates a system for tokenization and named entity recognition of ideographic language.

The research on Chinese NWI is also one of the most critical issues in Chinese NLP. The Chinese NWI research is closely related to Chinese NER and Chinese word segmentation research. As Sproat and Emerson [16] find that inefficient new word extraction causes over 60% of the word segmentation errors. From then on, many innovative algorithms such as statistical information based, class-based LM, user behavior based and collaborative methods have been brought forward to improve the accuracy of the Chinese NWI. Wu [19] presents a mechanism of new word identification in Chinese text where probabilities are used to filter candidate character strings and assign part-of-speech (POS) to the selected strings in a ruled-based system. Li [14] also uses a statistical learning approach based on SVM classifier employing different features such as the in-word probability of a character, the analogy between new words and lexicon words, the anti-word list and frequency documents to achieve the state-of-the-art performance. However, it is very time consuming given the complexity of the features. Fu proposes [8] a modified class-based LM approach by turning the problem into a classification problem with the part-of-speech information to classify each unknown word. Zheng[23] adds collaborative filtering to incorporate user behaviors into their New Word extraction system. However, these algorithms focus on Chinese text data only and its performance on other languages is unclear. In addition, some works mentioned above use features with high computation complexity thus are not suitable for large scale data.

3   Strategy Details

The semantic entity extraction task is executed as a scan procedure: the detector scans the input text string sequentially with a window, it outputs true when the sub string in the scan window is a semantic entity and false otherwise. In this paper, the string in the scan window is named as $s = \{x^1, ..., x^n\}$, where x^i is a segmented word andslies in the sentence $\zeta = \{..., a_{-2}, a_{-1}, s, a_1, a_2...\}$, where $a_i$is a context word. Then, the semantic entity extraction algorithm extracts features from ζ and feed to the classifier.

To achieve accurate semantic entity extraction, there must be a set of features carrying abundant information of semantic entities and they can be fast extracted from large volume of data. To resolve this problem, we propose several features that are suitable for semantic entity extraction.

3.1   Features Extraction

To decide whether s is a semantic entity or not, several types of features must be extracted from ζ and feed to the classifier. As discussed above, the feature extraction must be of low complexity since there are extremely large number of different s. Therefore, we propose the following features which can be fast obtained while still carrying abundant information of semantic entities.

First, the words or phrases composing a semantic entity must have frequent co-occurrence rather than random. Therefore, any statistical quantities measuring the correlations, closeness, or similar properties among the words and phrases of a semantic entity can be helpful for our target. Since these features are extracted from the internal components of the semantic entity, they are called "inner statistical features" in this paper.

Second, for the context words $\{..., a_{-2}, a_{-1}\}$ and $\{a_1, a_2...\}$, especially a_{-1} and a_1, may carry important information regarding the boundary of a semantic entity. For example, a_{-1} may have a high chance to be an article if s is a semantic entity. As a result, extracting statistical quantities from the context words as features may improve the accuracy of semantic entity extraction. Since they are computed by using words outside s, they are called "outer statistical features" in this paper.

Third, some novel semantic entities may appear in new data more frequently while some old semantic entities may gradually disappear. For most applications, those novel semantic entities are more important than general entities. However, they are more difficult to detect because they never occur in the training data. To resolve this problem, we propose a novelty feature to measure the novelty of a semantic entity.

In the following subsections, we will introduce the three types of features in detail. To simplify annotations, $p(x)$ and $p_{doc}(x)$ are used to represent the probability that word x may occur at any position and any document. They can be approximated as follows:

$$p(x) = \frac{F(x)}{\sum_{i=1}^{n} F(x_i)}, p_{doc}(x) = \frac{\|\{d: x \in d\}\|}{\|D\|}$$

Where $F(x)$ represents the times x appearing in the whole dataset. $\|D\|$ is the number of document of the dataset D and $\|\{d: x \in d\}\|$ the number of documents carrying x.

3.1.1   Inner Statistical Features

The information content $I^l(s)$, mutual information $M^l(s)$[13], correlation $C^l(s)$, TF-IDF [11] $T^l(s)$, cosine index $K^l(s)$, E index $E^l(s)$ and dice index $D^l(s)$[23] of s are computed as the most important inner statistical features.

The information content $I^l(s)$ is computed as the entropy of s:

$$I^l(s) = -\sum_{i=1}^{n} p(x_i) \log p(x_i)$$

Where n is the number of word s has. $I^l(s)$ can reflect how much information content s carries and the importance of s to the current news. Unlike the contribution of mutual information and correlation, the information content is calculated to determine the information that the whole semantic entity s includes and the degree of confusion it reflects.

The mutual information $M^I(x_i, x_j)[14]$ of two words x_i, x_j is defined as:

$$M^I(x_i, x_j) = \log\frac{p(x_i, x_j)}{p(x_i) * p(x_j)}$$

Where $p(x_i, x_j)$ means the joint term probability of x_i and x_j. If $M^I(x_i, x_j)$ is close to 0, it means x_i and x_j are just like independent random variables and have little connection. Therefore, only with high $M^I(x_i, x_j)$ values can the two words have correlation. Since the words or phrases composing a semantic entity must appear at the same time, they may have higher mutual information than words or phrases co-occurred randomly. In this paper, the traditional mutual information of two variables is extended to measure the mutual information over multiple variables as:

$$M^I(s) = \ln\left\{\frac{p(s)}{\prod_{i=1}^{n} p(x_i)}\right\}$$

The mutual information is a feature to measure correlations between two random variable from the view point of information theory. From the view point of statistical theory, the dependence has similar effect. The dependence of sis defined as:

$$C^I(s) = p(s) - \prod_{i=1}^{n} p(x_i)$$

If $C^I(s)$ is larger than 0, then words in s may not be independent and s has higher chance to be a semantic entity.

The TF-IDF $T^I(s)$ is a statistical quantity that shows the significance of a word or phrase to a document. Therefore, it may also help our semantic entity extraction task. In this paper, we use the normalized TF-IDF value so that the feature value can be comparable cross document boundary:

$$T^I(s) = \frac{p(s)}{idf(s)}$$

Where $idf(s)$ is the inverse document frequency of s[12]. In addition to the TF-IDF value of the whole entity s, the TF-IDF values of its components $(x_1, \ldots, x_n)$ may also carry useful information of semantic entities. Therefore, the sum, variance and median of TF-IDF values of $(x_1, \ldots, x_n)$ are also computed as the features.

Moreover, the cosine index $K^I(s)$, E index $E^I(s)$ as well as dice index $D^I(s)[22]$ are computed as:

$$K^I(s) = \frac{n * p(s)}{\sum_{i=1}^{n} \sqrt{p(x_i)}}$$

$$E^I(s) = \prod_{i=1}^{n} \frac{p(s)}{p(x_i)}$$

$$D^I(s) = \frac{n * p(s)}{\sum_{i=1}^{n} p(x_i)}$$

Please note that all the formulas use term probability p(x) are also applicable to the document probability $p_{doc}(x)$. So the mentioned features over $p_{doc}(x)$ are also computed in our strategy.

All of the above features only use $p(x)$ and $p_{doc}(x)$. If there is a table storing $p(x)$ and $p_{doc}(x)$ for all x_i and s, these features can be computed with constant complexity by look-up in the table. In addition, the table can be computed via term frequency of x_i and s, which can be computed by a sequential scan on the whole dataset with a scan window of max length s. Therefore, all features can be computed at linear complexity with respect to the word length of dataset.

3.1.2   Outer Statistical Features

Even though the inner statistical features may identify words and phrases that are parts of semantic entities, they may not carry enough information regarding the boundary of semantic entities. As a result, features extracted from the context of a semantic entity are needed for semantic entity extraction. In theory, the above features can be used to identify the semantic entity boundary if they are computed at the context of s. However, the computation of the above features needs the term frequency table. If they are computed at the context of s, one term frequency table must be computed for each potential semantic entity string s. As there are too many distinct s, it means either extremely large memory space (PB or EB size) is needed or large number of scans (millions) on the whole data. This would clearly consume too much time and storage.

To resolve this problem, we propose several outer statistical features that can be computed fast enough yet with reasonable memory consumption. They are normalized term frequency mean $V^o(s)$, max probability $L^o(s)$, outer mutual information $M^o(s)$, outer dependance $C^o(s)$ and the expand versions of cosine index $K^o(s)$, e index $E^o(s)$ as well as dice index $D^o(s)$ of s.

First, the context word may have high diversity than semantic entity elements. To measure the diversity of a context position i, the normalized term frequency mean $V_i^o(s)$ of the context position is used:

$$V_i^o(s) = \frac{1}{p(s)} * \frac{\sum_{x \in \cap(a_i(s))} p(x)}{\|\cap(a_i(s))\|}$$

Where a_i is the context position, $x \in \cap(a_i(s))$ is the set of words that appear at the context position i of s.

Also, feature $L_i^o(s)$ measures probability of word that appear most at position a_i:

$$L_i^o(s) = \frac{\max_{x \in \cap(a_i)} p(x, s)}{p(s)}$$

$p(x, s)$ is the joint probability that words appear at position a_i together with s. The outer mutual information between a_i and string s is defined as:

$$M_i^o(s) = \sum_{x \in \cap(a_i)} p(x, s) \log \frac{p(x, s)}{p(x) * p(s)}$$

The outer dependence between a_i and s is defined as:

$$C_i^o(s) = \sum_{x \in \cap(a_i)} p(x, s) \log \frac{p(x, s) - p(x) * p(s)}{\|\cap(a_i)\|}$$

Also, the expand versions of cosine index $K^o(s)$, E index $E^o(s)$ as well as dice index $D^o(s)$ are computed as:

$$K^o(s) = \frac{n * p(s)}{\sum_{x \in n(a_i)} \sqrt{p(x, s)}}$$

$$E^o(s) = \prod_{x \in n(a_i)} \frac{p(s)}{p(x, s)}$$

$$D^o(s) = \frac{n * p(s)}{\sum_{x \in n(a_i)} p(x, s)}$$

The regular form of above features needs computation intensive resources since there are a large number of various words appear at position a_i needed to be considered. However, only words at a_{-1} or a_1 are often believed really useful because they are the direct prefix and suffix of s. Moreover, if the scan window is large enough then the phrases as $\{a_{-1}, s\}$ or $\{s, a_1\}$ are included in the phrase term frequency tables. Therefore, only features with $i = \pm 1$ are used in our algorithm.

3.1.3   Novelty Statistical Features

New semantic entities may be repeated time after time and news after news during a period. So the frequencies of these semantic entities are often very large during at that time while they are scarcely appearing in the previous period. It implies that the novelty statistical features $N(s)$ is proportional to the occurrence probability $p_{current}(s)$ in current document and inverse proportional to the historical occurrence probability of s as $p_{history}(s)$:

$$\left(N(s) \propto p_{current}(s) \middle| N(s) \propto \frac{1}{p_{history}(s)} \right) \rightarrow N(s) = \gamma \frac{p_{current}(s)}{p_{history}(s)}$$

Where γ is for normalization. The historical occurrence probability of s as $p_{history}(s)$ can be calculated through many different methods, because it is believed that different time intervals will provide different effects. In this paper, several time intervals such as one day, one week, one month and one year are calculated as historical data. Meanwhile, several collective periods are also calculated. For example, if we have 3 days data, then data of first day, second day, third day, first day plus second day, first day plus second day and third day are all calculated as historical data. The historical data can be used to calculate several novelty features. These features can help to reflect the real novelty of semantic entities.

3.2   DT-SVM Classification Algorithm

With representative features, semantic entities can be detected by a sophisticated classifier. The SVM (Support Vector Machine) algorithm [6] is a widely used classification algorithm and can be adopted for semantic entity extraction. However, because our methods require the statistical features of all occurred adjacent string combinations in the dataset without considering whether it is a potential entity or not, the distribution of true semantic entities are so sparse (1~3%) and complicated. Therefore, traditional SVM algorithm may not be able to achieve acceptable performance given extremely imbalanced data.

To resolve this problem, we propose to use a decision tree to filter out most negative samples before the data are feed to the SVM classifier. The decision tree algorithm is chosen as the filter because it trains fast and is easy to tune between precision and recall. In our algorithm, we need to tune the filter to achieve almost 100% recall on semantic entities. To do so, we use the following filter training step: (1) train a decision tree model via C4.5 algorithm; (2) check all leaf nodes of the tree, mark leaf nodes covering only positive samples as "+" state, leaf nodes covering only negative nodes as "-" state and other leaf nodes as "0" state. With this filter, the proposed DT-SVM works as shown in Fig. 2.

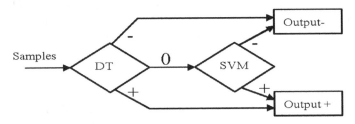

Fig. 2. DT-SVM framework

Algorithm 1 DT-SVM Classification Algorithm
Input:
Training Set $DA = \{X_k, Y_k\}$;
Output:
Semantic entity detection model: $DT - SVM$;
1: Divide training set DA into two parts, i.e. $DA1$ and $DA2$;
2: Apply decision tree algorithm on $DA1$, get model $DT1$;
3: Initialize a decision tree output array DTO;
4: for each $X_k \subset DA2$ do
5: Use $DT1$ to determine the class of X_k and get output Y_{kdt};
6: if $Y_{kdt}! = Y_k$
7: Put $DB_k[X_k, Y_k]$ into DTO;
8: end if
9: end for
11: Train DTO using svm algorithm, get model $SVM1$;
12: return model $DT1$ and $SVM1$ as $DT - SVM$;

Fig. 3. DT-SVM Classification Algorithm

The DT-SVM-based Classification algorithm plays a critical role in our strategy. It is designed and employed here for several reasons: 1) decision tree as a basic way of classification can ensure reasonable performance on those datasets which are easily classified; 2) the SVM methods are ineffective when there is too much noise in the unbalanced dataset, and the results will be greatly influenced. Meanwhile, the training time will also increase dramatically. Therefore, the decision tree methods with high recall rate (nearly 100%), acceptable precision rate and very low computation complexity are used to preprocess the dataset. Since a lot of negative data can be filtered and only small part of complex data is left to be treated by SVM, both the efficiency and effectiveness will be improved. 3) SVM is based on the maximum-margin rule so that the data with simple properties will contribute little to the whole process and usually be treated as distractors. Thus, there is no need to worry about whether the deleted negative samples will be useful to the classification algorithm.

Here, Fig. 3 presents the details of the DT-SVM classification algorithm. In the algorithm, X_k represents the feature vector composed of novel features, inner features and outer features. Y_k is the classification of X_k. After applying the DT filtering algorithm to remove most useless negative samples and feeding the remaining samples to the final SVM classification algorithm. The classifier model can be finally trained.

4   Experiments

In this section we evaluate the effectiveness of our strategy. We use both Chinese and English text data for comparison, so that we can evaluate the effectiveness of the strategy on different languages. The data are news web pages downloaded from Internet. There are 4.1 million Chinese pages and 690 thousand English pages covering 13 month in the dataset. The dataset is segmented to two parts. First 150 days of data are used for training and others are used for testing. For each dataset, the last day's semantic entities are manually annotated for quantitative evaluation. Other datasets containing 210 days of data are used as the background statistical material.

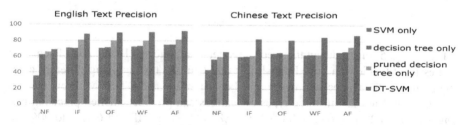

Fig. 4. Precision Rate

We first evaluate the effectiveness of inner statistical features, outer statistical features and novelty features. Then, our DT-SVM algorithm along with the standalone decision tree algorithm and SVM algorithm are also evaluated and compared to illustrate the characteristics of each algorithm. Criteria including precision, recall and F-One rate [26] are used for comparison. The results are shown in Fig. 4, Fig. 5, Fig. 6. In these figures, "NF" stands for the results using only novelty features, "IF" stands for the results using only inner statistical features, "OF" stands for the results using only outer statistical features, "WF" stands for the results using word features, viz. inner and outer statistical features, and "AF" stands for the results using all features, viz. inner, outer and novelty statistical features.

In our experiments, we use C4.5 to implement the DT algorithm. One of the great advantages of decision tree is that we do not need to set the parameters manually. Therefore, C4.5 can be easily implemented and applied. Meanwhile, grid searching which applies 5-fold cross validation is used to find the optimal parameters for the SVM algorithm. Specifically, c and gamma are two key parameters for the SVM algorithm where c is the penalty-factor that shows the acceptable error rate and gamma determines the distribution of data when it is mapped to the new feature space. We use the grid searching to find the best match of c and gamma. The searching ranges for c and gamma are set to a large number N (as 50) and the step is set to a small number M (as 0.5).

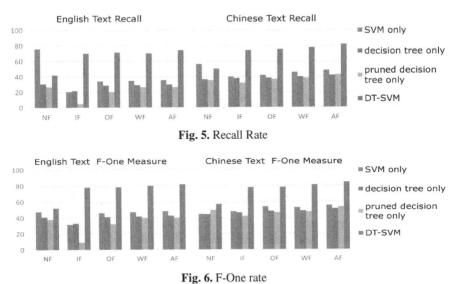

Fig. 5. Recall Rate

Fig. 6. F-One rate

Fig. 4, Fig. 5, Fig. 6 depict the precision rate, recall rate and F-One rate of semantic entities identification on both English text and Chinese text applying different algorithms such as SVM, decision tree, pruned decision tree and DT-SVM. From the figures, one can find that all three types of features can provide satisfied precision rate ranges from 35.28% to 68.3% for NF, 60.2% to 88.2% for IF and 63.7% to 90% for OF, recall rate ranges from 27% to 75.98% for NF, 4.8% to 74% for IF and 20.2% to 75.6% for OF, and F-One rate ranges from 38.18% to 57.16% for NF, 9.06% to 78.2% for IF and 32.25% to 72% for OF. By using the WF and AF features, the precision rate, recall rate and F-One rate can improve 5%-25% than using standalone features such as NF, IF and OF. In the meantime, the precision rate, recall rate and F-One rate of AF can improve 5%-10% than using WF. In comparison, DT-SVM strategy can provide precision rate ranges from 87.3% to 92.3% with an average of 89.3%, recall rate ranges from 73.5% to 81.8% with an average of 77.6%, F-One rate ranges from 81.53% to 84.56% with an average of 82.54% on both English and Chinese text using AF features which is better than other strategies such as SVM, DT and pruned-DT. It is because the highest precision rate, recall rate and F-One rate of those strategies are 81.7%, 50.5% and 57% which are lower than DT-SVM strategy provides. Specifically, the average increase on precision rate, recall rate and F-One rate are 10%, 27%, 25% respectively.

Based on these results we can conclude that: (1) all the 3 types of features are effective for the semantic entity extraction task; (2) the combination of these features can provide much better results than using standalone features; (3) our proposed DT-SVM strategy outperforms standalone decision tree algorithm and SVM algorithm, the proposed two-step classification strategy is very effective for handling imbalanced classification.

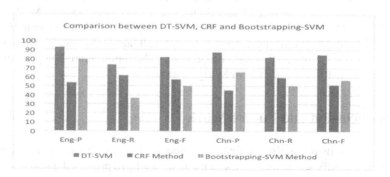

Fig. 7. Comparison between DT-SVM, CRF and Bootstrapping-SVM

To further evaluate the effectiveness of our strategy, we have compared with CRF-based semantic entity extraction approach [8, 12] which is regarded as one of the most effective NER approaches. In this paper, the experiments are conducted on CRF++ [12] toolkit. The three key parameters are '-a', '-c' and '-f'. Specifically, '-a' is used to select the type of algorithm to conduct the experiments. In our experiments, we choose 'CRF-L2' because it is proved to be better that 'CRF-L1'[25]. '-c' is used to set the hyper-parameter which determines the balance between 'overfitting' and 'underfitting'. We use the cross validation methods to find the optimal '-c'. '-f' determines the cut-off threshold of features. We use the simple parameter searching methods which is to set a range and apply the CRF methods to get the results of every '-f'. Then, we select the '-f' that produces the best result. To further demonstrate the effectiveness of our strategy in dealing with imbalanced dataset, we have also compared our strategy with Bootstrapping-SVM [24] since the bootstrapping method is both efficient and effective to solve imbalanced data problems. The simple bootstrap method involves first taking the original data set of N heights, and then sampling from it to form a new sample (called a 'resample' or bootstrap sample) which is also of size N. Therefore, our process is repeated for a large number of times (typically 1,000 or 10,000 times), and for each of these bootstrap samples we compute its mean (each of these samples are called bootstrap estimates).The precision, recall and F-One rate results are shown in Fig. 7 where "Eng-P" stands for the precision rate on English Text, "Eng-R" stands for the recall rate on English Text, "Eng-F" stands for the F-One rate results on English Text, "Chn-P" stands for the precision rate on Chinese Text, "Chn-R" stands for the recall rate on Chinese Text and "Chn-F" stands for the F-One rate results on Chinese Text.

Fig. 7 depicts the comparison between DT-SVM, CRF-based approach and the Bootstrapping-SVM on the precision rate, recall rate and F-One rate of semantic entities identification on both English text and Chinese text. From the figure, we can easily find that our DT-SVM strategy outperforms CRF-based approach and

Bootstrapping-SVM significantly in both Chinese and English dataset. The precision, recall and F-One rate of DT-SVM on English dataset and Chinese dataset ranges from 87.3% to 92.3%, 73.5% to 81.8% and 81.83% to 84.56% while the highest precision, recall and F-One rate of CRF-based algorithms are 53.2%, 61.8% and 57.718%. In the meantime, DT-SVM is better than Bootstrapping-SVM in dealing with imbalanced dataset. The best precision, recall and F-One rate of Bootstrapping-SVM algorithms are 79.3%, 50.5% and 56.992% respectively, and they are much lower than the mean values of DT-SVM. On average, the precision, recall and F-One rate of DT-SVM are about 23.6%, 21.5% and 25.8% higher than that of the representative approaches such as CRF-based and Bootstrapping-SVM algorithms.

5   Conclusion and Future Work

In this paper, a novel statistical features based machine learning strategy has been proposed to identify semantic entities in the text. We have proposed a set of statistical features which are sensitive to new semantic entities and can be obtained efficiently. We have also proposed a two-step approach to integrate decision tree and SVM algorithm to handle extremely imbalanced classification problem. Comprehensive experimental results have shown that our proposed features can identify semantic entities effectively and our proposed two-step classification approach can achieve high performance on imbalanced data. The results have also shown that our proposed strategy can significantly outperform representative approaches such as CRF and Bootstrapping-SVM on semantic entity extraction as a whole.

In the future, our research interests will focus on the integration of relationship among the semantic entities. Therefore, connected semantic entities can be discovered and their relationship can be displayed.

References

[1] Altun, Y., Tsochantaridis, I., Hofmann, T., et al.: Hidden Markov Support Vector Machines. In: Machine Learning-International Workshop Then Conference, vol. 20 (2003)

[2] Arndt, R., Troncy, R., Staab, S., Hardman, L., Vacura, M.: COMM: Designing a Well-Founded Multimedia Ontology for The Web. The Semantic Web, 30–43 (2007)

[3] Bai, S., Wu, H.J.P., Li, H., Loudon, G.: System for Chinese Tokenization and Named Entity Recognition, Google Patents. US Patent 6,311,152 (2001)

[4] Berger, A.L., Pietra, V.J.D., Pietra, S.A.D.: A Maximum Entropy Approach to Natural Language Processing. Computational Linguistics 22(1), 39–71 (1996)

[5] Chen, A., Peng, F., Shan, R., Sun, G.: Chinese Named Entity Recognition with Conditional Probabilistic Models. In: 5th SIGHAN Workshop on Chinese Language Processing, Australia (2006)

[6] Cortes, C., Vapnik, V.: Support-Vector Networks. Machine learning 20(3), 273–297 (1995)

[7] Finkel, J.R., Grenager, T., Manning, C.: Incorporating Non-Local Information into Information Extraction Systems by Gibbs Sampling. In: Proceedings of the 43rd Annual Meeting of the Association for Computational Linguistics, pp. 363–370 (2005)

[8] Fu, G., Luke, K.K.: Chinese Unknown Word Identification using Class-Based LM. Natural Language 2004, 704–713 (2005)

[9] Gao, J., Li, M., Wu, A., Huang, C.N.: Chinese Word Segmentation and Named Entity Recognition: A Pragmatic Approach. Computational Linguistics 31(4), 531–574 (2005)

[10] Hunter, J.: Adding multimedia to the semantic web: Building an mpeg-7 ontology. In: International Semantic Web Working Symposium, SWWS (2011)

[11] Jones, K.S.: A Statistical Interpretation of Term Specificity and Its Application in Retrieval. Journal of Documentation 28(1), 11–21 (1972)

[12] Kudo, T.: CRF++: Yet Another CRF Toolkit, http://crfpp.sourceforge.net (accessed on March 1, 2012)

[13] Latham, P., Roudi, Y.: Mutual information. Scholarpedia 4(1), 16–58 (2009)

[14] Li, H., Huang, C.N., Gao, J., Fan, X.: The use of SVM for Chinese New Word Identification. Natural Language 2004, 723–732 (2005)

[15] Sekine, S., Grishman, R., Shinnou, H.: A Decision Tree Method for Finding and Classifying Names in Japanese Texts. In: Proceedings of the 6th Workshop on Very Large Corpora (1998)

[16] Sproat, R., Emerson, T.: The First International Chinese Word Segmentation Bakeoff. In: Proceedings of the 2nd SIGHAN Workshop on Chinese Language Processing, vol. 17, pp. 133–143 (2003)

[17] Takeuchi, K., Collier, N.: Use of Support Vector Machines in Extended Named Entity Recognition. In: Proceedings of the 6th Conference on Natural Language Learning, vol. 20, pp. 1–7 (2002)

[18] Tsai, T.H., Wu, S.H., Lee, C.W., Shih, C.W., Hsu, W.L.: Mencius: A Chinese Named Entity Recognizer Using the Maximum Entropy-Based Hybrid Model. International Journal of Computational Linguistics and Chinese Language Processing 9(1) (2004)

[19] Wu, A., Jiang, Z.: Statistically-Enhanced New Word Identification in a Rule-Based Chinese System. Proceedings of the 2nd Workshop on Chinese Language Processing: Held in Conjunction with the 38th Annual Meeting of the Association for Computational Linguistics 12, 46–51 (2000)

[20] Wu, Y., Zhao, J., Xu, B.: Chinese Named Entity Recognition Combining a Statistical Model with Human Knowledge. In: ACL 2003, vol. 15, pp. 65–72 (2003)

[21] Wu, Y., Zhao, J., Xu, B., Yu, H.: Chinese Named Entity Recognition based on Multiple Features. In: Proceedings of the Conference on Human Language Technology and Empirical Methods in Natural Language Processing, pp. 427–434 (2005)

[22] Zhao, Y., Cui, L., Yang, H.: Evaluating Reliability of Co-citation Clustering Analysis in Representing the Research History of Subject, 80(1), 91–102 (2009)

[23] Zheng, Y., Liu, Z., Sun, M., Ru, L., Zhang, Y.: Incorporating User Behaviors in New Word Detection. In: Proceedings of the 21st International Joint Conference on Artificial Intelligence, pp. 2101–2106 (2009)

[24] Niu, C., Li, W., Ding, J., et al.: A Bootstrapping Approach to Named Entity Classification using Successive Learners. In: Proceedings of the 41st Annual Meeting on Association for Computational Linguistics, vol. 1, pp. 335–342. Association for Computational Linguistics (2003)

[25] Tellier, I., Eshkol, I., Taalab, S., et al.: Pos-tagging for Oral Texts with Crf and Category Decomposition. Natural Language Processing and its Applications 46, 79–90 (2010)

[26] Goutte, C., Gaussier, É.: A Probabilistic Interpretation of Precision, Recall and F-score, with Implication for Evaluation. In: Losada, D.E., Fernández-Luna, J.M. (eds.) ECIR 2005. LNCS, vol. 3408, pp. 345–359. Springer, Heidelberg (2005)

Efficient Computation of Multiple XML Keyword Queries

Liang Yao, Chengfei Liu, Jianxin Li, and Rui Zhou

Swinburne University of Technology, Melbourne, VIC 3122, Australia
{liangyao,cliu,jianxinli,rzhou}@swin.edu.au

Abstract. Answering keyword queries on XML data has been extensively studied. Current XML keyword search solutions primarily focus on single query setting where queries are answered individually. In many applications for searching information such as jobs and publications, an application server often receives a large number of keyword queries in a short period of time and many of them may share common keywords. Therefore, answering keyword queries in batches will significantly enhance the performance of these applications. In this paper, we investigate efficient approaches for computing multiple XML keyword queries. We first propose an approach that maximizes the sharing among keyword queries. We then consider useful data information and propose two data-aware algorithms: a short eager algorithm and a log based optimal algorithm. We evaluate the proposed algorithms on real and synthetic datasets and the experimental results demonstrate their efficiencies.

1 Introduction

Recently, answering keyword queries on XML data has drawn the attention of web and database communities, because the success of this research will relieve users from learning complex XML query languages and knowing the underlying schema of the XML data. Unlike the traditional keyword search for querying text data, which returns whole documents as search results, more fine-grained XML fragments are expected to be returned from an XML keyword query due to the structure and rich semantics of XML data. As such, a family of LCA (Lowest Common Ancestor) based approaches have been proposed, such as, SLCA (Smallest LCA)[12,9] and ELCA (Exclusive LCA)[5,13,15]. We notice that most XML keyword search solutions proposed so far primarily focus on single query setting where queries are answered individually. In many domain specific applications for searching information such as jobs, publications, properties and goods, the application servers often receive a large number of keyword queries in a short period of time. As the queries are issued for information within a specific domain, the chance that some queries share a subset of keywords is high.

For example, four keyword queries $q_1=\{$"keyword", "search", "probabilistic", "XML"$\}$, $q_2 = \{$"rank", "keyword", "search", "XML"$\}$, $q_3 = \{$"structured", "query", "probabilistic", "XML"$\}$ and $q_4 = \{$"relational", "database", "keyword", "search"$\}$ are submitted to DBLP XML dataset within a short time

X. Lin et al. (Eds.): WISE 2013, Part I, LNCS 8180, pp. 368–381, 2013.

period by different users searching publications. One may choose to compute each keyword query individually. However, this is not time efficient. Another solution is to cache results for some repeatedly issued keyword queries, which can save processing time. However, it is not efficient in terms of space consumption. In addition, due to large amount of results to store, secondary storage may have to be used, which causes extra I/O cost and will degrade the performance. Then can we provide efficient solutions for computing multiple XML keyword queries?

In this paper, we aim to answer the above question by exploring the keyword overlapping information of multiple queries and information about the XML data. We propose how to efficiently compute results in terms of SLCA - a widely accepted semantics to model XML keyword search results (see Section 2.1 for the definition). We explore some nice properties of SLCA and use them for efficient computation of SLCA results of multiple XML keyword queries. For example, we have the property that the intermediate SLCA results of a keyword query on a subset of keywords are used to further compute a keyword query of the full set of keywords. As a result, we may reuse the SLCA results of keyword query { "keyword", "search"} to compute q_1, q_2 and q_4, and even the SLCA results of keyword query { "keyword", "search", "XML"} to compute q_1 and q_2. Similarly, we may reuse the SLCA results of keyword query { "probabilistic", "XML"} to compute q_1 and q_3. We also have the property that the SLCA results of keyword queries on keyword sets K_1 and K_2 are merged to compute a keyword query on keyword set K, where $K = K_1 \cup K_2$. Therefore, to efficiently compute q_1, in addition to the approach of using the intermediate SLCA results on { "keyword", "search", "XML"}, we may also choose to merge the intermediate SLCA result sets on { "keyword", "search"} and { "probabilistic", "XML"}, or even the SLCA results of { "keyword", "search", "XML"} and { "probabilistic", "XML"}. These properties of SLCA are only applicable to keyword search over XML data.

There is a counterpart study on sharing work in keyword queries on relational databases [6], however, our work substantially differs from that work in that we use different evaluation scheme due to different data types (XML vs. relational) and result models (SLCA vs. candidate network). In XML keyword query processing, [3] proposes to process multiple keyword queries over XML data stream, where the traversal of the whole XML document is needed to answer multiple queries and no index can be used to improve the performance. These weaknesses will be addressed by our work. Furthermore, we will also explore the usage of sharing information and available data information to efficiently compute multiple keyword queries over XML database.

We summarize our contributions in this paper as follows: (1)We study the keyword overlapping among multiple XML keyword queries and its relationship to the properties of SLCA results and propose a basic algorithm that maximizes computations of shared keywords among XML keyword queries (Section 3). (2) We propose a short eager heuristic algorithm and an optimal algorithm by exploring useful data information (Section 4). (3) We demonstrate the efficiency of the proposed algorithms through experimental evaluation (Section 5).

In Section 2, we discuss SLCA semantics and useful properties for XML keyword queries and formally define the problem of computing multiple XML keyword queries. Related work is discussed in Section 6, and conclusions are given in Section 7.

2   Preliminaries and Problem Definition

2.1   SLCA Semantics

We model XML documents as trees using the conventional labelled ordered tree model. Each node of an XML tree corresponds to an XML element, an attribute or a text string. The leaf nodes are all text strings. A keyword may appear in element names, attribute names or text strings. If a keyword w appears in the subtree $T_{sub}(v)$ rooted at a node v, we say the node v contains keyword w. If w appears in the element name or attribute name of v, or w appears in the text value of v when v is a text string, we say node v directly contains keyword w. Node v is regarded as an *SLCA result* if the subtree $T_{sub}(v)$ contains all the keywords, and there does not exist a descendant node v' of v such that $T_{sub}(v')$ contains all the keywords. The following two properties of SLCA are useful for our work of computing multiple XML keyword queries.

Property 1. (Order free)

$$slca(w_1, ..., w_k) = slca(slca(w_{i_1}, ..., w_{i_m}), slca(w_{j_1}, ..., w_{j_n}))$$

where $\{w_{i_1}, ..., w_{i_m}\}$ and $\{w_{j_1}, ..., w_{j_n}\}$ are any two subsets of $\{w_1, ..., w_k\}$ and $\{w_{i_1}, ..., w_{i_m}\} \cup \{w_{j_1}, ..., w_{j_n}\} = \{w_1, ..., w_k\}$.

This property allows different ways of SLCA computation. We can first generate intermediate results using part of keywords (especially the overlapped set of keywords) and then process the others. As a special case, we have the recursiveness property.

Property 2. (Recursiveness)

$$slca(w_1, ..., w_k) = slca(slca(w_1, ..., w_{k-1}), w_k), where\ k > 2$$

This property can be used to compute a keyword query by merging the SLCA results of part of its keywords that have already been computed with the keyword node lists of other keywords which have not been processed.

2.2   Problem Definition

Before we study the problem of computing the SLCA results for a given set of XML keyword queries at the minimum cost, we first look how SLCA computation cost is modeled in a single keyword query evaluation setting. Following the discussion in [12], the basic approach of SLCA computation is to scan and merge the query keyword node lists in which the nodes are encoded by Dewey encoding scheme. The problem of SLCA computation can be transformed to find the longest common prefix of the Dewey IDs in the different keyword node lists, which can be computed in a single pass over all keyword node lists. Generally, the time complexity of SLCA keyword query computation can be dominated by $d\sum_{i=1}^{k} l_i$ where k is the number of keywords in a query $q = \{w_1, ..., w_k\}$, d is the depth of XML document and l_i is the size of the node list of keyword w_i. Therefore, we can model the computation cost C as $C(q) = d\sum_{i=1}^{k} l_i$. Given a set of keyword queries $\{q_1, ..., q_n\}$, if we separately compute these queries one by one, the computation cost could be represented as $C(q_1, ..., q_n) = d\sum_{j_1=1}^{k_1} l_{1j_1} + ... + d\sum_{j_n=1}^{k_n} l_{nj_n} = d\sum_{i=1}^{n} \sum_{j_i=1}^{k_i} l_{ij_i}$, where $w_{ij_i} (1 \leq j_i \leq k_i)$ is a keyword in q_i and k_i is the number of keywords in q_i.

One intuitive approach for reducing the cost of answering multiple keyword queries is to share the computations of overlapped keywords among keyword queries. To make this happen, selecting the right processing sequence of operations on keywords of the query is important. Therefore we first define a processing sequence below.

Definition 1. *(Processing Sequence of Single Query) Given a keyword query q, a processing sequence s of q refers to a set of operations that output the final SLCA results of q. Following the Property 1 and Property 2 of SLCA, each operation computes the intermediate SLCA results by merging two or more keyword node lists or intermediate result node lists.*

For example, $q = \{w_1, w_2, w_3, w_4\}$, we can first compute $\{w_1, w_2\}$ and $\{w_3, w_4\}$ by merging keyword node lists w_1 and w_2, and w_3 and w_4, respectively, and then compute $\{w_1w_2, w_3w_4\}$ by merging the node lists w_1w_2 and w_3w_4 to get the SLCA results of q. Here w_1w_2 and w_3w_4 are the intermediate SLCA results of $\{w_1, w_2\}$ and $\{w_3, w_4\}$, respectively.

Definition 2. *(Execution Plan of Multiple Queries) Given a set of keyword queries $Q = \{q_1, q_2, ..., q_n\}$, an execution plan p of Q is defined as a set of processing sequences $S = \{s_1, ..., s_n\}$, where s_i is a processing sequence of q_i $(1 \leq i \leq n)$ in Q.*

Figure 1 shows a keyword query set $\{q_1, q_2\}$, where $q_1 = \{a, b, c, d\}$, $q_2 = \{b, c, d, e\}$. The list of processing sequences for each query are given. The execution plan $p_1 = \{s_{11}, s_{21}\}$ has no shared computations, while the execution plan $p_2 = \{s_{12}, s_{21}\}$ has shared operations $\{b, c\}$ and $\{bc, d\}$. From the execution order for p_2, we can see that the shared computations are computed only once.

According to Definition 2, the evaluation cost of an execution plan depends on the processing sequence of each individual query and their shared computations with other keyword queries. In this paper, given multiple keyword queries $Q = \{q_1, q_2, ..., q_n\}$, our problem is to find the optimal execution plan by minimizing the cost for evaluating Q.

Q= {q₁, q₂}, q₁ : {a, b, c, d}, q₂ : {b, c, d, e}

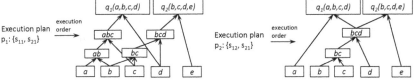

sequence id	operation(1)	operation(2)	operation(3)
s_{11}	{a, b}	{ab, c}	{abc, d}
s_{12}	{b, c}	{bc, d}	{bcd, a}
s_{13}	{c, d}	{cd, a}	{acd, b}
...	...	...	...

sequence id	operation(1)	operation(2)	operation(3)
s_{21}	{b, c}	{bc, d}	{bcd, e}
s_{22}	{b, c}	{bc, e}	{bce, d}
s_{23}	{b, c}	{d, e}	{bc, de}
...	...	...	...

Fig. 1. Example of Execution Plan

Definition 3. *(Optimal Execution Plan) Given multiple keyword queries $Q = \{q_1, q_2, ..., q_n\}$, our problem is to find the optimal execution plan $p^* = \{s_1^*, ..., s_n^*\}$ where s_i^* is a processing sequence of q_i such that the total cost $\sum_{i=1}^{n} C(s_i^*)$ is minimal.*

3   A Basic Approach

In this section, we propose a basic approach to find an efficient execution plan by analyzing the keyword sharing information among the given set of keyword queries $Q = \{q_1, q_2, ..., q_n\}$. As explained in the previous section, sharing operations on overlapped keywords among queries is useful for reducing the cost of answering Q. Therefore, we first introduce the concept of a sharing factor.

Definition 4. *(Sharing Factor) Consider a set of keyword queries $Q = \{q_1, ..., q_n\}$ ($n > 1$) that contain m distinct keywords $K = \{w_1, ..., w_m\}$. A sharing factor K_s is a subset of K (i.e., $K_s \subseteq K$) such that $\exists q_i, q_j \in Q$ ($i \neq j$), $K_s \subseteq q_i$ and $K_s \subseteq q_j$.*

Definition 5. *(Maximal Sharing Factor) Given any two queries q_i and q_j in Q, their maximal sharing factor is a set K_{maxs} of keywords such that $K_{maxs} \subseteq q_i$ and $K_{maxs} \subseteq q_j$, and there does not exist K'_{maxs} such that $K_{maxs} \subset K'_{maxs}$, $K'_{maxs} \subseteq q_i$ and $K'_{maxs} \subseteq q_j$.*

Our basic approach is based on finding and using the maximal sharing factors in Q. Without extra information about the data, making the full use of maximal sharing factors in Q allows us to maximally reduce the computation cost of Q. We now briefly describe the basic algorithm as shown in Algorithm 1.

First, for each pair of queries in Q, we compute their maximal sharing factor, get the set of maximal sharing factors of Q, and then sort this set of maximal sharing factors in the ascending order of their length and put them in a queue (Line 2 - Line 6). Next, for each query q in Q, we divide all keywords in q

Algorithm 1. Query Aware Algorithm

Input: keyword queries set: $Q:\{q_1, \dots , q_n\}$
Output: SLCA result for all queries
1: $\{K\} \leftarrow \emptyset$
2: **for** $i = 1 \rightarrow n - 1$ **do**
3:     **for** $j = i + 1 \rightarrow n$ **do**
4:         compare q_i and q_j, $K \leftarrow$ maximal common keywords of q_i and q_j;
5:         add K into $\{K\}$;
6:         $\forall K_i, K_j \in \{K\}$, K_i is placed before K_j if the size of K_i is smaller than that of K_j;
7: **for** $i = 1 \rightarrow n$ **do**
8:     **for** $\forall K$, $K \in \{K\}$, $K \subseteq q_i$ **do**
9:         put keywords of K into group $G_s(q_i)$;
10:    $G_u(q_i) = q_i - G_s(q_i)$;
11: $\{R\} \leftarrow \emptyset$
12: **while** $K \leftarrow \{K\}.next()$, $K \neq \emptyset$ **do**
13:    $r_K \leftarrow$ SLCA result of K;
14:    add r_K into R
15: **for** $i = 1 \rightarrow n$ **do**
16:    merge $r_K, r_K \in R, K \in K_{q_i}$ and $G_u(q_i)$ to compute SLCA result r of q_i;
17:    output r;

---

into two subgroups: the shared group $G_s(q)$ and the unshared group $G_u(q)$. $G_s(q)$ can be obtained by merging the set of maximal sharing factors of q, and $G_u(q)$ can be obtained by $G_u(q) = q - G_s(q)$ (Line 7 - Line 10). After that, we compute the SLCA results. We take a maximal sharing factor from the queue at a time and compute and store the intermediate SLCA results for this maximal sharing factor until the queue becomes empty (Line 12 - Line 14). Finally, for each query q, we compute its SLCA results by merging the intermediate results for the maximal sharing factors of q and the keyword node lists for unshared keywords in $G_u(q)$ (Line 15 - Line 17).

4   Data-Aware Approaches

In the basic approach, we discuss the problem of finding the optimal query execution plan by maximally sharing the computations of operations on keywords among queries. The basic approach can reach the best performance under the assumption that all keyword node lists have the same size. In some situations, the size of different keyword node lists may be significantly different, which may affect the performance of the basic approach. In this section, we propose two data-aware approaches which take the advantage of some data information. The first is a short eager heuristic approach, in which the sizes of keyword node lists are known. The second is a log-based optimal approach which is able to find the optimal solution when the size information of both the keyword node lists and the intermediate result node lists are available.

4.1   Short Eager Approach

In this section, we introduce a short eager algorithm to compute multiple keyword queries on XML data. In the basic approach, after we find all the maximal sharing factors among the multiple keyword queries and group all keywords into the shared group and the unshared group, we compute the result of these maximal sharing factors and then these keyword queries. During the procedure, we process a batch of queries without knowing the size of keyword node lists. If the keyword size information is available, we can enhance the performance of the basic approach.

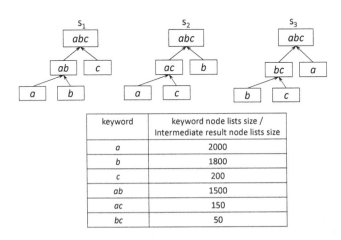

keyword	keyword node lists size / Intermediate result node lists size
a	2000
b	1800
c	200
ab	1500
ac	150
bc	50

Fig. 2. Different processing sequences cause different cost

Figure 2 shows keyword query $q : \{a, b, c\}$, the sizes of the keyword node lists and the intermediate result node lists, and three processing sequences of q. If we choose processing sequence $s_1 : (\{a, b\}, \{ab, c\})$, the cost can be measured by the sizes of the involved node lists, which is $2000 + 1800 + 1500 + 200 = 5500$. Similarly, the cost of processing sequence $s_2 : (\{a, c\}, \{ac, b\})$ is $2000 + 200 + 150 + 1800 = 4150$, and the cost of processing sequence $s_3 : (\{b, c\}, \{bc, a\})$ is $1800 + 200 + 50 + 2000 = 4050$.

One observation from the above example is that evaluating a keyword that has a short keyword node list first tends to be less expensive compared with evaluating a keyword that has a long keyword node list first. This heuristic rule naturally suggests us a short eager approach, which is described in Algorithm 2.

We first generate the set of maximal sharing factors (Line 1) and then divide keywords into groups (Line 2). Next, we choose to compute the maximal sharing factor that includes the keyword with the shortest keyword node list at a time (Line 3 - Line 5). After all maximal sharing factors have been computed, we compute and output the SLCA results for each query q by merging the intermediate results for the maximal sharing factors of q and the keyword node lists for unshared keywords in $G_u(q)$ (Line 7 - Line 10).

Algorithm 2. Short Eager Algorithm

Input: n multiple keyword queries: $Q:\{q_1, \ldots, q_n\}$
Output: SLCA result for all queries

1: generate maximal sharing factors set; (Line 2 to Line 5 of Algorithm 1)
2: divide keywords into shared and unshared groups; (Line 7 to Line 10 of Algorithm 1)
3: $\{R\} \leftarrow \emptyset$
4: **while** $K \leftarrow \{K\}.next()$ **do**
5: $\quad r_K \leftarrow$ SLCA result of K;
6: $\quad$ add r_K into R
7: **for** $i = 1 \rightarrow n$ **do**
8: $\quad$ **if** $\forall K_{q_i}, K_{q_i} \subseteq q_i, K_{q_i}$ is already computed **then**
9: $\quad\quad$ merge $r_K, r_K \in R, K \in K_{q_i}$ and $G_u(q_i)$ to compute SLCA result r of q_i;
10: $\quad\quad$ output r;

4.2 Log Based Optimal Approach

In the short eager approach, we use the size information of keyword node lists to help decide the processing sequence for each query. The advantage of this approach will be weakened if there is no significant difference between the sizes of keyword node lists. In fact, the real size of an intermediate result node list may not be necessarily always large from merging large keyword node lists. Sometimes, the size information of intermediate results can be found from the system log. As such, we come up with a log based optimal approach.

4.2.1 Bounding Minimal Cost of Processing Sequences

Given a keyword query q in the query set Q, there are many possible processing sequences to compute the SLCA results of q according to Definition 1. As the keyword node lists of unshared keywords in $G_u(q)$ can be processed in a fixed cost (i.e., $\sum_{i=1}^{k_u} l_i, 1 \leq i \leq k_u$ and $k_u = |G_u(q)|$) together with the intermediate keyword result node lists of maximal sharing factors, we only need to consider processing sequences for those operations that only contain the shared keywords, i.e., the keywords in $G_s(q)$. In general, q can be evaluated by choosing any processing sequence with a specified cost. Given a processing sequence $s_1 : \{o_1, o_2, ..., o_{m_1}\}$ of q, the cost of s_1 is $\sum_{i=1}^{m_1} C(o_i)$ where $C(o_i)$ is the computation cost of the o_i.

Note, $\sum_{i=1}^{m_1} C(o_i)$ is the maximal cost of s_1. If we consider the sharing factors when we measure the cost of s_1, the cost can be decreased when any of its operations o_i is shared with other queries in Q. If o_i is shared with other queries by t_{o_i} times, then its computation cost can be reduced to $\frac{C(o_i)}{t_{o_i}}$ because the intermediate results of o_i can be served to many other queries without computing from scratch. We can compute the cost of each shared operation. As a result, the total cost of s_1 can be minimized with regards to Q, i.e., $\sum \frac{C(o_i)}{t_{o_i}}$ where operation o_i is shared with other queries by t_{o_i} times. The minimal cost can be considered as the lower bound cost of processing s_1. The real cost of processing s_1, however, could be larger because the actual sharing count of o_i may be smaller than t_{o_i}.

To find the optimal execution plan, for each query q in Q, we generate a list of processing sequences with their corresponding lower bound cost values. They are sorted in the ascending order of their lower bound cost values. Then the problem is translated to that for each query q, we choose the first processing sequence of q in the list and constitute an execution plan to answer all queries with the optimal aggregated lowest computation cost. We now introduce how to find the optimal query execution plan in the following section.

4.2.2   Optimal Query Plan

Consider a set of queries $Q = \{q_1, q_2, ..., q_n\}$, each with a set of processing sequences $S_i = \{s_{i1}, s_{i2}, ..., s_{im_i}\}$ where $1 \le i \le n$ and m_i is the number of processing sequences of q_i. Finding the optimal execution plan of Q is to find a set of processing sequences $\{s_{1x_1}, s_{2x_2}, ..., s_{nx_n}\}$ over $\{S_1, S_2, ..., S_n\}$, which may approach the minimal total cost, i.e., $\sum_{i=1}^{n} C(s_{ix_i})$. However, the actual cost of q_i may be higher than the lower bound $C(s_{ix_i})$ because the shared times of some computations in s_{ix_i} may not be able to reach the maximal sharing times that have been used to compute $C(s_{ix_i})$ for q_i. Therefore, the score function of overall execution plan is not monotonic, i.e., given a set of keyword queries $\{q_1, ..., q_n\}$, we have an execution plan p_1 which contains a set of processing sequences $\{s_{11}, ..., s_{1n}\}$ for each query. Given another execution plan p_2, the overall real cost of p_2 may be less than that of p_1 even for all q_i, the lower bound score of s_{1i} is less than the lower bound score of s_{2i}. Therefore, existing methods to find out the top solution from multiple sorted lists cannot be applied because most of them deal with monotonic scoring functions. In this section, we introduce our log-based optimal algorithm on how to efficiently find the optimal execution plan from multiple processing sequence lists sorted by their lower bound costs.

As shown in Algorithm 3, we first compute and sort the list of processing sequences for each keyword query (Line 1), and initialize the execution plan pool P with the plan that has the lowest aggregated lower bound cost, which is formed from s_{i1} of each keyword query q_i (Line 2). Then, we start to process and expand the execution plan pool P by $P.next()$ to get the next plan p with the next lowest aggregated lower bound cost in P, and calculate and store its lower bound cost in σ and its real cost in t (Line 3 - Line 5). After that, we check whether $t = \sigma$, which is the *first stopping criteria* of our algorithm. The *second stopping criteria* is satisfied when the aggregated lower bound cost of the next plan to be considered is already larger than the real cost of the current plan stored in p. If the aggregated lower bound cost and real cost of a plan is the same, we compute the SLCA result for this plan and terminate the algorithm (Line 6 - Line 8). If $t > \sigma$, we remove p from P, and expand P (explained later) based on p and current real cost t (Line 9 - Line 11). Then we continue the process and get the next plan p' in P with the next lowest aggregated lower bound cost, and calculate and store its aggregated lower bound cost in σ' and real cost

---

Algorithm 3. Log Based Optimal Algorithm

Input: n keyword queries: $q_1, \ldots, q_n$
Output: SLCA result set R

1: compute n sorted processing sequence lists for n queries: $S_1, \ldots, S_n$;
2: $P \leftarrow \{\{s_{11}, ..., s_{n1}\}\}$; // P is initialized with the first processing sequence for each query. P is sorted in ascending order by aggregated lower bound cost.
3: $p \leftarrow P.next()$;
4: $\sigma \leftarrow$ lower bound cost of p;
5: $t \leftarrow compute\_real\_cost(p)$;
6: **if** $t = \sigma$ **then**
7:      R $\leftarrow$ SLCA result of p ;
8:      return R; // stopping criteria 1
9: **else**
10:     $P \leftarrow P - p$;
11:     $P.expand(p, t)$;
12: **while** $P \neq \emptyset$ **do**
13:     $p' \leftarrow P.next()$;
14:     $\sigma' \leftarrow$ lower bound cost of p';
15:     $t' \leftarrow compute\_real\_cost(p')$;
16:     **if** $\sigma' >= t$ **then**
17:         R $\leftarrow$ SLCA result of p ;
18:         return R; // stopping criteria 2
19:     **if** $t' = \sigma'$ **then**
20:         R $\leftarrow$ SLCA result of p' ;
21:         return R; // stopping criteria 1
22:     **if** $t' < t$ **then**
23:         $t \leftarrow t', \sigma \leftarrow \sigma', p \leftarrow p'$;
24:     $P \leftarrow P - p'$;
25:     $P.expand(p', t')$;
26: R $\leftarrow$ SLCA result of p
27: return R

---

in t' (Line 13 - Line 15). We check whether the stopping criteria 2 or criteria 1 is satisfied (Line 16 - Line 21). If so, we compute and output the SLCA results of the previous plan (criteria 2) or the current plan (criteria 1), and terminate the algorithm. Otherwise, if the new real cost t' is smaller than existing t, we keep p' in p by replacing p, t, σ with p', t', σ' (Line 22 - Line 23). We remove p' from P and expand P based on p' and current real cost t' (Line 24 - Line 25). We continuously verify the plans in P until P is empty, and no new valid plan could be generated. At this point, we compute and output the SLCA result for the plan p and terminate the process (Line 26 - Line 27).

In Algorithm 3, the function $compute\_real\_cost(p)$ is to compute the real cost of an execution plan, we identify their actual sharing times of each shared operation. We use the example shown in Figure 1 to describe how to compute real cost for an execution plan. Independent from the actual execution plans, the aggregated lower bound cost of s_{11} is $o_{\{a,b\}} + o_{\{ab,c\}} + o_{\{abc,d\}}$ because none of ab and abc is shared; the aggregated lower bound cost of s_{12} is $\frac{o_{\{b,c\}}}{2} + \frac{o_{\{bc,d\}}}{2} + o_{\{bcd,a\}}$

because the maximal sharing times for both bc and bcd are 2; the aggregated lower bound cost of s_{21} is $\frac{o_{\{b,c\}}}{2} + \frac{o_{\{bc,d\}}}{2} + o_{\{bcd,e\}}$ because the maximal sharing times for both bc and bcd are 2; the aggregated lower bound cost of s_{23} is $\frac{o_{\{b,c\}}}{2} + o_{\{d,e\}} + o_{\{bc,de\}}$ because only the maximal sharing time for bc is 2. For p_1, the real cost of s_{11} is the same as its aggregated lower bound cost, while the real cost of s_{21} is $o_{\{b,c\}} + o_{\{bc,d\}} + o_{\{bcd,e\}}$, which is different from its aggregated lower bound cost. For p_2, both the real cost of s_{12} and the real cost of s_{21} are the same as their aggregated lower bound costs. For $p_3 : \{s_{12}, s_{23}\}$, the real cost of s_{12} is $\frac{o_{\{b,c\}}}{2} + o_{\{bc,d\}} + o_{\{bcd,a\}}$, while the real cost of s_{23} is the same as its aggregated lower bound cost.

The procedure $expand(p, t)$ is used to expand current execution plan pool P. We first generate execution plan candidates. Each candidate is obtained by replacing the current processing sequence of one keyword query at a time with its next processing sequence. To reduce the size of execution plan pool and avoid unnecessary computations, for each candidate, we add it to P as a potential execution plan only when its aggregated lower bound cost is lower than the current real cost t of p.

5   Experiments

In this section, we present experimental evaluation of the proposed three algorithms: the basic algorithm (BA), the short eager algorithm (SE) and the log-based optimal algorithm (LBO). We also implement the algorithm for sequentially processing the queries without considering sharing (NS). All these algorithms are implemented in Java on a PC with 3.0 GHz CPU, 3.0 GB memory and Windows XP SP3.

5.1   Dataset and Query

We test the algorithms on both the real XML dataset - DBLP[1] and the synthetic XML dataset - XMark[2]. The size of DBLP is 909MB and the size of XMark (the scale factor is set as 6.0) is 698MB. The keywords in these keyword queries are randomly selected from high frequency keywords so that the probability of keyword overlapping among these queries becomes higher, and the keyword node lists with different sizes are maintained. We investigate the performance of the proposed algorithms by varying the query batch size of the set of queries and the frequency of sharing keywords.

5.2   Varying the Query Batch Size

We evaluate the algorithms when the query batch sizes are set from 10, 20 to 50.

For algorithm LBO, we do not need to keep the list of all possible processing sequences and the aggregated lower bound cost values for each query. From our

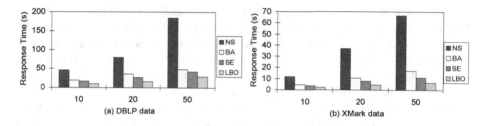

Fig. 3. Vary batch size on DBLP and XMark data

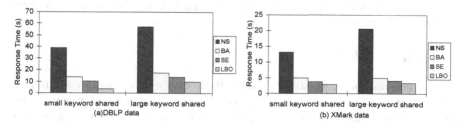

Fig. 4. Varying the sharing keywords frequency on DBLP and XMark data

experiment, we only need to use the first several processing sequences as the list is sorted in the ascending order of the aggregated lower bound cost values.

Figure 3 shows the performance of NS, BA, SE and LBO algorithms on DBLP and XMark datasets. If we set the processing time of the NS algorithm as a base, e.g., 100%, then it is easy for us to see the reduced time costs of other algorithms. For the DBLP dataset, when the batch size is set to 10, LBO saves 78% of processing time compared with NS, while SE saves 64% and BA saves 58% of processing time. When the batch size is 20, LBO can save 80% of processing time, and SE and BA can save 66% and 55% of processing time respectively compared with the NS algorithm. When the batch size is 50, the saving ratios of processing time for LBO, SE and BA are 84%, 77% and 74% respectively. Similarly, for the XMark dataset, the saving ratios of LBO, SE and BA are (90%, 88% and 81%), (83%, 79%, 69%), and (74%, 71%, 62%) respectively when the batch sizes are set to (10, 20 and 50). Under all settings, the LBO algorithm performs better than the others, followed by SE and BA algorithms. On average, LBO algorithm can save about 81% of processing time of the NS algorithm, while SE and BA can save about 74% and 66% respectively.

5.3   Varying the Shared Keyword Frequency

We vary the frequencies of shared keywords and investigate the performance under two situations. In the first situation, the keywords shared have large sizes of keyword node lists, and in the second situation, the keywords with the small sizes of the keyword node lists are shared between queries. Figure 4 shows the performance under these two sharing situations. The batch size is set to 10.

We also take the processing time of the NS algorithm as a base, e.g., 100%. When the shared keywords have larger sizes of keyword node lists (in other words, these keywords are "popular"), the LBO algorithm can save 83% of processing time compared with the NS algorithm on the DBLP dataset while the the SE and BA algorithms can save 76% and 69% respectively. On the XMark dataset, the saving ratios of the LBO, SE and BA algorithms are 84%, 80% and 76% respectively. When the sharing keywords have smaller sizes of keyword node lists, the saving ratios LBO, SE and BA algorithms are 91%, 74% and 64% on DBLP data set, and 78%, 72% and 62% on XMark data set. On both datasets, the LBO algorithm achieves the best performance among all algorithms, followed by the SE and BA algorithms. The LBO, SE and BA algorithms all outperform the NS algorithm significantly.

6   Related Work

In relational database, [6] studies the multiple query optimazation techniques. The evaluation of keyword queries in relational database is based on generating and evaluating Candidate Network(CN), which is an execution plan generated from the schema. It reduces the computation cost by reusing and combining the overlapped computations among the large number of CNs. However, the CN-based approaches in [6] cannot be applied to XML data. The "order free computation" property of SLCA allows keyword queries to be splitted and computed separately, then the intermediate results of the splitted queries can be merged to produce the result of original query. This property is implemented in all our proposed algorithms.

In XML, there are many works discussing efficient keyword search algorithms on single query setting. Most of existing approaches represented results using LCA model [12,5,13,15,10,7,14,8]. To compute SLCA, [12] introduces Indexed Lookup Eager Algorithm, Scan Eager Algorithm and Stack Algorithm. The complexity of Indexed Lookup Eager Algorithm is $O(kd|S_1|log|S|)$, where k is the keyword count in a query, d is the depth of XML document, $|S_1|$ is the length of the shortest keyword inverted list and $|S|$ is the total length of all keywords inverted lists. The complexity of both Scan Eager Algorithm and Stack Algorithm is $O(kd|S|)$. [11] improves the algorithm for specific tree patterns by looking for "Anchor" nodes. To compute ELCA, the complexity of Dewey Inverted List algorithm[5] is $O(kd|S|)$. [13] introduces the Indexed Stack (IS) with complexity of $O(kd|S_1|log|S_{max}|)$, where $|S_{max}|$ is the length of the longest keyword inverted list. [15] then saves the $|S_{max}|$ factor by introducing Hash Count method. [4] introduces a bottom-up approach to accelerate top-k results generation. [14] proposes efficient XML keyword query algorithms based on the set intersection operation for SLCA and ELCA semantics.

7   Conclusions

In this paper, we have studied the problem of answering multiple keyword queries on XML data. Based on the properties of SLCA and the analysis of the keyword overlapping among multiple keyword queries, we have designed three algorithms: the basic algorithm, the short eager algorithm and the log-based optimal algorithm. Through the extensive experimental evaluations over real and synthetic datasets, we have shown that our proposed algorithms can efficiently answer a batch of keyword queries, and all of them significantly outperform the algorithm that does not consider the sharing of computations in terms of processing time.

Acknowledgments. This work was supported by ARC Discovery Project DP110102407.

References

1. The dblp xml record, `http://dblp.uni-trier.de/xml/`
2. The xmark xml record, `http://www.xml-benchmark.org/index.html`
3. Hummel, F.C., da Silva, A.S., Moro, M.M., Laender, A.H.F.: Multiple keyword-based queries over xml streams. In: CIKM, pp. 1577–1582 (2011)
4. Chen, L.J., Papakonstantinou, Y.: Supporting top-k keyword search in xml databases. In: ICDE, pp. 689–700 (2010)
5. Guo, L., Shao, F., Botev, C., Shanmugasundaram, J.: Xrank: Ranked keyword search over xml documents. In: SIGMOD Conference, pp. 16–27 (2003)
6. Jacob, M., Ives, Z.G.: Sharing work in keyword search over databases. In: SIGMOD Conference, pp. 577–588 (2011)
7. Koloniari, G., Pitoura, E.: Lca-based selection for xml document collections. In: WWW, pp. 511–520 (2010)
8. Li, J., Liu, C., Zhou, R., Wang, W.: Suggestion of promising result types for xml keyword search. In: EDBT, pp. 561–572 (2010)
9. Li, J., Liu, C., Zhou, R., Wang, W.: Top-k keyword search over probabilistic xml data. In: ICDE, pp. 673–684 (2011)
10. Liu, Z., Chen, Y.: Reasoning and identifying relevant matches for xml keyword search. PVLDB 1(1), 921–932 (2008)
11. Sun, C., Chan, C.Y., Goenka, A.K.: Multiway slca-based keyword search in xml data. In: WWW, pp. 1043–1052 (2007)
12. Xu, Y., Papakonstantinou, Y.: Efficient keyword search for smallest lcas in xml databases. In: SIGMOD Conference, pp. 537–538 (2005)
13. Xu, Y., Papakonstantinou, Y.: Efficient lca based keyword search in xml data. In: EDBT, pp. 535–546 (2008)
14. Zhou, J., Bao, Z., Wang, W., Ling, T.W., Chen, Z., Lin, X., Guo, J.: Fast slca and elca computation for xml keyword queries based on set intersection. In: ICDE (2012)
15. Zhou, R., Liu, C., Li, J.: Fast elca computation for keyword queries on xml data. In: EDBT, pp. 549–560 (2010)

Soft Cardinality Constraints on XML Data
How Exceptions Prove the Business Rule

Flavio Ferrarotti[1], Sven Hartmann[2], Sebastian Link[3], Mauricio Marin[4],
and Emir Muñoz[5]

[1] Victoria University of Wellington
[2] Clausthal University of Technology
[3] The University of Auckland
[4] Yahoo! Research
[5] DERI, National University of Ireland Galway
`flavio.ferrarotti@vuw.ac.nz`

Abstract. We introduce soft cardinality constraints which need to be satisfied on average only, and thus permit violations in a controlled manner. Starting from a highly expressive but intractable class, we establish a fragment that is maximal with respect to both expressivity and efficiency. More precisely, we characterise the associated implication problem axiomatically and develop a low-degree polynomial time decision algorithm. Any increase in expressivity of our fragment results in coNP-hardness of the implication problem. Finally, we extensively test the performance of our algorithm. The performance evaluation provides first-hand evidence that reasoning about expressive notions of soft cardinality constraints on XML data is practically efficient and scales well. Our results unleash soft cardinality constraints on real-world XML practice, where a little more semantics makes applications a lot more effective in contexts where exceptions to common rules may occur.

1 Introduction

Cardinality constraints are a very natural class of constraints that can be observed easily and can express a lot of semantics important to applications such as consistency management, data integration, query optimization, view maintenance and cardinality estimation. Generally speaking, cardinality constraints capture information about the frequency with which certain data items occur in particular contexts.

Example 1. Suppose we use XML to store data about teams involved in projects within a research institute. Figure 1 shows an XML tree representing a small fragment of such an XML document. The nodes are annotated by their type: E for element nodes, A for attribute nodes, and S for text nodes. Of course, in reality there would be far more data stored in the XML document. We use the simplified example to illustrate how cardinality constraints can express important semantic properties of XML data. We assume that each project has a manager and that several research teams (RTeam) and support teams (STeam) can be involved in a given project. Technicians (Tech) belong to support teams,

X. Lin et al. (Eds.): WISE 2013, Part I, LNCS 8180, pp. 382–395, 2013.

Scientists (Sci) belong to research teams and Engineers (Eng) can belong to both, support and research teams. Some of the semantic properties which can be expressed by means of cardinality constraints are:

1. Every scientist is a member of 2, 3, or 4 research teams.
2. Every technician can work in up to 4 different support teams.
3. Every engineer is in 4 different support teams and 1 or 2 research teams.
4. A project cannot have more than one manager.
5. Every support team is involved in 1, 2, or 3 projects.
6. Every research team is involved in up to 2 different projects.
7. A maximum of 2 support teams and 2 research teams should be involved in a given project.
8. In every team, there should be two employees for each expertise level.
9. Within a given project, an employee cannot belong to more than one group.
10. No more than 5 employees of a given expertise level can be involved in the same project.   □

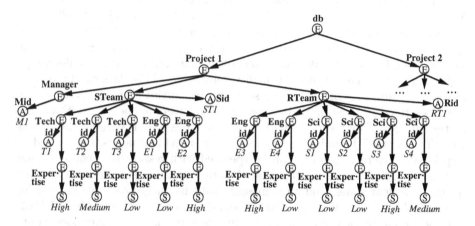

Fig. 1. Fragment of an XML tree with information on projects and teams

For the effectiveness of XML applications, it is an important problem to identify useful classes of cardinality constraints. This is challenging, since for semistructured XML data more so than for structured data, the more semantics of application domains can be captured the less efficient reasoning about these constraints usually becomes (see [1,2,8] among others).

In addition, exceptions to business rules are the common norm for (XML) data, and are therefore difficult to handle by strictly enforced constraints, which makes them often impractical to exploit. Indeed, it is often difficult to specify exact bounds for cardinality constraints. In many cases we only have a good idea of what rules should apply to most data.

Example 2. The constraints in Example 1 express rules that apply to most projects in the research institute, but clearly exceptions may occur. For instance, according to the first cardinality constraint in that example, every scientist is a member of 2, 3, or 4 research teams. However, it is likely that some scientists participate in 5 research teams or more. Nevertheless this will be an exception to most cases.   □

The previous example motivates the idea to treat cardinality constraints as *soft* constraints, which do not strictly enforce the constraint on all data, but on most data. Under this interpretation, soft cardinality constraints express an ideal or a preferred situation while still allowing room for violations of the strict constraints.

Example 3. The first cardinality constraint in Example 1 is better interpreted as a soft cardinality constraint. We can expect that scientists *on average* participate in 2 to 4 research teams, or that the number of researchers working in less than 2 and more than 4 research teams is considerably small. Similar observations apply to the other cardinality constraints in Example 1, too.   □

Paper Organization. In Section 2 we assemble useful technical notations on XML trees. In Section 3, we introduce soft cardinality constraints to effectively deal with the problem that exceptions confirm the rule in practice. Soft cardinality constraints need to be satisfied on average only, and thus permit exceptions to the general business rules in a controlled manner. We start by defining a highly expressive class of soft cardinality constraints. This class can be used to capture a wide range of interesting semantic properties of XML data, such as those described in Example 1. It allows data engineers to specify "soft" bounds on the number of nodes in an XML tree that are equal on some of their subnodes. These "soft" bounds can be violated by individual nodes, but should be respected on average. Bounds can be specified with respect to a context node. The soft constraints use a very general notion of value equality which is not restricted to leaf nodes, and an expressive path language to select nodes using single-label wildcards, child navigation, and descendant navigation from XPath.

In Section 4 we focus on a particular fragment of soft cardinality constraints, and characterise the associated implication problem axiomatically. In Section 5 we develop a low-degree polynomial time algorithm for deciding implication. The fragment under investigation is still expressive, allowing for instance to express most of the cardinality constraints in Example 1. Any increase in expressivity of our fragment is likely to cause intractability of the implication problem. There is great potential for practical uses of the proposed decision algorithm. For example, the process of checking XML data integrity against soft cardinality constraints can benefit a lot from the ability to decide implication efficiently. Clearly, if a set Σ of soft cardinality constraints implies a soft cardinality constraint φ and we have already checked that an XML data tree satisfies Σ then there is no need to test φ anymore, thus saving considerable resources. Section 5 reports on the outcomes of a performance anaylsis that provides first-hand evidence that reasoning about our fragment of soft cardinality constraints is practically efficient and scales well. Our results unleash cardinality constraints to application contexts where a little more semantics makes applications a lot more effective and where exceptions can be tolerated.

Related Work. The topic of XML constraints has attracted much attention over the last decade (see [5,10,11,17] among others). However, as far as we know, this article is the first to explore the concept of *soft* constraints in the context of XML. It is not the first time that cardinality constraints for XML data are studied [6], and they have a long and successful history in the field of database design (see [16] for a recent survey). Soft constraints are not new in the context of database design, where deontic logic has long been used as a tool to model soft constraints, see [12] for a survey. More recently, soft constraints have also been studied in the context of the constraint satisfaction problem (see [4,9,13] among others) where constraints are often soft in the sense that they do not have to be satisfied for a solution to be acceptable. None of these works, however, deals with the implication problem for soft constraints on XML data.

2 Basic Terminology

We use the common tree model of XML for our investigation. Let $\mathbf{E}$ denote a countably infinite set of element tags, $\mathbf{A}$ a countably infinite set of attribute names, and let S represent simple text data in XML (PCDATA). These sets are pairwise disjoint. The elements of $\mathcal{L} = \mathbf{E} \cup \mathbf{A} \cup \{S\}$ are called *labels*. An *XML tree* is a 6-tuple $T = (V, lab, ele, att, val, r)$ where V is a set of nodes, and lab is a mapping $V \to \mathcal{L}$ assigning a label to every node in V. A node $v \in V$ is an *element node* if $lab(v) \in \mathbf{E}$, an *attribute node* if $lab(v) \in \mathbf{A}$, and a *text node* if $lab(v) = S$. Moreover, ele and att are partial mappings defining the edge relation of T: for any node $v \in V$, if v is an element node, then $ele(v)$ is a list of element and text nodes, and $att(v)$ is a set of attribute nodes in V. If v is an attribute or text node, then $ele(v)$ and $att(v)$ are undefined. The partial mapping val assigns a string to each attribute and text node: for each node $v \in V$, $val(v)$ is a string if v is an attribute or text node, while $val(v)$ is undefined otherwise. Finally, r is the unique and distinguished root node.

A *path* p of T is a finite sequence of nodes $v_0, \ldots, v_m$ in V such that (v_{i-1}, v_i) is an edge of T for $i = 1, \ldots, m$. The path p determines a word $lab(v_1) \cdots \cdot lab(v_m)$ over the alphabet $\mathcal{L}$, denoted by $lab(p)$. For navigation in the XML tree, we use the path language $PL^{\{\cdot, \_, \_^*\}}$ consisting of words given by the following grammar: $Q \to \ell \mid \varepsilon \mid Q.Q \mid \_ \mid \_^*$. Here $\ell \in \mathcal{L}$ is any label, ε denotes the empty path expression, "." denotes the concatenation of two path expressions, "$\_$" denotes the *single-label* wildcard, and "$\_^*$" denotes the *variable length don't care* wildcard. Let P, Q be words from $PL^{\{\cdot, \_, \_^*\}}$. P is a *refinement* of Q, denoted by $P \lesssim Q$, if P is obtained from Q by replacing variable length wildcards in Q by words from $PL^{\{\cdot, \_, \_^*\}}$ and single-label wildcards in Q by labels from $\mathcal{L}$. Let Q be a word from $PL^{\{\cdot, \_, \_^*\}}$. A path p in the XML tree T is called a Q-*path* if $lab(p)$ is a refinement of Q. For a node v of T, $v[\![Q]\!]$ denotes the set of nodes in T that are reachable from v following any Q-path.

We use $[\![Q]\!]$ as an abbreviation for $r[\![Q]\!]$ where r is the root node. For $\mathcal{S} \subseteq \{., \_, \_^*\}$, $PL^{\mathcal{S}}$ denotes the subset of $PL^{\{\cdot, \_, \_^*\}}$ expressions restricted to the constructs in $\mathcal{S}$. $Q \in PL^{\{\cdot, \_, \_^*\}}$ is *valid* if it does not have labels $\ell \in \mathbf{A}$ or $\ell = S$ in a position other than the last one. Let P, Q be words from $PL^{\{\cdot, \_, \_^*\}}$. P is *contained* in Q, denoted by $P \subseteq Q$, if for every XML tree T and every node v of T we have $v[\![P]\!] \subseteq v[\![Q]\!]$.

If a node u lies on the path from a node v to the root, then u is an *ancestor* of v, and v a *descendent* of u. An *independent set* J of an XML tree T is a set of pairwise incomparable nodes, i.e., no node in J is an ancestor of any other node in J. Every path from a leaf to the root is a *branch* of T. An independent set intersects a branch at most once. For a node u of T, a *u-independent set* J of T is a set of descendents of u such that each pair of distinct nodes in J has u as their lowest common ancestor. Clearly, u-independent sets are independent.

Two nodes u, v are *value equal*, denoted by $u =_v v$, if the subtrees rooted at u and v are isomorphic by an isomorphism that preserves string values. For nodes v and v' of an XML tree T, the *value intersection* of $v[\![Q]\!]$ and $v'[\![Q]\!]$ is given by
$$v[\![Q]\!] \cap_v v'[\![Q]\!] = \{(w, w') \mid w \in v[\![Q]\!], w' \in v'[\![Q]\!], w =_v w'\}.$$

3   Soft Cardinality Constraints

Now we define a highly expressive class of soft cardinality constraints. The first source of expressivity comes from the ability to specify soft upper bounds (soft-max) as well as soft lower bounds (soft-min) on the number of nodes (target nodes) in an XML tree that are value-equal on some of its subnodes (field nodes). These soft bounds can be violated by some individual nodes, but they should be respected in average. There is also the possibility of specifying the soft bounds w.r.t. a context node. The second source of expressivity results from the generality of the path language $PL^{\{.,-,*\}}$ used for the selection of nodes. The final source of expressivity is due to the use of the robust notion of value-equality defined in the previous section, which is *not* restricted to leaf or attribute nodes.

Definition 1. *We define a* soft cardinality constraint *φ for XML as an expression of the form soft-card$(Q, (Q', \{Q_1, \ldots, Q_k\})) = (soft\text{-}min, soft\text{-}max)$ where k is a non-negative integer, where $Q, Q', Q_1, \ldots, Q_k \in PL^{\{.,-,*\}}$ such that $Q.Q'$, $Q.Q'.Q_1, \ldots, Q.Q'.Q_k$ are valid, and where soft-min $\in \mathbb{N}$ and soft-max $\in \mathbb{N} \cup \{\infty\}$ with soft-min $\leq$ soft-max. Herein, Q is called the* context path, *Q' is called the* target path, *$Q_1, \ldots, Q_k$ are called the* field paths, *soft-min is called the* soft lower bound, *and soft-max the* soft upper bound *of φ. If $Q = \varepsilon$, we call φ* absolute; *otherwise φ is called* relative.

In the sequel, for a soft cardinality constraint φ, we denote its context path as Q_φ, its target path as Q'_φ, its field paths as $Q_1^\varphi, \ldots, Q_{k_\varphi}^\varphi$ and its soft lower and upper bounds as soft-min$_\varphi$ and soft-max$_\varphi$, respectively.

Definition 2. *Consider a soft cardinality constraint φ, an XML tree T, a context node $q \in [\![Q_\varphi]\!]$ and a target node $q'_0 \in q[\![Q'_\varphi]\!]$. We set $f_T^\varphi(q, q'_0)$ as the maximum of $|\{q' \in q[\![Q'_\varphi]\!] \mid \exists y_1, \ldots, y_k. \forall i = 1, \ldots, k. y_i \in q'[\![Q_i^\varphi]\!] \wedge x_i =_v y_i\}|$ where $x_1, \ldots, x_k$ ranges through all $x_i \in q'_0[\![Q_i^\varphi]\!]$ (with $i = 1, \ldots, k$). That is, $f_T^\varphi(q, q'_0)$ is the maximum number of target nodes q' in the sub-tree of T rooted* *at the context node q that share with q'_0 the same information on their field paths. We say that T satisfies φ as a soft cardinality constraint if*

$$soft\text{-}min_\varphi \leq \frac{1}{|U|} \sum_{q'_0 \in U} f^\varphi_T(q, q'_0) \leq soft\text{-}max_\varphi$$

holds for every context node $q \in [\![Q_\varphi]\!]$ and every maximal q-independent set $U \subseteq q[\![Q'_\varphi]\!]$. If there is no target node $q'_0 \in q[\![Q'_\varphi]\!]$ in T for which for all $i = 1, \ldots, k_\varphi$, field nodes $x_i \in q'_0[\![Q^\varphi_i]\!]$ exists in T, then T satisfies the soft cardinality constraint φ by default since it does not apply to T.

Example 4. Following the discussion in Examples 2 and 3 above, the following expressions formalise the cardinality constraints in Example 1 when interpreted as soft cardinality constraints over trees of the form illustrated in Figure 1.
1. soft-card$(\varepsilon, (\_.RTeam.Sci, \{id\})) = (2, 4)$ or equivalently
   soft-card$(\varepsilon, (\_^*.RTeam.Sci, \{id\})) = (2, 4)$.
2. soft-card$(\varepsilon, (\_.STeam.Tech, \{id\})) = (1, 4)$ or equivalently
   soft-card$(\varepsilon, (\_^*.STeam.Tech, \{id\})) = (1, 4)$
3. soft-card$(\varepsilon, (\_.STeam.Eng, \{id\})) = (4, 4)$ and
   soft-card$(\varepsilon, (\_.RTeam.Eng, \{id\})) = (1, 2)$.
4. soft-card$(\_, (Manager, \emptyset)) = (1, 1)$.
5. soft-card$(\varepsilon, (\_.Steam, \{Sid\})) = (1, 3)$.
6. soft-card$(\varepsilon, (\_.Rteam, \{\_^*.Rid\})) = (1, 2)$.
7. soft-card$(\_, (STeam, \{\_^*.Sid\})) = (1, 2)$ and
   soft-card$(\_, (RTeam, \{Rid\})) = (1, 2)$.
8. soft-card$(\_\_\_, (\_, \{\_^*.S\})) = (2, 2)$ or equivalently
   soft-card$(\_\_\_, (\_, \{Expertise.S\})) = (2, 2)$.
9. soft-card$(\_, (\_\_\_, \{id\})) = (1, 1)$.
10. soft-card$(\_, (\_\_\_, \{Expertise.S\})) = (1, 5)$.

Note that the soft cardinality constraints in point 1–3, 5 and 6 are absolute while the soft cardinality constraints in the remaining points are relative. □

Let $\Sigma \cup \{\varphi\}$ be a finite set of (soft) constraints in a class $\mathcal{S}$. We say that Σ *finitely implies* φ, denoted by $\Sigma \models \varphi$, if every finite XML tree T that satisfies all $\sigma \in \Sigma$ also satisfies φ. The *finite implication problem* for the class $\mathcal{S}$ is to decide whether $\Sigma \models \varphi$. By Σ^* we denote the *(finite) semantic closure* of Σ, i.e., the set of all (soft) constraints finitely implied by Σ.

If we want to take advantage of the proposed soft cardinality constraints in real-world XML applications, then we must be able to reason about them efficiently. Central to this task is the finite implication problem describe above. Unfortunately the implication problem for the general class of soft cardinality constraints introduced in Definition 1, is likely intractable. In fact, as stated in the next theorem, there are at least three different sources of intractability:
   i. the simultaneous use of both soft lower and soft upper bounds (as permitted in $\mathcal{S}_1$ in Theorem 1),
  ii. the complete absence of field paths (as permitted in $\mathcal{S}_2$ in Theorem 1), and
 iii. the simultaneous use of arbitrary length wildcards in both target and field paths (as permitted in $\mathcal{S}_3$ in Theorem 1).

Theorem 1. *The finite implication problem for each of the following fragments of soft cardinality constraints is coNP-hard.*

$$S_1 = \{soft\text{-}card(\varepsilon, (P', \{P_1, \ldots, P_k\})) = (soft\text{-}min, soft\text{-}max)$$
$$| \ P', P_1, \ldots, P_k \in PL^{\{\cdot\}}, k \geq 1, soft\text{-}max \leq 5\},$$
$$S_2 = \{soft\text{-}card(\varepsilon, (P', \{P_1, \ldots, P_k\})) = (1, soft\text{-}max)$$
$$| \ P', P_1, \ldots, P_k \in PL^{\{\cdot\}}, k \geq 0, soft\text{-}max \leq 6\},$$
$$S_3 = \{soft\text{-}card(\varepsilon, (Q', \{Q_1, \ldots, Q_k\})) = (1, soft\text{-}max)$$
$$| \ Q', Q_1, \ldots, Q_k \in PL^{\{\cdot, \cdot^*\}}, k \geq 1, soft\text{-}max \leq 4\}.$$

We note that for each of the classes considered in Theorem 1, the 3-colorability problem over graphs can be polynomially transformed to the complement of the implication problem for soft cardinality constraints, but due to space limitations we omit the formal proof.

To avoid the sources of intractability pointed out in Theorem 1, we will consider a fragment of soft cardinality constraints that provides an optimal balance with respect to expressivity and efficiency.

Definition 3. *We define the fragment $\mathfrak{M}^{\text{soft}}$ of soft-max cardinality constraints as follows. $\mathfrak{M}^{\text{soft}} = \{soft\text{-}card(Q, (Q', \{Q_1, \ldots, Q_k\})) = (1, soft\text{-}max) \mid Q, Q', Q_1, \ldots, Q_k \in PL^{\{\cdot, \cdot^*\}}$ but s.t. Q' or $Q_1 \cdots .Q_k \in PL^{\{\cdot\}}\}$. Since soft-min is always set to 1, we use the abbreviation $soft\text{-}card(Q, (Q', \{Q_1, \ldots, Q_k\})) \leq soft\text{-}max$ to denote the soft constraints in $\mathfrak{M}^{\text{soft}}$.*

The fragment of soft-max cardinality constraints is still expressive, allowing for instance to express most of the cardinality constraints in Example 1.

Example 5. The soft cardinality constraints in points 2, 4–7, 9 and 10 in Example 4 belong to $\mathfrak{M}^{\text{soft}}$. Also the second soft cardinality constraint in point 3 belongs to $\mathfrak{M}^{\text{soft}}$. The remaining three soft cardinality constraints can still be partially expressed as soft-max cardinality constraints if we change the lower bound soft-min to 1. □

Note that by Theorem 1, any increase in expressivity of the fragment of soft-max constraints results in coNP-hardness of the implication problem.

4   Axiomatization

Table 1 shows a set of inference rules which constitutes a finite axiomatization for the implication of soft-max cardinality constraints. Each inference rule has the form $\frac{premises}{conclusion}$ *condition* with premises from $\mathfrak{M}^{\text{soft}}$. That is, the path expressions used in the premises are always chosen such that the respective soft cardinality constraint lies in $\mathfrak{M}^{\text{soft}}$.

Example 6. Let us define the soft-max cardinality constraints $\sigma_1 = soft\text{-}card(\_, (RTeam, \{Eng.\_^*.S\})) \leq 2$ and $\sigma_2 = soft\text{-}card(\_\_, (Eng, \{\_^*.S\})) \leq 2$, which are

Table 1. A finite axiomatization for soft cardinality constraints in $\mathfrak{M}^{\text{soft}}$

$$\frac{}{\text{soft-card}(Q, (Q', S)) \leq \infty} \quad \begin{array}{l} Q' \in PL^{\{\cdot, \cdot\text{-}\}} \text{ or} \\ \emptyset \neq S \subseteq PL^{\{\cdot, \cdot\text{-}\}} \end{array} \qquad \frac{}{\text{soft-card}(Q, (\epsilon, S)) \leq 1}$$
$$\text{(infinity)} \qquad\qquad\qquad \text{(epsilon)}$$

$$\frac{\text{soft-card}(Q, (Q'.Q'', S)) \leq \text{soft-max}}{\text{soft-card}(Q.Q', (Q'', S)) \leq \text{soft-max}} \qquad \frac{\text{soft-card}(Q, (Q', S)) \leq \text{soft-max}}{\text{soft-card}(Q, (Q', S)) \leq \text{soft-max} + 1}$$
$$\text{(target-to-context)} \qquad\qquad \text{(weakening)}$$

$$\frac{\text{soft-card}(Q, (Q', S \cup \{\epsilon, P\})) \leq \text{soft-max}}{\text{soft-card}(Q, (Q', S \cup \{\epsilon, P.P'\})) \leq \text{soft-max}} \qquad \frac{\text{soft-card}(Q, (Q', S)) \leq \text{soft-max}}{\text{soft-card}(Q, (Q', S \cup \{P\})) \leq \text{soft-max}} \quad \begin{array}{l} Q' \text{ or} \\ P \in PL^{\{\cdot, \cdot\text{-}\}} \end{array}$$
$$\text{(prefix-epsilon)} \qquad\qquad\qquad \text{(superfield)}$$

$$\frac{\text{soft-card}(Q, (Q'.P, \{P'\})) \leq \text{soft-max}}{\text{soft-card}(Q, (Q', \{P.P'\})) \leq \text{soft-max}} \quad \begin{array}{l} \text{at least 2 of} \\ Q', P, P' \in PL^{\{\cdot, \cdot\text{-}\}} \end{array} \qquad \frac{\text{soft-card}(Q, (Q', S)) \leq \text{soft-max}}{\text{soft-card}(Q'', (Q', S)) \leq \text{soft-max}} \quad Q'' \subseteq Q$$
$$\text{(subnodes)} \qquad\qquad\qquad \text{(context-path-containment)}$$

$$\frac{\text{soft-card}(Q, (Q'.P, \{\epsilon, P'\})) \leq \text{soft-max}}{\text{soft-card}(Q, (Q', \{\epsilon, P.P'\})) \leq \text{soft-max}} \quad \begin{array}{l} \text{at least 2 of} \\ Q', P, P' \in PL^{\{\cdot, \cdot\text{-}\}} \end{array} \qquad \frac{\text{soft-card}(Q, (Q', S)) \leq \text{soft-max}}{\text{soft-card}(Q, (Q'', S)) \leq \text{soft-max}} \quad Q'' \subseteq Q'$$
$$\text{(subnodes-epsilon)} \qquad\qquad\qquad \text{(target-path-containment)}$$

$$\frac{\begin{array}{l}\text{soft-card}(Q, (Q', \{P.P_1, \ldots, P.P_k\})) \leq \text{soft-max,} \\ \text{soft-card}(Q.Q', (P, \{P_1, \ldots, P_k\})) \leq \text{soft-max}'\end{array}}{\text{soft-card}(Q, (Q'.P, \{P_1, \ldots, P_k\})) \leq \text{soft-max} \cdot \text{soft-max}'} \qquad \frac{\text{soft-card}(Q, (Q', S \cup \{P\})) \leq \text{soft-max}}{\text{soft-card}(Q, (Q', S \cup \{P'\})) \leq \text{soft-max}} \quad P' \subseteq P$$
$$\text{(multiplication)} \qquad\qquad\qquad \text{(field-path-containment)}$$

applicable to XML documents that are structured in the way schematised by the tree in Figure 1. The soft constraint σ_1 states that in a given project, it is rare that more than two research teams have engineers of a same expertise level. The soft constraint σ_2 states that it is unusual that there is more than two engineers of a same expertise level within a given team. By applying the *context-path-containment* rule to σ_2 we derive $\sigma_3 = \text{soft-card}(\_.RTeam, (Eng, \{\_^*.S\})) \leq 2$. Then, by applying the *multiplication* rule to σ_1 and σ_3 we derive $\varphi = \text{soft-card}(\_, (RTeam.Eng, \{\_^*.S\})) \leq 4$, which expresses that it is infrequent to find more than 4 engineers of a same expertise level if we look at all the engineers in all the research teams involved in a given project. $\qquad\qquad\square$

We omit the tedious, but not very difficult proof of the soundness of the inference rules. Our next goal is to demonstrate that the set $\mathfrak{R}$ of inference rules in Table 1 is complete for the implication of soft-max constraints in the class $\mathfrak{M}^{\text{soft}}$. Completeness means we need to show that for an arbitrary finite set $\Sigma \cup \{\varphi\}$ of soft-max constraints in the class $\mathfrak{M}^{\text{soft}}$, if φ is not derivable from Σ by $\mathfrak{R}$, then there is some XML tree T that satisfies all members of Σ but violates φ. That is, T is a counter-example tree for the implication of φ by Σ.

In a first step, we represent φ as a finite node-labeled tree $T_{\Sigma,\varphi}$, which we call the φ-tree.

Definition 4. *(φ-tree).* *Let $\Sigma \cup \{\varphi\}$ be a finite set of soft-max constraints in the class $\mathfrak{M}^{\text{soft}}$. Let $\mathcal{L}_{\Sigma,\varphi}$ denote the set of all labels $\ell \in \mathcal{L}$ that occur in path expressions of members in $\Sigma \cup \{\varphi\}$, and fix a label $\ell_0 \in \boldsymbol{E} - \mathcal{L}_{\Sigma,\varphi}$. First we transform the path expressions occurring in φ into simple path expressions in $PL^{\{\cdot\}}$. For that purpose we replace each single-label wildcard "_" by ℓ_0 and each variable-length wildcard "_*" by a sequence of $l + 1$ labels ℓ_0, where l is the* *maximum number of consecutive single-label wildcards that occurs in any soft constraint in* $\Sigma \cup \{\varphi\}$. *This transformation turns* Q_φ *into* O_φ, Q'_φ *into* O'_φ, *and each* Q_i^φ *into* O_i^φ *for* $i = 1, \ldots, k_\varphi$. *The path expressions after the transformation do not contain any more wildcards (neither single-label nor variable-length ones). Let* p *be an* O_φ-*path from a node* r_φ *to a node* q_φ, *let* p' *be an* O'_φ-*path from a node* r'_φ *to a node* q'_φ *and, for* $i = 1, \ldots, k_\varphi$, *let* p_i *be a* O_i^φ-*path from a node* r_i^φ *to a node* x_i^φ, *such that the paths* $p, p', p_1, \ldots, p_{k_\varphi}$ *are mutually node-disjoint. From the paths* $p, p', p_1, \ldots, p_{k_\varphi}$ *we obtain the* φ-*tree* $T_{\Sigma,\varphi}$ *by identifying the node* r'_φ *with* q_φ, *and by identifying each of the nodes* r_i^φ *with* q'_φ.

The *marking* of the φ-tree $T_{\Sigma,\varphi}$ is a subset $\mathcal{M}$ of the node set of $T_{\Sigma,\varphi}$: if for all $i = 1, \ldots, k_\varphi$ we have $Q_i^\varphi \neq \varepsilon$, then $\mathcal{M}$ consists of the leaves of $T_{\Sigma,\varphi}$, and otherwise $\mathcal{K}$ consists of all descendant nodes of q'_φ in $T_{\Sigma,\varphi}$.

We use φ-trees to calculate the impact of soft-max constraints in Σ on a possible counter-example tree T for the implication of φ by Σ. To distinguish soft-max constraints that have an impact from those that do not, we introduce the notion of *applicability*. Intuitively, when a soft-max constraint is not applicable, then we do not need to satisfy its soft upper bound in a counter-example tree as it does not require all its field paths.

Definition 5. **(Applicability).** *Consider a* φ-*tree* $T_{\Sigma,\varphi}$, *and let* $\mathcal{M}$ *be its marking. A soft-max constraint* σ *is said to be* applicable *to* φ *if there are nodes* $w_\sigma \in [Q_\sigma]$ *and* $w'_\sigma \in w_\sigma[Q'_\sigma]$ *in* $T_{\Sigma,\varphi}$ *such that* $w'_\sigma[P_i^\sigma] \cap \mathcal{M} \neq \emptyset$ *for all* $i = 1, \ldots, k_\sigma$. *We say that* w_σ *and* w'_σ witness *the applicability of* σ *to* φ.

Then, we reverse the edges of the φ-tree and add to the resulting tree downward edges for the applicable members of Σ. Finally, each upward edge receives a label of 1 and each downward edge resulting from $\sigma \in \Sigma$ a label of soft-max$_\sigma$. This final directed graph $G_{\Sigma,\varphi}$ is called the *cardinality network*. A downward edge resulting from σ tells us that under each source node there can be at most soft-max$_\sigma$ target nodes.

Definition 6. *(Cardinality Network).** *We define the* cardinality network $G_{\Sigma,\varphi}$ *of* φ *and* Σ *as the node-labeled directed graph obtained from* $T_{\Sigma,\varphi}$ *as follows: the nodes and node-labels of* $G_{\Sigma,\varphi}$ *are exactly the nodes and node-labels of* $T_{\Sigma,\varphi}$, *respectively. The edges of* $G_{\Sigma,\varphi}$ *consist of the reversed edges from* $T_{\Sigma,\varphi}$. *Furthermore, for each soft-max constraint* $\sigma \in \Sigma$ *that is applicable to* φ *and for each pair of nodes* $w_\sigma \in [Q_\sigma]$ *and* $w'_\sigma \in w_\sigma[Q'_\sigma]$ *that witness the applicability of* σ *to* φ *we add a directed edge* (w_σ, w'_σ) *to* $G_{\Sigma,\varphi}$. *We refer to these additional edges as* witness edges *while the reversed edges from* $T_{\Sigma,\varphi}$ *are referred to as* upward edges *of* $G_{\Sigma,\varphi}$. *This is the case since for every witness* w_σ *and* w'_σ *the node* w'_σ *is a descendant of the node* w_σ *in* $T_{\Sigma,\varphi}$, *and thus the witness edge* (w_σ, w'_σ) *is a downward edge or loop in* $G_{\Sigma,\varphi}$. *We now introduce weights as edge-labels: every upward edge* e *of* $G_{\Sigma,\varphi}$ *has weight* $\omega(e) = 1$, *and every witness edge* (u, v) *of* $G_{\Sigma,\varphi}$ *has weight* $\omega(u, v) = \min\{soft\text{-}max_\sigma \mid (u, v) \text{ witnesses the applicability of some } \sigma \in \Sigma \text{ to } \varphi\}$.

The *weight* of a path t in the cardinality network is defined as the product of the weights of its edges, i.e., $\omega(t) = \prod_{i=1}^{n} \omega(v_{i-1}, v_i)$, or $\omega(t) = 1$ if t has no edges. The *distance* $d(v, w)$ from a node v to a node w is the minimum over the weights of all paths from v to w, or ∞ if no such path exists. When the target node q'_φ of constraint φ can be reached from its context node q_φ along a path of weight at most soft-max$_\varphi$ in the cardinality network $G_{\Sigma,\varphi}$ then there exists no counter-example tree T.

Example 7. Figure 2 shows the cardinality network $G_{\Sigma,\varphi}$ obtained for $\Sigma = \{\sigma_1, \sigma_2\}$ and φ, where σ_1, σ_2, and φ are the soft-max constraints used in Example 6. Note that the distance $d(q_\varphi, q'_\varphi) = 4$ and soft-max$_\varphi = 4$. Thus, there is no counter-example tree T, which is correct since φ is indeed implied by Σ.   □

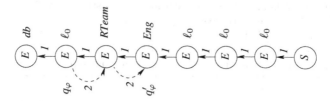

Fig. 2. Cardinality Network

The result below prove the following crucial observation. If φ is not derivable from Σ by $\mathfrak{R}$, then every path from q_φ to q'_φ in $G_{\Sigma,\varphi}$ has distance at least soft-max$_\varphi + 1$.

Lemma 1. *Let $\Sigma \cup \{\varphi\}$ be a finite set of soft-max cardinality constraints in the class $\mathfrak{M}^{\mathrm{soft}}$. If the distance $d(q_\varphi, q'_\varphi) \leq$ soft-max$_\varphi$ in the cardinality network $G_{\Sigma,\varphi}$, then φ is derivable from Σ by $\mathfrak{R}$.*

The strategy to prove this lemma is to encode an inference by $\mathfrak{R}$ by witness edges of the cardinality network. We omit this proof as it is technical and lengthy.

If $\Sigma \cup \{\varphi\}$ is a finite set of soft-max constraints in the class $\mathfrak{M}^{\mathrm{soft}}$ such that φ is not derivable from Σ by $\mathfrak{R}$, then the previous lemma allows us to construct a finite XML tree T which satisfies all soft-max constraints in Σ but does not satisfy φ. This fact proves the following important result.

Theorem 2. *The inference rules in Table 1 are complete for the implication of soft-max constraints in $\mathfrak{M}^{\mathrm{soft}}$.*

5   An Algorithm for Deciding Implication

Our Algorithm 1 for deciding the implication of soft-max constraints is similar to the corresponding algorithms in [11,6] for deciding the implication of the strictly less expressive classes of numerical keys and (strict) cardinality constraints, respectively. However, the construction of the cardinality network $G_{\Sigma,\varphi}$, which is central to the algorithms, requires considerably more effort for (strict) cardinality constraints as studied in [6] as well as for the soft-max cardinality constraints as studied here. This effort results in an increase in the worst-case time complexity of the algorithm compared to numerical keys. Nevertheless, Algorithm 1 enables us to conclude that the implication of soft cardinality constraints in $\mathfrak{M}^{\text{soft}}$ can be decided in low-degree polynomial time in the worst case.

The correctness of Algorithm 1 is due to the fact that for $\Sigma \cup \{\varphi\}$ a finite set of soft-max cardinality constraints in $\mathfrak{M}^{\text{soft}}$, $\Sigma \models \varphi$ holds if and only if $d(q_\varphi, q'_\varphi) \leq \text{soft-max}_\varphi$ in the cardinality network $G_{\Sigma,\varphi}$. This fact can be easily proved from the results in the previous section.

Algorithm 1. Soft-max constraint implication

Input: a finite set $\Sigma \cup \{\varphi\}$ of soft-max cardinality constraints in $\mathfrak{M}^{\text{soft}}$
Output: yes, if $\Sigma \models \varphi$; no, otherwise
1: Construct $G_{\Sigma,\varphi}$ for Σ and φ;
2: Find the shortest path P from q_φ to q'_φ in $G_{\Sigma,\varphi}$;
3: **if** $\omega(P) \leq \text{soft-max}_\varphi$ **then return**(yes); **else return**(no).

Interestingly, the algorithm has the same worst-case complexity as the algorithm for the class of cardinality constraints in [6], which is clearly less expressive since it does not allow variable length wildcards to appear in the field paths, and does not cater for soft bounds. In fact, if $\Sigma \cup \{\varphi\}$ is a finite set of soft-max cardinality constraints in $\mathfrak{M}^{\text{soft}}$, then the implication problem $\Sigma \models \varphi$ can be decided in time $\mathcal{O}(|\varphi| \times l \times (||\Sigma|| + |\varphi| \times l))$, where $|\varphi|$ is the sum of the lengths of all path expressions in φ, $||\Sigma||$ is the sum of all sizes $|\sigma|$ for $\sigma \in \Sigma$, and l is the maximum number of consecutive single-label wildcards that occur in Σ.

It is important to note the blow-up in the size of the counter-example with respect to φ. This is due to the occurrence of consecutive single-label wildcards. If the number l is fixed in advance, then Algorithm 1 establishes a worst-case time complexity that is quadratic in the input. In particular, if the input consists of (numerical) keys, as studied in [10,11], then the worst-case time complexity of Algorithm 1 is that of the algorithm dedicated to (numerical) keys only [10].

6 Experimental Evaluation

We have amply tested our decision algorithm and analysed its performance. We compare the performance against the implementation presented in [7] which is optimised for deciding implication of the strictly less expressive class of XML keys from [10]. The performance results were obtained in a fairly modest Intel Core i7 2.8 GHz machine, with 4 GB of RAM, running a Linux kernel 2.6.32. We compiled our C++ implementation of the algorithms using the standard g++ compiler from the GNU Compiler Collection 4.6.3.

Test Cases. To generate realistic sets of soft constraints to test our algorithm, we generated soft-max constraints applicable to large XML documents from [15]: *321gone.xml* and *yahoo.xml* (auction data), *dblp.xml* (bibliographic information on CS), *nasa.xml* (astronomical data), *SigmodRecord.xml* (articles from SIG-MOD Record), and *mondial-3.0.xml* (world geographic db).

We started by writing, for each document in the collection, a corresponding set of around 10 appropriate (in the context of the document) soft-max cardinality constraints. On adapting the strategy from [7], we generated large sets of soft-max cardinality constraints as follows. Firstly, using the manually defined sets of soft-max constraints as seeds, we computed new implied soft-max constraints by successively applying the inference rules from the axiomatization of soft-max cardinality constraints shown in Table 1. Each constraint generated by this method was added to the original set. We applied the *multiplication, target-to-context, prefix-epsilon, subnodes, subnodes-epsilon, superfield, context-path containment, target-path containment,* and *field-path-containment* rules whenever possible, since those are the rules which can produce implied soft-max cardinality constraints with corresponding non trivial cardinality networks. Secondly, we defined some non-implied (by the soft constraint defined previously) soft-max cardinality constraints. We did that by taking non-implied soft cardinality constraints φ, building their corresponding cardinality networks $G_{\Sigma,\varphi}$, adding several witness edges to them while keeping the weights $\omega(P) > $ soft-max$_\varphi$, for P the shortest path from q_φ to q'_φ, and finally defining new non-implied soft-max cardinality constraints corresponding to those witness edges. This process gave us a robust collection of soft-max cardinality constraints to thoroughly test the performance of the implication algorithm[1].

Tests Results. To have a base line for determining how much the increase in expressivity of the considered class of constraints affects the performance of the decision algorithm, we first measured the performance of the algorithm in [7] which is optimised for deciding implication of the strictly less expressive class of XML keys from [10]. For this we used the same sets of XML keys with simple non empty field paths in $PL^{\{\cdot\}}$ and context and target paths in

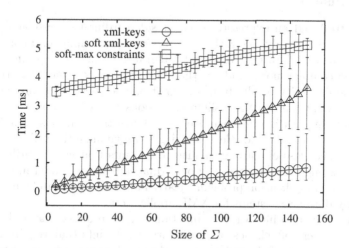

Fig. 3. Performance of the Decision Algorithms for the Implication Problem

---

[1] All sets of constraints generated to test the performance of the decision algorithms as well as the full set of results from those experiments and the binary codes, can be downloaded from http://emir-munoz.github.com/xml-constraints

$PL^{\{\ldots^*\}}$ than in [7]. We then ran our algorithm for soft-max constraints using the same sets of XML keys but treating them as soft-max cardinality constraints with soft-max = 1 (i.e., "soft keys"). This allowed us to quantify the gain produced by the optimization of the algorithm for XML keys presented in [7] and produced a base line to measure the effect of introducing single label wildcards in the soft-max cardinality constraints. Finally we measured the performance of the algorithm for soft-max cardinality constraints over the set of full soft-max constraints obtained with the process described at the beginning of this section.

The results are shown in Figure 3. The x-axis corresponds to the number of (soft) constraints in Σ, and the y-axis corresponds to the *average* running time required to decide whether Σ implies a given (soft) constraint φ. More precisely, let $time(\Sigma, \varphi)$ be the running time required to decide $\Sigma \models \varphi$ and let Φ be a set of (soft) constraints such that $\Sigma \cap \Phi = \emptyset$, the running time shown in Figure 3 corresponds to $\left(\sum_{\varphi_i \in \Phi} time(\Sigma, \varphi_i) \right) / |\Phi|$. In our experiments the sets Φ were composed of 20 fixed (soft) constraints. We tested the scalability of the algorithms by adding, in each iteration, 5 new (soft) constraints to the corresponding Σ sets. Each of the experiments was executed 5 times. The resulting error bars are include in the graph. They are consistent with time variations commonly produced by the scheduling of the operating system and the use of the $time()$ function to measure the experiments [14].

From the experiments it is clear that the implication algorithm is practically efficient and scales well in all three cases. Notably, the extra price to pay for the added expressivity provided by the class of soft-max cardinality constraints is in the order of just 5 milliseconds for a considerable big set of 150 constraints.

7   Conclusion

We have introduced an expressive class of *soft* cardinality constraints that is sufficiently flexible to advance XML data processing in important areas of XML application such as data exchange and integration, where exceptions to strict rules are the common norm and are therefore difficult to handle by strict constraints. The flexibility results from the right balance between expressivity and efficiency of maintenance. While slight extensions result in the intractability of the associated implication problem, we have shown that our class is finitely axiomatizable, robust and decidable in low-degree polynomial time. Thus, our class forms a precious class of soft cardinality constraints that can be utilised effectively by data engineers. Indeed, the performance tests presented in this paper for its associated implication problem, clearly indicate that it can be maintained efficiently by database systems for XML applications.

Future work can go into various directions. XML practice may well warrant the study of other soft classes of cardinality constraints that require different paradigms to specify soft bounds and to select and compare nodes. It would be interesting to investigate soft cardinality constraints with regard to data cleaning, where one of the most important questions is how to model the consistency of the data; for instance exploring conditional XML constraints in connection with the idea of conditional functional dependencies [3]. Finally, it would also be interesting to explore practical applications of the decision algorithm for the implication problem in areas such as cardinality estimation and optimization of XPath queries, XML constraint mining and validation of XML documents.

References

1. Arenas, M., Fan, W., Libkin, L.: What's Hard about XML Schema Constraints? In: Hameurlain, A., Cicchetti, R., Traunmüller, R. (eds.) DEXA 2002. LNCS, vol. 2453, pp. 269–278. Springer, Heidelberg (2002)
2. Arenas, M., Fan, W., Libkin, L.: On the Complexity of Verifying Consistency of XML Specifications. SIAM J. Comput. 38(3), 841–880 (2008)
3. Bohannon, P., Fan, W., Geerts, F., Jia, X., Kementsietsidis, A.: Conditional Functional Dependencies for Data Cleaning. In: ICDE, pp. 746–755. IEEE (2007)
4. Brown, K.: Soft consistencies for weighted csps. In: Proceedings of Soft 2003: 5th International Workshop on Soft. Constraints, Kinsale, Ireland (2003)
5. Buneman, P., Davidson, S.B., Fan, W., Hara, C.S., Tan, W.C.: Keys for XML. Computer Networks 39(5), 473–487 (2002)
6. Ferrarotti, F., Hartmann, S., Link, S.: A Precious Class of Cardinality Constraints for Flexible XML Data Processing. In: Jeusfeld, M., Delcambre, L., Ling, T.-W. (eds.) ER 2011. LNCS, vol. 6998, pp. 175–188. Springer, Heidelberg (2011)
7. Ferrarotti, F., Hartmann, S., Link, S., Marin, M., Muñoz, E.: Performance Analysis of Algorithms to Reason about XML Keys. In: Liddle, S.W., Schewe, K.-D., Tjoa, A.M., Zhou, X. (eds.) DEXA 2012, Part I. LNCS, vol. 7446, pp. 101–115. Springer, Heidelberg (2012)
8. Franceschet, M., Gubiani, D., Montanari, A., Piazza, C.: From Entity Relationship to XML Schema: A Graph-Theoretic Approach. In: Bellahsène, Z., Hunt, E., Rys, M., Unland, R. (eds.) XSym 2009. LNCS, vol. 5679, pp. 165–179. Springer, Heidelberg (2009)
9. Hartmann, S.: Soft Constraints and Heuristic Constraint Correction in Entity-Relationship Modelling. In: Bertossi, L., Katona, G.O.H., Schewe, K.-D., Thalheim, B. (eds.) Semantics in Databases 2001. LNCS, vol. 2582, pp. 82–99. Springer, Heidelberg (2003)
10. Hartmann, S., Link, S.: Efficient reasoning about a robust XML key fragment. ACM Trans. Database Syst. 34(2) (2009)
11. Hartmann, S., Link, S.: Numerical constraints on XML data. Inf. Comput. 208(5), 521–544 (2010)
12. Meyer, J.J.C., Wieringa, R., Dignum, F.: The Role of Deontic Logic in the Specification of Information Systems. In: Logics for Databases and Information Systems, pp. 71–115. Kluwer (1998)
13. Preece, A.D., Chalmers, S., McKenzie, C., Pan, J.Z., Gray, P.M.D.: A semantic web approach to handling soft constraints in virtual organisations. Electronic Commerce Research and Applications 7(3), 264–273 (2008)
14. Stewart, D.B., Khosla, P.K.: Mechanisms for Detecting and Handling Timing Errors. Commun. ACM 40(1), 87–93 (1997)
15. Suciu, D.: XML Data Repository, University of Washington (2002), http://www.cs.washington.edu/research/xmldatasets/www/repository.html
16. Thalheim, B.: Integrity Constraints in (Conceptual) Database Models. In: Kaschek, R., Delcambre, L. (eds.) The Evolution of Conceptual Modeling. LNCS, vol. 6520, pp. 42–67. Springer, Heidelberg (2011)
17. Yu, C., Jagadish, H.V.: XML schema refinement through redundancy detection and normalization. VLDB J. 17(2), 203–223 (2008)

A Framework for Processing Uncertain RFID Data in Supply Chain Management

Dong Xie[1], Quan Z. Sheng[2], Jiangang Ma[2], Yun Cheng[1],Yongrui Qin[2], and Rui Zeng[3]

[1] Department of Computer Science and Technology
Hunan University of Humanities, Science and Technology, China
`dong.xie@hotmail.com, chy6677@163.com`
[2] School of Computer Science, The University of Adelaide, SA 5005, Australia
`{michael.sheng,jiangang.ma,yongrui.qin}@adelaide.edu.au`
[3] The School of Information Science and Technology, Yunnan Normal University, China
`zengruyn@126.com`

Abstract. Radio Frequency Identification (RFID) is widely used to track and trace objects in supply chain management. However, massive uncertain data produced by RFID readers are not suitable for directly use in RFID applications. Following our thorough analysis of key features of RFID objects, this paper proposes a new framework for effectively and efficiently processing uncertain RFID data, and supporting a variety of queries for tracking and tracing RFID objects. In particular, we propose an adaptive cleaning method by adjusting size of smoothing window according to various rates of uncertain data, employing different strategies to process uncertain readings, and distinguishing different types of uncertain data according to their appearing positions. We propose a comprehensive data model, which is suitable for a wide range of application scenarios. In addition, a path coding scheme is proposed to significantly compress massive data by aggregating the path sequences, the positions and the time intervals. Experimental evaluations show that our approach is effective and efficient in terms of the compression and traceability queries.

1 Introduction

Radio-frequency identification (RFID) is a wireless communication technology that is increasingly useful for identifying objects. RFID uses radio-frequency waves to transfer identifying information between tagged objects and readers without line of sight, thus enabling automatic identification [1,2]. In recent years, RFID technology has been widely used in supply chain management [3,4], where products are distributed through multi-enterprises such as manufacturers, dealers, wholesalers, retailers, and customers. With RFID technology, tracking applications of products can automatically analyze recorded identification events to discover the current location of an individual item. They can also retrieve historical information, such as previous locations, time of travel between locations, and time spent in storage. Such technological advances will revolutionize our ability to monitor the world around us, allowing critical decisions and required interventions to be made in a timely fashion.

X. Lin et al. (Eds.): WISE 2013, Part I, LNCS 8180, pp. 396–409, 2013.

While RFID provides promising benefits in many applications, there remains significant challenges to be overcome before these benefits can be realized. Central to these challenges is the *uncertainty* of the data collected by the underlying RFID networks. As RFID devices are intrinsically sensitive to environmental factors such as signal interference and malfunction of reading components, RFID data are typically incomplete, imprecise, and even misleading [5,6]. Obviously, when such data are used directly in monitoring and tracking applications (e.g., product recall), the quality of the applications can be a significant concern. Currently, most of existing approaches use different types of data cleaning strategies to process uncertain RFID data, such as the window-based smoothing filtering, rule-based inference algorithms, probabilistic model-based approaches, and path-oriented queries over RFID data [3,7,4,8]. However, these approaches do not consider features of RFID applications such as temporal and spatial relationships of objects. For example, most of current path-oriented queries cannot be directly processed on uncertain RFID data without pre-processing in complex cases of RFID environments. As a result, it is difficult to employ these methods for storing and processing uncertain data for tracking RFID objects' movements.

In this paper, we propose a framework to process uncertain RFID data. According to key features of RFID applications, we present a comprehensive data model for storing uncertain RFID data, and employ an adaptive strategy to adjust sizes of smoothing windows for capturing suitable rates of different RFID readings. We further develop inference rules for different types of uncertain data, and propose a path coding scheme for compressing massive data by efficiently aggregating the path sequence, the time interval and the position. In addition, we propose an approximation approach for query processing. The main contributions of our work are summarized as follows. Firstly, we classify different types of uncertain RFID data, including ghost, missing, redundant, inconsistent, and incomplete data. Based on such classification, we employ an adaptive strategy to determine suitable rates of different uncertain RFID readings by adjusting the size of smoothing windows. In addition, we develop a set of inference rules for processing different uncertain RFID data. Secondly, we propose a path coding scheme to significantly compress massive data by aggregating the path sequence, the position and the time interval. The scheme also expresses object movements with cyclic or long paths, and comprehensively integrates objects with group and individual movements. Thirdl, we conduct extensive experimental evaluations for our model and the proposed algorithms. Experimental evaluations show that our approach is effective and efficient in terms of compression and object traceability.

The rest of the paper is organized as follows. Section 2 presents the related backgrounds. Section 3 introduces an adaptive RFID data cleaning model. Section 4 describes an inference algorithm for RFID data streams, and the experimental evaluation is reported in Section 5. Finally, Section 6 is dedicated to related work and Section 7 concludes the paper and discusses some future research directions.

2   Background

In this section, we first describe a scenario and data characteristics in a supply chain management. We then propose a system framework and show how to use our new data models and architecture to efficiently manage, query and analyze uncertain data collected from RFID data sources.

2.1   RFID-based Supply Chains

Figure 1 shows the whole process of an RFID-based supply chain. Each product item is tagged with an electronic product code (EPC) in the production line and related product specifications are written into an RFID tag. Products with tags are then packed into cases, Which are packed into pallets at the supplier warehouses. After that, pallets are loaded onto trucks, which then depart to dealers. At zones of retail stores, all pallets are unloaded from the trucks, and all cases are unpacked from the pallets. Eventually, product items are purchased by consumers.

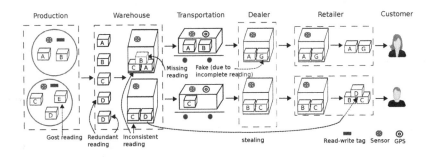

Fig. 1. An RFID-enabled supply chain

2.2   Uncertain RFID Data

From the scenario introduced in the above, we can have several observations. First, tracking and tracing objects in a supply chain generates a huge amount of data containing the movement information associated with the tagged objects. Second, due to the impact of context factors, such information is often imprecise and unreliable. For example, it has been proved that the communication between a tagged object and an RFID reader is sensitive to environment settings such as the distance between the tag and the reader, as well as the orientation between the tag and the reader antenna. We can further summarize the basic characteristics of RFID readings in the following.

Temporality and Spatiality. RFID data is generated dynamically and is associated with timestamp when readings are made. In addition, tagged objects are typically mobile and go through different locations during their life cycle. Temporal and spatial information is important for tracking and monitoring RFID objects. For example, using timestamp, it is possible to track how long it takes an aircraft part to move from the warehouse to the maintenance venue.

Implicit Inferences. RFID data always carries implicit information such as changes of state and containment relationships among objects. To use RFID data, applications require other information, such as environmental situations to make proper inferences.

Limited Active Lifespan. RFID data normally has a limited active life period, during which a given system actively updates, tracks, and monitors the data. For example, in a supply-chain management system this period starts when the manufacturer delivers the products and ends when customers purchase the products.

We represent a raw RFID data record as a tuple for supporting data cleaning. Our model incorporates RFID data's application information, including tagID, time, and location. These original data tuples form a spatial-temporal data streams **DS**.

Definition 1. *(RFID reading) An observed raw RFID reading is defined as a tuple $ds_i = (d^i, t^i, l^i)$, where d^i, t^i, l^i denotes the 'tagID', 'timestamp', 'location'.*

Definition 2. *(RFID data streams) An RFID data stream is a spatial-temporal sequence of tuples $DS =< ds_1, ds_2, ..., ds_k >$, where each tuple $ds_i \in$ **DS** is represented as $ds_i = (d^i, t^i, l^i)$, where d^i, t^i, l^i denotes the 'tagID', 'timestamp', 'location'.*

Definition 3. *(Problem statement) Given a stream of raw RFID readings $DS =< ds_1, ds_2, ..., ds_k >$, which could be noisy, we aim to derive a clean data steam to support path-oriented queries.*

2.3   Overview of System Framework for Processing Uncertain RFID Data

Our data processing framework is shown in Figure 2, where RFID readers read EPC tags from multi-streams. A smoothing window is first specified by analysing the uncertainties of existing RFID readings. Next, missing readings are inferred by related algorithms proposed in this paper. What is more, redundant readings need to be temporarily stored in databases because they may be helpful to infer missing readings. Apart from incomplete data (generated by fake or stolen object) and missing data, uncertain data also includes ghost data that needs to be cleaned. To distinguish ghost and inconsistent readings, we need to get and store more readings in RFID database because these data record EPC, position and timestamp of RFID objects. Repairing processing modifies inconsistent data by aggregating time interval, position and path information. Finally, query processing focuses on tracking and tracing objects according to data lineage.

3   Adaptive RFID Data Cleaning Model

In this section, we first introduce an adaptive smoothing window to process uncertain data, and then define the basic cleaning rules to detect and correct error readings.

3.1   Smoothing Window

Data smoothing is a data cleaning technology to correct error readings in data streams due to interference and noise from the network environment. More precisely, smoothing applies cleaning operations to data streams through a time window. If there is some data lost within the widow, new data will be interpolated to compensate the missing readings. On the other hand, if there are some abnormal readings appearing in the window, they

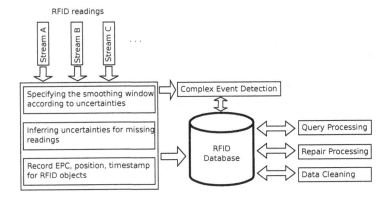

Fig. 2. A system view of our proposed framework to process RFID data

will be removed by cleaning operations like spatial aggregation. To illustrate the idea of the adjustable strategy on the data streams, we show a simplified diagram in Figure 3. Stream A (small window) should have a large window whenever missing readings frequently happen (e.g., from point p_1 to point p_2). However, stream C (large window) should have a smaller window whenever ghost, redundant and inconsistent readings frequently happen (e.g., from point p_2 to point p_3). Stream B (medium window) might be "neutral" such that it tends to balance the two types of readings (e.g., point p_2).

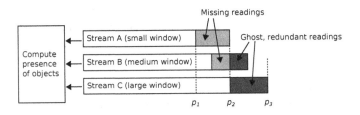

Fig. 3. Three different smoothing windows for reading RFID objects

According to the rate of missing data from a certain data stream with a small window (e.g., stream A), the small window may be increased to a suitable size. Similarly, the window of stream C may be decreased to a suitable size according to the rate of ghost, redundant and inconsistent data from a certain data stream with large window. The strategy is to use different sizes of windows, which can adjust the rate of uncertain data over multi-streams obtained by different antennae configurations of RFID readers. Here we employ a formula to adjust the size of smoothing windows in the following: $sw = adj * (m * (acra - ra_1) - n * (acra - ra_2))$, where sw is the size of smoothing windows for reading EPC tags at the current position; $acra$ is the acceptable rate of uncertain data; ra_1 is the rate of the latest referable ghost and redundant data at the current position; ra_2 is the rate of the latest inconsistent and missing data at the current position; adj is the adjustable interval time; m and n are parameters to adjust ra_1 and ra_2 for the size of smoothing windows respectively.

3.2   Inference Rules for Processing Uncertain Data

Our cleaning rule consists of three parts: *pattern, condition* and *action*. First, a pattern represents an ordered data list in the input data stream. For example, two consecutive readings A and B in a data stream are represented as (A, B), which means A and B are adjacent within the data stream. Second, a condition specifies the existential semantics of the patterns. For example, if two duplicate readings A and B in a data stream are coming, it is true that A's location is the same as that of B. Finally, action part of the cleaning rule indicates the designed operations to be performed when the condition is satisfied. For example, if data stream B is duplicate with respect to A, then a Delete operation will be applied to remove data stream B.

Definition 4. *(Cleaning rule) A cleaning rule is defined as Pattern-Condition-Action (PCA) rule:*
CLEANING RULE<*rule-id*>
PATTERN <*pattern* >**CONDITION** <*condition*>
ACTION<*action* >
If some patterns appearing in data streams and conditions are satisfied, then an action will be performed to add or drop data.

With our rules-based cleaning approaches, observed raw RFID data could be converted into application logic data. We now apply the cleaning rules to process erroneous data in data streams. The erroneous data are characterized into four types of data: (1) missing data—no tags are detected; (2) inconsistent data—multiple possible readings are recorded; (3) ghost data—impossible readings are recorded; and (4) redundant data—a reading is recorded for several times. We assume that data is represented by a standard pattern (o, p, t_1, t_2), where o denotes an object's EPC, p denotes position, t_1 and t_2 denote time interval $[time\ 1, time\ 2]$ when object o is read at position p. Due to the constraint of space, we here discuss how to deal with missing data, inconsistent data and redundant data.

- **Missing data.** Figure 4(a) shows a scenario of inferring missing data. Suppose an object o_1 moves through three consecutive positions (where readers are installed) $p_1 \rightarrow p_2 \rightarrow p_3$. However, the reader at p_2 does not detect object o_1 due to noise. To infer the position of o_1, we can make use of spatial relationship among readers or the containment information of objects. For example, if object o_1 moves from p_1 to p_3, it must pass p_2. Similarly, if items are contained in a box and the box is contained in a pallet, these items would move with the pallet together. Thus it is possible to infer missing data in certain positions if other readings in the same containment are recorded.
- **Inconsistent data.** In Figure 4(b), o_1 goes from p_1 to p_3 passing through p_2 but it is also read by a near reader placed in another position p_2'. Similarly, we may infer missing data, which refers the relationships of the readers or the position information of related objects. Another rule will store the parent position p_a of p_2 and p_2' if o_1 cannot refer related information. As a result, o_1 should be stored as (o_1, p_a, t_2, t_3).

– **Redundant data.** In Figure 4(c), an object may be frequently read within several smoothing windows, but the object does not change its position during the period. For example, o_1 is read twice within different smoothing windows from the timestamp t_2 to the timestamp t_2'. In general, we may simply clean the redundant data. However, o_1 may be missed at the next position, so inferring the position of o_1 needs the redundant readings within the nearest smoothing window.

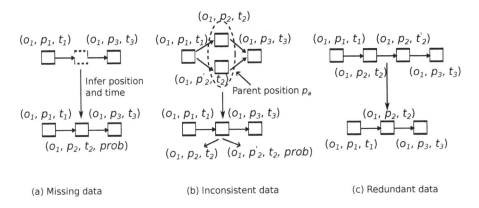

(a) Missing data   (b) Inconsistent data   (c) Redundant data

Fig. 4. Inferring missing and redundant data

4   Inference Algorithm

In this section, we first present a temporal-spatial-based data modeling of RFID data, and then introduce an inference algorithm for RFID data streams.

4.1   Data Model

Figure 5 presents a model of uncertain RFID data. The proposed data model extends the traditional ER model, and introduces the following new entities for RFID data: READER, OBJECT, BUSINESS, POSITION, SEQ and PATH. From this entity model, we can induce dynamic relationships between entities. For example, there are three dynamic relationships that can be generated from the model: READING can be inferred from entities READER and OBJECT; relation STAY can be acquired from entities OBJECT, PATH and SEQ; relation REA_BUS can be generated from entities READER, BUSINESS and POSITION.

One of the main advantages in our model is that the model abstraction integrates the practical RFID application logic into the data model and can effectively support query operations such as RFID object tracking and monitoring. For example, RFID readers read EPC-tagged objects and store raw data as $(EPC, reader\_id, timestamp)$, where EPC is a unique ID of a tag and $reader\_id$ is a unique RFID reader ID at the position of EPC and timestamp is the time instant of reading EPC. With the model, we obtain the position of EPC by connecting REA_BUS, which expresses positions and businesses (e.g., unloading situation) of readers. We can further transfer raw data into the STAY table for expressing an RFID tag moving through a position within a time interval.

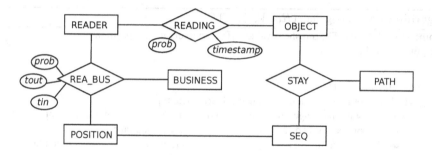

Fig. 5. Data Model

4.2   RFID Data Path Coding Scheme

A path coding scheme represents the movement of a tag. With current data models, the movement information including RFID readings at each position is stored in the database in terms of *stay records*. In particular, a stay record takes the form of $(EPC, position, tin, tout)$, where tin and $tout$ are the timestamps of the first detection and the final detection of the tag. The existing coding techniques called path (sequence) store and compress data into $SEQ(sqid, postion_id, tin, tout)$ by aggregating the path sequence, the position and the time interval. To effectively support path queries and reduce the data volume, we extend this method by employing $(EPC, sqid)$ instead of $(EPC, position, tin, tout)$ in STAY. As our new aggregation approach is able to resolve two main problems, namely the cyclic path and long path, it significantly reduces redundant data and the volume of RFID data storage.

Aggregating Time Intervals. Some users may not concern about concrete timestamps of objects. For example, users may only want to know that the stay time of object a is from t_1' to t_2'. Moreover, users may increase the time interval from t_1' to t_3'. As a result, though a is read at the timestamp t_1 and object b is read at the timestamp t_2, their stay times may have the same the interval $[t_1', t_2']$.

Aggregating Positions. Similarly, some users may not concern about concrete positions of objects. For example, if a customer buys a computer, it is enough for the customer to know the logistic position storing the computer rather than the concrete warehouse of the logistic position. Therefore, an object at a position can be aggregated into its parent's position as a single reading, in spite of multi-readings at different positions. For example, although in a warehouse w_1, object a is read at position p_1 and object b is read at position p_2, their positions can be represented as warehouse w_1.

Aggregating Path Sequences. As a group of objects has the same path, we can further compress the path expression as the main path by aggregating the same child path sequences. For example, if object b is moving from the path sequence sq_1 to sq_3, we denote the path sequence sq_{13} to express from sq_1 to sq_3. As a result, the path sequence of b can be denoted as (b, sq_{13}) by aggregating sq_1 and sq_2. In general, the aggregation is used to compress paths of group of objects. If a large number of objects do not follow the same path sequences, it is unnecessary to aggregate path sequences.

Algorithm 1. Cleaning data (R, TV, AT, AP)

Input: R: READING; TV: threshold value; AT: time interval; AP: position
Output: Cleaned data

1. **for** $\forall$ record $r^i \in \mathbf{R}$ **do**
2.   **if** $r^i \in$ root node without references **then** delete tuple r^i
3.   **else if** r^i with unchanged position **then** delete tuple r^i with later timestamp
4.   **if** $r^i \notin$ root node **then** remark tuple r^i as a non-cloned fake
5.   **else if** r^i with complete path at root node **then** remark tuple r^i as a cloned fake
6.   **if** $r^i \notin$ a node **then** remark tuple r^i as a missing object
7.   **for** $i = 1$ to number of r^i_paths **do**
8. $W \leftarrow$ weight(the ith path from parent to child of missing node)
9.   **if** $W > TV$ **then** insert r^i missing node to STAY
10.   **else** aggregate higher AT and AP
11.   **end for**
12.   **if** $r^i \in$ two nodes with a same timestamp **then** remark tuple r^i as an inconsistent object

13. **end for**

4.3 The Algorithm

We propose an algorithm for dealing with uncertain data according to our analysis from Figure 4. The algorithm 1 first visits records in table READINGS, and cleans ghost and redundant data according to appearing positions of records' EPC. Then, the algorithm 1 further processes missing and inconsistent data with the given threshold value according to the weights of paths from parent nodes to child nodes in the directed graph. Finally, EPC's node is inserted into table STAY.

5 Experimental Evaluation

In this section, we present our experiments to empirically validate the effectiveness and efficiency of the proposed framework.

5.1 Experimental Setting

All the experiments were conducted on a computer with Core i2 2.00GHz, 2GB RAM, running Windows 7. Since there is no well-known RFID data set, we generate synthetic data sets and formulate 7 queries (1 tracking query, 4 tracing queries, and 2 aggregate queries). The related parameters are listed in the Table 1.

We first generate stay records instead of raw RFID data to reflect the environment for a real-life food distribution. The objects are generated at root nodes such as import markets and agricultural input suppliers. Then, they move to the next position by a group of objects or single object in RFID applications. We consider the grouping factor to generate stay records using a directed graph for object movements, and analyze how

Table 1. Experimental parameters

Parameter	Statement
G	The size of the groups at each level of the directed graph
M	The rate of missing data in the whole data
RG	The rate of redundant and ghost data in the whole data
L	The path length
P	The level of the position aggregate
AT	The level of the time interval aggregate
AP	The level of the position aggregate
S	The data size

the query performance is affected by the grouping factor. Each node in the directed graph represents a set of objects in a position, and an edge represents the movement of an object between positions. In general, objects at locations near the root of the directed graph move in larger groups, while objects near the leaves move in smaller groups. The size of the groups at each level of the directed graph is denoted as G, where G is the number of objects that move together in the directed graph, and SEQ corresponds to the object movements indicated by the edges in the directed graph. We randomly generated data according to a set of directed graphs with a given level of G.

Since the majority of uncertain data is missing data, we consider the effect of the rate of missing data on the whole RFID data collected. We also consider the influence of the rate of redundant and ghost data. In addition, users need to specify aggregate levels of the position and the time interval. Moreover, we test the performance over different sizes of data.

5.2 Data Cleaning and Compression

We evaluate the data cleaning and compression under M = (0.5, 0.25, 0.1, 0.5 and 0.025), RG = (0, 0.333, 0.5, 0.666 and 0.75), and S = (3.8k, 31.7k, 135k, 427.5k, and 975k). The tuples are generated by 1k, 2.5k, 5k, 7.5k, 10k objects respectively.

Since the rate of missing data is opposite with the rates of redundant and ghost data, we set several sets of rates denoted as (M = 0.5, RG = 0), (M = 0.25, RG = 0.333), (M = 0.1, RG = 0.5), (M = 0.05, RG = 0.666), and (M = 0.025, RG = 0.75). The rate of missing data declines from 0.5 to 0.025, and the rates of redundant and ghost data raise from 0 to 0.75. We generate data for 10k objects. Figure 6 shows that the sizes of non-cleaned data linearly increase. The linear results show that the high RG might significantly increase the size of READING, and this will increase system overloads. However, the smaller RG might indicate that missing data are too much and it is difficult to infer what is missing according to related data. Hence, RG should be adjusted to a smaller value for better query performance, and M should be adjusted to a suitable value.

The above experiments do not consider aggregation factors. Since our algorithm automatically aggregates the path sequences, the levels of the positions and the time interval need to be specified by users. Here we need to consider the following scenario in a typical supply chain: moving an object within a short time interval from a position

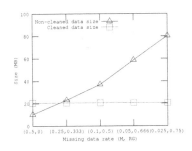

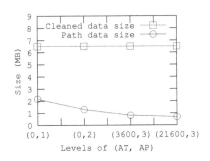

Fig. 6. Missing, redundant, ghost data

Fig. 7. Aggregate levels of positions and time intervals

to another position is likely impossible (i.e., objects normally travel between different organizations). In this case, a level of a position should correspond to a suitable level of time intervals. Figure 7 shows that path data sizes decrease with increasing AT and AP under S = 10k, P = 3-22, M = 0.1, and RG = 0.5. Decreasing path data sizes also efficiently decreases the overloads for path queries.

5.3 Query Processing

Since non-cleaned data are unsuitable for comparisons of query processing performance, we employ our algorithm to clean data, then aggregate data to execute the queries. We formulate 7 queries to test various features which are shown in Table 2. Q1 tests the performance of tracking query; Q2-Q3 and Q6-Q7 are tracing queries; and Q4-Q5 are aggregate queries. Specially, Q6 queries incomplete data, and Q7 queries objects with a cyclic and long path.

As EPC and path sequences of objects are stored in the table STAY (STAY only stores object path sequences), Q1 and Q2 do not need to join multi-tables to obtain path information. Though Q3 needs to join two tables to obtain path information, SEQ (SEQ only stores the *check-in* and *check-out* time intervals of positions in spite of object EPCs) may be significantly compressed. Queries Q4 and Q5 need to join two tables to aggregate the object total numbers at certain positions, but similar data in two tables can be compressed via aggregation. Therefore, the compression rate may efficiently improve overloads of aggregate queries. Since fakes are related to three tables STAY, SEQ and PATH (PATH only stores the path graph for all objects with respect to all positions), Q6 needs to join the three tables to distinguish non-cloned fakes and cloned fakes.

Since readings of EPC-tagged objects are within continuous time intervals without the same timestamps in RFID applications, our method proposes rough time intervals and positions to ignore concrete timestamps and positions by aggregating time intervals and positions. This can significantly improve efficiencies of storage and queries. Figure 8 shows the execute time under S = 10k, G = 200, and P = 3-22. We set four sets of different aggregate levels (AT = 0 and AP = 1; AT = 60 and AP = 2; AT = 3600 and AP = 3; AT = 21600 and AP = 4) to evaluate Q1-Q7.

Table 2. Test queries

Query	Description
Q_1	Tracking an object
Q_2	Tracing an object without conditions
Q_3	Tracing an object with conditions
Q_4	Counting objects at a position without a time interval
Q_5	Counting objects at a position within a time interval
Q_6	Tracing fakes
Q_7	Tracing objects with a cyclic and long path

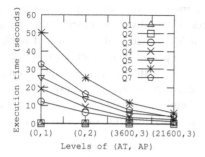

Fig. 8. Positions and time intervals

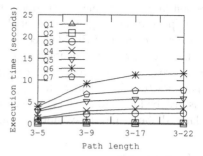

Fig. 9. Different path lengths

Figure 9 shows the execute time of Q1-Q7 under G = 200, S = 10K, AT = 3600, and AP = 3, and different path lengths. The execute time of Q3-7 increases when the path length increases. This is because shorter paths include a large number of same path information, which is compressed by aggregating. Longer paths include a large number of different path information, which cannot be compressed by aggregation. Since the objects with 10-22 length are rare (see discussions in Section 5.2), the trends for increasing the execute time shows a slow growth. Similarly, different path lengths have little influence for Q1 and Q2.

6   Related Work

In this section, we briefly discuss some of the research work related to managing uncertain data in RFID supply chain management.

Previous works process raw RFID data by resolving inconsistencies. Inconsistencies have two major aspects: database repairing and consistent query answer (CQA) [9]. The existing repairing models can be classified into (a) the minimal distance-based repair of the inconsistent database generates the maximum consistent subset of the inconsistent database [10]; (b) the priority-based repair prefers more reliable data sources according to different reliabilities of conflicting data sources [11]; (c) the attribute-based repair minimizes attribute modifications [11]. Probabilistic consistent query answering (PCQA) [12] utilizes all-possible-repairs semantics and is extended to process uncertain data over Internet of Things (IoT) [13]. However, these works cannot be directly used in complex cases of an RFID supply chain environment.

Inference rules are widely used to process uncertain data. For example, Spire system [14] uses a time-varying graph model to estimate the most likely location and containment of an object from raw RFID data streams. In addition, the work in [15] proposes a scalable and distributed stream processing system to estimate an object's location and a probabilistic data stream system called Claro [8] uses continuous random variables to process uncertain data streams. However, these works do not consider all types of data such as inconsistent and incomplete data.

Oriented-path queries require performing multi-self-joining of the table involving many related positions. Gonzalez et al. [4,16] focus on groups of objects and uses compression to preserve object transition relationships for reducing the join cost of processing path selection queries. Lee et al., [7] focus on individual objects and proposes a movement path of an EPC-tagged object, which is coded as the position of readers and the order of positions denoted as a series of unique prime number pairs. However, their approaches do not code longer paths and cyclic paths. Ng [2] uses finite continued fraction (rational numbers) to represent respective positions and their orders, whose paths may be long and cyclic. However, rational numbers notably increase the volume of data. All the methods have not adequately considered uncertainties of RFID data.

7   Conclusion

Effectively processing uncertain RFID data still remains a challenge. In this paper, we have studied the main problems related to cleansing uncertain data collected in a supply chain network and particularly focused on processing ambiguous and imprecise RFID data. We have designed and implemented a novel framework, which improves the existing techniques for modeling, cleansing and querying RFID data. Based on the proposed model, data can be efficiently stored and compressed by aggregating positions and time intervals. Experimental evaluations show that our model and algorithms are effective and efficient in terms of the storage, tracking and tracing. Our future work aims to construct the main path for moving objects tracking.

Acknowledgment. This work is supported by the Research Foundation of Education Committee of Hunan Province, China (11B104 , 12C 0 999), the Hunan Provincial Natural Science Foundation of China (12JJ3057, 12JJ2040), the Hunan Science and Technology Plan of China (2011FJ3033, 2010TT2055, 2012GK3073), the Construct Program of the Key Discipline "Computer Application Technology" in Hunan Province, China.

References

1. Wu, Y., Sheng, Q.Z., Ranasinghe, D.C.: Facilitating efficient object tracking in large-scale traceability networks. The Computer Journal 54(12), 2053–2071 (2011)
2. Ng, W.: Developing rfid database models for analysing moving tags in supply chain management. In: Jeusfeld, M., Delcambre, L., Ling, T.-W. (eds.) ER 2011. LNCS, vol. 6998, pp. 204–218. Springer, Heidelberg (2011)

3. Alexander, I., Thomas, A., Florian, M.: Increasing supply-chain visibility with rule-based rfid data analysis. IEEE Internet Computing 13(1), 31–38 (2009)
4. Gonzalez, H., Han, J., Li, X., Klabjan, D.: Warehousing and analyzing massive rfid data sets. In: Proceedings of the 22nd International Conference on Data Engineering (ICDE 2006), pp. 83–83. IEEE (2006)
5. Sheng, Q.Z., Li, X., Zeadally, S.: Enabling next-generation rfid applications: solutions and challenges. Computer 41(9), 21–28 (2008)
6. Ma, C., Zhang, R., Lin, X., Chen, G.: Duowave: Mitigating the curse of dimensionality for uncertain data. Data and Knowledge Engineering 76-78, 16–38 (2012)
7. Lee, C.H., Chung, C.W.: Rfid data processing in supply chain management using a path encoding scheme. IEEE Transactions on Knowledge and Data Engineering 23(5), 742–758 (2011)
8. Tran, T.T., Peng, L., Diao, Y., McGregor, A., Liu, A.: Claro: modeling and processing uncertain data streams. The International Journal on Very Large Data Bases (VLDB Journal) 21(5), 651–676 (2012)
9. Arenas, M., Bertossi, L., Chomicki, J., He, X., Raghavan, V., Spinrad, J.: Scalar aggregation in inconsistent databases. Theoretical Computer Science 296(3), 405–434 (2003)
10. Arenas, M., Bertossi, L., Chomicki, J.: Consistent query answers in inconsistent databases. In: Proceedings of the Eighteenth ACM SIGMOD-SIGACT-SIGART Symposium on Principles of Database Systems, pp. 68–79. ACM (1999)
11. Staworko, S., Chomicki, J., Marcinkowski, J.: Preference-driven querying of inconsistent relational databases. In: Grust, T., et al. (eds.) EDBT 2006. LNCS, vol. 4254, pp. 318–335. Springer, Heidelberg (2006)
12. Lian, X., Chen, L., Song, S.: Consistent query answers in inconsistent probabilistic databases. In: Proceedings of the 2010 International Conference on Management of Data, pp. 303–314. ACM (2010)
13. Chen, L., Tseng, M., Lian, X.: Development of foundation models for internet of things. Frontiers of Computer Science in China 4(3), 376–385 (2010)
14. Nie, Y., Cocci, R., Cao, Z., Diao, Y., Shenoy, P.: Spire: Efficient data inference and compression over rfid streams. IEEE Transactions on Knowledge and Data Engineering 24(1), 141–155 (2012)
15. Cao, Z., Sutton, C., Diao, Y., Shenoy, P.: Distributed inference and query processing for rfid tracking and monitoring. Proceedings of the VLDB Endowment 4(5), 326–337 (2011)
16. Gonzalez, H., Han, J., Cheng, H., Li, X., Klabjan, D., Wu, T.: Modeling massive rfid data sets: a gateway-based movement graph approach. IEEE Transactions on Knowledge and Data Engineering 22(1), 90–104 (2010)

An Ontology-Based Approach to Context-Aware Access Control for Software Services

A.S.M. Kayes, Jun Han, and Alan Colman

Faculty of Information and Communication Technologies
Swinburne University of Technology, VIC 3122, Australia
{akayes,jhan,acolman}@swin.edu.au

Abstract. In modern communication environments, the ability to provide access control to services in a context-aware manner is crucial. By leveraging the dynamically changing context information, we can achieve context-specific control over access to services, better satisfying the security and privacy requirements of the stakeholders. In this paper, we introduce a new *Context-Aware Access Control (CAAC) Framework* that adopts an ontological approach in modelling dynamic context information and the corresponding CAAC policies. It includes a *context model* specific to access control, capturing the relevant low-level context information and inferring the high-level implicit context information. Using the context model, the *policy model* of the framework provides support for specifying and enforcing CAAC policies. We have developed a prototype and presented a healthcare case study to realise the framework.

Keywords: Context-Awareness, Context-Aware Access Control, Context Model, High-Level Context, Access Control Policy.

1 Introduction

The rapid advancement of computing technologies has led to the computing paradigm shift from fixed desktop to context-aware environments, as described by Weiser [12]. Such a shift brings with it opportunities and challenges. On the one hand, users demand access to services in an anywhere, anytime fashion. On the other hand, such access has to be carefully controlled due to the additional challenges coming with the dynamically changing environments, so as not to compromise the relevant security and privacy requirements. Such new challenges require new Context-Aware Access Control (CAAC) approaches. In general, the information about the changing environment, called context information [7], needs to be taken into account when making access control decisions.

A number of context-sensitive access control approaches have highlighted the importance and use of context information in the access control processes (e.g., [3],[5],[6],[8],[11]). However, the existing approaches to access control have only considered specific types of context information. There is still a lack of a general context model specific to access control, to capture the basic context information, and infer richer information from other information in a comprehensive manner. In addition, an appropriate access control policy model that incorporates contexts into access control decision making is required.

X. Lin et al. (Eds.): WISE 2013, Part I, LNCS 8180, pp. 410–420, 2013.

Towards this goal, in this paper we introduce a new Context-Aware Access Control (CAAC) approach to provide context-aware access control support for software services. It makes the following key contributions.

First, we propose a *context model*, in order to represent and capture different types of context information in a systematic way; and to reason about high-level implicit context information that is not directly available but can be derived from other information. The context model uses the ontology language OWL, extended with SWRL for inferring implicit context with user-defined rules.

Second, we introduce a *policy model* for defining and enforcing access control policies that take into account relevant contexts from the context model. The policy model also uses OWL and SWRL. The main novel point in our policy model is that it provides context-aware access control decisions. In particular, it has reasoning capability which grants the right access to the appropriate parts of a resource (fine-grained access control to resource) by the appropriate users, considering the different granularity levels of the resource. Other than the above two major contributions, we have developed a *prototype framework* implementing the approach and carried out a *case study* in the healthcare domain.

The rest of this paper is organized as follows. Section 2 presents an application scenario to motivate our work. Section 3 discusses the design of our context model for access control. Section 4 presents a policy model for context-aware access control. The applicability of our approach is investigated in Section 5. Section 6 discusses related work. Finally, Section 7 concludes the paper.

2   Motivation and General Requirement

Motivating Scenario. Let us consider an application scenario from the healthcare domain, requiring context-aware access control.

Scene #1: The scenario begins with patient Bob who is in the emergency room due to a heart attack. While not being Bob's usual treating physician, Jane, a medical practitioner of the hospital, is required to treat Bob and needs to access Bob's emergency medical records from the emergency room.

Scene #2: After getting emergency treatment, Bob is shifted from the emergency room to the general ward of the hospital and has been assigned a registered nurse Mary, who has regular follow-up visits to monitor his health condition. Mary needs to access several types of Bob's records (daily medical records and past medical history) from the general ward.

Basic Analysis. The context information describing the context entities relevant to access control in the above scenario includes: the *identity/role* of the Users, the *location* of the Users, the *relationship* between the User and Patient, the *health status* of the Patient, etc. To provide fine-grained access control and grant the right access to the appropriate parts of a resource by the appropriate users, the resource (patient medical records) needs to be considered in a hierarchical manner. Furthermore, access to the resource and its components at different levels of granularity can be managed in a service oriented manner, i.e., service wrapper operations can be defined them and invoked by the users as

Table 1. An Informal Access Control Policy for Our Example Application

No	Policy
1	A medical practitioner, who is a treating physician of a patient, is allowed to read/write the patient's emergency medical records in the hospital. However, in an emergency situation (like the above), all medical practitioners should be able to access the emergency medical records from the emergency room of the hospital.

permission allows. For example, the write operation on the emergency medical records can be defined as *writeEMR()*. In this way, fine-grained access control to resources can be easily realized by managing the access to the service operations.

Concerning the above two scenes, one of the relevant access control policy is shown in Table 1. The policy is based on a set of constraints on the user role and service (a *service* can be seen as a *pair <a, r>* with r being a requested resource and a being the action/operation on the resource), and the policy refers to need to be evaluated in conjunction with the relevant contexts.

General Requirement. As different types of contexts are integrated into the access control processes, some important issues arise.

(R1) *Representation of context entities and information: What access control-specific context entities and information should be considered as part of building a context model specific to CAAC? Furthermore, how to model and capture the context entities and information in an effective way?*

(R2) *Inferring high-level context information: How to express user-defined reasoning rules, in order to capture knowledge which is only implicitly present, thereby extending the context model?*

(R3) *Enforcement of access control policies: How to specify and evaluate the access control policies based on the relevant contexts to realize a flexible and dynamic access control scheme?*

3   Context Model - COAC Context Ontology

Many researchers have attempted to define the concept of context. According to Dey [4], the context entities are *Person, Place* and *Object* and he defines context as *any information that can be used to characterize the situation of an entity.* We specialize Dey's definition of context to cover access control applications.

Definition: Context Information and Context-Awareness. Context Information used in an access control decision is defined as any relevant information about the state of a relevant entity or the state of a relevant relationship between different relevant entities at a particular time. *Context-awareness* relates to the use of this context information for access control decision making.

Experience from existing research (e.g., [9]) shows that ontologies are very suitable for modelling context for pervasive computing applications. To meet requirements (1) and (2), we in this section propose an ontology-based context model, named Context Ontology for Access Control (COAC).

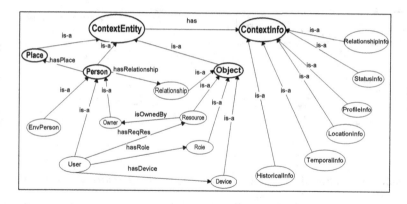

Fig. 1. COAC Upper Context Ontology - *Context Entity* and *Context Information*

Design Considerations. Our COAC ontology, representing *context entity* and *information* as ontology elements, is capable of: *representing general (upper) concepts, representing domain-specific concepts,* and *supporting reasoning according to user-defined rules (to obtain high-level context information).*

To model context information, we adopt the OWL language, which has been the most practical choice for most ontological applications because of its considered trade-off between computational complexity and expressiveness [9]. Ontology-based reasoning using Description Logic (DL) may not always be sufficient to infer the high-level implicit contexts. The expressivity of OWL can be extended by adding SWRL rules to an ontology. We express the user-defined reasoning rules using the SWRL rule language which provide the ability to infer additional information in our COAC context ontology.

Upper Context Ontology. Our *COAC Upper Ontology* is the main ontology having *classes* modelling *context entities* and *context information* and *relationships* between classes (*object* and *data type* properties).

The main part of the upper ontology is shown in Figure 1. We classify the access control-specific *context entities* into two groups: *core* and *environmental.* *User, Resource,* and resource *Owner* are the *core entities* as they are core concepts of access control. To offer the advantages of the RBAC role, which regulates access to services based on user roles rather than individual users, our model also has *Role* as a core entity. A *hasRole* object property is used to relate *User* and *Role* classes for representing the fact that a user has a role. *User* is linked to *Resource* by the *hasReqRes* object property for capturing a user's access interest in a resource. The relationship between a *Resource* and its *Owner* is captured by the property *isOwnedBy*. The *Relationship* between *Persons* is another class of core entity, has its own characterizing context information. An object property *hasRelationship* is used to link the *Persons* and the *Relationship*. Following Dey's general context entities [4], we consider *Role* and *Relationship* as special *Objects*.

The *environmental entities* are the other entities that are relevant to the access request. They include *EnvPerson* (a person who is neither a *User* nor an *Owner*

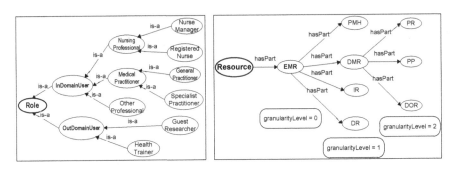

Fig. 2. Domain-Specific (a) *Role Ontology* (left), (b) *Resource Ontology* (right)

but relevant), *Place*, and *Device*. The *Device* is the communication device that the *User* uses to issue the access request, and consequently the *User* is linked to *Device* through *hasDevice*. Furthermore, a *Person* is at a particular place and therefore is connected to *Place* with an object property *hasPlace*.

Focusing on the *context information* types that are relevant to making access control decisions, we have classified the context information into six categories, represented by six context information types (see Figure 1)): *RelationshipInfo*, *StatusInfo*, *ProfileInfo*, *LocationInfo*, *TemporalInfo*, and *HistoricalInfo* classes.

Based on the motivating scenario, we further classify relationship information into different types. For example, the *Person-Centric* relationship is a relationship information type, which contains a data type property (*xsd:string* type) named *interRelationship*. It indicates the interpersonal relationship between the *Persons* concerned such as "doctor-patient". Due to page limit, the complete descriptions of other context types are not included in this paper.

Domain-Specific Context Ontologies. In the context of the motivating scenario, we define the *domain-specific ontologies* for the healthcare domain on the basis of the *COAC upper ontology*. In particular, we focus on the representation of the relevant *Role* and *Resource* ontologies due to space limitation.

The Role can be categorized as *InDomainUser* or *OutDomainUser* (Figure 2(a) shows a part of the *Role* ontology). This categorization is to facilitate different fine-grained control for different types (roles) of users. Additionally, this improved structure is beneficial from the reasoning viewpoint, as some services only relevant to the introduced generalizations of these roles, i.e., *InDomainUser* and *OutDomainUser* in the example. In the healthcare application, we model *InDomainUser* roles, relying on the Australian Standard Classification of Occupations (ASCO) of the health professionals. The *NursingProfessional*, *MedicalPractitioner*, and *OtherProfessional* are subclasses of *InDomainUser*; the *GeneralPractitioner* and *SpecialistPractitioner* are in turn subclasses of *MedicalPractitioner*; etc. Besides, some other members such as *GuestResearcher*, *HealthTrainer*, etc, are not healthcare members but need to access some of the patient information, and therefore classified under *OutDomainUser*. This superclass-subclass hierarchy can facilitate access control in a way similar to the RBAC's senior-junior role concept [10]. For example, a user playing the role *MedicalPractitioner* can access a patient's daily

Table 2. User-Defined Reasoning Rules

No	Rule ($C_1 \wedge C_2 \wedge ... \wedge C_n \rightarrow C_1 \wedge C_2 \wedge ... \wedge C_r$, where each C_i is a rule concept.
Rule #1(a)	**Owner**(?o) $\wedge$ **StatusInfo**(?hs) $\wedge$ has(?o, ?hs) $\wedge$ **ProfileInfo**(?hp) $\wedge$ has(?o, ?hp) $\wedge$ **bodyTemperature**(?hp, "normal") $\wedge$ **heartRate**(?hp, "abnormal") $\rightarrow$ **healthStatus**(?hs, "critical") // **Rule #1(b)** ... $\wedge$ ... $\rightarrow$ **healthStatus**(?hs, "normal")
Rule #2(a)	**User**(?u) $\wedge$ **Role**(?rol) $\wedge$ hasRole(?u, ?rol) $\wedge$ swrlb:equal(?rol, "MedicalPractitioner_1") $\wedge$ **Resource**(?r) $\wedge$ hasReqRes(?u, ?r) $\wedge$ **Owner**(?o) $\wedge$ isOwnedBy(?r, ?o) $\wedge$ **Relationship**(?re) $\wedge$ hasRelationship(?u, ?re) $\wedge$ hasRelationship(?o, ?re) $\wedge$ **RelationshipInfo**(?rel) $\wedge$ has(?re, ?rel) $\wedge$ **ProfileInfo**(?pp) $\wedge$ has(?u, ?pp) $\wedge$ **userIdentity**(?pp, ?uID) $\wedge$ roleIdentity(?pp, ?rolID) $\wedge$ **ProfileInfo**(?sp) $\wedge$ has(?o, ?sp) $\wedge$ **conPeopleID**(?sp, ?cpID) $\wedge$ **conPeopleRoleID**(?sp, ?cpRID) $\wedge$ swrlb:notEqual(?cpID, ?uID) $\wedge$ swrlb:notEqual(?cpRID, ?rolID) $\rightarrow$ **interRelationship**(?rel, "nonTreatingPhysician") // **Rule #2(b)** ... $\wedge$ ... $\rightarrow$ **interRelationship**(?rel, "treatingPhysician")

medical records, which means a user playing the role *GeneralPractitioner* or *SpecialistPractitioner*, is also permitted to access the patient's daily medical records. However, the converse is not true.

The different components at various granularity levels of a patient's medical record (*Resource*) are individually identifiable, so as to achieve fine-grained control over access to them. As such, there is the healthcare *Resource* (patient data) hierarchy in the domain ontology (Figure 2(b) shows a part of the *Resource* ontology). In formulating the structure of the patient medical record, we follow the Health Level Seven (HL7) standard. Emergency Medical Records (*EMR*, at *granularityLevel* 0) includes a patient's complete medical records: a patient's Daily Medical Records (*DMR*, at *granularityLevel* 1), which includes Physiological Records (*PR*), Physician Prescriptions (*PP*), Daily Observations Reports (*DOR*), etc., at *granularityLevel* 2; a patient's Past Medical History (*PMH*); Identification Records (*IR*); Demographic Records (*DR*); and so on. The *granularityLevel* is an important data type property (*xsd:int* type) in our model that regulates access to different parts of a resource individually. In the motivating scenario (Scene #1), for example, Jane can access Bob's *EMR*, i.e., including all other sub-components of the *EMR*, while Mary can only access *DMR*.

Context Reasoning. Our COAC ontology is extended with user-defined reasoning rules (specified in the SWRL language) to infer high-level implicit contexts (requirement (2)). This aims to improve the request for specific services through automated inference of implicit contexts from limited contexts.

A *user-defined reasoning rule* has the form: $C_1 \wedge C_2 \wedge ... \wedge C_n \rightarrow C_1 \wedge C_2 \wedge ... \wedge C_r$, where C_1, C_2, ... C_n are the input concepts (*body* of the rule) and C_1, C_2, ... C_r are the resultant/derived concepts (*head* of the rule).

Example Rules. The SWRL Rule #1(a) in Table 2 states that *if* the patient's *bodyTemperature* is "normal" and its *heartRate* is "abnormal" *then* his *healthStatus* is "critical". Rule #2(a) in Table 2 states that *if* a User (playing the *MedicalPractitioner* role) has requested access to a *Resource* which is owned by an *Owner* (a Patient) and the *Owner* is not connected with the *MedicalPractitioner* through a *interRelationship* *then* the association between the *User* and *Owner* is a "nonTreatingPhysician" relationship. Note that this rule has used the required (basic) information, the user's *personal profile* information and the

Table 3. An Example Access Control Policy

AccessPolicy (hasSubject, hasObject, hasAccessPerm, hasAction, hasAccessConds)
{
Subject.Role.roleID = "MP00X"; & // MedicalPractitioner's role identity
Object.granularityLevel = 0; & **Object**.privacyAttribute = 1; & // Emergency Records (EMR)
    AccessCond.intRelationship(User, Owner) = "anyPhysician"; &
    AccessCond.healthStatus(Owner) = "critical"; &
    AccessCond.locAddress(User) = **AccessCond**.locAddress(Owner) = "ER00X"; &
    → **AccessPermission**.permission = "granted"; &
        Action.actionType = "read" *or* "write".
}

---

patient's *social profile* information. For the scenario (Scene #1), based on the context ontology, this rule can determine that Jane and Bob has a "nonTreating-Physician" relationship, because Jane is not assigned as the treating physician of patient Bob. Similarly, the SWRL rules can be used to derive *Location-Centric* relationships (e.g., *colocatedRelationship* between *User* and *Owner*).

4   Policy Model - CAPO Policy Ontology

The access control policies specify whether a *subject* is permitted or denied access to a set of target *objects* for their specific *action* or sequence of *actions*, when a set of *conditions* are satisfied. We refer to the user's role rather than explicitly name each user to reflect the way users can be grouped as the subject.

Various policy languages have been proposed in the literature. Our goal in this research is to provide a way in which CAAC policies can be specified by incorporating dynamic contexts. To be of practical use, it must be expressive enough to specify the policies in an easy and natural way. In particular, it needs to use the concepts from the context model, to specify under which conditions the requested resource is accessible. To do so, we use the same ontology-based languages (the OWL and SWRL) as the policy language. In the following, we present the two main parts of our policy model, supporting the requirement (3).

Policy Specification. Our *context-aware policy ontology (CAPO)*, as depicted in Figure 3, has the following concepts: *AccessPolicy, Subject, Role, Object, AccessPermission, Action*, and access conditions (i.e., *AccessCond*). Our policy model also uses the concepts (*ContextInfo*, shown in shaded ellipse) from the COAC context model, to identify and capture the relevant contexts at runtime.

A *CAAC Policy* captures the *who/what/when* dimensions which can be read as follows: an *AccessPolicy* specifies that a *Subject* (who is playing a *Role*) has *AccessPermission* ("granted" or "denied") to which parts (*privacy attributes*) of an *Object* (at which *levels of granularity*) for a specific *Action* ("read" or "write") or sequence of actions under which context-dependent *access conditions*.

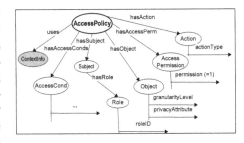

Fig. 3. CAPO Policy Ontology

Table 4. A Simplified Version of an Access Query

AccessPolicy(?policy) ∧ Subject(?subject) ∧ Role(?role) ∧ Object(?resource) ∧ Action(?action) ∧ AccessPermission(?permission) ∧ AccessCond(?condition) ∧ ContextInfo(?context) ∧ swrlb:equal(?condition, ?context) → **sqwrl:select(?role, ?resource, ?permission, ?action)**

Table 5. Access Query Result

?role	?resource	?permission	?action
1 MP00X(MedicalPractitioner)	EMR	granted	read
2 MP00X(MedicalPractitioner)	EMR	granted	write
3 MP00X(MedicalPractitioner)	DMR	granted	read

Let us consider an access control policy for the medical practitioners (Policy #1 in our application scenario). The access decision is based on the following policy constraints: **who** the user is (subject's *role*), **what** resource being requested (object's *privacy attribute* and *granularity level*), and **when** the user sends the request (the *interpersonal relationship* between user and owner, their *locations*, the *health status* of the patient). The template of the policy in a readable form is shown in Table 3 and the specific policy states that all medical practitioners can access a patient's complete medical records when his health status is critical.

One of the main features of our policy model is its ability to specify access permissions at different levels of resource granularity. For example, in the above policy, all medical practitioners can access a patient's emergency records (EMR) at granularity level 0 (*highest* level), which means they also can access all other records at the *lower* granularity levels (i.e., all the sub-components of the EMR).

Policy Evaluation. During the evaluation phase, an *access query* is used to process the user's *access request*. We use the SWRL rules to evaluate the policies. In particular, the query language SQWRL, which is based on OWL and SWRL, is adopted to process service access requests.

To determine the access permissions on the requested services, an *Access Query* is formulated based on the access control policies, by capturing the *policy constraints* and *relevant contexts* that are currently in effect.

A *Service Access Request* is defined as a tuple, <*pass, service*>, where *pass* is the user's access pass *(identity, password)*, and *service* is the *(action, resource)* pair. When a *service access request* comes, the user is first identified based on his provided *pass*. Then, the policy ontology identifies the relevant access control policies. It also identifies the low-level and high-level context information using the context ontology. Then the defined access query is used to process the user's request to access the services using both the policy constraints and the relevant contexts. For the application example (*Scene #1*), a simplified version of the access query (see Table 4) is used to match the relevant policy constraints against the relevant contexts, in order to determine whether the access is *granted* (if matching result is *true*) or *denied* (otherwise). Table 5 presents query results.

5   Prototype Implementation and Case Study

Framework Prototype. We have developed a prototype implementing CAAC approach in J2SE. We have used the Protégé-OWL API to implement the ontologies. We have used Java and the Jess Rule Engine to implement a context reasoner for executing the SWRL rules. We have implemented a set of *APIs*, which can support the software engineers to develop CAAC applications using this framework. The *ContextManager* provides functionalities to manipulate the context ontologies (capturing low-level context facts and inferring high-level contexts). We have developed a number of *context providers* and the *context reasoner* as parts of the *ContextManager*. The *PolicyManager* allows CAAC application developers to add, edit and delete access control policies. The *CAACDecisionEngine* checks the user's request to access the software services/resources and makes access control decisions using the relevant context information. In addition, the *query manager* (part of the *CAACDecisionEngine*) allows application developers to add, edit and delete the access queries.

Application Prototype. Following the motivation scenario, we have developed a demo CAAC application, called *Patient Medical Record Management (PMRM)*, as an example of how to realise CAAC decisions. The application allows different users to invoke different operations on the requested services to access specific patient records in a context-aware manner.

For Scene #1, when Jane wants to access the requested service *writeEMR()*, a service *AccessRequest* is submitted to the *CAACDecisionEngine* for evaluation. The defined query (see Table 4) is used to retrieve the access control decision (access *permission* is "granted" or "denied").

The COAC ontology contains the relevant low-level context facts (from the *context providers*). It also captures the relevant high-level implicit contexts using the *context reasoner* based on the basic information in the ontology and the reasoning rules (e.g., Rule #1(a) and Rule #2(a) in Table 2). The CAPO ontology captures the relevant policy constraints (e.g., the policy in Table 3) applicable to the requested service, i.e., (*action, resource*) pair. The relevant contexts and the policy constraints, captured at the time of access request, are provided to the *CAACDecisionEngine* as part of the request processing. The current contexts are then matched against the constraints of the policy as part of making the access control decision. Based on this information, the *CAACDecisionEngine* returns an access control decision for the submitted access request. One of the entries in the access query results (Line #2 of Table 5) satisfies Jane's *AccessRequest*. That is, *if* Jane and Bob both are *located in* the "emergency room" of the hospital *and* Bob's current *health status* is "critical" *and* the *interpersonal relationship* between Jane and Bob is "any physician", *then* Jane is authorized to access the service *writeEMR()*, because the available contexts and policy constraints indicate that the access conditions are satisfied.

6   Related Work and Discussion

Several research efforts (e.g., [1],[5],[8],[11]) have adopted and extended the Role-based Access Control (RBAC) approach for access control to software services. Some of these efforts (e.g., [1]) incorporate specific types of contexts such as location and time. Kulkarni et al [8] have proposed a context-aware RBAC (CA-RBAC) model for pervasive applications. They consider user and resource attributes as the context constraints. He et al [5] have considered access control for Web service based on the user role and presented a CAAC policy model considering the user, resource and environment concepts. These approaches consider specific types of contexts which are not general enough in dynamic environments. In contrast, we have proposed a general and extensible ontology-based context model, in which we introduce several additional general concepts for context modelling, including resource owner and relationship between different persons. Toninelli et al [11] have proposed a semantic CAAC approach which provides resource access permission on the basis of context (resource availability, roles of user, location and time). It includes an ontology-based framework with context and policy models. Similar to the above approaches, however, its context model does not consider several concepts which are important for access control in today's dynamic environments, including owner, relationship between user and owner, the derived entity status (e.g., health status) information, etc. In addition to these concepts, our approach also supports resource hierarchy, and consequently user's access to resources at different levels of granularity.

A number of further research efforts (e.g., [2],[3],[6]) have extended the Attribute-based Access Control (ABAC) approach to provide access control to software services in a context-aware manner. Corradi et al [2] have proposed a CAAC model for ubiquitous environments, where permissions are directly associated with contexts: user location, user activities, user device, time, resource availability and resource status. Hulsebosch et al [6] have proposed a context-sensitive access control framework based on the user's location and access history. These approaches also have limitations in considering a limited set of contexts. A recent CAAC framework for the Web of data is grounded on two ontologies which deal with the core access control policy concepts and the context concepts [3]. Even though it does consider three important dimensions of context: user, device and environment, it does not have the capability of inferring high-level implicit contexts. These attribute-based approaches have major limitations when applied in large-scale domains because of the huge number of attributes involved. In addition, they do not directly treat *roles* as first class entity.

7   Conclusion

In this paper, we have introduced a new ontology-based approach to context-aware access control for software services. It includes an extensible context model specific to access control, and a reasoning model for inferring high-level implicit contexts based on user-defined rules, and an access control policy model incorporating context information from the context model. In order to demonstrate the practical applicability of our approach, we have developed a prototype framework implementing the approach. We have also developed a context-aware access control application in the healthcare domain and have presented a healthcare case study. The case study has shown that our framework is effective in practice.

Acknowledgment. Jun Han is partly supported by the Qatar National Research Fund (QNRF) under Grant No. NPRP 09-069-1-009. The statements made herein are solely the responsibility of the authors.

References

1. Chandran, S.M., Joshi, J.B.D.: *IoT-RBAC*: A location and time-based RBAC model. In: Ngu, A.H.H., Kitsuregawa, M., Neuhold, E.J., Chung, J.-Y., Sheng, Q.Z. (eds.) WISE 2005. LNCS, vol. 3806, pp. 361–375. Springer, Heidelberg (2005)
2. Corradi, A., Montanari, R., Tibaldi, D.: Context-based access control management in ubiquitous environments. In: NCA, pp. 253–260 (2004)
3. Costabello, L., Villata, S., Gandon, F.: Context-aware access control for rdf graph stores. In: ECAI, pp. 282–287 (2012)
4. Dey, A.K.: Understanding and using context. Personal and Ubiquitous Computing 5(1), 4–7 (2001)
5. He, Z., Wu, L., Li, H., Lai, H., Hong, Z.: Semantics-based access control approach for web service. JCP 6(6), 1152–1161 (2011)
6. Hulsebosch, R.J., Salden, A.H., Bargh, M.S., Ebben, P.W.G., Reitsma, J.: Context sensitive access control. In: SACMAT, pp. 111–119 (2005)
7. Kayes, A.S.M., Han, J., Colman, A.: ICAF: A context-aware framework for access control. In: Susilo, W., Mu, Y., Seberry, J. (eds.) ACISP 2012. LNCS, vol. 7372, pp. 442–449. Springer, Heidelberg (2012)
8. Kulkarni, D., Tripathi, A.: Context-aware role-based access control in pervasive computing systems. In: SACMAT, pp. 113–122 (2008)
9. Riboni, D., Bettini, C.: Owl 2 modeling and reasoning with complex human activities. Pervasive and Mobile Computing 7(3), 379–395 (2011)
10. Sandhu, R.S., Coyne, E.J., Feinstein, H.L., Youman, C.E.: Role-based access control models. IEEE Computer 29(2), 38–47 (1996)
11. Toninelli, A., Montanari, R., Kagal, L., Lassila, O.: A semantic context-aware access control framework for secure collaborations in pervasive computing environments. In: Cruz, I., Decker, S., Allemang, D., Preist, C., Schwabe, D., Mika, P., Uschold, M., Aroyo, L.M. (eds.) ISWC 2006. LNCS, vol. 4273, pp. 473–486. Springer, Heidelberg (2006)
12. Weiser, M.: Some computer science issues in ubiquitous computing. Communications of the ACM 36(7), 75–84 (1993)

A Novel Method for Finding Similarities
between Unordered Trees Using Matrix Data Model

Israt Jahan Chowdhury and Richi Nayak

School of Electrical Engineering and Computer Science, Science and Engineering Faculty,
Queensland University of Technology, Brisbane, Australia
{israt.chowdhury,r.nayak}@qut.edu.au

Abstract. Trees are capable of portraying the semi-structured data which is common in web domain. Finding similarities between trees is mandatory for several applications that deal with semi-structured data. Existing similarity methods examine a pair of trees by comparing through nodes and paths of two trees, and find the similarity between them. However, these methods provide unfavorable results for unordered tree data and result in yielding NP-hard or MAX-SNP hard complexity. In this paper, we present a novel method that encodes a tree with an optimal traversing approach first, and then, utilizes it to model the tree with its equivalent matrix representation for finding similarity between unordered trees efficiently. Empirical analysis shows that the proposed method is able to achieve high accuracy even on the large data sets.

Keywords: Semi-structured Data, Unordered Tree, Similarity Measure, Matrix Representation.

1 Introduction

The web domain consists of heterogeneous data in various forms such as HTML, XML, image, videos and text. Some of these data are naturally represented as tree data structures. Comparing the tree-structured data is important as it enable searching for interesting information among the abundant data efficiently. Many researchers confirm the significance of unordered tree data representation and their comparisons [1, 2]. An unordered tree does not have left-to-right fixed order among siblings node and only preserves the ancestor-descendant or parent-child relationship. Especially in the web domain where the data source is heterogeneous, the unordered tree representation gives more freedom for flexible matching and concise representation.

A large number of tree mining methods have been developed for finding similarities [3]. Majority of them are for ordered trees and very few are available for unordered trees due to the complexities involved with the unordered tree processing. Existing similarity methods examine a pair of trees by comparing through nodes and paths of two trees, and aggregate the similarity between them [4]. Some similarity measure methods use tree level information by considering their common nodes in the corresponding levels and giving different weight in different levels, but it fails to reserve the child-parent relationship among tree nodes [5]. Higher order models such

X. Lin et al. (Eds.): WISE 2013, Part I, LNCS 8180, pp. 421–430, 2013.

as Tensor Space Model (TSM) have also been used for representing tree data and finding similarities, though these techniques suffer from high dimensionality as well as complexity problems [6]. Tree edit distance methods are also commonly used in measuring similarity between the tree data. These methods measure the distance between two trees in terms of minimum cost to transform one tree into another tree by applying edit operations such as deletion, insertion and substitution [3]. The edit distance computing algorithms for ordered tree data are known to exhibit $O(n^3)$ complexity, where n is the maximum number of nodes in two input trees [7]. The tree-edit distance based methods for unordered trees show NP-hard complexity [8, 9]. A few methods have been developed by reducing the tree edit distance problem to the maximum clique problem [10, 11] or proposing variants of the tree edit distance problem [12]. However, they still suffer from high complexity for large unordered tree structure [10]. Other examples of unordered tree matching methods are tree pattern matching [13], maximum agreement subtree [14], largest common subtree [15], and smallest common supertree. These methods also suffer from the complexity problem. In summary, existing methods provide unfavorable results for unordered tree data and result in yielding high complexity.

We propose a novel idea of representing the trees with matrix data structure using tree encoding, and then comparing two matrix structures using efficient cosine similarity measure. An optimal traversing adapting a well known optimization problem called "Simple Assemble Line Balancing" is used to provide tree encoding for unordered tree data. A matrix based representation called "Augmented Adjacency Matrix" is proposed to represent the tree data based on the encoding information. The empirical analysis shows that the proposed method performs well with high accuracy and outperforms benchmarking methods for the large size data. The proposed method is able to achieve $O(n^2)$ complexity due to its incorporation of matrix data for comparison. This is remarkable as the existing similarity methods for unordered trees mostly give NP-hard and MAX-SNP hard complexity [8, 9].

2   The Proposed Similarity Measure Method

The proposed unordered tree similarity method includes three steps. Firstly, the tree data is encoded with an optimal traversing approach. Secondly, an equivalent matrix representation is obtained for each tree structure utilizing the tree encoding with other tree information. Thirdly, cosine similarity measure is used to calculate the similarity between two matrices representing unordered trees.

2.1   Step 1: Tree Encoding Using an Optimal Traversal Approach

Tree Traversal. A tree traversal is a systematic approach of visiting each node once in a tree by following certain strategy and returns a list containing the node sequence traversed along the way. The depth-first search (DFS) and breadth-first search (BFS) are two commonly used traversing algorithms that rely on the fixed ordering among sibling nodes. A DFS algorithm starts from root node and explores each branch as far as possible before backtracking. They can be classified as pre-order, in-order and post-order, based on the sequence of visiting nodes on right or left order. A BFS algorithm, also known as level order traversal algorithm, starts visiting a tree from its root node and then follows a strategy for traversing other nodes in the order of their level from left to right [16]. These strategies are able to represent ordered trees efficiently; however, they face challenges when applied for unordered tree traversal as there is no fixed order among sibling nodes. To our best knowledge, these are the only two strategies that have been used for representing and canonization of unordered trees [17].

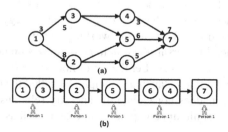

Fig. 1. The Simple Assemble line balancing problem, first diagram replicates an assembly line (a), second one representing optimal sequence of operations on various machines (b)

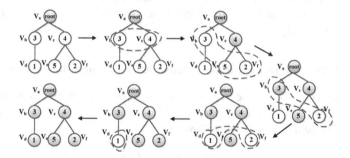

Fig. 2. Optimal Tree Traversal

Optimal Tree Traversal. In this paper we introduce an optimal tree traversal method for representing unordered tree. This method is inspired by a well-known optimization problem known as "Simple Assemble line balancing" from the "Operation Research" paradigm [18]. In manufacturing, the line balancing problem is used to minimize the cost of production by balancing the machine sequences of an assembly line based on their operating time and finds the optimal sequence that will support minimum operation or cycle time. Fig. 1(a) illustrates a scenario where the nodes are representing various machines in an assembly line and the numerical values outside the nodes stand for the operation time requiring for each machine. Fig. 1(b) shows the optimal sequence of completion tasks according to the assembly line problem. In the proposed method we metaphor the assembly line as the unordered tree; a machine as a tree node; and the optimal sequence as the optimal tree traversal. The weight of a node is calculated by counting the number of occurrences of each node under its parent node. The traversal process begins at the root node. The children nodes are visited only after their parent nodes are visited. This is done to ensure that the ancestral ordering constraint is preserved. The objective of the traversal approach is to minimize the overall traversal time and return an optimal node sequence for the unordered tree.

Problem Definition. Let tree $T = (V, E)$ be an unordered labeled tree where $V = (v_0, v_1, ..., v_n)$ denotes the set of nodes that presumes a partial order ρ due to the ancestral relation (i.e., $i \, \rho \, j \rightarrow i > j$ where i and j are node indices and i is ancestor of j). If function $tr: T \rightarrow T^*$ that passes over the tree, listing all nodes that met along the way, then it is called tree traversal. T^* is n-dimensional vector, representing the list of nodes in the order of traversal according to the specified traversal strategy, $(v_0, v_1, ..., v_n) = V \in T^*$, where, v_0 is the root node. By using the working principle of line balancing problem, we define the general traversal function to an optimization problem for achieving the optimal node traversal sequence. Let the set of nodes $V = (v_0, v_1, ..., v_n)$, traversed in a sequence by using the line balancing principle, be called the optimal tree traversal if the traversal function tr does not violate the ancestry relationship given by the unordered tree and ensures minimum computational cost as well as traversal time.

Tree Encoding. After receiving the optimal sequence for traversing all tree nodes, each node will be encoded according to its order in this sequence. For instance, in Fig. 2, the traversal will start from the root node V_a and the optimal sequence is V_a-V_c-V_e-V_b-V_f-V_d. The encoded values for the nodes in the tree will be 1-2-3-4-5-6 for V_a-V_c-V_e-V_b-V_f-V_d respectively.

2.2   Step 2: Tree Modeling with the Augmented Adjacency Matrix Representation

Adjacency matrix has been used for representing trees and graphs by modeling the adjacency information regarding parent-child relationship [19]. Let the adjacency matrix A model the tree $T(V, E)$ as followings.

$$A_{ij} = \begin{cases} \text{true,} & v_i, v_j \in E(T) \\ \text{false,} & \text{otherwise} \end{cases} \tag{1}$$

A tree data is a hierarchical representation that includes the inherent implicit relationships and semantics of various nodes. The traditional adjacency matrix fails to represent the label information, level information, encoding information, and ancestry relationships. To overcome these limitations the following Augmented Adjacency matrix is proposed to model tree data more accurately.

Augmented Adjacency Matrix. This is a square matrix that utilizes the level, encoding and weight information of a tree to represent the cell values.

Encoding information: By using the optimal traversing sequence, we obtain the encoding values of the tree nodes according to the order they are visited. The root node becomes the first row and column to be represented in the matrix and the other nodes are arranged in the optimal order achieved by the optimal traversal. This encoding value also integrates with the level value between two nodes.

Level information: The level information in a tree represents the ancestry relationships of the nodes. This structural information is important for finding similarity between trees [5]. The nodes appearing high on tree carry more influence than nodes appearing near the leaf nodes. Consequently, the level assignment is bottom-up; the lowest leaf node is assigned the level 1 and the higher value is assigned to the root node level. The following rules are applied to assign a value to two nodes, V_i and V_j, incorporating the level information.

1. If an ancestor-descendant relationship exists between two nodes V_i and V_j, where V_i is the ancestor of V_j, or if the encoding value of V_i is less than the encoding value of V_j then the level value of cell C_{ij} is: $\dfrac{level(V_j)}{level(V_i)}$. The function *level* outputs the level value of a node.
2. If an ancestral relationship does not exists between two nodes V_i and V_j, or if the encoding value of V_i is greater than the encoding value of V_j then the level value for cell C_{ij} will be 0.

Weight information: In this method, nodes carry a weight displaying how frequently the node occurs under its parent node. The node weight is added to the corresponding level value. Additionally, a value of 1 is added to each diagonal cell of the adjacency matrix to represent the existence of corresponding node on that tree.

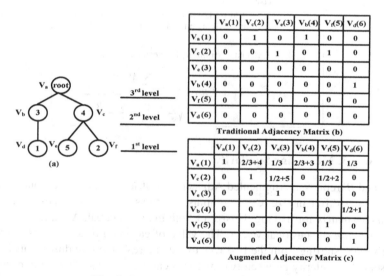

	$V_a(1)$	$V_c(2)$	$V_e(3)$	$V_b(4)$	$V_f(5)$	$V_d(6)$
V_a (1)	0	1	0	1	0	0
V_c (2)	0	0	1	0	1	0
V_e (3)	0	0	0	0	0	0
V_b (4)	0	0	0	0	0	1
V_f (5)	0	0	0	0	0	0
V_d (6)	0	0	0	0	0	0

Traditional Adjacency Matrix (b)

	$V_a(1)$	$V_c(2)$	$V_e(3)$	$V_b(4)$	$V_f(5)$	$V_d(6)$
V_a (1)	1	2/3+4	1/3	2/3+3	1/3	1/3
V_c (2)	0	1	1/2+5	0	1/2+2	0
V_e (3)	0	0	1	0	0	0
V_b (4)	0	0	0	1	0	1/2+1
V_f (5)	0	0	0	0	1	0
V_d (6)	0	0	0	0	0	1

Augmented Adjacency Matrix (c)

Fig. 3. Augmented Adjacency Matrix

We illustrate the process of modeling the tree with the augmented adjacency matrix and populating the matrix values. Fig. 3 illustrates the traditional adjacency matrix and the augmented adjacency matrix for a given tree. The example tree has three levels, and the root node level is considered as the highest one. The encoding value of nodes is received from Fig. 2 by using the optimal traversal. The traversal sequence is V_a-V_c-V_e-V_b-V_f-V_d and the encoding values for these nodes are 1-2-3-4-5-6 respectively. The level information of corresponding nodes is calculated, and the node weights are added to the level values. For instance, consider the calculation of the cell value, C_{23}, showing the relation between V_a-V_c. The encoding value of $V_a = 1$ which is less than the encoding value of $V_c = 2$ that means V_a is ancestor of V_c. According to rule 1, the level value of C_{23} is $\dfrac{level(V_c)}{level(V_a)} = \dfrac{2}{3}$. The weight of V_c is 4. The final cell value will be 2/3+4. The rest of the cell values are being calculated in the same way.

Table 1. The proposed similarity measure algorithm

Algorithm : Measuring Similarity
Input: Unordered trees T_a and T_b
Output: Measurement similarity between tree pair
1. Model the tree T_a with the Augmented Adjacency Matrix A';
2. Model the tree T_b with the Augmented Adjacency Matrix B';

3. **if** $\lvert B'\rvert > \lvert A'\rvert$ **then**
        Add $(\lvert B'\rvert - \lvert A'\rvert)$ rows and columns of zeros at the right end and bottom of the matrix A';
    else
        Add $(\lvert A'\rvert - \lvert B'\rvert)$ rows and columns of zeros at the right end and bottom of the matrix B';
    end if

4. Calculate similarity between two trees using $$Cos(A',B') = \frac{\sum_{x=1}^{n}\sum_{y=1}^{n} A'_{xy} B'_{xy}}{\sqrt{\sum_{x=1}^{n}\sum_{y=1}^{n} A'^{2}_{xy}}\sqrt{\sum_{x=1}^{n}\sum_{y=1}^{n} B'^{2}_{xy}}}$$

2.3   Step 3: Measuring Similarity

Let A' and B' represent Augmented Adjacency Matrices of the corresponding trees. If the two trees differ in size, extra columns and rows with zero elements are added to the smaller matrix for making the size of both matrixes equal. A matrix can be considered as a $n \times n$ dimensional vector. The value of each cell of a matrix is a dimension of the vector, starting from the first row to the end row; the $n \times n$ dimensional vector is represented. Similarity between two matrices can be calculated by using cosine similarity. Table 1 illustrates the similarity process.

It is expected to achieve a polynomial time complexity with the proposed method detailed in Table 1. The method consists of three steps. The complexity of the first step is $O(n^2)$, same as the line balancing optimisation problem. The complexity of the second step is known to be $O(n^2)$ for modelling the adjacency matrix based on tree encoding information. The final step comprises cosine similarity calculation, too small to count in; consequently it can be ignored during complexity analysis. The overall complexity is $O(n^2)$ where n is the maximum number of nodes in the input trees pair.

3   Experimental Results

The proposed similarity measure method is evaluated on two datasets including the bill of material (BOM) data that has the similar structure as XML documents [20] and the Glycan structures obtained from the KEGG/Glycan database [21]. The proposed method is implemented on Matlab and experiments are performed on a PC with RAM size 8.00 GB and a processor Intel Core i7.

Performance on the BOM Data: The BOM data set consists of 404 sample BOMs with 50,000 nodes and 12,000 unique nodes. The dataset includes trees with maximum and minimum depth of 8 and 4 respectively, whereas the maximum and minimum breadth is 10 and 6 respectively. The well known evaluation metrics such as precision, recall, F-score and AUC are calculated. To calculate these measures, positive and negative samples were needed. For this purpose, a tree pair in the data set is regarded as positive if the distance score is smaller than a given threshold. Otherwise it is regarded as negative. The threshold value is determined empirically. Fig. 4(a) and (b) show the performance of the matrices with varied threshold values. As expected, data in Fig. 4(a) shows that with the increase in threshold, matching accuracy is improved yielding the best matches showing increase in precision, however it reduces the number of matches resulting the fall in recall. Considering the tradeoff between precision and recall, the proposed method produces the best result when the threshold is set in the range between 0.6~0.65 (Fig. 4(a)). For thresholds below the value of 0.3, AUC score is less than 0.5, indicating the random classification (Fig. 4(b)). The threshold value higher than 0.5 gives a good quality solution yielding higher AUC.

We performed a scalability test by varying the BOM data set of different size reporting the CPU time and memory usage. Fig. 4(c) reveals that the method is able to provide the $O(n^2)$ complexity, confirming the theoretical complexity analysis. The memory usage does not change with the increased data size, as the proposed method just needs to keep a pair of trees in the memory at a time

Performance on the Glycan Structures: We used the Glycan data for comparing scalability of the proposed method with the state-of-the-art similarity measure methods such as CliqueEdit, UwCliqueEdit, and DpCliqueEdit [11]. It is to be noted that none of these available methods empirically analysis their accuracy. They conduct the CPU time analysis to show the complexity. We compare our proposed method based on CPU time with these methods. For analysis, tree pairs are selected randomly from the data set with a specified range of the total number of nodes (i.e., sum of the numbers of nodes in two trees) and the average CPU time per pair is measured.

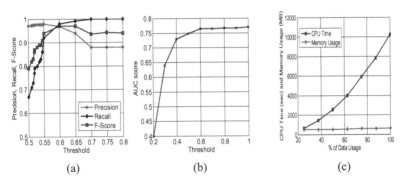

Fig. 4. Evaluation metrics with varied thresholds (a, b) and scalability test (c)

Results in Table 2 show that our proposed method performs well for almost all sizes of trees. Although the proposed method does not give best result for the smaller tree node sizes, between the ranges of 55~59 and 75~79, but several other methods perform worse than our method. After reaching the range 80~84, our method outperforms others due to the use of optimal traversal. Overall the average performance of all subsets of datasets (the last row) indicates that our method outperforms all methods, some with very large margin. The CliqueEdit, UwCliqueEdit, and DpCliqueEdit [11] methods implement several heuristics to cut the CPU expense, but provides no results about accuracy of the matching process. We provide the accuracy test for our proposed method on the BOM dataset. Results ascertain that the proposed method is able to achieve high accuracy and polynomial complexity.

Table 2. Average CPU time (sec) per glycan pair is shown for each case. Boldface indicates the best results for each case and the highlighted cell indicates the worst results for each case.

Total # nodes	Clique Edit	UwClique Edit	DpClique Edit-A	DpClique Edit-B	DpClique Edit-C	DpClique Edit-D	DpClique Edit-E	**Proposed Method**
55~59	1.987	0.433	8.968	0.108	**0.088**	0.086	0.096	0.374
60~64	2.746	4.949	1.78	0.167	0.163	**0.149**	0.177	0.47
65~69	64.29	9.303	39.46	0.381	0.364	**0.328**	0.357	1.513
70~74	58.69	**0.099**	1.337	0.545	0.436	0.463	0.501	1.517
75~79	2.441	0.918	4.051	0.953	**0.752**	0.754	0.781	1.547
80~84	7.150	6.570	44.63	2.516	2.268	1.620	1.653	**1.55**
85~89	237.7	28.03	21.11	3.205	3.205	2.413	2.490	**1.641**
90~94	303.2	1211	1710	38.81	26.30	8.165	9.475	**1.761**
Average	84.78	157.66	228.92	5.84	1.75	1.75	1.94	**1.29**

4   Conclusion

The unordered tree data represents information inherent in many domains naturally. This presses the need of developing an efficient method of measuring similarity between trees especially when we are living in the big data era. This paper proposes an efficient method of measuring similarity between unordered trees. The proposed method introduces an augmented adjacency matrix structure for modeling the tree data. The matrix representation enables efficient computation of pair of trees for finding similarity. An optimal traversal of the tree is obtained using a line of balance optimization problem. The encoding values of the nodes with this optimal traversal are utilized in representing the tree with the matrix structure.

Empirical analysis shows that the proposed method is able to achieve improved complexity in comparison to existing methods even for large datasets. Results also showed that an improved complexity is achieved with high accuracy. The proposed method is able to achieve polynomial complexity whereas the existing methods for calculating similarity amongst unordered trees suffer from the high complexity problem and are NP-hard or MAX-SNP hard.

Our future plan is to work on the detail of the optimal traversal approach to improve the overall performance. We plan to apply heuristics to improve the scalability further. We also plan to do more experiments to analyze effectiveness and versatility of the proposed method.

Acknowledgements. We would like to thank Tatsuya Akutsu and Tomoya Mori for providing us the objective codes of their algorithms [22], [11] and Carol J.Romonowski, Rakesh Nagi for providing the BOM data set [20]. Also thanks to Noor Ifada for the helpful editing.

References

1. Yamamoto, Y., Hirata, K., Kuboyama, T.: On Computing Tractable Variations of Unordered Tree Edit Distance with Network Algorithms. In: Okumura, M., Bekki, D., Satoh, K. (eds.) JSAI-isAI 2012. LNCS, vol. 7258, pp. 211–223. Springer, Heidelberg (2012)
2. Pawlik, M., Augsten, N.: RTED: a robust algorithm for the tree edit distance. Proceedings of the VLDB Endowment 5(4), 334–345 (2011)
3. Bille, P.: A survey on Tree Edit Distance and Related Problems. Theoretical Computer Science 337(1-3), 217–239 (2005)
4. Shasha, D., Wang, J.T.L., Kaizhong, Z., Shih, F.Y.: Exact and approximate algorithms for unordered tree matching. IEEE Transactions on Systems, Man and Cybernetics 24(4), 668–678 (1994)
5. Nayak, R.: Fast and effective clustering of XML data using structural information. Knowledge and Information Systems 14(2), 197–215 (2008)
6. Kutty, S., Nayak, R., Li, Y.: XML Documents Clustering Using a Tensor Space Model. In: Huang, J.Z., Cao, L., Srivastava, J. (eds.) PAKDD 2011, Part I. LNCS, vol. 6634, pp. 488–499. Springer, Heidelberg (2011)
7. Demaine, E.D., Mozes, S., Rossman, B., Weimann, O.: An optimal decomposition algorithm for tree edit distance. ACM Transactions on Algorithms 6(1), 1–19 (2009)
8. Zhang, K., Statman, R., Shasha, D.: On the Editing Distance between Unordered Labeled Trees. Information Processing Letters 42(3), 133–139 (1992)

9. Hirata, K., Yamamoto, Y., Kuboyama, T.: Improved MAX SNP-Hard Results for Finding an Edit Distance between Unordered Trees. In: Giancarlo, R., Manzini, G. (eds.) CPM 2011. LNCS, vol. 6661, pp. 402–415. Springer, Heidelberg (2011)

10. Fukagawa, D., Tamura, T., Takasu, A., Tomita, E., Akutsu, T.: A Clique-based Method for the Edit Distance between Unordered Trees and Its Application to Analysis of Glycan Structures. BMC Bioinformatics 12(1), 1–9 (2011)

11. Mori, T., Tamura, T., Fukagawa, D., Takasu, A., Tomita, E., Akutsu, T.: A clique-based method using dynamic programming for computing edit distance between unordered trees. Journal of Computational Biology 19(10), 1089–1104 (2012)

12. Torsello, A., Hancock, E.R.: Computing approximate tree edit distance using relaxation labeling. Pattern Recognition Letters 24(8), 1089–1097 (2003)

13. Chen, Y., Cooke, D.: Unordered Tree Matching and Strict Unordered Tree Matching: The Evaluation of Tree Pattern Queries. In: The 2010 International Conference on Cyber-Enabled Distributed Computing and Knowledge Discovery, pp. 33–41. IEEE Computer Society, Huangshan (2010)

14. Zhang, S., Wang, J.T.L.: Discovering Frequent Agreement Subtrees from Phylogenetic Data. IEEE Transactions on Knowledge and Data Engineering 20(1), 68–82 (2008)

15. Akutsu, T., Fukagawa, D., Takasu, A.: Improved approximation of the largest common subtree of two unordered trees of bounded height. Information Processing Letters 109(2), 165–170 (2008)

16. Valiente, G.: Algorithms on trees and graphs. Springer, Heidelberg (2002)

17. Chi, Y., Yang, Y., Muntz, R.R.: Canonical Forms for Labelled Trees and Their Applications in Frequent Subtree Mining. Knowledge and Information System 8(2), 203–234 (2005)

18. Scholl, A.: Balancing and Sequencing of Assembly Lines, 2nd edn. Physica-Verlag, Heidelberg (1999)

19. Cormen, T.H., Stein, C., Rivest, R.L., Leiserson, C.E.: Representations of graphs. In: Introduction to Algorithms, 3rd edn., pp. 524–531. MIT Press and McGraw-Hill, Cambridge (2009)

20. Romanowski, C.J., Nagi, R.: On Comparing Bills of Materials: A Similarity/Distance Measure for Unordered Trees. IEEE Transactions on System, Man, and Cybernets 35(2), 249–260 (2005)

21. Kanehisa, M.: KEGG for representation and analysis of molecular networks involving diseases and drugs. Nucleic acids research. 38(suppl 1), D355–D360 (2010)

22. Akutsu, T., Mori, T., Tamura, T., Fukagawa, D., Takasu, A., Tomita, E.: An Improved Clique-Based Method for Computing Edit Distance between Unordered Trees and Its Application to Comparison of Glycan Structures. In: International Conference on Complex, Intelligent and Software Intensive Systems (CISIS 2010), pp. 536–540 (2011)

Semantic Inversion in XML Keyword Search with General Conditional Random Fields

Shu-Han Wang and Zhi-Hong Deng*

Key Laboratory of Machine Perception (Ministry of Education),
School of Electronic Engineering and Computer Science, Peking University
forsona@pku.edu.cn, zhdeng@cis.pku.edu.cn

Abstract. Keyword search has been widely used in information retrieval systems, such as search engines. However, the input retrieval keywords are so ambiguous that we can hardly know the retrieval intent explicitly. Therefore, how to inverse keywords into semantic is meaningful. In this paper, we clearly define the *Semantic Inversion* problem in XML keyword search and solve it with *General Conditional Random Fields*. Our algorithm concerns different categories of relevance and provides the alternative label sequences corresponding to the retrieval keywords. The results of experiments show that our algorithm is effective and 12% higher than the baseline in terms of precision.

1 Introduction

As a widely accepted tool, **Keyword Search** has been extensively used to search information from all kinds of databases, such as document corpus, relation databases, and semi-structure databases. However, the main weakness of Keyword Search is its ambiguity. Given a Keyword Search that consists of only keywords, it is hard to know what the users really want to search. For example, a user wants to find a paper published by IJCAI. The paper is written by Pineau and is about some technique based on point. For the above information requirement, a proper keyword search may be 'Point based Pineau IJCAI'. However, most all existing works compute results by statistical information of keywords without understanding the semantics under these keywords. In fact, if the system knows that 'Point based' is part of the paper's title(sometimes users may hardly remember the whole title, so only type in part of it), 'Pineau' is the author, and 'IJCAI' is the title of a proceeding. As we see, if we can recognize the inner semantic of users' input, user's search intent will be much more explicit.

Recent work have started to help users to construct semantic queries. [1] extracts structural semantic in graph data, and provide the top-k matching subgraphs according to the query. [2] proposes an automatic keyword query reformulation approach which extracts information in dataset offline and generates semantic of query online. [3] presents a novel system guiding users through a process of increasing semantic to specify their query intention. Pandey and Punera analyze user's search intent by extracting template structure of search queries[4].

---

* Corresponding author.

X. Lin et al. (Eds.): WISE 2013, Part I, LNCS 8180, pp. 431–440, 2013.

These work fully proved that deeper semantic respect to keywords can greatly help our retrieval. However, different from all methods above, our algorithm mainly concentrate on tagging the keywords with labels in XML database, and gives users top-k matching label sequence according to keyword sequence. Since XML labels can be well semantical, users can affirm what they really need by selecting proper labels. Here we present our main contributions to keyword search on XML database.

The Semantic Inversion for Keyword Search

Given keyword sequence, our algorithm recognize it into label sequence, so as to understand the semantic of the keywords. We call this recognition '**Semantic Inversion**'. In our algorithm, alternative label sequences are provided after keywords are typed in. Users select the best labels matching their keywords, in order to clarify their retrieval intention. As semantic becomes so important in retrieval, Semantic Inversion can be a promising way to optimize the keyword search.

Model the Semantic Inversion with CRF

If the Semantic Inversion problem were difficult, it could hardly be useful for retrieval. Fortunately, we find out that the Semantic Inversion is similar with the Part of Speech Tagging(POS) and other sequential learning problems. So existing models may be useful. Conditional Random Fields has been proved efficient in sequential learning and results of our experiments also prove that CRFs can solve the Semantic Inversion problem outstandingly.

Quantize the Relevance by Weighing Diverse Features

Existing algorithms aiming to compute the relations in keyword field always concentrate on one or few factors (LCA of keywords, co-occurrence between keywords, etc.). In our algorithm, keyword-keyword, label-label, keyword-label, different categories of relevance are weighed to quantize the relevance between keyword sequence and label sequence jointly. As we will discuss in later part of this paper, our learning algorithm is to find the best parameters for optimizing the weight of various features.

In this paper, we discuss keyword search only in XML domain. The rest of the paper is organized as follows. Section 2 presents the definition of Semantic Inversion for keyword search. Section 3 introduces general CRFs which we have applied to our problem. Details of features and the algorithm are provided in Section 4. The following two parts shows the experiments and several related work. Finally, we close with conclusion in Section 7.

2   Semantic Inversion

In this section, we provide the concept of Semantic Inversion in XML domain, and why Semantic Inversion is able to improve keyword search.

Definition 1 (*Label*): Given a set of XML files and a word, a tag is the label of the term if the content of the tag contains the word. A single word may have many probable labels.

For instance: the label of word 'Pineau' is 'author', and the label of words (appearing together) 'Torran Dubh' can be 'conflict' or 'caption'.

In XML files, the label can be regarded as the semantic of its content, so we will not distinguish 'semantic' and 'label' in later parts.

Definition 2 (*Label Sequence*): Given a search keyword sequence S consists of a sequence of words, the corresponding label sequence is composed of labels respect to each word in keyword sequence. Apparently, a label sequence can be recognized as various probable label sequences, which express diverse semantic.

For instance: the label sequence of the word sequence 'Point, based, Pineau, IJCAI' can be 'title, title, author, booktitle'.

Definition 3 (*Semantic Inversion*): Given a set of XML files and search keyword sequence $S = \{w_1, w_2, \cdots, w_k\}$, the problem of semantic inversion is to find sequential label(s) $L = \{l_1, l_2, \cdots, l_k\}$ which maximizes $Sim(S, L)$, where $Sim(S, L)$ is a function to evaluate the fitness or relevance of S and L. The answer sequences (label sequences) are given in descending order of $Sim(S, L)$. In our algorithm, the CRF model uses conditional probability $Pr(\mathbf{y}|\mathbf{x})$ as the relevance function $Sim(S, L)$.

Semantic Inversion is quite useful in keyword search. We state it in three aspects:

Semantic Inversion Can Help the Search Engine to Improve Accuracy. Traditional search engine may also recognize the word 'Pineau' in the query 'Point, based, Pineau, IJCAI' as a person's name, but after Semantic Inversion, 'Pineau' can be recognized as a author's name, rather than director's, politician's or others'. As a result, the misunderstanding of the search engine can be greatly reduced, so the search accuracy is improved.

Semantic Inversion Can Help Users Prevent Ambiguity. If the words in query have diverse semantic, Semantic Inversion can provide alternative label sequence. Since tags in XML is semantical and easy to understand, users can clarify their needs by just selecting the proper label sequence.

Semantic Inversion Can Reduce the Search Time. When the label sequence is selected, the search range has been greatly narrowed. Search engines only need to search from pages relevant to labels so the search time is greatly reduced.

3 General CRFs

CRFs(Conditional Random Fields) have been widely used by sequential algorithms, especially in sequential tagging problems, CRFs outperform other models. This section will briefly review CRFs and the General CRFs which have been applied in our algorithm.

3.1 Conditional Random Fields

Assume $\mathbf{x} = x_1, x_2, \cdots, x_n$ is the input keyword sequence and $\mathbf{y} = y_1, y_2, \cdots, y_n$ is the label(the semantic) sequence. $\mathbf{x}$ and $\mathbf{y}$ have the same length. CRF(Conditional Random Fields)[5] models the conditional probability $Pr(\mathbf{y}|\mathbf{x})$ by using a Markov random field for the structured $\mathbf{y}$, and find the best $\mathbf{y}_i$ to maximize $Pr(\mathbf{y}_i|\mathbf{x})$.

For the keyword sequence $\mathbf{x}$ and the semantic sequence $\mathbf{y}$, the *global feature vector* of CRF is the sum of all the *local feature functions*:

$$\mathbf{F}(\mathbf{y}, \mathbf{x}) = \sum_{i=1}^{n} \mathbf{f}(\mathbf{y}, \mathbf{x}, i) \tag{1}$$

CRF computes the conditional probability with parameter vector $\mathbf{w}$ by

$$Pr(\mathbf{y}|\mathbf{x}, \mathbf{w}) = \frac{e^{\mathbf{w} \cdot \mathbf{F}(\mathbf{y}, \mathbf{x})}}{Z_{\mathbf{w}}(\mathbf{x})} \tag{2}$$

where

$$Z_{\mathbf{w}}(\mathbf{x}) = \sum_{\mathbf{y}'} e^{\mathbf{w} \cdot \mathbf{F}(\mathbf{y}', \mathbf{x})} \tag{3}$$

For the given keyword sequence, the most probable label sequence maximize the conditional probability, Since $Z_{\mathbf{w}}(\mathbf{x})$ does not depend on $\mathbf{y}$, we can also say:

$$\hat{\mathbf{y}} = \arg\max_{\mathbf{y}} \mathbf{w} \cdot \mathbf{F}(\mathbf{y}, \mathbf{x}) \tag{4}$$

3.2 Learning Algorithm

Training a CRF is to learn λ for maximizing the log-likelihood of a given training set $T = \{(\mathbf{x}_k, \mathbf{y}_k)\}_{k=1}^{N}$. Meanwhile, we need to penalize the likelihood with a spherical Gaussian weight prior[6] to prevent from overfitting. So the gradient is:

$$\nabla L_{\mathbf{w}} = \sum_{k} [\mathbf{F}(\mathbf{y}_k, \mathbf{x}_k) - E_{Pr(\mathbf{y}'|\mathbf{x}_k, \mathbf{w})} \mathbf{F}(\mathbf{y}', \mathbf{x}_k)]$$
$$- \frac{\mathbf{w}}{\sigma^2} \tag{5}$$

The Learning Algorithm seeks the zero of the gradient. In other words, $\nabla L_{\mathbf{w}} = 0$.

3.3 Linear-Chain CRFs and General CRFs

Linear-chain CRFs performs well in sequential learning problems such as NP chunking[7], Part of Speech tagging[5], Opinion Expression Identification[8] and Named Entity Recognition[9]. To solve this kind of problems, the Markov field of $\mathbf{y}$ should be a linear chain, and the transition features are just between the adjacent y_i. In those applications, we suppose labels are sequential and use Linear-chain CRFs to concentrate on the dependence of the adjacent labels(or adjacent segments, Semi-Markov CRF[10]).

However, the problems of retrieval are quite different. Several labels appear together to express one subject jointly. Of course, labels are not sequential. We need to structure $\mathbf{y}$ to describe the relevance of all pairs of labels, rather than only the adjacent ones. That is general CRF. In general CRFs, the structure of labels can be a complete graph. For its complexity, general CRFs are not so commonly used as Linear-chain CRFs.

4   The Approach

Semantic Inversion can be naively solved by the random select algorithm and the greedy algorithm. Random select algorithm gives the answer randomly selected from all candidate answers. Greedy algorithm recognize each keyword x_i as the label which x_i most frequently appears in. However, both algorithms fail to consider the relevance between labels and the relevance between adjacent keywords.

To fully considerate all categories of relevance, we employ the general CRFs to model the relevance and then proposed an algorithm to solve Semantic Inversion. The algorithm weighs keyword-label relevance, label-label relevance and adjacent keywords' relevance, and uses Gradient Descent algorithm to learn best parameters for the model.

4.1   Features

In this section, we concentrate on features used in the general CRF. We need to extract textual features for quantize of relevance keywords and labels beforehand. In our algorithm, there are three categories of features for one keyword sequence-label sequence pair$(\mathbf{x}, \mathbf{y})$: f for keyword-label relevance, g for label-label relevance, h and h' for the relevance between adjacent keywords.

Feature $\mathbf{f}(\mathbf{x_i}, \mathbf{y_i})$ expresses the dependence between the keyword and the label in position i. They appear in the same position, which indicates we try to recognize the keyword x_i as the content of label y_i. How frequently x_i appears under the label y_i should be our first consideration. We measure this kind of dependence like:

$$f(x_i, y_i) = \frac{p(x_i, y_i)}{p(x_i)} \tag{6}$$

where $p(x_i, y_i)$ is the frequency that keyword x_i appear in the content of label y_i, and $p(x_i)$ denotes the frequency of keyword x_i. Since the frequency of the labels could be different one another, we will not consider this factor in feature f.

Feature $\mathbf{g}(\mathbf{y_i}, \mathbf{y_j})$ expresses the relevance of two labels. Existing methods(such as SCLA[11]) measures it mainly based on XML files' tree-like structure. We measure this relevance based on co-occurrence. The knowledge base can be seen as a set of instances(people, cities, films, etc.), and labels which often appear commonly to describe the same instances should be deeper relevant.

$$g(y_i, y_j) = \frac{p(y_i, y_j)}{p(y_i)p(y_j)} \tag{7}$$

where $p(y_i, y_j)$ is the frequency that label y_i and label y_j appear together in one instance, $p(y_i)$ denotes te frequency of label y_i and $p(y_j)$ for y_j, respectively. In general CRFs, this transform feature should be calculated between each pair of labels, which differs from the Linear-chain CRFs.

Feature $\mathbf{h}(\mathbf{x_i}, \mathbf{x_{i+1}}, \mathbf{y_i}, \mathbf{y_{i+1}})$ **and** $\mathbf{h'}(\mathbf{x_i}, \mathbf{x_{i+1}}, \mathbf{y_i}, \mathbf{y_{i+1}})$ measure the relevance of the adjacent keywords. Here we also use the co-occurrence of keywords to measure it:

$$h_0(x_i, x_{i+1}) = \frac{p(x_i, x_{i+1})}{p(x_i)p(x_{i+1})} \tag{8}$$

where $p(x_i, x_{i+1})$ denotes keyword x_i and keyword x_{i+1} appear in the content of one label.

Adjacent keywords could probably express the similar semantic. $h(x_i, x_{i+1}, y_i, y_{i+1})$ describes the contribution for recognizing adjacent keywords into the same label:

$$h(x_i, x_{i+1}, y_i, y_{i+1}) = \begin{cases} h_0(x_i, x_{i+1}), & y_i = y_{i+1} \\ 0, & y_i \neq y_{i+1} \end{cases} \tag{9}$$

If an alternative label sequences inverses the adjacent keywords into different labels, we also need to penalize. $h'(x_i, x_{i+1}, y_i, y_{i+1})$ is the penalty according to the relevance of the keywords.

$$h'(x_i, x_{i+1}, y_i, y_{i+1}) = \begin{cases} 0, & y_i = y_{i+1} \\ h_0(x_i, x_{i+1}), & y_i \neq y_{i+1} \end{cases} \tag{10}$$

All the features can be extracted quite easily. For Semantic Inversion, the joint information is contained in feature g and sequential information is concluded in feature h and h'. Feature f is the basic and the most natural features for our problem. We put these three sorts of features into general CRF, then learn the parameters.

The total weighed features are

$$\begin{aligned} \mathbf{w} \cdot \mathbf{F}(\mathbf{y}, \mathbf{x}) = & w_1 \sum_i f(x_i, y_i) + w_2 \sum_i \sum_j g(y_i, y_j) \\ & + w_3 \sum_i h(x_i, x_{i+1}, y_i, y_{i+1}) \\ & + w_4 \sum_i h'(x_i, x_{i+1}, y_i, y_{i+1}) \end{aligned} \tag{11}$$

where the parameter vector $\mathbf{w} = [w_1, w_2, w_3, w_4]$ is what we need to learn from the general CRF. Then we can use the equation (2) and (3) to calculate the probabilities of each alternative label sequence.

4.2   Parameter Learning

First, we find out all the keywords and the labels in the Training Data and find out all probable labels of each keyword. Then

5   Experiments

In our experiments, the Test Algorithm gives the best 10 probable label sequences for each keyword sequence.

Algorithm 1. The Learning Algorithm

1: **Learn**($TrainingData = \{< \mathbf{x}, \mathbf{y} >\}$)
2: Find out all the keywords and the labels in $TrainingData$, and all probable labels for each keyword.
3: Calculate the features f, g, h, h'
4: Initialize CRF
5: **repeat**
6:     Calculate ∇L by Equation (5)
7:     Modify $\mathbf{w}$ by ∇L: $\mathbf{w}' = \mathbf{w} + \nabla L$
8: **until** $\|\nabla L\| < threshold$
9: **return** CRF
10: **End Learn**

5.1 Data Source and Extraction

We use Wikipedia dataset for our experiments. Wikipedia dataset contains over 1,000,000 XML documents, involving all fields of knowledge. We randomly select 50,000 documents of them, and no fields are selected particularly.

We extract the 'keyword sequence - label sequence' pairs from the *infobox* of each XML file. The infobox of Wikipedia is the ideal source of our experiments for its neat 'attribute - content' format. We randomly select attributes as the labels, and extract part of the respective content as the keywords. For each label, selected keywords will not be more than 4. The total length of the label sequence will not be more than 6.

5.2 Evaluation

As we discussed before, our algorithm learns from the training set $T = \{(\mathbf{x}_1, \mathbf{y}_1),$ $(\mathbf{x}_2, \mathbf{y}_2), \cdots\}$ which consisting of several 'keyword sequence - label sequence' pairs. The algorithm modified the four parameters (w_1, w_2, w_3 and w_4) and imply them into the test set. We evaluate how the **answer**(label sequences) our algorithm gives resemble the **correct** (label) sequence. Here the correct sequence is the sequence of the labels, the content of which keywords are exactly contained in.

The test contains only the keyword sequence. For each keyword sequence, we generate all probable label sequences. Label sequences with too low $\sum_i f(x_i, y_i)$ feature (in other words, keywords appear too few times in these labels) will be taken out. The algorithm will grade all the probable label sequences by computing the conditional probability $Pr(\mathbf{y}'|\mathbf{x})$.

For each $\mathbf{x}$ in the test set S, we concentrate on the best label sequence(with the highest conditional probability) $\hat{\mathbf{y}}$, and compare it to the correct sequence $\mathbf{y}$. The accuracy is calculated by:

$$Acc = \frac{\sum\limits_{\mathbf{x} \in S} Match(\hat{\mathbf{y}}, \mathbf{y})}{\sum\limits_{\mathbf{x} \in S} Length(\mathbf{x})} \qquad (12)$$

and

$$Match(\hat{\mathbf{y}}, \mathbf{y}) = \sum_i eq(\hat{\mathbf{y}}, \mathbf{y}, i) \tag{13}$$

Another concentration is the accuracy of the best sequence within top-N sequences(In our experiments, $N = 4, 7, 10$). In the real occasion, the search engine should provide users with several alternative results within the first page so that users can choose the best one for their own. Since some input keyword sequences are originally ambiguous, the Top-N accuracy can sometimes be more convincing.

Algorithms sorts the label sequences by the conditional probability, and the set $TopN(\mathbf{x})$ contains the top-N sequences. The Top-N accuracy is calculated by:

$$AccN = \frac{\sum\limits_{\mathbf{x} \in S} \max\limits_{\mathbf{y}' \in TopN(\mathbf{x})} Match(\mathbf{y}', \mathbf{y})}{\sum\limits_{\mathbf{x} \in S} Length(\mathbf{x})} \tag{14}$$

In the other hand, we want to know how the algorithm ranked the real correct sequence. If the algorithm is efficient, the rank of correct label sequence should be small. We also evaluate the algorithm by how frequently the correct label sequence appear in Top-N($N = 1, 4, 7, 10$) answers(label sequences). This could show whether the correct sequence has been highly scored.

5.3   Results and Discussion

We use the cross-validation to evaluate the experiments. All the 'keyword sequence - label sequence' pairs are split into 10 parts. At each time, 9 parts are used for learning and one part for testing. The code is written by C++. The programs are performed on a server with 4 core processors and 16GB memory.

Accuracy of the First Answer. Our algorithms(73.2%) outperforms the baseline algorithms(38.1% for Random Select Algorithm and 61.2% for Greedy Algorithm), which confirms our assumption that fully consideration of various categories of relevance can improve the quality of semantic reversion.

Accuracy of Top-N Answers. Figure (a) shows the accuracy of the Top-N answers, $N = 1, 4, 7, 10$. This accuracy is calculated by Equation (11). Users can select the best answer from the N answers our algorithm gives. The best accuracy of Top10 answers can be over 90%! That is to say: Users can lead the search engine to understand the precise semantic by no more than 10% extra work. That is quite exciting.

The Rank of the Correct Answer. Figure (b) shows how our algorithm ranks the correct answer. Nearly half of the correct answers are ranked at the first place. Over 80% of cases, the real correct sequence has been ranked before 10 and will be shown within the first page of the results. If so, the only thing users need to do is selecting.

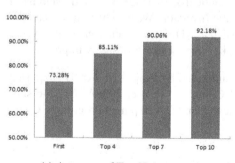

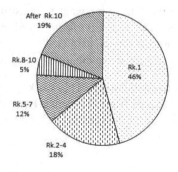

(a) Accuracy of Top-N Answers   (b) The Rank of the Correct Answer

6 Related Work

In this section, we will present some related work around the utilization of Conditional Random Fields and algorithms for the keyword search on structured database.

Semi-Markov Conditional Random Fields[10] split the sequence into several segments. Their goal is to find the best segmentation maximizing the conditional probability which is defined by the CRF model. Semi-CRFs perform very well in NER problems[12]. For our problem, Semi-CRF can easily find the phrases and the names contained in input sequence, but will not fully describe the joint semantic of the labels because Semi-CRFs are mainly based on the Linear-chain CRFs.

On structured data, there is a work focusing on Keyword Query Reformulation[2]. The reformulated queries provide alternative descriptions of original input, so as to better capture users information need and guide users to explore related items in the target structured data. The data are modeled with a heterogenous graph, and a probabilistic generation model is utilized for query reformulation. Its aim is to help users to claim the semantic clearly.

In the field of RDF, recent work provides QUICK[3], a novel system for helping users to construct semantic queries in a given domain. QUICK works with the schema graph and the query templates. Users can conveniently express their search intent by increasingly selecting the semantic and the structure provided by QUICK. What is more, an online system with a user-friendly interface has been established based on QUICK.

7 Conclusions and Future Work

In nowadays keyword search, good search systems should understand users' intent deeply. How to recognize and represent the keyword semantic becomes more and more important. In XML data, labels are naturally semantical, so recognizing the keywords into XML labels is what we need to concentrate on. In this paper we define this process as Semantic Inversion and model it with general conditional random fields. From our experiments, our algorithms can efficiently recognize the keywords into labels and top-k label sequences also provide users with the chance to reclaim their real search intent.

In the future, we want to construct a semantical knowledge base and establish a better XML retrieval system based on Semantic Inversion. We also hope to expand the Semantic Inversion to other structured data, such as RDF. If semantic can be simply and accurately inversed into other explicit forms, keyword search will surely improve.

Acknowledgement. This work is partially supported by Project 61170091 supported by National Natural Science Foundation of China and Project 2009AA01Z136 supported by the National High Technology Research and Development Program of China (863 Program).

References

1. Tran, T., Wang, H., Rudolph, S., Cimiano, P.: Top-k exploration of query candidates for efficient keyword search on graph-shaped (rdf) data. In: International Conference on Data Engineering - ICDE 2009, pp. 405–416 (2009)
2. Yao, J., Cui, B., Hua, L., Huang, Y.: Keyword Query Reformulation on Structured Data. In: International Conference on Data Engineering, ICDE (2012)
3. Zenz, G., Zhou, X., Minack, E., Siberski, W., Nejdl, W.: From keywords to semantic queries - Incremental query construction on the semantic web. Journal of Web Semantics 7(3), 166–176 (2009)
4. Pandey, S., Punera, K.: Unsupervised Extraction of Template Structure in Web Search Queries. In: International World Wide Web Conference - WWW (2012)
5. Lafferty, J.D., McCallum, A., Pereira, F.C.N.: Conditional Random Fields: Probabilistic Models for Segmenting and Labeling Sequence Data. In: International Conference on Machine Learning, ICML 2001, pp. 282–289 (2001)
6. Chen, S.F., Rosenfeld, R.: A Gaussian Prior for Smoothing Maximum Entropy Models. Technical Report CMU-CS-99-108, Carnegie Mellon University (1999)
7. Sha, F., Pereira, F.C.N.: Shallow parsing with conditional random fields. In: North American Chapter of the Association for Computational Linguistics, NAACL (2003)
8. Breck, E., Choi, Y., Cardie, C.: Identifying expressions of opinion in context. In: International Joint Conference on Artificial Intelligence, IJCAI, pp. 2683–2688 (2007)
9. McCallum, A., Li, W.: Early Results for Named Entity Recognition with Conditional Random Fields, Feature Induction and Web-Enhanced Lexicons. In: Proceedings of the Seventh Conference on Natural Language Learning - CoNLL (2003)
10. Sarawagi, S., Cohen, W.W.: Semi-Markov Conditional Random Fields for Information Extraction. In: Neural Information Processing Systems, NIPS (2004)
11. Xu, Y., Papakonstantinou, Y.: Efficient keyword search for smallest LCAs in XML databases. In: International Conference on Management of Data - SIGMOD, pp. 527–538 (2005)
12. Okanohara, D., Miyao, Y., Tsuruoka, Y., Tsujii, J.: Improving the Scalability of Semi-Markov Conditional Random Fields for Named Entity Recognition. In: Meeting of the Association for Computational Linguistics. ACL (2006)

Automatically Training Form Classifiers

Mauricio C. Moraes[1], Carlos A. Heuser[1], Viviane P. Moreira[1],
and Denilson Barbosa[2]

[1] Instituto de Informática, Universidade Federal do Rio Grande do Sul
{mcmoraes,heuser,viviane}@inf.ufrgs.br
[2] Department of Computing Science, University of Alberta
denilson@ualberta.ca

Abstract. The state-of-the-art in domain-specific Web form discovery relies on supervised methods requiring substantial human effort in providing training examples, which limits their applicability in practice. This paper proposes an effective alternative to reduce the human effort: obtaining high-quality domain-specific training forms. In our approach, the only user input is the domain of interest; we use a search engine and a focused crawler to locate query forms which are fed as training data into supervised form classifiers. We tested this approach thoroughly, using thousands of real Web forms from six domains, including a representative subset of a publicly available form base to validate this approach. The results reported in this paper show that it is feasible to mitigate the demanding manual work required by some methods of the current state-of-the-art in form discovery, at the cost of a negligible loss in effectiveness.

Keywords: deep web, hidden web, domain-specific search, query form discovery.

1 Introduction

Form discovery is an essential step in many important applications, such as Deep Web crawling [21, 25], Web information integration [12, 18, 26] and vertical search engines [1, 5]). Form discovery can be divided into two main tasks: (i) *locating* HTML pages containing forms on the Web, and (ii) *identifying* among those forms which are relevant to the application at hand, for instance, based on relevance to a domain of interest. The state-of-the-art in form discovery relies on supervised machine-learning methods, and thus require considerable involvement from the human expert in order to operate. More automatic methods could opportunely be employed as direct replacement or as complement for existing methods of form discovery, which heavily depend on the human expert.

It could be argued that one could leverage publicly available form bases [3, 4] as training examples. However, such bases are composed of forms that belong to a predefined set of domains, written mainly in English. They are of no avail for other domains and/or languages, for instance. Moreover, there are no guaranties that these bases will remain publicly available in the future. Finally, as we show

X. Lin et al. (Eds.): WISE 2013, Part I, LNCS 8180, pp. 441–453, 2013.

in this paper, the form bases are not perfect, and a small fraction of the forms in them are misclassified, which would lead to poor training examples.

In this paper, we combine and evaluate several heuristics for pre-query discovery of domain-specific structured forms on the Web. In other words, we are interested in rule-based strategies that do not require form submission (i.e., that use the pre-query approach [22]) for the discovery of domain-specific forms that contain an implicit schema providing hints about the structure of their underlying data sources (i.e., structured forms). Domain-independent form discovery [21] and the discovery of unstructured forms are outside the scope of this paper.

Individually, each heuristic we evaluate is either borrowed or inspired by previous work. The novel aspect in this paper is the way in which we combine the referred heuristics and the purpose to which we employ them. The final goal of this work is to mitigate the demanding manual work required by the current state-of-the-art in form discovery, and thus relieve the human expert from tiresome work while discovering domain-specific structured HTML forms. We achieve our goal through the automatic creation of training bases for two state-of-the-art form classifiers. To the best of our knowledge, this is the first work on automatically training Web form classifiers. The main contributions of this paper are:

- The proposal, combination and evaluation of several heuristic-based methods for pre-query discovery of domain-specific structured HTML forms on the Web; and
- Substantial empirical evidence that it is feasible to automatically train two supervised form classifiers that compose the current state-of-the-art in form identification, at the cost of a relatively small loss in effectiveness.

The remainder of the paper is organized as follows: Section 2 discusses the related work and Section 3 explains the two form bases we use in our experiments: one of which was collected by us, while the other is a publicly available form base. Sections 4, 5 and 6 propose, combine and evaluate several heuristic-based methods for the automatic creation of training examples for two state-of-the-art form classifiers. Section 7 concludes this paper.

2   Related Work

Many methods for domain-specific form discovery were developed in the last years. These methods can be functionally organized into five groups, namely: (i) *Deep Web crawlers*, (ii) *form classifiers*, (iii) *form crawlers*, (iv) *form rankers*, and (v) *form clusterers*. One of the main distinctions among these broad groups is their primary goal: while form crawlers aim at locating forms, form classifiers, form clusterers, and form rankers aim at identifying forms. Deep Web (DW) crawlers aim at providing both location and identification of forms. Refer to [22] for an in-depth discussion on the subject.

In short, all of these groups of related work directly or indirectly rely on significant human intervention to locate and/or identify domain-specific relevant forms on the Web. DW crawlers require knowledge bases that are hard

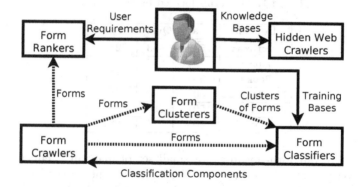

Fig. 1. Dependency of domain-specific structured form discovery groups of techniques on the human expert and on each other

to be automatically built; form classifiers demand laborious manual training; form crawlers are built on top of form classifiers; finally, form rankers and form clusterers indirectly, although not necessarily, depend on form crawlers.

Figure 1 illustrates how the referred groups of related work depend on the human expert and on each other. The arrows denote the dependencies, pointing to the dependent on the relations. The dashed arrows represent optional (but desirable) dependencies.

There are two exceptions to this "heavy" dependency scenario: (i) the DW crawler created by Bergholz and Chidlovskii [11], which requires from the human expert only a small set of keywords related to the *domain of interest* (i.e., the domain in which the discovery should occur); and (ii) the form ranker created by Lin and Chen [20], which requires from the human expert only the identification of Google Directory categories related to the domain of interest. Nevertheless, both of these works apply only to unstructured or keyword-based forms, which are outside the scope of this paper. Moreover, the DW crawler of Bergholz and Chidlovskii is not fully reproducible, since it is based on the Xelda Server [6], a proprietary solution from Xerox for which there are no published details.

3   Experimental Setting

We start with our form locator, which was inspired by that of Bergholz and Chidlovskii [11]. The main difference between their solution and ours is that theirs uses a Web Directory as the source of relevant URLs to crawl, while ours uses a general-purpose Web search engine (e.g., Google, Yahoo!, Bing, etc.). Like theirs, our form locator performs limited-depth in-site crawling[1], since there is evidence that the vast majority of query forms can be found up to three links in depth from the home page of a Web site [11–13].

---

[1] In-site crawlers are restricted to (i.e., follow only) the links that point to pages that belong to the same site.

Table 1. The forms located by our proposal

Domain	Located	Query	On-topic
Airfares	502	353	240
Autos	504	286	155
Books	504	344	161
Car Rentals	501	318	183
Hotels	501	398	280
Jobs	503	331	282

The input to our form locator is a small set of manually defined keywords that describe the domain of interest. This input is sent directly as a query to a general-purpose Web search engine and the resulting URLs are used as seed URLs (i.e., the initial set of URLs) by our limited-depth in-site crawler, which collects some of the forms it can find in a Web site.

The Small Test Set. Table 1 provides the details about the experimented domains, their descriptions and their respective located forms, which were manually classified by function and by domain. The keywords sent to the search engine were, literally, the names of the domains of interest, exactly as they are shown in Table 1. The *Query* column indicates the number of query forms found among the ones located for each domain. The *On-topic* column indicates the number of all forms (i.e., query and non-query forms) that belong to their respective domains. Notice that we considered as on-topic only the forms that were written in English (i.e., forms from the domains of interest but written in any languages other than English were considered misidentified).

For clarity, hereafter we refer to the set of forms we collected and verified as the *small test set*, to distinguish it from the random sample of the publicly available set of forms discovered by the DeepPeep search engine [2, 3].

The Big Test Set. The big test set is a manually classified random sample containing more than 10,000 forms from the DeepPeep form base after filtering for the removal of replicas. We identified replicated forms observing two *renderable properties* of forms (i.e., properties that can be rendered by a web browser). More specifically, we considered replicas the forms that present: (i) identical *renderable texts* (i.e., the form itself minus its HTML tags); and (ii) the same number of *renderable fields* (i.e., all fields except the hidden ones). There is no point in allowing replicas in the test base, since it is likely that the more disparate the forms of the test base, the more accurately is measured the effectiveness of the form identifier tested against it.

The big test set is detailed in Table 2. The *Total* column indicates the total number of forms present in the DeepPeep base as it was downloaded from the Web site [2] of one of its authors. The *Distinct* column indicates the number of distinct (i.e., non-replicated) forms found on the same base. The *Sample* column indicates the sizes of the random samples, which were built according to a confidence level of 95% and a confidence interval of 2%. The semantics of the columns *Query* and *On-topic* are the same of Table 1.

Table 2. Statistics about the Big Test Set

Domain	Total	Distinct	Sample	Query (Error)	On-topic (Error)
Airfares	3,641	1,955	1,307	1,264 (3.3%)	1,240 (5.2%)
Autos	9,972	7,565	2,111	1,846 (12.6%)	956 (54.8%)
Books	2,035	1,809	1,305	1,266 (3.0%)	1,130 (13.5%)
Car Rentals	2,515	1,995	1,316	1,140 (13.4%)	906 (32.2%)
Hotels	20,303	13,417	2,361	2,172 (8,1%)	1,490 (36.9%)
Jobs	14,958	11,573	2,261	1,896 (16.2%)	1,088 (51.9%)

The form discoverer of the DeepPeep Project is supposed to retrieve only domain-specific query forms [10]. In other words, no non-query form was expected to be found in the big test set. Also, among the forms attributed to a domain in the big test set, only the ones that actually belong to that domain were expected to be found. Therefore, since we found both non-query forms and forms that do not belong to their respective domains in the big test set, it was possible to attribute an error-rate to the DeepPeep form discoverer with respect to both identifications by function and by domain. These error rates are shown in Table 2 between parentheses in the cells of the columns *Query* and *On-topic*, respectively. We did not analyze in details the forms that the DeepPeep's discoverer misidentified, but we are convinced that such bad behavior was caused by two main reasons. First, it had to identify forms from domains that present vocabulary overlap (i.e., they share words of their characteristic vocabularies [9]). Second, it had to identify forms from the domain of interest, but that were written in a language other than English.

Evaluation Metrics. The measures used to assess the quality of the methods were the same described by Barbosa and Freire [9], namely: Recall (R), Precision (P), F1, Accuracy (A) and Specificity (S), which are defined in terms of the number of true positive (TP), true negative (TN), false positive (FP), and false negative (FN) classifications produced by a method:

$$R = \frac{TP}{TP + FN} \qquad P = \frac{TP}{TP + FP} \qquad S = \frac{TN}{TN + FP}$$

$$A = \frac{TP + TN}{TP + TN + FP + FN} \qquad F1 = \frac{2 * R * P}{R + P}$$

4   Identifying Forms by Function

As customary in this field [11, 14, 15, 23, 24], we worked with three form features in order to perform form identification by function: (i) the use of the HTTP `Get` submission method; (ii) the absence of password fields; and (iii) the presence of

Table 3. Effectiveness of GFC on the big test set with automatic training

Scenario	R	P	F1	S	A
Baseline (manual training)	0.907	0.986	0.945	0.867	0.904
Automatic (trained for `Airfares`)	0.858	0.959	0.906	0.677	0.840
Automatic (trained for `Autos`)	0.855	0.963	0.906	0.713	0.841
Automatic (trained for `Books`)	0.851	0.967	0.905	0.744	0.840
Automatic (trained for `Car Rentals`)	0.858	0.959	0.906	0.677	0.840
Automatic (trained for `Hotels`)	0.849	0.968	0.905	0.750	0.839
Automatic (trained for `Jobs`)	0.849	0.968	0.905	0.750	0.839

the word *"search"* in the body of a form. These features were combined in the following seven heuristics, which identify any given form as a query form if it:

1. uses the HTTP `Get` submission method;
2. does not present a password field;
3. presents the word *"search"*;
4. is identified as a query form by heuristics #1 and #2;
5. is identified as a query form by heuristics #1 and #3;
6. is identified as a query form by heuristics #2 and #3;
7. is identified as a query form by heuristics #1, #2 and #3.

4.1 Reducing the Manual Effort in Identifying Forms by Function

We now report on an experiment to assess whether our proposed heuristic-based method for form identification by function can mitigate the manual work required from the human expert while discovering forms, using the state-of-the-art form classifier GFC [9]. It should be noted that GFC requires dozens to hundreds of training examples.

Direct classification against the small and the big test sets showed that heuristic #6 is the most effective, so this is the one we use. As positive examples, we used forms from the small test set identified as query forms, while the negative examples were other forms of the same test not classified as query forms. As the baseline, we manually train GFC using randomly chosen query forms from all domains of the big test set. We used 220 positive examples and 220 negative examples in each execution (c.f. [9]), and we report the average scores of 100 executions. In all our experiments, there is no overlap between the training and test sets.

The evaluation was done on the big test set (recall Table 2), and the results are shown in Table 3. We report the results of the automatic method by domain, since a different classifier was used in each case. The results provide empirical evidence that GFC is virtually as effective when trained using our heuristic-based method, relatively to when it receives manual training. Indeed, the losses summed up to approximately 4% in F1 measure.

5   Identifying Forms by Domain

We developed three heuristics for form identification by domain. The first heuristic is able to recognize at least some of the words that characterize the attribute labels frequently present in the forms of the domain of interest [16, 17], by means of measuring the frequencies of the renderable words present in domain-specific query forms and selecting the most frequent ones.

Algorithmically, the main steps of the first heuristic are detailed as follows, where $0 \leq \alpha \leq 1$ is a manually defined threshold:

1. Locate a set of forms from the domain of interest using the method presented in Section 3;
2. Identify the query forms among the ones located in step #1, using the method presented in Section 4;
3. Compute occurrence statistics for the words that compose the renderable texts of the query forms identified in step #2;
4. Return the set of words that occur in at least α of the query forms identified in step #2 (i.e. the vocabulary that characterizes the domain of interest).

Once the vocabulary for the domain of interest is defined, the second heuristic is used to identify any given form as belonging to the desired domain or not. The rationale of the second heuristic is straightforward: a form is considered as belonging to the domain of interest if it contains at least N of the words present in the vocabulary that characterizes the domain of interest, where N is a manually defined input parameter.

Finally, the third heuristic ranks forms (inspired by Kabra et al. [19]), and uses the top-F most relevant ones as training examples for a form classifier. The ranking of forms by relevance is based on the presence of the top-W most frequent words in the domain vocabulary domain (as per heuristic 1). The rationale that substantiates the ranking-based training is straightforward: the more relevant words a form contains in its renderable text, the more relevant it will be as a training example.

We experimented with different values for α, N, F and W, and set them to $\alpha = 0.29$, $N = 2$, $F = 100$ and $W = 5$, based on empirical evidence.

5.1   Experiments

In order to test if our proposed heuristic-based method for form identification by domain is effective in reducing the training effort for form discovery, we performed experiments with the state-of-the-art form classifier DSFC [9], which also requires dozens to hundreds of manually labeled forms as training examples. We report the effectiveness of DSFC for binary classifications for each of the domains. We use two scenarios, differing in the mix of positive and negative training examples.

In the first scenario, which we call *pure-domain*, all negative examples come from an extraneous domain (biology in this case). In the second scenario, which we call *mixed-domain*, the negative examples come from any of the domains

Table 4. DSFC's results in the pure-domain scenario

Domain	Manual Training					Automatic Training				
	R	P	F1	S	A	R	P	F1	S	A
Airfares	0.991	0.438	0.608	0.767	0.801	0.867	0.759	0.809	0.938	0.925
Autos	0.988	0.483	0.649	0.868	0.882	0.910	0.826	0.866	0.969	0.960
Books	0.982	0.591	0.738	0.891	0.903	0.937	0.907	0.922	0.980	0.973
Car Rentals	0.988	0.279	0.435	0.703	0.732	0.772	0.433	0.555	0.845	0.835
Hotels	0.982	0.333	0.497	0.531	0.618	0.590	0.660	0.623	0.915	0.844
Jobs	0.973	0.479	0.642	0.839	0.857	0.700	0.938	0.802	0.991	0.944

Table 5. DSFC's error rates in the pure-domain scenario

Manual Training						
	Airfares	Autos	Books	Car Rentals	Hotels	Jobs
Airfares	0.8%	2.7%	1.7%	65.3%	39.2%	6.6%
Autos	7.5%	1.1%	6.3%	40.1%	4.8%	15.3%
Books	4.4%	5.6%	1.7%	8.6%	5.7%	31.7%
Car Rentals	72.3%	21.0%	1.5%	1.1%	30.3%	16.8%
Hotels	93.0%	22.8%	11.2%	78.3%	1.7%	26.0%
Jobs	20.5%	11.2%	11.5%	31.3%	9.5%	2.6%
Automatic Training						
	Airfares	Autos	Books	Car Rentals	Hotels	Jobs
Airfares	13.2%	0.1%	0.9%	14.2%	12.3%	1.5%
Autos	1.1%	8.9%	2.2%	4.1%	2.8%	5.6%
Books	0.3%	0.1%	6.2%	0.1%	0.0%	9.3%
Car Rentals	5.5%	1.6%	1.7%	22.7%	8.6%	5.9%
Hotels	27.9%	0.4%	1.5%	7.1%	40.9%	1.6%
Jobs	0.4%	0.5%	2.4%	1.1%	0.1%	29.9%

in Table 1 (e.g., the negative examples for the `Airfares` domain would come from `Autos`, `Books`, and so on). Biology was chosen as the extraneous domain for the first scenario as its vocabulary has a relatively small overlap with the other domains, which are all business-related. Thus, the two scenarios emulate the best-case and worst-case for our method, in a sense.

The Pure-Domain Scenario. For this experiment, the results for the manually trained baseline are the average of ten runs. In each run, 220 random forms from the domain of interest and 220 random forms from the `Biology` domain present in the DeepPeep form base (c.f. the methodology in Barbosa and Freire [9]). As for the automatic training, the positive and negative examples of forms employed to automatically train DSFC came from the small test set. More concretely, for each domain of interest, the positive training examples were the forms that the method identified as belonging to the domain of interest and the

Table 6. DSFC's results in the mixed-domain scenario

Domain	Manual Training					Automatic Training				
	R	P	F1	S	A	R	P	F1	S	A
Airfares	0.971	0.898	0.933	0.978	0.977	0.867	0.790	0.827	0.948	0.933
Autos	0.993	0.937	0.964	0.991	0.991	0.884	0.964	0.922	0.994	0.979
Books	0.991	0.981	0.986	0.996	0.996	0.986	0.978	0.982	0.995	0.994
Car Rentals	0.970	0.805	0.880	0.971	0.971	0.843	0.835	0.839	0.974	0.957
Hotels	0.973	0.870	0.919	0.964	0.965	0.710	0.918	0.801	0.982	0.922
Jobs	0.975	0.930	0.951	0.988	0.986	0.927	0.992	0.958	0.998	0.987

Table 7. Manually trained DSFC's error rates in the mixed-domain scenario

Manual Training						
	Airfares	Autos	Books	Car Rentals	Hotels	Jobs
Airfares	2.8%	0.0%	0.3%	7.3%	2.7%	0.4%
Autos	0.1%	0.6%	0.1%	1.9%	0.8%	1.4%
Books	0.0%	0.1%	0.8%	0.0%	0.0%	1.4%
Car Rentals	6.6%	2.3%	0.2%	2.9%	2.6%	1.8%
Hotels	0.8%	0.9%	0.3%	3.9%	2.6%	2.8%
Jobs	0.3%	0.7%	2.7%	1.0%	0.9%	2.5%
Automatic Training						
	Airfares	Autos	Books	Car Rentals	Hotels	Jobs
Airfares	13.2%	0.7%	0.2%	6.8%	13.9%	0.5%
Autos	0.0%	11.5%	0.0%	1.4%	0.5%	0.8%
Books	0.0%	0.8%	13.2%	0.0%	0.0%	1.4%
Car Rentals	7.8%	2.8%	0.0%	15.6%	13.4%	0.4%
Hotels	5.0%	0.0%	0.0%	2.4%	28.9%	0.7%
Jobs	0.0%	0.0%	0.2%	0.0%	0.2%	7.2%

negative examples were the forms that the method identified as not belonging to the domain of interest.

Table 4 report the effectiveness of DSFC in this experiment. A closer look at shows that although the automatic training has lower recall values (losing by 5% to 40%, for an average loss of 19%), it has far superior precision (winning by 53% to 98%, for an average improvement of 74%). Consequently, the gains in F1 (28% on average) and Accuracy (15% on average) are significant.

Table 5 shows error rates of the identifications based on the vocabularies that characterize the domains named in the first column. The columns present the error rates of the identifications of the on-topic forms of the big test set related to the domains named in the first line of each column. More concretely, the results of manual training presented in Table 5 shows that the classifier misclassified: (i) 0.8% of the forms of the Airfares domain when it used the vocabulary of the Airfares domain; (ii) 2.7% of the forms of the Autos domain when it used the vocabulary of the Airfares domain; (iii) 7.5% of the forms of the Airfares domain when it used the vocabulary of the Autos domain; and so on.

The Mixed-Domain Scenario. For this experiment, the results of the manually trained classifier correspond to the average of ten runs executed for each domain. In each run, 220 random forms from the domain of interest and 45 (i.e., 220/5) random forms from each of the other five domains present in the big test set composed the positive and negative training examples, respectively. As for the automatic training of DSFC, all examples came from the small test set, with the positive examples belonging to the domain of interest and the negative examples belonging to all the domains, with equal distributions.

Tables 6 and 7 report the effectiveness of DSFC in this scenario. As with the pure-domain case, the automatic training has worse recall than the manual training approach (varying from 0.5% to 27%, for an average loss of 11.2%) but better precision overall (up to 6.7% better), except for `Airfares` (loss of 12% in precision). In the end, the F1 score for the automatic approach is on average 5.5% worse, and the overall accuracy is almost 2% worse. Nevertheless, the automatic approach slightly outperformed the manual approach for the `Jobs` domain both in the F1 and Accuracy scores.

6   Further Discussion

This section discusses some of the harder questions raised during the development of this work. It provides interesting information that more completely characterizes the contributions of this paper.

What is the minimum number of forms that have to be located through the search engine so that our proposal presents its best results? Our proposal relies on the ability of the search engine to return a significant number of relevant URLs in response to a keyword-based query consisting of descriptive keywords for the domain of interest. Since forms are sparsely distributed on the Web [7, 8], it is reasonable to expect that there are domains for which search general engines may be unable to find a significant number of relevant URLs.

We simulated such a pessimistic scenario by restricting the number of relevant forms used in the pure-domain scenario of the experiment discussed in Section 5.1. For each domain, we executed several slightly modified executions of that experiment: the first one used only 10 relevant forms, the second one used 20, the third one used 30, and so on, up to the total of relevant forms located by our form locator. In all executions the number of irrelevant forms was left intact (i.e., the number of irrelevant forms used in each run was the total number of irrelevant forms located by the form locator).

The results showed that our method for form identification by domain achieves its best results when at least approximately 150 relevant forms are found among the approximately 500 forms located per domain. This is empirical evidence that our discoverer works reasonably well in the domains where it can locate at least approximately one relevant form for each three located forms.

What is the impact on the effectiveness of the discovery process when different keywords are sent to the search engine? Arguably, the results of our approach are highly dependent on the query sent to the search engine. In order to asses how the results vary depending on the keywords sent to the search engine, we performed two experiments that were executed in the first scenario for form identification, again on the pure-domain setting described in Section 5.1.

The first experiment assessed the impact of the use of "special terms" as keywords to be sent to the search engine together the domain name. The experimented special terms (i.e., words supposed to give a hint to the search engine that Web sites that present query forms are desired) were: *"online"* and *"search"*. The results show that the use of special terms do not consistently affect the effectiveness of our method for form identification by domain: in most cases, the results are similar, but in some cases the method become completely ineffective. We omit the numbers here in order to save space.

In the second experiment, we sent to the search engine different descriptive keywords for the experimented domains (e.g., *"car"* and *"vehicles"* for the Autos domain, *"employment"* and *"careers"* for the Jobs domain, etc). The Car Rentals domain was not experimented because we could not define reasonable alternative keywords for it. The results (omitted here in order to save space) show that different descriptive keywords lead our method to achieve significantly different results, often worse than the ones achieved using the names of the experimented domains, as described in the previous sections.

7   Conclusion

The very nature of the Web (i.e., it is huge, dynamic and decentralized in a way that pages are freely and autonomously added, deleted, or modified) compels Web processing methods to evolve from manual to completely unsupervised. Approaches for domain-specific structured form discovery are no exception, but the state-of-the-art in form discovery, directly or indirectly, still relies on significant human intervention. Consequently, novel methods that mitigate the demanding manual work required by the current state-of-the-art in form discovery are welcomed.

This paper proposes, combines and evaluates a group of heuristic-based domain-specific structured form discovery methods that require almost no involvement from the human expert in order to operate, pushing the state-of-the-art in form discovery towards the evolutionary goal of complete unsupervised execution. The only manual input needed by our methods is the definition of a few keywords that appropriately describe the domain of interest (i.e., the domain in which the form discovery should occur).

Our experiments ran on six domains and employed real Web data composed of thousands of forms, including a representative subset of the publicly available DeepPeep form base [2, 3]. Somewhat surprisingly, the automatic training in the pure-domain scenario led to better results than manual training. In the mixed-domain scenario, automatic training led to classifiers that were very effective

($F1 >= 0.8$) but poorer than manual training for the domains that present vocabulary overlap. Of course, in a real application, the actual mix of domains will sit somewhere between the two scenarios described above. Thus, overall, it is fair to expect that automatic training will be slightly less effective, but requiring no human effort in training.

Our form locator is able to locate a significant number of relevant forms from all the experimented domains, providing enough data to our heuristic-based form identification methods. Moreover, the form locator relies on a commercial search engine, and the best results were not dependent on using obscure or highly specific search keywords. Also, the number of hits needed for the method to work was fairly small. Thus, the observed results show that the heuristic approach described in this paper is able to effectively replace manual training for the cutting-edge form classifiers GFC and DSFC.

There are opportunities for future work. To start with, it would be interesting to further investigate the performance of the classifiers in detail. Also, it would be good to understand why adding some keywords hurts the performance of the form locator, and whether there are ways to find other keywords that would actually help. Finally, our approach could be extended into a "query-by-example" method, where the user could give a sample of some data of interest, and the system would locate forms containing more similar data.

Acknowledgments. This work has been partially supported by the Brazilian National Institute of Science and Technology for the Web (CNPq Grant No. 573871/2008-6), by CNPq projects No. 480283/2010-9 and 478979/2012-6, and of the Natural Sciences and Engineering Research Council of Canada. Mauricio C. Moraes received a scholarship from CNPq while working on this paper. He also received a scholarship from Capes during his visit to the University of Alberta, Canada, where he worked on this paper.

References

[1] Cazoodle, http://www.cazoodle.com/ (last access in January 2011)

[2] DeepPeep repository, http://www.cs.utah.edu/~lbarbosa/forms/forms.tar.gz (last access in January 2011)

[3] DeepPeep. http://www.deeppeep.org/ (last access in October 2009)

[4] UIUC web integration repository, http://metaquerier.cs.uiuc.edu/repository (last access in October 2009)

[5] Trulia, http://www.trulia.com/ (last access in January 2011)

[6] Xelda, http://www.xrce.xerox.com/Research-Development/ Historical-projects/XeLDA/ (last access in January 2011)

[7] Barbosa, L., Freire, J.: Searching for hidden-web databases. In: Proceedings of the Eight International Workshop on the Web and Databases, vol. 5, pp. 1–6. ACM (2005)

[8] Barbosa, L., Freire, J.: An adaptive crawler for locating hidden-web entry points. In: Proceedings of the 16th International World Wide Web Conference, pp. 441–450. ACM, New York (2007)

[9] Barbosa, L., Freire, J.: Combining classifiers to identify online databases. In: Proceedings of the 16th International World Wide Web Conference, pp. 431–440. ACM, New York (2007)

[10] Barbosa, L., Nguyen, H., Nguyen, T., Pinnamaneni, R., Freire, J.: Creating and exploring web form repositories. In: Proceedings of the ACM SIGMOD International Conference on Management of Data, pp. 1175–1178. ACM (2010)

[11] Bergholz, A., Chidlovskii, B.: Crawling for domain-specific hidden web resources. In: Proceedings of the 4th International Conference on Web Information Systems Engineering, pp. 125–133 (2003)

[12] Chang, K., He, B., Zhang, Z.: Toward large scale integration: Building a meta-querier over databases on the web. In: Second Biennial Conference on Innovative Data Systems Research, pp. 44–55 (2005)

[13] Chang, K.C.-C., He, B., Li, C., Patel, M., Zhang, Z.: Structured databases on the web: observations and implications. SIGMOD Rec. 33(3), 61–70 (2004)

[14] Doorenbos, R.B., Etzioni, O., Weld, D.S.: A scalable comparison-shopping agent for the world-wide web. In: International Conference on Autonomous Agents, pp. 39–48. ACM, New York (1997)

[15] Gong, Z., Zhang, J., Liu, Q.: Hidden-Web Database Exploration. In: Sixth International Conf. on Intelligent Systems Design and Applications, pp. 838–843. IEEE (2006)

[16] He, B., Tao, T., Chang, K.: Organizing structured web sources by query schemas: a clustering approach. In: Proceedings of the 2004 ACM CIKM International Conference on Information and Knowledge Management, pp. 22–31. ACM (2004)

[17] He, B., Tao, T., Chang, K.C.-C.: Clustering structured web sources: A schema-based, model-differentiation approach. In: Lindner, W., Fischer, F., Türker, C., Tzitzikas, Y., Vakali, A.I. (eds.) EDBT 2004. LNCS, vol. 3268, pp. 536–546. Springer, Heidelberg (2004)

[18] He, H., Meng, W., Yu, C., Wu, Z.: Wise-integrator: an automatic integrator of web search interfaces for e-commerce. In: Proceedings of 29th International Conference on Very Large Data Bases, pp. 357–368 (2003)

[19] Kabra, G., Li, C., Chang, K.: Query routing: Finding ways in the maze of the DeepWeb. In: International Workshop on Challenges in Web Information Retrieval and Integration, pp. 64–73 (2005)

[20] Lin, K., Chen, H.: Automatic information discovery from the invisible Web. In: International Conference on Information Technology: Coding and Computing, pp. 332–337 (2002)

[21] Madhavan, J., Ko, D., Kot, L., Ganapathy, V., Rasmussen, A., Halevy, A.: Google's deep web crawl. PVLDB 1(2), 1241–1252 (2008)

[22] Moraes, M.C., Heuser, C.A., Moreira, V.P., Barbosa, D.: Prequery discovery of domain-specific query forms: A survey. IEEE Trans. Knowl. Data Eng. 25(8), 1830–1848 (2013)

[23] Peng, Q., Meng, W., He, H., Yu, C.: WISE-Cluster: Clustering e-commerce search engines automatically. In: Sixth ACM CIKM International Workshop on Web Information and Data Management, pp. 104–111. ACM (2004)

[24] Raghavan, S., Garcia-Molina, H.: Crawling the hidden web. In: Proceedings of 27th International Conference on Very Large Data Bases, pp. 129–138 (2001)

[25] Wu, P., Wen, J., Liu, H., Ma, W.: Query selection techniques for efficient crawling of structured web sources. In: ICDE, pp. 47–56. ACM (2006)

[26] Wu, W., Yu, C., Doan, A., Meng, W.: An interactive clustering-based approach to integrating source query interfaces on the deep web. In: Proceedings of the 2004 ACM SIGMOD International Conference on Management of Data, pp. 95–106. ACM (2004)

Page-Level Wrapper Verification for Unsupervised Web Data Extraction

Chia-Hui Chang[1], Yen-Ling Lin[1], Kuan-Chen Lin[1], and Mohammed Kayed[2]

[1] National Central University, Chungli, Taiwan
chia@csie.ncu.edu.tw, {xaononlin,deeper.fox}@gmail.com
[2] Faculty of Science, Beni-Suef University, Egypt
mskayed@yahoo.com

Abstract. Unsupervised information extraction has been studied a lot in the past decade. However, not much attention has been paid to its wrapper maintenance. In this paper, we study wrapper construction and verification problem based on the given schema and template which is induced from unsupervised page-level wrapper induction system. We model the verification problem as a constraint satisfaction problem (CSP) for leaf node label assignment with respect to constraints specified by a finite state machine (FSM) which is constructed from previous learned schema and template. If there exists no solution to the CSP, i.e. no valid label sequence exists, we say the test page fails the verification; otherwise, we rank all valid label sequences by measuring the fitness of each label sequence for extraction. We evaluate the FSM based approach with XML validation via false positive rate and false negative rate and measure the extraction performance through extraction accuracy. The experimental result shows the proposed method can effectively filter invalid pages (zero false positive rate) and rank the correct label sequence with the highest score with 96.5% accuracy.

Keywords: Unsupervised Information Extraction, Wrapper Induction, Wrapper Verification, Extractor.

1 Introduction

Wrapper Induction (WI) which aims to generate rules for Web data extraction is a key for information integration systems. Many WI approaches have been surveyed in [2] with different extraction targets (field-level, record-level or page-level), with various automation degree (supervised, unsupervised), and with different wrapper representations (HTML tokens, DOM tree paths, etc.). Although many researches focused on WI, the development of tools for wrapper maintenance that following up the correctness of the generated wrapper over time has received less attention.

For supervised WI, wrapper maintenance can be divided into two main tasks: wrapper verification and wrapper re-induction. Wrapper verification remedies the cases when the wrapper seems to work normally but the extracted data are

X. Lin et al. (Eds.): WISE 2013, Part I, LNCS 8180, pp. 454–467, 2013.

invalid, while wrapper re-induction repairs the wrapper by gaining new labeled training data in order to learn new extraction rules. Example of such wrapper verification systems are RAPTURE [5] and DataProlog [7].

For unsupervised WI, the first thought is that we do not need wrapper verifier because unsupervised induction can be used to learn new wrappers each time. However, unsupervised web data extraction usually takes longer time (several seconds) than pure extraction module based on the induced schema and template (several milliseconds). Meanwhile, schema matching is quite challenging for two schemas without labels. If no good schema-matching algorithm is available, we cannot use the extracted data directly. Thus, extraction modules for unsupervised IE systems are still a necessity.

Nevertheless, the construction of an extraction module for page-level WI systems (e.g. RoadRunner [3], EXALG [1] and FivaTech [4]) is not an easy task and is much sophisticated than record-level extraction module (e.g. WIEN [5] and STALKER [8]). For record-level WI systems, the job of the extractor is to repeatedly match all the occurrences of the record template in the input testing page. As for page-level WI systems, the job of the extractor needs to align the whole page with the complete page template that contains sets and optional. The difficulty of matching optional and sets with similar templates in a new page makes the alignment especially complex and requires not just template information but also the data content to assist the comparison.

In this paper, we propose a page-level schema/template verification approach that detects changes of a website and extracts the embedded data from new pages based on existing schema and template induced by the unsupervised WI system FivaTech. Instead of dealing with every node in the DOM tree of a new page, only leaf nodes (with their respective DOM tree paths) are processed to verify if the new page complies with the finite state automata built from the existing schema. That is, there exists a sequence of label assignments for all leaf nodes such that transitions among labels comply with the state transitions of the FSM. Finally, we find the best label sequence by measuring the fitness score of every possible sequence of label assignments such that the data field will be extracted respectively.

The paper is organized as follows. Section 2 presents background of this research. Section 3 introduces some preliminaries for the definitions of page-level wrapper induction. Section 4 describes our proposed wrapper verifier. Section 5 presents the experimental results. Finally, section 6 concludes our work.

2 Background

While WI has been widely studied from various aspects, wrapper maintenance has not been well studied yet. In fact, most wrapper maintenance researches focus on record-level wrappers. Their wrapper verifiers only check the validity of extracted data to remedy the situations when the wrappers work normally while the extracted data are invalid. For example, based on WIEN approach [5], Kushmerick [6] introduced RAPTURE which uses nine features: digit density, letter density, upper-case density, lower-case density, punctuation density, HTML density, length, word count, and mean word length to measure the similarities between data observed by the wrapper and that expected.

Similarly, Lerman et al. introduced DATAPROG [7] for the verification of STALKER [8]. They use examples of previously extracted data in order to acquire semantic descriptions of the data. Two types of features are used to describe the data: extraction patterns learned by DATAPROG and global numeric features. The wrapper verifier checks the correctness of the data extracted for each data field by statistically comparing two distributions that describe the field in both the training examples and the new test examples. If the distributions are statistically the same, the wrapper is judged to be correct; otherwise, it is failed.

For both RAPTURE and DATAPROG, if the extraction process failed due to template change, the verification of the semantic of data is not necessary. Therefore, Pek et al. [9] combine template and content verification for record-level data extraction and verification. The method uses three features for wrapper validation: the extraction path based on HTML tree structure, the number of children per parent node, and the number of possible data nodes that could be extracted by this extraction path. The DOM-tree based feature is considered as a complementary to RAPTURE to advance the verification during extraction.

In spite of these researches for record-level wrappers, there is no literatures or studies on page-level wappers construction or verification to our knowledge. Most wrapper induction focus on record-level extraction, where the wrappers can rely on pattern match for data extraction. As for page-level data extraction, the wrappers need to align the whole test page with the induced page template and schema, which is almost as complex as wrapper induction. In addition, some researchers might consider unsupervised wrapper induction and schema matching as a solution for maintaining the consistency of the extracted data. However, due to limited label information with schema matching, alternative approach based on page-level wrapper is more reliable.

3   Problem Definition

Deep Web pages are generated by a CGI program which embeds a data instance x (taken from a database) into a predefined template T where all data instances conform to a common schema Ω. In this paper, we follows the definition of the structured data used in EXALG [1] where basic data are allocated in leaf nodes while composite data (e.g. tuple, set, option, etc.) are allocated in internal nodes. Figure 1(a) shows an example schema with 5 basic data and 6 composite data, including 1 set (denoted by {}), 2 options (denoted by ()?) and 3 tuples (denoted by <>).

The page generation process can be viewed as an encoding process (denoted by $\lambda_\Omega(T, x)$) which embeds a data instance x into the template T, while wrapper induction is a reverse engineering which decodes from the pages to its template and schema. Figure 2 shows a page template tree with embeded schema information (e.g. *schemaType, virtual_optional, sid template_id*, etc.) from Figure 1(a) to

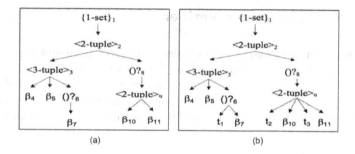

Fig. 1. (a) A sample schema and (b) The same schema tree with augmented leaf templates t_1, t_2 and t_3 defined by Figure 2

```
1    <?xml version="1.0" ?>
2    <html schemaType="tuple">
3      <body>
4        <table schemaType="set" sid="1" order="1">
5          <tr schemaType="tuple" sid="2" order="2">
6            <td schemaType="tuple" sid="3" order="3">
7              <b><text schemaType="basic" sid="4"/></b>
8              <text schemaType="basic" sid="5"/>
9              <br>
10             <virtual_optional sid="6" order="1">
11               <text template_id="t1">
12                 <![CDATA[ Avg. Customer Review: ]]></text>
13               <b><text schemaType="basic" sid="7"/></b>
14             </virtual_optional>
15           </td>
16           <virtual_optional sid="8" order="1">
17             <td schemaType="tuple" sid="9" order="2">
18               <b><text template_id="t2">
19                 <![CDATA[ List Price:]]></text></b>
20               <span colspan="2">
21                 <text schemaType="basic" sid="10"/></span>
22               <b><text template_id="t3">
23                 <![CDATA[ You save:]]></text></b>
24               <font color="#990000">
25                 <text schemaType="basic" sid="11"/></font>
26             </td>
27           </virtual_optional>
28         </tr>
29       </table>
30     </body>
31   </html>
```

Fig. 2. A page template tree with embeded schema from Figure 1(a)

define the generation of HTML pages. In principle, all HTML tags are template patterns, while leaf text can be either template (embraced by CDATA in Figure 2; e.g. t_1="Avg. Customer Review:", t_2="List Price:" and t_3="You save:" at lines 12, 19 and 23, respectively) or data strings (embraced by <text> tags of basic types).

Given a page template tree with embedded schema induced from page-level WI system as shown in Figure 2, the task of wrapper verification is to decide whether an input page follows the template and schema, and to extract the embedded data if the answer is "yes".

4   Page-Level Wrapper Verifier

Our page-level template and schema verifier contains three modules: path-guided semantic comparison, finite-state-machine (FSM) construction for structure verification and most possible path selection for extraction. Note that if no path can be found, the test page is considered invalid for the current template and schema. In such a case, wrapper re-induction can then be executed to learn new template and schema.

4.1   Path-Guided Semantic Comparison

To extract data from a test page, we need to match the test page against the template tree. To avoid recursive comparison and backtracking for mismatch, we use a path-guided approach and compare leaf texts in the new page with data items or leaf templates that have identical path in the template tree. To be more specific, each leaf text x with path p is a candidate instance of basic identifiers and leaf template that have the same path p.

For example, given a text string "A A CD" with path $/html/body/table/tr/td$, we will compare it with leaf identifiers β_5 and t_1 according to the template tree in Figure 2. As another example, for the leaf text "Toddler Favorites" with path $/html/body/table/tr/td/b$', we will compare it with basic identifiers β_4 and β_7 as well as leaf template t_2 and t_3. Note that if a leaf text has no candidate identifiers, it means the test page does not comply with the current page template tree and there is no need to continue the verification process.

We define node similarity between a leaf text in the test page and a candidate leaf template or basic identifier as follows. For leaf template candidate, the node similarity is 1 if the leaf text is identical with the text of the leaf template or 0 otherwise. For basic candidates, we use five features to characterize text strings for similarity measure of the given leaf text and its candidate basic identifier. These features are defined and illustrated with string $s = $ "Op 123-900" as follows:

- LetterDensity: density of letters in the string; e.g. LetterDensity(s)=0.2.
- DigitDensity: density of digits in the string, e.g. DigitDensity(s)=0.6.
- PunDensity: density of punctuations in the string, e.g. PunDensity(s)=0.1.
- CapitalStartTokenDensity: density of tokens that begins with upper letter, e.g. CapitalStartDensity(s)=1/2.
- IsHttpStart: whether the string begins with "http" or not, e.g. IsHttpStart(s) =0.

Similar to Rapture, we assume that each instance of a basic identifier $\beta_c \in S$ follows a multivariate normal distribution with mean vector μ_c and variance vector σ_c estimated from training examples. Therefore, we can estimate the probability of a leaf node $x = (x_1, \ldots, x_d)$ to be an instance of this basic identifier as follows:

$$p(\boldsymbol{x}|\beta_c) = N(\boldsymbol{x}|\boldsymbol{\mu}_c, \boldsymbol{\sigma}_c) = \frac{1}{\sqrt{(2\pi)^d \Pi_{j=1}^d \sigma_j}} \exp\{-\frac{1}{2}\Sigma_{j=1}^d (\frac{x_j - \mu_j}{\sigma_j})^2\} \quad (1)$$

where d is the number of features. We also use this probability for candidate selection. If the probability is greater than a threshold θ, we consider the basic ID as a candidate for this leaf node; otherwise we remove this basic ID β_c from the CandidateSet. The selection of the threshold is studied in the experiment.

Note that instead of finding the best id for each leaf node, we only filter impossible IDs for each leaf node. The reason is that if we choose the id with the maximum probability for each leaf value, we would take only content into consideration; yet the relation between leaf nodes is ignored. Therefore the best id for each leaf node will be decided by the schema as shown next.

4.2 Finite State Machine Construction

To verify whether a test page complies with the schema induced from the old pages, we need to ensure the occurrence order of assigned identifiers/labels for the leaf nodes follow the schema. Therefore, we make use of the schema tree with augmented template leaf nodes (see Figure 1(b)) and transform it into a finite state machine (FSM) with only leaf nodes. The FSM can then be used to guide the assignment of identifiers/labels for all leaf texts in the new page. As an example, Figure 3 is the FSM for the augmented schema tree in Figure 1(b).

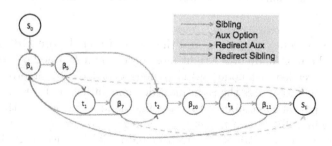

Fig. 3. A finite automata generated from Figure 1(b)

Constructing an FSM from an augmented schema includes 3 main steps:

Step 1: Create a directed binary tree of basic transitions where each left link points to the first child and each right link points to the next sibling (Figure 4(a)).

Step 2: Create auxiliary transitions from leaf nodes to internal nodes for set and optional data types as follows.

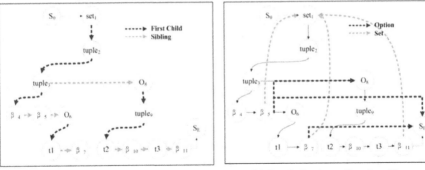

(a) Step 1: Create basic transitions.   (b) Step 2: Create auxiliary transitions.

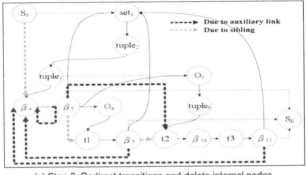

(c) Step 3: Redirect transitions and delete internal nodes.

Fig. 4. Generating a finite automata from the augmented schema tree in Figure 1(b)

1. For each "option" node x, find the last leaf node before x, denoted by y, and the next node of x's right most descendant leaf node, denoted by z. The so called last node and next node is ordered by preorder traversal such that y denotes the right most descendent leaf node of x's left sibling or x's parent's left sibling if x is the first child and z denotes the first node following all the leaf nodes in x's subtree. For example, the transition from β_5 to O_8 is added due to optional node O_6 since β_5 is the last leaf before O_6 and O_8 is the next node of β_7, the last descendant leaf of O_6. Similarly, the transition from β_7 to S_E is added since S_E is the next of the last descendant leaf β_{11} (See deep-blue dotted lines in Figure 4(b)). Note that if z is also optional, then additional transition will be added from y to the next node of z's rightmost descendant leaf node until non-optional node is encountered. For example, the transition from β_5 to S_E is added since O_8 is optional.

2. For a "set" type node x, we need to add a transition from the right most leaf node b of x to the set node x to allow repetition. If any ancestor of b is an option type node, an additional transition from l, the last leaf node before the ancestor node, to x needs to be added. The node l denotes the

right most leaf node of the option node's left sibling[1]. This is similar to the skip link added for option nodes, but this transition is added to allow repetition for set nodes. Again, if any ancestor of l is an option node, we will repeat the process to add proper links from leaf nodes to some internal nodes. For example, the set node in Figure 4 will generate three transitions from β_{11}, β_7 (the last leaf node before O_8) and β_5 (the last leaf node before O_6) to set_1 (See light-blue dotted lines in Figure 4(b)).

Step 3: Delete internal nodes such that only leaf nodes of basic type and template nodes are reserved. This will be done as follows.

1. For each node v with auxiliary transitions from some leaf f created at step 2, we create a transition from f to the first leaf under v or v itself if it is leaf. For example, transitions from β_5, β_7, and β_{11} to set_1 will be directed to β_4 (the deep-blue dotted lines in Figure 4(c)).

2. For each sibling link from v to r: (i) if v is a leaf and t is an internal, we create a transition from v to the first leaf under r; or (ii) if v is an internal and r is a leaf, we create a transition from r's previous node (i.e. the right most leaf under v) to r; or (iii) if both are internal, we instead create a transition from r's previous node (i.e. the right most leaf under v) to the first leaf under r. For example, sibling link from β_5 to O_6 demonstrates case 1 with a new link from β_5 to t_1, the first child under O_6. Also, sibling from $tuple_3$ to O_8 illustrates case 3 with the new link from β_7 (the previous node of O_8) to t_2, the first child under O_8.

3. Remove all internal nodes and their transitions. Just keep leaf nodes.

Once the FSM is constructed, we then model the task of assigning labels to all leaf nodes as a constrained satisfaction problem where each leaf node is a variable with a set of basic/template identifiers as its domain and the transitions among states in the generated FSM are binary constraints that enumerate possible value pairs for two consecutive variables (leaf nodes). Therefore, we apply constraint propagation to reduce the number of candidate IDs for each node. For example, the only candidate β_{10} for node 6 with path "/html/body/table/tr/td/span" implies the previous label could only be t_2 and the next label could be t_3. Such arc consistency checking can reduce the number of candidates for each leaf node and improve the efficiency of label sequence enumeration without backtracking. In this particular example, one complete sequence of label assignment after arc consistency checking is β_4, β_5, t_2, β_{10}, t_3, β_{11} as shown on the right most column of Table 1.

4.3   Finding the Most Possible Label Assignment

If there exists more than one label sequence that complies with the generated FSM, we would need to decide which label sequence can be used to extract the

---

[1] If the option node is the first child, then l will be the right most leaf node of its parent's left sibling.

Table 1. Data extraction and verification for new page

No.	Leaf Node	Path	CandidateSet	State
0				S_0
1	Toddler Favorites	$/html/body/table/tr/td/b$	β_4, β_7	β_4
2	A A CD	$/html/body/table/tr/td$	β_5	β_5
5	List Prices:	$/html/body/table/tr/td/b$	β_4, β_7, t_2	t_2
6	6.29	$/html/body/table/tr/td/span$	β_{10}	β_{10}
7	You save:	$/html/body/table/tr/td/b$	β_4, β_7, t_3	t_3
8	0.62	$/html/body/table/tr/td/font$	β_{11}	β_{11}
9	EOF			S_E

data[2]. Two possible measures are considered here. The first one is based on node similarity as discussed in section 4.1. The second one is based on transition probability between labels. The transition probability can be obtained based on the extracted data at earlier time. For example, given the extracted data sequence:

"$\beta_4, \beta_5, t_1, \beta_7, \beta_4, \beta_5, t_1, \beta_7, \beta_4, \beta_5, t_2, \beta_{10}, t_3, \beta_{11}$",

we could accumulate the transition counts for each state pair. For state pairs without transition occurence, we give a small value η to differentiate it from zero probability which denotes no transition is allowed. For example, β_7 has possible transitions to β_4, t_2, and S_E with 2, 0, 0 counts in the above sequence, respetively. By adding a small value η, the transition probabilities are $2/(2+2\eta)$, $\eta/(2+2\eta)$, $\eta/(2+2\eta)$, respectively.

Now to make use of the above transition probabilities and the node similarity defined above, we enumerate every possible label and evalute the sequence probability $SP(X, L)$ for a page X with n leaf nodes $x_1, x_2, ..., x_n$ and label assignment $L = S_0, l_1, l_2, ..., l_n, l_{n+1}(S_E)$ as follows:

$$\log SP(X, L) = \log TP(S_0, l_1) + \Sigma_{i=1}^{n} TP(l_i, l_{i+1}) + \Sigma_{i=1}^{n} \log P(\boldsymbol{x}_i | \beta_{l_i}) \quad (2)$$

5 Experiments

In the following, we make use of Rapture dataset for the experiment. For each website with n collected pages, we choose two web pages with different keywords to build the schema and template using FivaTech. We then test these n webpages to see if they pass or fail the FSM-based verification, and manually examine each webpage to see if they comply with the induced schema/template (valid) or not (invalid). The process is repeated $n/2$ times for each website.

We evaluate the performance via false positive rate (i.e. the percentage of invalid pages which pass through the verifier) and false negative rate (i.e. the percentage of valid pages that fail the verification) to compare the proposed method with XML validation as described below.

---

[2] If some leaf node of the test page is left with no candidate label, then the schema is considered not complete and needs to be re-induced.

5.1   Baseline - XML Validation

We use XML validation as the baseline for comparison. XML validation is a module to check whether an XML file conforms to some XML DTD. We first transform the training HTML pages into XML files, and then summarize an XML schema representation, i.e. XSD (XML Schema Description) from these XML files through the XML2XSD function supported by .NET Framework. For validation, the XML validation module takes the XSD file of the template and the XML file of the new page as inputs, and checks whether the XML file is valid with respect to the XSD file.

Note that XSD considers sibling tags with the same tag name as repeated elements even if they contain different subtrees, leading to incorrect schema. In some sense, XML validation only ensures that every path of the test page is a valid one in the old pages. Thus, we might expect a high false positive rate and low false negative rate (≈ 0) for XML validation.

5.2   Verification Effectiveness Evaluation

We first compare the false positive rate from invalid pages (i.e. type I error) for the proposed methods on Rapture dataset. In our experiment, we did not find any invalid page that can pass through our FSM verification even with a zero threshold θ which is used to control the least probability of a leaf node to be an instance of a basic identifier. All invalid pages would be left with no candidate during arc consistency checking as shown in Fig. 5. Therefore, we obtain zero type I error rate for all possible threshold θ. On the other hand, XML validation has constant type I error 20.2% (i.e. 79.8% specificity) since the parameter θ is used only for FSM verification.

Next, we compare the false negative rate from valid pages (i.e. type II error). The result is shown in Fig. 6. As expected, XML validation has zero type II error, i.e. all valid pages will pass through the validation, while FSM-based approaches also has zero type II error if no threshold θ is set. As the threshold θ increases,

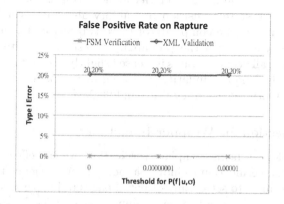

Fig. 5. False positive rate for FSM-based and XML-based approaches

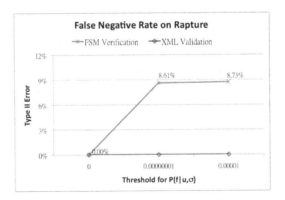

Fig. 6. False negative rate for FSM-based and XML-based approaches

type II error increases (around 8.6% to 8.7%) since larger θ will filter more candidate basic identifiers, leaving less chance for valid pages to pass through the verification. Since the threshold θ is a mechanism to filter the number of candidates for each leaf node in the DOM-tree, we can ignore this filtering if verification time is not too long.

5.3 Extraction Effectiveness

Next, we evaluate the effectiveness of data extraction for pages that pass through the FSM verification. This is equivalent to check whether the correct label sequence can be ranked highest when more than one valid label sequence are available. We compare the proposed label sequence evaluation in Eq. 2 with individual effect based state transition probability (the first two terms of Eq. 2) and leaf node similarity (the last term of Eq. 2). The performance is measured by accuracy, the number of correctly extracted page divided by the total number of test pages.

As shown in Fig. 7, leaf node similarity alone performs the best with 99.73% accuracy rate, while state transition probability also has a good performance with 96.53% accuracy. The composite of these two factors has a less accuracy at 99.54%. Meanwhile, when the probability threshold for filtering candidate labels is increased, all accuracy rates increased slightly except for state transition probability. For leaf node similarity, the accuracy is 99.89%, while for state transition probability the accuracy is around 96.23%.

5.4 Page Selection for Wrapper Induction

With this page-level schema/template verification, we can effectively add more pages that fail the verification to induce a more general schema/template. That is, we can use verifiers to select only essential pages to induce schema/template, reducing the time cost for induction on unnecessary pages. In some situations, the verifier can also be used to detect bugs in the induced schema/template.

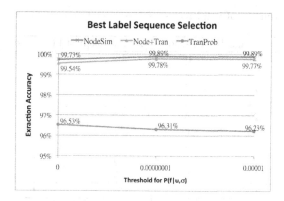

Fig. 7. Best label sequence selection for extraction effectiveness

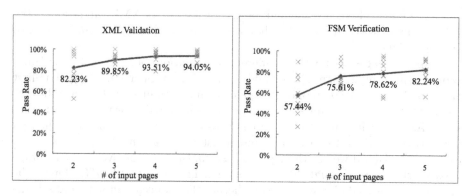

Fig. 8. Percentage of pages that pass through verification with various training pages for wrapper induction

To demonstrate this idea, we collect search result pages for nine web sites (as shown in Table 2) by submitting the same query to each search form and obtain an average of 20 pages for each website. We use both randomly selected pages and specific pages for training set, respectively.

For randomly selected pages, we choose k input pages for each web site for wrapper induction (k=2, 3, 4, 5) and test the generated wrapper to see how many pages from the same website can pass through the generated extractor. The process is repeated five times for each web site for a given k. For XML validation, the coverage is 82% when two pages are used for wrapper induction. The ratio increases to 94% when five pages are used. This shows that although all pages are generated by the same CGI program, there are still many schema variations that are not detected using five pages. Figure 8 shows the percentage of pages that comply with the induced wrapper for schema verification. If we use only two pages to induce the wrapper, only 57% pages can pass through FSM verification. The ratio increases to 82% when five pages are used for wrapper induction.

Table 2. Number of valid pages with respect to specific training pages

Website	# of Pages	first+second	first+midst	first+last	first+second+last
BingS	17	16	16	16	16
BookD	23	2	15	15	15
iWatch	18	2	13	13	14
Acm	11	2	3	2	4
YahooS	21	4	10	7	18
CuisineNet	35	32	32	3	34
Dogtoy	17	16	16	2	17
Tradekey	21	3	20	20	20
Tradingpost	19	2	15	15	15
Avg # pages	20.2	8.8	15.6	10.3	17.0
Pass rate	N/A	40.1%	72.57%	53.3%	81.17 %

This experiment also echos the high false positive rate of XML validation as the XSD schema generated by XML allows more (both valid and invalid) pages to pass the validation.

For specific pages, we try three different combinations of input including {first, second}, {first, midst} and {first, midst, last} for wrapper induction. The first and second pages are commonly chosen for training input since they are the earliest pages one could obtain and often contain additional information than other pages. As shown in Table 2, the percentages of pages that could pass through verification based on two the first and midst pages are higher (72.57%) than average (57.44%). If we use three pages first, midst, last, the average percentage of pages that pass through FSM verification increases to 81.17%. Hence, the wrapper verifier could be used to test whether the induced template and schema are complete or not.

6   Conclusions

In this paper, we solve the problem of page-level wrapper maintenance, which is a problem that has not been studied thoroughly. To solve this problem, we reduce the full schema tree verification to label assignment problem for leaf nodes and exploit finite state machine and constraint satisfaction problem (arc consistency) to verify if the order of label assignments for leaf nodes complies with the schema.

For performance evaluation, we use XML validation which induces a general XSD as a baseline for comparison. The experiments show that the proposed FSM-based approaches have much better effectiveness than XML validation. XML validation, with high pass rate, has zero false negative rate for valid pages but high false positive rate for invalid pages; while the proposed FSM-based approaches guarded by leaf node evaluation and transition probability, with low pass rate, has zero false positive rate for invalid pages and acceptable false negative rate for valid pages.

References

1. Arasu, A., Garcia-Molina, H.: Extracting Structured Data from Web Pages. In: ACM SIGMOD International Conference on Management of Data, pp. 337–348 (2003)
2. Chang, C.-H., Kayed, M., Girgis, M.R., Shaalan, K.F.: A survey of web information extraction systems. IEEE Trans. on Know. and Data Eng. 18(10), 1411–1428 (2006)
3. Crescenzi, V., Mecca, G., Merialdo, P.: RoadRunner: Towards Automatic Data Extraction from Large Web Sites. In: Proceedings of the 27th International Conference on Very Large Databases (2001)
4. Kayed, M., Chang, C.-H.: Page-level web data extraction from template pages. IEEE Trans. on Know. and Data Eng. 22(2), 249–263 (2010)
5. Kushmerick, N.: Wrapper induction: Efficiency and expressiveness. Artificial Intelligence 118(1-2), 15–68 (2000)
6. Kushmerick, N.: Wrapper verification. World Wide Web Journal 3(2), 79–94 (2000)
7. Lerman, K., Minton, S.N., Knoblock, C.A.: Wrapper maintenance: A machine learning approach. Journal of Artificial Intelligence Research 18(1), 149–181 (2003)
8. Muslea, I., Minton, S., Knoblock, C.A.: Hierarchical wrapper induction for semistructured information sources. Autonomous Agents and Multi-Agent Systems 4(1-2), 93–114 (2001)
9. Pek, E.-h., Li, C.-M., Liu, M.-l.: Web wrapper validation. In: Zhou, X., Zhang, Y., Orlowska, M.E. (eds.) APWeb 2003. LNCS, vol. 2642, pp. 388–393. Springer, Heidelberg (2003)

Co-Learning Ranking for Query-Based Retrieval

Min Peng*, Jiajia Huang, Jiahui Zhu, Li Zhou, Hui Fu, Yanxiang He, and Fei Li

Computer School, Wuhan University, Wuhan, China
{pengm,huangjj,zhujiahui,zhouli_88,fuhui,yxhe}@whu.edu.cn,
{kevin.lifei}@gmail.com

Abstract. In this paper, we propose a novel blending ranking model, named Co-Learning ranking, in which two ranked results produced by two basic rankers interact with each other adequately and are combined linearly with a pair of appropriate weights. Specifically, in the interaction process, a *reinforcement strategy* is proposed to boost the performance of each ranked results. In addition, an automatic combination method is designed to detect the better-performance ranked result and assign a higher weight to it automatically. The Co-Learning ranking model is applied to the document ranking problem in query-based retrieval, and evaluated on the TAC 2009 and TAC 2011 datasets. Experimental results show that our model has higher precision than basic ranked results and better stability than linear combination.

Keywords: Co-Learning Ranking, Interactive Learning, Automatic Combination, Ranked Result.

1 Introduction

Ranking, which is to sort a group of objects with certain feature, is applied widely in natural language process and information retrieval. In recent years, kinds of supervised and unsupervised ranking methods have been widely studied. However, neither supervised nor unsupervised methods are able to work well for all types of ranking tasks. Therefore, it opens a new idea that build a blending ranking model by integrating different ranking results produced by different basic rankers. In fact, this issue has already attracted some researchers. One alternative strategy is *learn-then-combine* methods [1, 2, 3], which composites different ranked results after their interactive learning with each other. In this paper, we are interested in this strategy which includes two ideas as follows.

First, different ranked results can learn with each other. Namely, each result adjusts itself and improves its performance by providing reliable and valuable information to others. This idea has been successfully applied in the multi-documents automatic summarization problem [1, 2, 3]. Second, top N instances from each ranked result are more valuable than the rest. This idea was both employed in

---

* This material is based on the work supported by National Science Foundation of China (NSFC) under Award 61070083 as well as the Key Technologies R&D Program of Wuhan under Award 201210421135.

X. Lin et al. (Eds.): WISE 2013, Part I, LNCS 8180, pp. 468–477, 2013.

learn-then-combine methods [1, 2, 3] and other methods[4, 5]. For example, Zheng Chen *et al.* utilized the best k collaborators of a query sentence to finish the interaction ranking process [4]. The user study report [10] also supported this idea. It pointed out that 62% of search engine users only click on the first page's items, and 90% of users click on the first three pages' items.

In this paper, we think both the interactive learning process and the top N instances are important for blending ranking. However, some drawbacks in the learn-then-combine methods proposed in [1, 2, 3] made them unfit for all kinds of ranking tasks. In detail, similar values between different documents are involved into the interactive learning process to adjust each basic ranker's score. This method may work well on document ranking tasks, but is not fit for other nun-textual ranking tasks, such as ranking people based on social relationship. Therefore, we design a novel interactive learning strategy, which is not based on the ranking target's type(document or others). In addition, it is not appropriate to combine two ranked results linearly with a given weight after their interactive learning with each other. The reason is obvious: the performance of each result after learning is changing with ranking task. Therefore, one method is necessary to automatically detect the performance of different ranked results after learning.

Along these lines, we proposed a novel blending ranking model, which tries to blend any two kinds of ranked results within two steps as follows. This ranking model is called Co-Learning ranking.

- **Interactive Learning Process:** two ranked results interact with each other and adjust each one's result with guidance of information provided by the top N instances from the other.
- **Automatic Combination Process:** based on the interactive learning process, detect the performance of each ranked result and grant a matching weight to each one.

In step (1), Spearman Ranking Correlation Coefficient (SRCC) [6] is employed to guide the interactive learning process. Its basic idea is that higher rank correlation score between two sequences denotes closer relationship between them. In this paper, we employ the basic idea of SRCC but improve it (called iSRCC) for measuring the relationship between two top N ranked instances sequences. In step (2), we design a simple and effective detecting strategy to detect the performance of different ranked results.

Compared with the basic rankers and linear combination methods (used as the ranking criterion), the Co-Learning ranking model only concerns about the basic ranked score of each instance, and ignore the instance's category (i.e., document, picture or people). Therefore, it can be widely used in various ranking tasks, such as document retrieval, meta-search engine, pattern recognition and so on. Furthermore, this model can detect the performance of different ranked results without supervise. Therefore, it will be work well on different pairs of basic ranking algorithms by adjusting combination weight automatically. These pairs could be supervise-unsupervise, like Profile-ListNet [4], or query dependent-query independent, like cosine-LexRank [1].

In this paper, we apply this model into the task of query-based document retrieval and design a novel Co-Learning ranking algorithm. In this algorithm, two ranked results interactively learning with each other and then to be automatically combined. At last, we show its efficacy on TAC datasets.

The remainder of this paper is organized as follows. Section 2 describes the approach in detail. Section 3 demonstrates evaluation methodology by related experimental results. Section 4 discusses conclusion and future work.

2   Co-Learning Ranking Framework

Set $I = \{i_1, i_2, ..., i_n\}$ as the instances to be ranked. R_1 and R_2 are two ranked results of I, which are independently produced by two different basic rankers f_1 and f_2. $R_i = \{(Id_j, value_j)|0 < j <= n\}, (i = 1, 2)$ is *two-tuples*. Id_j denotes the jth instance in I, and $value_j$ denotes the jth instance's score calculated by the basic ranker.

The goal of Co-Learning ranking is to seek a blending ranking model which is able to boost the final performance by considering the two ranked results.

2.1   Co-Learning Ranking Framework

We develop a Co-Learning ranking framework. For an instances set I, from a global view, two ranked results R_1 and R_2 produced by f_1 and f_2 are not good enough. However, this pair of results can improve themselves by exchanging information with each other. Especially, the top N instances in each result may be more important because they often attract much more attention than the rest in real ranking tasks [5]. Therefore, in the Co-Learning ranking framework, R_1 and R_2 can adjust themselves for many times with the guidance of iSRCC value u between top N instances in two results. The iSRCC value u is produced by a ranking reinforcement strategy. This process is called Interactive Learning Process. Until nothing can be exchanged between them, a blending result is produced with a *weight-decision function* h. This function h is formed by automatically detecting each result's efficacy. This process is called Automatic Combination Process. The Co-Learning ranking framework is as follows:

The termination condition $I(O)$ can be defined depending on different application occasions. For instance, the top N instances in two results are not changed any more or the interaction times touch the upper limit.

2.2   Interactive Learning Process

Spearman Ranking Correlation Coefficient. We aim to conduct a framework which could ignore the mechanism of the basic ranking algorithms. Thus, the basic ranked results and the correlativity between them are very important. The SRCC is employed to measure the correlation between two results. Formula (1) presents a simple way to calculate the SRCC value between two sequences :

$$\rho = 1 - \frac{6\sum_{i=1}^{n}(x_j - y_j)^2}{n(n^2 - 1)} \tag{1}$$

---

1 f_1 and f_2 are employed to **rank** the instance set I to get ranked results R_1, R_2;
2 Normalize R_1 and R_2;
3 **repeat**
4   Select the top N instances I_2 from R_2, calculate the iSRCC value u_1 as
     guidance to adjust R_1: **Adjust** $R_1^* = F(u_1, R_1, R_2)$;
5   Select the top N instances I_1 from R_1, calculate the iSRCC value u_2 as
     guidance to adjust R_2: **Adjust** $R_2^* = F(u_2, R_2, R_1)$;
6     $R_1 = R_1^*$;
7     $R_2 = R_2^*$;
8 **until** $I(O)$;
9 Calculate the combination weights w_1 and w_2;
10 Combine the two ranked results R_1 and R_2 with automatic combination
   weights w_1 and w_2: $R(x_i) = w_1 R_1(x_i) + w_2 R_2(x_i), i = 1, 2$;

---

where x_j and y_j are the *rankings* of the jth variables in two sequences respectively.

The ranking is generated by sorting variables with their scores. SRCC between two sequences is denoted as ρ_s, where $\rho_s \in [-1, 1]$. (Here assuming there are no same score for all variables in each group.)

Ranking Reinforcement Strategy. We improve the SRCC to work better on the interactive learning process. The ranking reinforcement strategy is designed with following considerations.

- In the framework, the top N instances from one result could guide the adjustment of the other. Hence, we set an *interaction degree* between two results as the sum of ranking differentia between the top N instances from one result and the instances with same Id from the other, other than the differentia between all pairs of instances.
- The interaction degree between the top N instances in R_1 and the same ones in R_2 is considered as an indicator which states how many information R_2 can learn from R_1, and vice versa.
- In each interactive learning process, both ranked results should keep more than 50% ranking results of themselves.

Based on the considerations above, formula (2) evolved from formula (1) is employed to measure the interaction degree between the two ranked results:

$$u_i = 1 - \frac{12 \sum_{i=1}^{N}(x_j - y_j)^2}{n(n^2 - 1)} \tag{2}$$

where N denotes the number of top N instances, x_j and y_j denote the ranking of jth instance in R_1 and R_2, n denotes the size of each ranked result.

u_i is called as iSRCC value. When $N \in [0, 0.1n]$, $u_i \in [0.5, 1]$.

Based on the iSRCC value u_1 and u_2, R_1 and R_2 adjust themselves as follows:

$$R_1^*.value_j = u_1 * R_1.value_j + (1 - u_1) * R_2.value_j \tag{3}$$

$$R_2^*.value_j = u_2 * R_2.value_j + (1 - u_2) * R_1.value_j \qquad (4)$$

According to formulas (2) (3) (4), a reinforcement strategy based on the improvement of SRCC is designed.

Through the interactive learning process, the two cooperators will come into agreement gradually. Until nothing exchanges with each other, this process is ended. Because the adjustment is guided by the positive instances from two sides, both ranked results are boosted after the process.

2.3   Automatic Combination Process

Though the performance of each ranked result is improved by the interactive learning process, the final result generated by linearly combining the two results with appointed weights is still not the best option. Reasons are obvious: different ranking tasks employ different ranking functions, but well knowing about every pair's performance is difficult. Moreover, it is still not sure about how much each ranked result is improved through the interactive learning process. So it is also hard to assign them appropriate weights. In addition, we note that the ranked result with slighter adjustment implies better performance because it only needs less information to improve itself. Along these lines, we design an automatic combination strategy, which can detect the performance of each ranked result via the interactive learning process. Explicit description of the automatic combination strategy is presented as follows.

The iSRCC value $u_i(i = 1, 2)$ between two ranked results are changing with the interaction process, so a stronger fluctuate of u_i denotes a bigger adjustment of ith result. Usually it further implies worse performance of ith ranked result compared with the other one.

Upon the discussion above, a simple but effective detecting method is proposed as follows:

$$w_i = max(u_i) - min(u_i) \qquad (5)$$

where $max(u_i)$ and $min(u_i)$ are the maximum value and the minimum value of u_i. w_1 and w_2 are a pair of weights to combine the two ranked results.

Combination weights $w_i \in [0, 1](i = 1, 2)$ are normalized as follows:

$$w_i^* = \frac{w_i}{w_1 + w_2} \qquad (6)$$

Then, the final result produced by linearly combining the two learning-after results with w_1^* and w_2^* is shown as follows:

$$R.value_j = w_1^* * R_1.value_j + w_2^* * R_2.value_j \qquad (7)$$

Besides, there are some smooth measures for w_1^* and w_2^*:

- If the two weights are equal, then their contributions are regarded as same, i.e., if $w_1 = w_2 = 0$, then $w_1^* = w_2^* = 0.5$.
- In order to let each ranked result hold some right in the final decision making, i.e., if $0 \le w_i^* \le 0.1$, then $w_i^* = 0.1$; if $0.9 \le w_i^* \le 1$, then $w_i^* = 0.9, i = 1, 2$.

Based on the detecting method above, a dynamic combination strategy is formed.

2.4 Application Scenario

Co-Learning ranking framework has many applications. How to using it will depend on the specific situation of task. Here we are interested in the task of query based document retrieval. The basic ranker can be a complex search engine or just a simple classifier [1] or a *learn-to-rank* function [8, 9].

Given a document set D, a query sentence q and a basic ranker f, the score of each document d in D can be calculated by the basic ranker f. As the basic ranker's performance is not satisfactory, it is needed to select a method to blending different ranked results. Here, two critical steps are involved, namely document ranking based on query sentence and the results blending of different basic rankers. In this paper, we focus on the blending step mainly, by employing Co-Learning ranking framework.

With two basic rankers f_1 and f_2, the Co-Learning ranking framework proposed above is instantiated as a Co-Learning ranking algorithm. It blends two basic ranked results in query-based document retrieval. The algorithm is detailed in Algorithm 1.

---

input : document set D and a query sentence q
output: Final ranked result R.

1   Rank D with f_1 and generate ranked result R_1;
2   Rank D with f_2 and generate ranked result R_2;
3   Normalize R_1 and R_2:
$$R_i.value_j = \frac{R_i.value_j - min(R_i.value_j)}{max(R_i.value_j) - min(R_i.value_j)}, i = 1, 2; j = 1, 2, \ldots, n;$$
4   **repeat**
5   Calculate the iSRCC value u_1 of R_1 with formula (2);
6   **for** $i =1$ *to* n **do**
7 | **Adjust** R_1 with formula (3);
8   **end**
9   Calculate the iSRCC value u_2 of R_2 with formula (2);
10   **for** $i =1$ *to* n **do**
11 | **Adjust** R_2 with formula (4);
12   **end**
13     $R_1 = R_1^*$;
14     $R_2 = R_2^*$;
15 **until** u_1, u_2 *don't change any more or the iteration times (M) touch the upper limit*;
16 **Detect** the performance of each ranked result, namely calculate w_1 and w_2 with formula (5);
17 Normalize w_1 and w_2 with formula (6);
18 **for** $i =1$ *to* n **do**
19 | Generate final result R with formula (7);
20 **end**

Algorithm 1. Co-Learning ranking

The most confidential documents number N can be defined upon the document set size n and the number of the documents that most relevant to query sentence q.

3   Experiment and Analysis

3.1   Datasets and Basic Rankers

We take TAC 2009 [11] and TAC 2011 [12] datasets as experimental analysis data. Both corpuses contain 880 documents and 44 query sentences. For each query sentence, 20 of 880 documents in each set are most relevant.

In this paper, each dataset is divided into two parts: training corpus and testing corpus. There are 660 documents and 220 documents in the two corpuses respectively. For each query sentence, 5 of 220 documents in the testing corpus are most relevant. The training corpuses are used for training the basic rankers. The testing corpuses are used for testing the performance of each basic rankers and different kinds of combination methods, including linear combination and Co-Learning ranking method. As the testing result is a binary judgement (i.e., relevant or irrelevant), we employ Mean Average Precision (MAP) [7] as the performance evaluation.

In this experiment, we select two kinds of unsupervised models as the weak basic rankers and list as follows.

- **Random Projection** (f_1): RP model allows to substantially reducing the n-dimensional document space into a random d-dimensional subspace where $d \ll n$, while still retaining a significant degree of its structure.
- **LDA** (f_2): LDA model is a topic mixture proportions, each document is drawn from some latent topics with multinomial distribution, and each topic is drawn from a list of words with multinomial distribution.

The two models are quite different from each other. Both can be used in query based retrieval task. When applied in the two datasets, their MAP values are poor, as shown in the third line of Table 1.

3.2   Impact of Interactive Learning Process

In order to evaluate the impact of interactive learning process on the performance of difference pairs of combination, the MAP of the two learned and ranked results are shown in Table 1. Here, we tried different values of N (i.e., N=5, 10, 15)and tested on the two datasets. The results are shown in Table 1.

As is shown in Table 1, both ranked results are boosted after the interactive learning process. Specially, the weaker result always can be significantly boosted, on both TAC 2009 and TAC 2011 sets. In addition, when setting $N = 15$, both basic rankers have the biggest improvement. This clearly proves our theoretical analysis.

For further evaluating how much each query task is improved through the interactive learning process, we show the before learning and after learning results

Table 1. MAP of two basic rankers and MAP of them after interactive learning process

N	TAC 09		TAC 11	
	RP	LDA	RP	LDA
Before learning	0.210	0.581	0.147	0.691
5	0.419	0.605	0.415	0.708
10	0.599	0.620	0.570	0.713
15	**0.647**	**0.627**	**0.642**	**0.714**

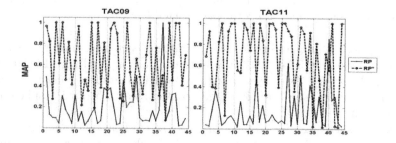

Fig. 1. MAP of RP and RP* in 44 queries over TAC 2009 and TAC 2011

produced by RP basic ranker. Here, we set N=10 and test it on TAC 2009 and TAC 2011 datasets respectively. The results are shown in Fig.1 (RP* denotes the RP result after learning, RP denotes its original result).

From Fig. 1, we obverse that, each query result has a significant improvement after the interactive learning process. Some of them even reach to 100%. That means, the effectiveness of our proposed reinforcement strategy for the weaker ranked result is evidently.

3.3   Stability of Co-Learning Ranking

In order to test the impact of automatic combination process on the final result, we applied the two ranked results on two corpuses and observe the variety of MAP with N changing. As comparison, we also measured the MAP of linear combination ranking (e.g., $R(x_j) = \lambda R_1(x_j) + (1 - \lambda)R_2(x_j)$) with varied λ. The results are shown in Fig.2(a) (b).

From Fig. 2, we can see, compared with the linear combination, the varieties of Co-Learning ranking result are much small (less than 3%). In the linear combination method, the performance greatly relies on the combination weight. Its MAP value shows widely changes (more than 40%). Therefore, the Co-Learning ranking method shows its stability significantly. That is to say, no matter what the value of N is and how much difference between two original ranked results, the final results are always very promising.

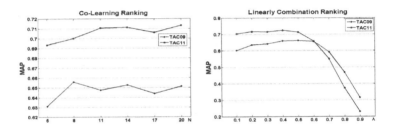

Fig. 2. (a) MAP of Co-Learning ranking with N changing from 5 to 20; (b) MAP of linear combination ranking with w changing from 0.1 to 0.9

In addition, the final results are also better than each basic ranker's result. For example, on data set TAC 2009, it is boosted 7% compared with LDA original result and boosted 44% compared with RP original result. It achieves a much better performance than the RP original ranked result when it is applied in the corpuses alone.

Generally speaking, the best performance of Co-Learning ranking is almost equal to the best of linear combination method, while the worst performance of Co-Learning is much better than the one of the later. In short, the Co-Learning ranking shows its strong stability and better performance, fitting for more universal and expandable scenarios.

4 Conclusions and Future Work

In this paper, we propose a Co-Leaning ranking framework, which can be applied into various applications. In the framework, first, the two ranked results learn interactively with each other and adjust themselves under the guidance of the proposed reinforcement strategy. Then, an automatic combination method is put forward to detect the performance of two ranked result and to assign a pair of matching weights to them. The proposed framework is applied into the query-retrieval task as a scenario. Experimental results show that the proposed framework can significantly outperform their original version and has much better stability than the linear combination method.

As future work, we plan to focus on the following issue: (1) Study the blending of supervised algorithm with unsupervised algorithm and applied it into other ranking tasks. (2) Try to employ other automatic combination strategies.

Acknowledgments. This material is based on work supported by National Science Foundation of China (NSFC) under Award 61070083 as well as the Key Technologies R&D Program of Wuhan under Award 201210421135. The authors are very grateful for these generous supports.

References

1. Wei, F., Li, W., Liu, S.: iRANK: A Rank-learn-combine Framework for Unsupervised Ensemble Ranking. J. Am. Soc. Inf. Sci. Tec. 61(6), 1232–1243 (2010)
2. Wei, F., Li, W., He, Y.: Co-feedback Ranking for Query-focused Summarization. In: The ACL-IJCNLP 2009 Conference Short Papers, pp. 117–120. ACL Press, Singapore (2009)
3. Wei, F., Li, W., Wang, W., He, Y.: iRANK: An Interactive Ranking Framework and Its Application in Query-focused Summarization. In: The 18th ACM Conference on Information and Knowledge Management, pp. 1557–1560. ACM Press, New York (2009)
4. Chen, Z., Ji, H.: Collaborative Ranking: a Case Study on Entity Linking. In: The Conference on Empirical Methods in Natural Language Processing, pp. 771–781. ACL Press, Edinburgh (2011)
5. Xia, F., Liu, T.Y., Li, H.: Statistical Consistency of Top-k Ranking. In: Advances in Neural Information Processing Systems, vol. 22, pp. 2098–2106 (2009)
6. Spearman, C.: 'Footrule' for Measuring Correlation. British Journal of Psychology, 1904-1920 2(1), 89–108 (1906)
7. Baeza-Yates, R., Ribeiro-Neto, B.: Modern Information Retrieval. ACM Press, New York (1999)
8. Cortes, C., Mohri, M., Rastogi, A.: Magnitude-preserving Ranking Algorithms. In: The 24th International Conference on Machine Learning, pp. 169–176. ACM Press, Corvallis (2007)
9. Qin, T., Zhang, X.D., Tsai, M.F., Wang, D.S., Liu, T.Y., Li, H.: Query-level Loss Functions for Information Retrieval. J. Inform. Process. Manag. 44(2), 838–855 (2008)
10. iProspect Search Engine User Behavior Study (April 2006), http://www.iprospect.com/
11. TAC 2009 Update Summarization Task, http://www.nist.gov/tac/2009/Summarization/
12. TAC 2011 Update Summarization Task, http://www.nist.gov/tac/2011/Summarization/

Fine-Grained Access Control
for RDF Data on Mobile Devices*

Owen Sacco[1], Matteo Collina[1,2], Gregor Schiele[1], Giovanni Emanuele Corazza[2],
John G. Breslin[1], and Manfred Hauswirth[1]

[1] DERI, National University of Ireland, Galway
{owen.sacco,gregor.schiele,john.breslin,manfred.hauswirth}@deri.org
[2] University of Bologna, Italy
{matteo.collina,giovanni.corazza}@unibo.it

Abstract. Existing approaches for fine-grained access control for RDF data suffer from high overhead, making them ill-suited for mobile devices. This makes it difficult to develop mobile applications that manage personal RDF data in a privacy preserving manner. In this paper we propose a new approach to realise fine-grained access control for mobile devices. We show how fine-grained privacy settings for personal information stored in mobile devices can be described using the Privacy Preference Ontology (PPO) – a light-weight vocabulary for defining fine-grained privacy preferences. Moreover, we introduce a two stage privacy preservation approach for efficient filtering of personal information on mobile devices. Our approach combines (1) an efficient query-based analysis stage with (2) a result filtering stage based on the privacy preferences described using PPO.

Keywords: Semantic Web, Access Control, Mobile, Privacy, PPO, PPM, Gambas.

1 Introduction

Web information systems are getting mobile. Due to more powerful mobile devices like smartphones and tablets, users increasingly let them manage and publish their personal data like social network information or sensor readings. This, in combination with the increasing popularity of Linked Data technologies like RDF and SPARQL, has led to the need to develop efficient data storage systems (i.e. RDF stores) for mobile devices. However, current storage systems such as RDF on the Go [6] do not offer efficient support for fine-grained access control for the data contained in them. This makes it difficult to develop mobile applications that manage personal RDF data in a privacy preserving manner. Existing approaches for access control for RDF data either suffer from high overhead,

* This work is funded by the Science Foundation Ireland (grant SFI/08/CE/I1380 (Líon 2)), by an IRCSET scholarship co-funded by Cisco systems and by the European Commission (FP7, contract FPF-2011-7-287661, GAMBAS).

X. Lin et al. (Eds.): WISE 2013, Part I, LNCS 8180, pp. 478–487, 2013.

making them ill-suited for mobile devices – or do not provide fine-grained control, i.e. the ability to control access at the level of individual triples (e.g. by filtering triples resulting from SPARQL queries).

In our previous work [10,8], we presented a Privacy Preference Ontology (PPO) – a light weight vocabulary for defining privacy preferences for RDF data – and a Privacy Preference Manager (PPM) [9] enforcing them. However, this system induces large overhead on mobile devices.

In this paper we propose a new approach for fine-grained RDF data access control that is specifically designed for mobile devices. Our approach allows users to specify privacy preferences and enforces them when RDF data is accessed. It is co-located with the data and executed entirely on the user's mobile device. No external server support is needed, giving the user full control over his/her data at any time without trusting an external party. Our approach is based on RDF and SPARQL, modelling privacy preferences with RDF and checking them with SPARQL queries. This allows us to reuse the full power of RDF/SPARQL support in the existing RDF store on the mobile device without the need to add an additional reasoner or specific parser to process language specific rules. At the same time we retain the expressiveness of access control policies. Our approach does not assume any special support from the RDF store and can be used on top of any RDF store that offers support for SPARQL. To filter RDF triples we introduce a novel two stage approach that combines (1) an initial efficient query analysis stage that extracts the necessary metadata about the query and the (2) filtering phase that filters the result set without having to access the store for additional metadata (about the query). Our evaluation shows that this improved filtering algorithm results in a 10 times increase in system performance compared to our previous approach.

The paper is structured as follows: Section 2 presents our scenario, our privacy preference ontology PPO and privacy preference manager PPM. Based on this, Section 3 presents the current PPM filtering algorithm which was not published in our previous work. Then, we introduce our new improved filtering algorithm in Section 4. Section 5 presents evaluation results. Finally, Section 6 gives an overview of related work before we wrap up the paper with future work and a conclusion in Section 7.

2   Access Control for RDF Stores on Mobile Devices

Consider two friends Alice and Bob who want to exchange personal data with their smartphone devices. Each device by default denies access unless otherwise instructed by its user. Alice uses her smartphone, contacts Bob's smartphone and asks for his location. Bob receives a notification that Alice has requested to access his location. Bob grants Alice access and this privacy preference is stored in his smartphone. Alice can now retrieve and view Bob's location on her smartphone. Other data is still not accessible. Next time Alice requests to view Bob's location, if the request matches Bob's stored privacy preference, then she is automatically granted (or denied) access. Otherwise, Bob is notified about Alice's new request and decides whether to grant her access or not.

To realise this example we propose an access control system for RDF stores on mobile devices. By storing the data directly on the users' mobile devices, users can have full control over their data without trusting any external server or provider. However, the access control algorithms must be executed on the mobile device, too and thus they must be very efficient to respond in a timely fashion and not waste battery life.

Our approach models access control policies for RDF data using the Privacy Preference Ontology (PPO). PPO is non-domain specific and can model privacy preferences for any RDF scenario. In this section, we provide an overview of PPO and we explain how we model privacy preferences using it. Subsequently, we describe how the Privacy Preference Manager (PPM) enforces such privacy preferences by filtering out RDF data based on them. The PPM is datastore independent and therefore can be easily customised to provide fine-grained access control to any datastore.

2.1 Privacy Preference Ontology (PPO)

PPO[1] [8,10] is a *light-weight* Attribute-based Access Control (ABAC) vocabulary that allows users to describe fine-grained privacy preferences for restricting or granting access to non-domain specific Linked Data elements, such as Social Semantic Data. Considering that PPO is described in RDF(S), it does not require a specific parser or reasoner but it retains the expressivity of fine-grained access control policies similar to rule-based approaches. Among other use-cases, PPO can be used to restrict part of FOAF[2] profile records to users that have specific attributes. It provides a machine-readable way to define settings such as "Grant read access to my location only to Alice".

As PPO deals with RDF(S)/OWL data, a privacy preference defines: (1) the resource, statement, named graph, dataset or context it must grant or restrict access to; (2) the conditions refining what to grant or restrict (for example defining which instance of a class as subject or object to grant); (3) the access control privileges (including **Create**, **Read**, **Write**, **Update**, **Delete** and **Append**); and (4) an **AccessSpace**, defined by either an agent or a SPARQL query that specifies a graph pattern that must be satisfied by the requesting user.

Example. Figure 1 illustrates Bob's privacy preference that restricts his location only to Alice. The location is modelled as an instance of type **SpatialThing**[3] which includes longitude and latitude. Hence the privacy preference is applied to any resource of this type – in our case, Bob's location. In this example Alice is granted the **read** access to Bob's location.

---

[1] PPO – http://vocab.deri.ie/ppo#

[2] Friend-of-a-Friend (FOAF) – http://www.foaf-project.org

[3] WGS84 – http://www.w3.org/2003/01/geo/wgs84_pos#

```
PREFIX ppo:      <http://vocab.deri.ie/ppo#> .
PREFIX wgs84: <http://www.w3.org/2003/01/geo/wgs84_pos#>
<http://bob.com/PrivacyPref#1> a ppo:PrivacyPreference ;
ppo:hasCondition [ ppo:classAsSubject wgs84:SpatialThing ];
ppo:assignAccess acl:Read;
ppo:hasAccessSpace [
    ppo:hasAccessAgent <http://alice.com/me> ].
```

Fig. 1. Bob's privacy preference to grant Alice his location

2.2   Privacy Preference Manager (PPM)

The PPM [9] is an access control manager that allows users to create privacy preferences for RDF data. The manager also filters the requested data by returning only a *subset* of the requested data containing only those triples that are granted access as specified by the privacy preferences. The PPM was developed as a Web application – either as a centralised Web application or in a federated Web environment. The privacy preferences are stored separately from the data and can only be accessed by the PPM.

Although the PPM is suited for Web environments, it is not originally designed for operating on mobile devices due to their limited resources – such as processing power, memory resources and battery life. To port the PPM to mobile devices we modified the enforcing algorithm substantially to reduce the number of querying operations needed for filtering. In addition we designed a new filtering algorithm that extends our previous one to further reduce the number of queries.

3   PPM Access Control Filtering Algorithm (PPF-1)

The original PPM access control filtering algorithm (called PPF-1 in this paper) consists of (1) a matching part which maps the triples in the requested result set to the specific privacy preferences that apply to the triple; and (2) a filtering part that filters the result set by checking which triples a requester is granted access. This algorithm was not published in our previous work and therefore in this section we provide a detailed overview.

Initially, PPF-1 expects a list of requested triples and a set of privacy preferences. With these, PPF-1 first matches the triples to their corresponding privacy preferences; then, it checks what the requester can access and grants the requester a filtered result set.

Algorithm 1 illustrates the matching between triples and privacy preferences. This part iterates through every triple in the result set and for every triple it checks all the privacy preferences to match which ones apply to the triple. The algorithm checks whether each privacy preference applies to: (1) the named graph in which the triple resides; (2) a resource in the triple; and (3) a rectified statement – i.e. the triple's subject, predicate and object.

Data: *resultSet* and *privacyPreferencesList*
Result: (1) *protectedTriplesList*; (2) *unprotectedTriplesList*;
            (3) *accessAgentsList*; and (4) *accessPrivilegesList*.
List<PrivacyPreference> pList ← privacyPreferencesList;
List<Triple> rs ← resultSet;
Triple t ← new Triple();
PrivacyPreference p ← new PrivacyPreference();
forall the $t \in rs$ **do**
    forall the $p \in pList$ **do**
        if *p.Match(t)* **then**
            pURI ← p.getPrivacyPreferenceURI();
            aURI ← p.getAgentURI();
            privilege ← getAccessPrivilege();
            protectedTriplesList.add(t, pURI);
            accessAgentsList.add(aURI, pURI);
            accessPrivilegesList.add(privilege, pURI);
        else
            unprotectedTriplesList.add(t);
        end
    end
end

Algorithm 1: Privacy Preferences and Triples Matching

The algorithm checks whether each privacy preference has a condition that specifies: (1) the resource must be the subject of the triple; (2) the resource must be the object of the triple; (3) the subject of the triple must be an instance of a certain class; (4) the object of the triple must be an instance of a certain class; (5) contains a particular predicate; and (6) contains a particular literal.

For most of these checks, the values in both the requested triples and in the privacy preferences are tested to check whether they are both the same. However, for testing whether a subject or object of the triple are instances of a particular class, the algorithm queries the store each time a privacy preference (for each triple) is tested. The algorithm constructs a query that gets the class type of the subject or object. If the class type matches with the restricted class then the algorithm returns true. Otherwise it fetches the endpoint URI of the datastore in which the class types for the subject or object are specified. The endpoint URIs are mapped to the subjects and objects. Once the class type is retrieved, the algorithm returns whether they match (true) or not (false). Once all the triples are iterated, the filtering part filters the protected triples.

The filtering algorithm checks that for each triple in the protected triples list, the agent has been granted access by matching the privacy preference URI bound to the triple with the URI bound to the agent. If these match, then the triple is added to the access triples list. Once completed, the filtering algorithm sends back the access triples list that represents the filtered result set.

4 Extended Access Control Filtering Algorithm (PPF-2)

PPF-1 has a major performance bottleneck in the privacy preference matching phase: for each restricted triple PPF-1 executes a query on the RDF store to test whether the subject or object is of a particular class type. This may result in a large overhead since executing a query can be expensive – specifically on mobile devices with restricted resources. To increase efficiency, the number of necessary store accesses for identifying the class of a resource must be reduced without losing PPF-1's fine-grained control over data access.

In this section we introduce an extended filtering algorithm (called PPF-2) that fulfils these requirements. The main idea of PPF-2 is to identify the class of a resource by analysing both the requested query and the ontologies used by the data. To reduce the effort of analysing the used ontologies, we perform an ahead-of-time indexing phase for the ontologies at the system start time. This index is later used to identify the given classes. With this ahead-of-time indexing in place, the actual filtering process becomes a two stage algorithm, as follows: (1) analysis of the query to derive the resources' classes (Stage 1); (2) filtering of the triples (Stage 2), using the knowledge derived in Stage 1.

In the following we describe how we realise Step 1. Stage 2 is similar to the filtering done in PPF-1 and thus not explained again.

4.1 Knowledge Extraction from the Ontology and Query

Our solution is based on a query analysis step that allows to identify the classes of each resource based on the attributes that are used in the query. The query analyser parses the SPARQL query and for each resource it extracts *inbound* and *outbound* properties. Inbound properties are extracted from the triples in which the resource is the object. Outbound properties are extracted from the triples in which the resource is the subject. Based on these properties it is possible to identify the classes of a resource by looking at the ontologies data. Our approach uses a closed-world assumption, i.e. we assume that the filtering algorithm knows every ontology on which a privacy preference can be defined. This assumption is valid because: if an ontology is unknown when the privacy preference is defined, then the PPM can retrieve it before any actual query is run.

Similarly to accessing the store, querying the ontologies is a slow process. This can be improved by indexing the ontologies (once) *before* any actual query is run. The index is built by leveraging two relationships for properties defined in RDF Schema [4]: rdfs:domain and rdfs:range. The first is used to state that any resource that has a given property is an instance of a class, while the second is used to state that the values of a property are instances of a class. Thus, both of them can be used to derive the actual class(es) of a resource.

Example. Figure 2 shows a SPARQL query usable to extract the location (given as latitude and longitude) of a given user and the noise level at this location. The ?user is modelled as a gambas:User, a subclass of foaf:Agent. The

---

[4] RDF Schema – http://www.w3.org/TR/rdf-schema/

```
PREFIX gambas: http://www.gambas-ict.eu/ont/
PREFIX wgs84: http://www.w3.org/2003/01/geo/wgs84_pos#
SELECT ?lat ?long ?noise
WHERE {?user <gambas:userLocation> ?location .
  ?location <wgs84:lat> ?lat .
  ?location <wgs84:long> ?long .
  ?location <gambas:noiseLevel> ?noise}
```

Fig. 2. A SPARQL query example where the resources' class can be uniquely determined by the query analysis step

?location is a gambas:Place, a subclass of dol:Location [5], which has an attached wgs84:lat (latitude) and wgs84:long (longitude). In order to derive the classes of the variables in the query of Figure 2, the algorithm proceeds as follows for the ?user resource: (1) extract the <gambas:userLocation> property; (2) access the index on rdfs:domain using the property as key; (3) access the linked classes set, which contains only the gambas:User class. A similar approach can be applied to the ?location resource.

In the following section we will show a comparison of the performances of this modification versus the base case.

5   Evaluation

In order to evaluate the performance gain achieved by our extended filtering algorithm, we conducted a number of experiments on a Google Nexus 7 device running Android 4.2.2. Our system is implemented in Java. We compared two configurations with a PPM running on top of an RDF On the Go data store [6]. In the first configuration the PPM is using our previous filtering algorithm PPF-1. In the second one, the PPM is using our new filtering algorithm PPF-2.

The evaluation dataset was composed of 15000 triples, containing data about seven real-world user profiles. Using this dataset we executed a sample query on a user's topic interests and filtered the intermediate results with both algorithms (PPF-1 and PPF-2). Since we are mainly interested in the overhead induced by access control instead of query execution, we measured the execution time for filtering, omitting the time needed to execute the sample query on the dataset. The latter time depends only on the underlying RDF store and thus is the same for both filtering algorithms. To characterise the filtering performance in scenarios with different complexity, we varied both the number of triples in the intermediate result and the number of checked privacy preferences. Each experiment was repeated ten times. We started measuring after an initial preheating phase consisting of ten filtering runs. This reduced the variance introduced by the Android Just-in-Time optimiser. Moreover, each experiment was executed

---

[5] DOLCE – http://ontologydesignpatterns.org/wiki/Ontology:DOLCE%2BDnS_Ultralite

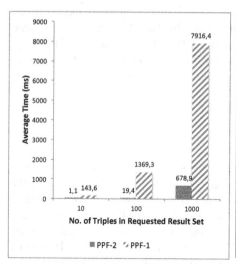
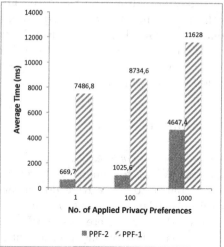

Fig. 3. Performance with varying size of result set

Fig. 4. Performance with varying number of privacy preferences

independently in a separate Android App, with no other running App and with all synchronisation services disabled – further reducing variances.

Figure 3 shows the execution time for filtering an intermediate result set of varying size (10, 100, and 1000 triples) using a single privacy preference. As can be seen, PPF-2 clearly outperforms PPF-1 by at least a factor of 10, confirming the effectiveness of the predefined index technique (see Section 4). Even for an intermediate result set of 1000 triples (representing the result of a query matching a comparatively large number of the 15000 triples in the RDF store), PPF-2 requires only approximately 0.7s to check access and filter the result set. In comparison, PPF-1 requires nearly 8s, making it unsuitable for many scenarios, e.g. interactive systems. The time required for filtering a mid size intermediate result set of 100 triples is around 0.02s for PPF-2 (compared to approximately 1.4s for PPF-1). Filtering a small intermediate result set of only 10 triples is nearly not measurable with both algorithms.

Figure 4 shows the execution time for filtering an intermediate result set of fixed size (1000 triples) using a varying number of privacy preferences (1, 100, and 1000 preferences). Again, PPF-2 clearly outperforms PPF-1 for all measurement points, reducing the absolute time for filtering triples with 100 privacy preferences to around 1s, down from 8.7s. Interestingly, the results for filtering with one privacy preference are quite similar (0.7s for PPF-2, down from 7.5s) due to fixed (i.e. size-independent) execution efforts. For 1000 privacy preferences, PPF-2 can still outperform PPF-1 by a factor of approximately 2.5 but both algorithms may still be too slow to be used in time critical scenarios (with PPF-1 requiring around 11.6s and PPF-2 around 4.6s).

Note that the presented results are only valid for situations in which the original query contains knowledge that can be used for filtering optimisation.

This may not always be the case. Therefore we also conducted experiments with an unbound query that requested all triples in the RDF store. This query contains no knowledge for PPF-2. In this case PPF-2 is reduced to PPF-1. It must access the store for each triple check and thus cannot perform better than PPF-1. This is confirmed by our measurements, since the results for PPF-1 and PPF-2 are the same in this case.

6   Related Work

Access control privileges for RDF data can be modelled using the Web Access Control (WAC) vocabulary[6]. However, this vocabulary is designed to specify access control to entire RDF documents rather than to specific data contained within the RDF document. *Privacy policies* can be modelled using the Platform for Privacy Preferences (P3P)[7]. It specifies a protocol that enables Web sites to share their privacy policies with Web users expressed in XML. P3P does not ensure that Web sites act according to their publicised policies and it does not enable end users to define their own privacy preferences. The authors in [5] propose a *privacy preference formal model* consisting of relationships between subjects and objects in Social Semantic Web applications. However, the proposed formal model does not provide fine-grain access control for RDF data. Similarly, the authors in [7] also propose an access control model for semantic networks. However, they do not cater for RDF data in mobile devices. RelBac [4] is a relational access control model that provides a formal model based on relationships amongst communities and resources. It is also not intended for RDF data stored in mobile devices.

The authors in [1] propose an *access control framework* for Social Networks by specifying privacy rules using the Semantic Web Rule Language (SWRL). However, this work does not support processing SWRL rules on mobile devices and requires a specific parser to process the SWRL syntax. The authors in [3] compare 12 rule-based languages for enforcing access control. Most of them require defining a large amount of rules for defining access control policies. Moreover, these require specific reasoners and parsers; apart from a system to enforce them.

Finally, the authors in [2] propose an access control vocabulary that is similar to our PPO and a manager similar to our PPM. However, their model applies only to named graphs. Although they provide support for mobile devices, the access control policies are sent to a central server and processed on this server. Our approach supports access control filtering directly on mobile devices.

7   Conclusion and Future Work

Access to personal data on mobile devices must be controlled tightly and efficiently. In this paper we presented our approach for fine-grained access control

---

[6] WAC — http://www.w3.org/ns/auth/acl
[7] P3P — http://www.w3.org/TR/P3P/

for RDF data on mobile devices. It allows users to fully control access to their data directly on their mobile devices, increasing their trust in the system. As we have shown, Linked Data technology like RDF and SPARQL can be used – even on mobile devices – to realise access control for RDF data. By using RDF to model our privacy preferences and a SPARQL engine to check them, no special rule language and reasoner components are necessary. Instead, the store managing the user data can be used to realise the access control on this data. Our experiments show that to be efficient such a system should combine multiple techniques, e.g. pre-indexing, query analysis as well as result filtering. Our work can be extended in several directions. Firstly, an evaluation of the impact of the proposed index on a combination of different types of privacy preferences is needed. Secondly, access space queries remain problematic, as they need to be tested on the store. It should be possible to address this in a similar manner as PPF-2 by analysing and building indexes for access space queries prior to executing the filtering algorithm.

References

1. Carminati, B., Ferrari, E., Heatherly, R., Kantarcioglu, M., Thuraisingham, B.: A Semantic Web Based Framework for Social Network Access Control. In: SACMAT 2009 (2009)
2. Costabello, L., Villata, S., Gandon, F., et al.: Context-aware access control for rdf graph stores. In: ECAI (2012)
3. De Coi, J.L., Olmedilla, D.: A review of trust management, security and privacy policy languages. In: International Conference on Security and Cryptography, SECRYPT (2008)
4. Giunchiglia, F., Zhang, R., Crispo, B.: Ontology Driven Community Access Control. In: Trust and Privacy on the Social and Semantic Web, SPOT 2009 (2009)
5. Kärger, P., Siberski, W.: Guarding a Walled Garden — Semantic Privacy Preferences for the Social Web. In: Aroyo, L., Antoniou, G., Hyvönen, E., ten Teije, A., Stuckenschmidt, H., Cabral, L., Tudorache, T. (eds.) ESWC 2010, Part II. LNCS, vol. 6089, pp. 151–165. Springer, Heidelberg (2010)
6. Le-Phuoc, D., Parreira, J.X., Reynolds, V., Hauswirth, M.: RDF on the Go: An RDF Storage and Query Processor for Mobile Devices. Posters and Demos of the ISWC 2010 (2010)
7. Ryutov, T., Kichkaylo, T., Neches, R.: Access control policies for semantic networks. In: POLICY (2009)
8. Sacco, O., Breslin, J.G.: PPO & PPM 2.0: Extending the privacy preference framework to provide finer-grained access control for the web of data. In: I-SEMANTICS 2012 (2012)
9. Sacco, O., Passant, A.: A Privacy Preference Manager for the Social Semantic Web. In: SPIM Workshop (2011)
10. Sacco, O., Passant, A.: A Privacy Preference Ontology (PPO) for Linked Data. In: Linked Data on the Web Workshop, LDOW 2011 (2011)

Overview of the WISE 2013 Challenge

Yueguo Chen[1], Lexi Gao[1], Xuan Ming[2], Weining Qian[3], and Yabo Xu[2]

[1] Renmin University of China
[2] Sun Yat-Sen University
[3] East China Normal University
{chenyueguo,gaolexi}@ruc.edu.cn, xuanming@mail2.sysu.edu.cn,
wnqian@sei.ecnu.edu.cn, xuyabo@mail.sysu.edu.cn

Abstract. WISE 2013 Challenge is designed to be a competition of technologies and systems for challenging problems of web information system engineering over real-life web data. The challenge has two tracks, namely *entity linking track* (T1) and *Weibo prediction track* (T2). The design of both tracks and results are summarized in this report.

1 T1: Entity Linking Track

WISE 2013 Challenge T1 task[1] is to label entities within plain texts based on a given set of entities. The set of entities used for labeling are extracted from the Wikilinks dataset[2]. In this challenge, the attendants are given a test corpus (in English) of plain texts, and a list of entities (distinct Wikipedia URLs extracted from the Wikilinks dataset). Attendees are required to automatically label the mentioned entities or concrete concepts within the given test corpus, using the given Wikipedia URLs. Results will be evaluated based on the recall and precision of entities and concrete concepts that have been correctly labeled.

The Wikilinks dataset contains around 3 million entity names (Wikipedia URLs) and their 40 million mentions. Since a certain percentage of Wikipedia URLs in the current Wikilinks dataset are in wrong format, a revised version of the dataset is used[1]. The attendants are however allowed to use any other datasets, as long as the linked entities in the submitted results are from the provided dataset.

Attendees need to find solutions to automatically detect mentions of entities, and link the detected mentions to entities in the given entity file. Detection of proper nouns such as *Shanghai, Google Translate, IBM*, is relatively easy. For concepts (that typically as classes of entities), attendants are suggested to ignore those common concepts such as trip, book, hotels, because they will be excluded from evaluation whether they are correctly labeled or not. However, concrete concepts such as *Chinese characters, guidebooks, online travel agencies* will be considered for evaluation. It is sometimes not that easy to distinguish between common concepts and concrete concepts. The attendants are suggested to label

---

[1] http://deke.ruc.edu.cn/wisechallenge/
[2] http://www.iesl.cs.umass.edu/data/wiki-links

X. Lin et al. (Eds.): WISE 2013, Part I, LNCS 8180, pp. 488–490, 2013.

a concept as concrete as possible. For example, the concept *online travel agencies* will be better than *travel agencies*. If a concrete concept has a linking entity, the label to a relatively general concept will be treated as a negative label.

1.1   Evaluation

Attendees are expected to submit results of a solution in a single file, with each line representing a label of a mention to an entity in the following format:

file_ID mention_start_pos entity_URL mention

The test files are labeled manually by experts, to generate a set of labels as ground truth. Note that both proper nous and concrete concepts are labeled in the ground truth label set. We further generate a subset of ground truth which contains only labels of proper nouns. It is called proper nouns label set. The submissions will be evaluated based on the percentage of the correct labels covering the full ground truth label set, and the percentage of the correct labels covering the proper nouns label set. Therefore, each run will be evaluated using two scores, one for the recall of both concrete concepts and proper nouns, and the other for the recall of proper nouns only. In addition, the precision of both concrete concepts and proper nouns is also considered.

1.2   Results

Finally, 13 runs from 6 teams are submitted. Among them, three teams achieved the best performance in 3 measures respectively. They are listed in Table 1. Details of the techniques of winners will be discussed in their papers respectively.

Table 1. Best results of the entity linking task

team	Recall of PN + DC	Recall of PN	Prec. of PN + DC
Northeast University, China	**0.475**	0.449	0.140
Beihang University	0.442	**0.456**	0.278
University of Brasilia	0.401	0.387	**0.425**

2   T2: Weibo Prediction Track

WISE 2013 Challenge T2 task[3] is based on a dataset collected from Sina Weibo, the dominant micro-blog service (http://weibo.com) in China. In this challenge, the attendees are required to build a system for predicting Weibo users' age range.

The original dataset was crawled from Sina Weibo via the open API Sina provided[4]. The dataset distributed in WISE 2013 Challenge is preprocessed by the

---

[3] http://www.wise2013.org/wise2013challenge.html#T2
[4] http://open.weibo.com/

following steps: 1) user IDs are anoymized; 2) user name mentioned in tweets is removed, and also re-tweet content; 3) original tweets are them segmented into words, and each Chinese word are mapped to a unique integer. After all pre-processing steps, there are only alphanumeric contents in the dataset. The final dataset released in the track consists of three files: 1) user Info: it includes basic information about users (anonymized user ID, tags, jobs, personal description, age, gender, education etc.); 2) tweets: it includes basic information about tweets (user ID, timestamp, segmented tweet content); 3) user birthday: It includes the user's birthday (only provided in training dataset). All the birthday information is obtained from Sina, and may contain noisy data inevitably (some user may fill in the incorrect birthday for privacy concern). It should also be noted that the dataset is not complete, yet is only a sample of the whole dataset in the Sina micro-blogging service.

Attendees are required to build a system for predicting Weibo users' age range. There are four age ranges (1, 2, 3, and 4) in total. Each represents by an integer number. Specifically, range 1 represents the age interval [0-18], range 2 represents the age interval (18-24], range 3 represents the age interval (24-35], and range 4 represents the ages above 35. Every user's age falls into one of the four ranges, and the goal of this task is to achieve high prediction accuracy.

2.1   Evaluation

Attendees are expected to submit results in a single plain text format with 17519 rows, in which each row contains 2 fields (separated by a tab): the user id and the predicted range value. A small dataset with the same setting (without user birth file) is provided for testing purpose. Each submission is evaluated by Micro-averaged F-Measure.

2.2   Results

Finally, submissions from 6 teams are received. Among them, two teams achieved the best performance in terms of F-Measure. Both teams tried diversed classification models, and the highest reported accuracy are tied on 83%. And we finally select the team from Renmin University of China as the winner as they tried a wider range of models that offer more insight into this task, and the runner-up is East China Normal University. Details of the techniques of these teams will be discussed in their papers respectively.

Acknowledgement. This work is partially supported by National 863 High-tech Program (Grant No. 2012AA011001).

Entity Extraction within Plain-Text Collections WISE 2013 Challenge - T1: Entity Linking Track

Carolina Abreu*, Flávio Costa, Laécio Santos, Lucas Monteiro,
Luiz Fernando Peres de Oliveira, Patrícia Lustosa, and Li Weigang

University of Brasilia, Brasilia, Brazil
carolabreu@unb.br,
{capregueira,laecio,lucasbmonteiro,patlustosa}@gmail.com,
luiz_peres10@hotmail.com, weigang@cic.unb.br

Abstract. The WISE 2013 conference proposed a challenge (T1 Track) in which teams must label entities within plain texts based on Wikilinks dataset which comprises 40 million mentions over 3 million existed entities. This paper describe a straightforward two-fold unsupervised strategy to extract and tag entities, aiming to achieve accurate results in the identification of proper nouns and concrete concepts, regardless the domain. The proposed solution is based on a pipeline of text processing modules that includes a lexical parser. The solution labelled 8824 texts, and the results achieved satisfying precision measures.

Keywords: Information extraction, Text analysis, Entity extraction, Wikilinks.

1 Introduction

With the rise of the Semantic Web, the need for built-in mechanisms for defining relationships between data becomes even more relevant [1]. Nevertheless, to identify relations within a myriad of information is still a challenge. For large corpus of data, it is impractical to manually label each text in order to define relations to extract information.

Automatic extraction of information from textual corpora is a well-known problem that has long been of interest in Information Extraction (IE) research. Named entity recognition (NER) is one of the first steps in many applications of IE and other applications of Nature Language Processing. A NER system allows the identification of proper nouns in unstructured text. The NER problem usually involves names, localizations, dates and monetary amount, but it can be expanded to involve the concrete concept identification as well [2]. After entity recognition, entity matching is the task of identifying to which entity from a knowledge base a given mention in free text refers [3].

This paper is focused on the Entity Linking Track (T1), a challenge to create an automatic system that identifies entities within simple texts and relate them

---

* This research has been partially supported by the CAPES and FINEP grants. The authors also thank the Electronic Warfare Education Center of Brazilian Army.

X. Lin et al. (Eds.): WISE 2013, Part I, LNCS 8180, pp. 491–496, 2013.

to the respective URLs of Wikipedia's set of entities provided by WISE. The Wikilinks dataset consists of over 3 million lines, where each line represents an entity and its over 40 million mentions. Attendees need to automatically detect proper nouns and concrete concepts. We present a straightforward two-fold unsupervised strategy to extract and tag entities, aiming to achieve accurate results in the identification of proper nouns and concrete concepts, regardless the domain. The proposed solution is based on a pipeline of text processing modules that includes a lexical parser. This general solution may be applied in any text corpus when an entity set is known. Even if there is no performance requirement in this track, the large number of entities and plain text to be processed in a short period is a constraint itself.

This paper is organized as follows: In Section 2, we provide a formal description of Data. After that, the problem definition and approach of the solution are described in Section 3. Section 4 reports on our experimental results and evaluation and Section 5 closes the paper with conclusions and a discussion of future work.

2   Formal Description of the Data

In order to supply a theoretical analysis of the data, we use a similar formalization as the one used in [5] and later in [4], for a Disambiguation to Wikipedia (D2W) problem.

Consider we are given a file f, containing a list of entities and its mentions, where each line represents a Wikipedia URL. Consider we are given a plain text document d with a set of named entities $N = \{n_1, \ldots, n_N\}$. We use the term *named entity* (NE) to denote the occurrence of a proper name or a concrete concept inside a plain text. Our purpose is to produce a mapping from the set of named entities to the set of Wikipedia entities (URLs) $W = \{e_1, \ldots, e_{|W|}\}$. It is possible for a NE to correspond to a entity that is not listed within f, therefore a *null* entity is added to the set W. Each Wikipedia entity has a set of mentions $M = \{m_1, \ldots, m_N\}$ associated to itself in the file f. According to [4], to match the named entities with the Wikipedia entities may be expressed as a problem of finding a many-to-one matching on a bipartite graph, with NEs forming one partition and Wikipedia entities the other partition. We denote the output matching as a N-tuple $\Gamma = (e_1, \ldots, e_N)$ where e_i is the more precise match, or the disambiguation, for NE n_i.

A local NE approach matches each named entity n_i apart from the others. The match is defined by some parameters. Let $\phi(n_i, e_j)$ be a score function reflecting the likelihood that the entity $e_j \in W$ is the more precise match for $n_i \in N$. Previous research considers that to identify all named entities based on a local approach can be expressed as optimization problem, as the one presented in the following:

$$\Gamma^*_{local} = argmax_\Gamma \sum_{i=1}^{N} \phi(n_i, e_i) . \tag{1}$$

Local approaches define a ϕ function to establish which one is the most accurate matching, by assigning higher scores to entities with content similar to that of the input document. Global approaches work in a more complex manner, by matching the entire set of NEs simultaneously, aiming to improve the coherence among the linked entities [4]. We intend to show that even a simpler approach, such as local matching, provides sound results that may be competitive with the state-of-the-art global oriented approaches. By keeping it simple, it is possible to reach very interesting results with limited resources.

3   Problem Definition and Approach

The entity linking method we propose here for entity recognition and extraction follows the basic framework for any IE system, summarized by [6]. The solution may be described in five main steps: **(1)** The entity file and the text corpus are pre-processed to load the database and to remove formatting issues, respectively; **(2)** Each plain text of the corpus is tokenized and grammatically annotated with its POS tag; **(3)** Each sentence is analyzed to detect proper nouns and concrete concepts, and the NEs are extracted using a set of rules; **(4)** The solution will identify to which URL from the Wikilinks entity file, those NEs in plain text refers; **(5)** All the chosen NEs and their respective entities are compiled in a result's file. A more detailed description of these steps is reported in the following topics and summarized in the overview of the architecture in Fig. 1.

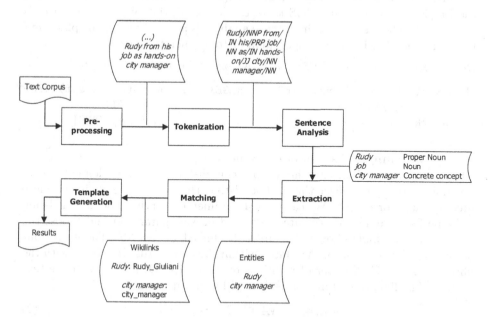

Fig. 1. Overview of the System Architecture. Each internal box represent a process. The external elements are an example of the solution applied to a single sentence.

Pre-processing. The Wikilinks entity file is processed and, as a result, database tables are created. All the entities and their respective mentions are organized in two separate tables. In addition, each entity is processed to create an auxiliary table, containing the tokens of the core words of the entity, which will be used as one of the matching parameters.

Tokenization. The pre-processed texts are divided into tokens and they go through an automatic grammatical tagging for its appropriate part-of-speech. An automatic tagger tool grammatically annotates each mention, and for each token it is assigned the most common Part-Of-Speech (POS), according to the Penn Treebank tag set[7]. The POS result is persisted in the database and will be used in the sentence analysis and extraction. The tokenization is done to the complete plain text.

Sentence Analysis and Entity Extraction. The Sentence Analysis step aims to identify proper noun groups and concrete concepts based on parsing. It is not the purpose of this research to perform a complete, detailed parse tree for each sentence. Instead, the system needs only to perform a partial parsing. We analyzed the grammatical structure of proper nouns and concrete concepts and proposed a couple of regular expressions, regex, to guide the solution to find these terms within the plain text. This technique identify these fragments deterministically based on purely local syntactic cues. For this reason, its coverage is limited. We defined a group of regular expressions based on the word level of the English Penn Treebank POS tag set. The description of the occurrences is shown next. The result of each sentence analysis is a set of NEs in the plain text that correspond either to a proper noun or to a concrete concept.

Regular Expressions to identify Proper Nouns and Concrete Concepts

```
(REGEX #1: Proper Nouns) p+n?     (REGEX #2: Concrete Concept) a?n+
    p:[/NNP, /NNPS]                             a:[/JJ]
    n:[/NN, /NNS]                             n:[/NN, /NNS]
```

Entity Matching. The goal of entity matching is to determine whether the NE refers to which Wikilinks entity. To accomplish that goal, we defined a score function $\phi(n_i, e_j)$ to reflect the likelihood that the entity $e_j \in W$ is the more precise match for $n_i \in N$. We calculate 4 parameters, varying from 0 to 1, being 1 the perfect match. We estimate the weight of each parameter and choose the entity with the highest overall score to be linked in the NE. There are four parameters for our solution:**A:** Match within the core words; **B:** Match with the phrase context; **C:** Number of mentions of the entity; **D:** Match with the text context. The final score function is defined by the equation

$$\phi(n_i, e_j) = \alpha A + \beta B + \gamma C + \delta D . \tag{2}$$

The parameter C is independent of the word received and gives a sense of the size of the entity. All the constant (α, β, γ, δ) weights were calibrated by specialists.

Design of Solution. The object of the present work is to achieve high precision level in labelling entities within plain texts. To achieve this goal, we developed a solution as illustrated in Fig. 1. The algorithms were implemented in Java, C# and Python. The processed data was loaded to a Postgrees database. The parser used by the algorithms was the extension of the Tree Tagger toolkit[1]. The texts were executed in a Dell PowerEdge R710 with 16 x Quad-core Intel Xeon E5630 2.53 GHz, Cache 12MB and 141 GB RAM.

4   Experiments and Evaluation

The test corpus dataset provided by WISE 2013 Challenge contain 8.824 files to be analyzed, with a total size of 19.2 MB of text. The smaller file is 573 bytes, whereas the biggest is 3.182 bytes. The number of lines of text within the corpus is 522.600. The average number of lines per text is 58. The minimum number of lines from a text has 6 lines. The maximum number of lines on a text is 250. We chose randomly a representative sample of the texts and a set of 3 specialists analyzed the samples and manually label proper nouns and concrete concepts that they think were relevant entities in each of the texts, in a process similar to the one executed by WISE 2013 Challenge Committee. The equation was calibrated according the comparison between the specialist answer and the solution answer.

The evaluation of the corpus text was made based on the WISE 2013 Guidelines, in which all proper nouns should be identified in the results. In addition, the evaluation ignored common concepts (usually common nouns) and focus on the concrete concepts. The results show a recall of proper nouns and concrete concepts of 40,1%. The recall of proper nouns only was 38,7%. The overall precision of the solution was 42,5%, the best results in the challenge. The f-measure was 41,27%, showing the strength of our solution.

In some of the texts, the automatic solution identified and linked some concrete concepts that were not predicted by the specialists. This results showed that our algorithm returned a large number of the relevant results (recall) and returned substantially more relevant than irrelevant results.

5   Conclusions

In the WISE 2013 Challenge T1: Entity Linking Track, we were asked to identify and link entities within plain text with the Wikilinks dataset. In this paper, a synthesis formalization of the problem and an overview of the architecture and the steps developed in the solution were presented. We successfully achieved the basic requirement to identify the proper nouns and concrete concepts from 8824 plain texts, with competitive precision and recall evaluation measures. The architecture is very scalable and we assumed that basic semantic relations can be inferred from matching given lexical-syntactic patterns. The contributions of

---

[1] http://www.cis.uni-muenchen.de/~schmid/tools/TreeTagger/

this paper can be noted as: a) developed an EI system to label automatically the English texts by WIKI entities with reasonable precision; b) proposed an algorithm to implement a local approach disambiguation with no training data requirements or restrict domain limitations.

During the development of the EI system, we overcame various difficulties such as the non-specific domain limitation, the ambiguity around concrete concepts and proper nouns and the lack of training data. As part of future works, we intend to implement the proposed solution to a specific domain, such as the energy industry, to evaluate if a controlled vocabulary it is possible to achieve even higher precision and recall measures.

Even the proposed solution is used for the entity extraction within plain-text collections, it still has great potential application in industry for example to extract the keywords from daily operation schedules and system log to avoid possible accidences in the important equipment etc. This opens a new line or research in both Nature Language Processing and Data Mining in energy industry.

References

1. Ruiz-Casado, M., Alfonseca, E., Okumura, M., Castells, P.: Information Extraction and Semantic Annotation of Wikipedia. In: Proceeding of the 2008 Conference on Ontology Learning and Population: Bridging the Gap between Text and Knowledge, June 16, pp. 145–169 (2008)
2. Shaalan, K., Raza, H.: NERA: Named Entity Recognition for Arabic. J. Am. Soc. Inf. Sci. Technol. 60(8), 1652–1663 (2009)
3. Singh, S., Subramanya, A., Pereira, F., McCallum, A.: Wikilinks: A Large-scale Cross-Document Coreference Corpus Labeled via Links to Wikipedia. CMPSCI Technical Report, UM-CS-2012-015, University of Massachusetts Amherst (2012)
4. Ratinov, L., Roth, D., Downey, D., Anderson, M.: Local and Global Algorithms for Disambiguation to Wikipedia. Computational Linguistics 1, 1375–1384 (2011)
5. Bunescu, R., Pasca, M.: Using Encyclopedic Knowledge for Named Entity Disambiguation. In: Proceedings of EACL, vol. 6, pp. 9–16. ACL (2006)
6. Cardie, C.: Empirical Methods in Information Extraction. AI Magazine 18(4), 65–79 (1997)
7. Marcus, M.P., Beatrice, S., Marcinkiewicz, M.A.: Building a large annotated corpus of English: the Penn Treebank. Computational Linguistics 19, 313–330 (1994)

ELS: An Efficient Entity Linking System

Chen Chen, Huilin Liu, Junchang Xin, Tiezheng Nie, and Zhiqiang Pang

College of Information Science and Engineering, Northeastern University, China

Abstract. WISE 2013 Challenge T1 is to label entities within plain texts based on a given set of entities. The set of entities used for labeling is extracted from the Wikilinks dataset. For the given test corpus (in English) and a list of entities, the challenge will automatically label the mentioned entities or concrete concepts within the given test corpus. In this report, we present ELS, which is an efficient entity linking system that can detect the entities effectively. Firstly,we convert each file into a list of tokens. In order to make sure the labeled concepts are concrete,the windows model is designed to extend the tokens. As the same entity may be linked by different URLs,we select the optimal linking via edit distance. Finally, we conduct extensive experiments to verify the efficiency of our proposed ELS.

Keywords: entity, entity linking, Wikilinks.

1 Introduction

The Wiki is a website which allows people to add, modify, or delete the content via a web browser usually using a simplified markup language or rich-text editor [1]. At present, the wiki website contains a large amount of entities and corresponding explanations. For the given list of entities which are extracted from the Wikilinks dataset, the WISE Challenge T1 is to label entities within plain texts using the given Wikipedia URLs. For example, for the test sentence "getting ready for a trip through China", we should first label China as an entity and then provide its linking URL in the Wikipedia, such as http://en.wikipedia.org/wiki/China.

The entities are usually nouns,and the most simple way to label entities is to detect the proper nouns such as Shanghai, Google Translate, IBM. However, this challenge suggests to ignore the common concepts such as trip, book, hotels. The concrete concepts such as Chinese character, guide books, online trave agencies are more significant. It is sometimes not that easy to distinguish between common concepts and concrete concepts, so the challenge requires to label a concept as concrete as possible. For example, the concept "online travel agencies" will be better than "travel agencies".

To achieve high performance of entity linking, the ELS system is built in the report. We use the Java programming language to write ELS for both efficient execution and reusability. All the given entities are stored in the MySQL [2] database. For the given test corpus, we can label the entities by matching them with the entities in the database. Furthermore, the windows model is designed to make the labeled entities more concrete.

X. Lin et al. (Eds.): WISE 2013, Part I, LNCS 8180, pp. 497–502, 2013.

The rest of this report is organized as follows. Section 2 is the preparatory stage which describes the storage of the given list of entities. Section 3 describes the design and implementation details of our system. A performance evaluation is presented in Section 4. And finally, Section 5 concludes the whole report.

2   Storage of the Given Entities

Based on the given list of entities, the most direct approach for entity linking is string matching. For example, if a single word or phrase exists in the given list of entities, we can see it as the new mention of the matched entity. Based on this consideration, it is necessary to organize the given entities in proper data structures to guarantee the match efficiency.

By parsing the given list of entities in the Wikipedia, 5,599,138 entities are extracted. It's important to note that the same entity may have multiple mentions. All these mentions are considered as the entities used for matching.

Due to the large amount of the entities, MySQL database is used to store and manage them. The most simple way is to store all the entities and corresponding Wikipedia URLs in the same data table. However, the big table is hard to maintain and the key word searching in it is time consuming. In the report, all the given entities are hashed to relatively small tables. The organization of the tables is shown in Figure 1.

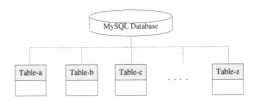

Fig. 1. The data tables used for maintaining the given entities

In the report, 27 tables are created. And according to the first letter of the entities , the entities are inserted into different tables. For example, we insert the entity whose first letter is "a" or "A" into Table-a shown in Figure 1. By the same way, other entities are inserted into the corresponding tables from Table-b to Table-z. Additionally, in the given list there are some entities which start with the number or special symbols, such as 1.98 Beauty Show,@Home. For this kind of entities, we insert them into the table of Tableother. Taking Table-a as an example, its table schema is shown in Table 1. The column of "entityname" stores the concrete entities in the given list and the column of "url" stores the corresponding URLs. The entities and corresponding URLs are stored as varchar in the MYSQL database and the engine type of the table is MYISAM.

Table 1. The table schema of Table-a

Field	Type	Null	Key
url	varchar(3100)	NO	
entityname	varchar(300)	NO	

3 System Design and Implementation

The goal of the report is to build a system which can label the entities in the test corpus. This section will give the concrete design and implementation of the proposed ELS entity linking system.

3.1 Preprocessing

For each test file in the test corpus, we will first convert it into a single string. Furthermore, we encapsulate each word in the string on the Class of Token. The class diagram of Token is shown in Figure 2. There are three member variables in the class. The variable of *startpos* is used to store the start pos of the word in the string; The variable of *name* is used to store the content of the word; and *partofspeech* is used to store the part of speech of the word. For example, for the test string "getting ready for a trip through China", the values of the word "China" are 33, China, and noun.

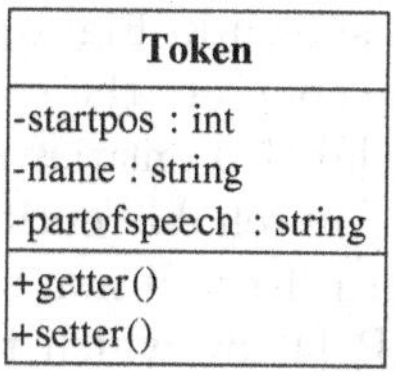

Token
-startpos : int
-name : string
-partofspeech : string
+getter()
+setter()

Fig. 2. The class diagram of Token

Following the encapsulation, the test file is converted into a list of Tokens.

3.2 The Windows Model for the ELS Entity Linking System

According to the requirement of the WISE Challenge, the concrete concept is better than the common ones. For the given test corpus, to make the labeled entities concrete, the windows model shown in Figure 3 is designed.

For the input list of tokens, the labeling starts from the first token, such as the token 1 in Figure 3. It is possible that several consecutive tokens stand for a single concept. For example, the concept "geography book" contains two tokens. We think the concept which contains more tokens will be more concrete.

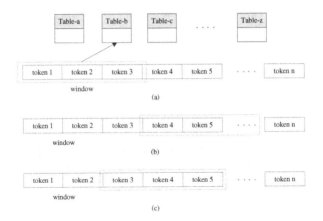

Fig. 3. The windows model for the ELS entity linking system

Based on this idea, we set a window to extend the tokens. Supposing that the current token is token 1, we do not match the content of token 1 with the entities in the database directly. Instead, more tokens involved in a window (the tokens in the red rectangle in Figure 3) are considered. For all the tokens in the window, we will find the matched longest token substring starting from token 1. For example, if the substring {token 1, token 2, token 3} is matched with an entity in the database, then the substring will be returned as the concrete concept. After the matching, the window will slide to the right. The current window is shown in Figure 3(b). If the matched longest substring is {token 1, token 2}, then the next window is shown in Figure 3(c). In the windows model, the bigger the window is, the more concrete the concept will be.

To improve the precision of the labeled entities in the test corpus. Two additional tools are used to filter some unreliable entities. The first one is the stop word list. All the words in the word list will not be labeled as an entity in any case. The second one is OpenNLP [4] by which the part of speech of a word can be easily labeled. For a single word, if the part of speech is not noun or verb-noun, we do not label it as an entity.

3.3   Entity Linking

The final step of the system is to link the labeled entity with the corresponding URL in the Wikipedia. According to the entity list given by WISE 2013, the same entity may be linked by different Wikipedia URLs. Therefore, it is necessary to select the optimal one as the final linking result.

For the given entity e and its corresponding URLs set $U = \{url_1, url_2, ..., url_n\}$, the edit distance $f(e, url_i)$ is used to measure the distance between e and url_i. The final linking URL of e is computed by Equation 1.

$$linkurl(e) = arm\min_{i \in U} f(e, url_i) \tag{1}$$

Here, the edit distance measures distance as the number of operations required to transform a string into another. More details are referred to [3].

3.4 The Overall Algorithm

The overall entity labeling and linking algorithm is shown in the following. We first convert file into a list of tokens *tokenarray* (line 1). For each window extended by tokens,we find the matched longest token substring as *entity* (line 4). If the *entity* is not *Null*,we can get its corresponding URLs set *US* (line 6). Finally we select the optimal one as the final linking *linkurl* and write the result to file (line 7,8).

Algorithm 1 The overall algorithm

Require: windowsize,file
 1: converted file into tokenarray
 2: **while** *tokenarray.hasNextWindow()* **do**
 3:    *windowtokens ← NextWindow(tokenarray, windowsize)*
 4:    *entity ← getmaxMatch(windowtokens)*
 5:    **if** *entity ≠ NULL* **then**
 6:      *US ← entity.getURLset()*
 7:      *linkurl ← entity.getoptimalone(US)*
 8:      *writetofile(entity, linkurl)*
 9:    **end if**
10: **end while**

---

4 Performance

In order to evaluate the performance of our system,1500 files were randomly selected from the test corpus for testing. These selected files are labeled manually by experts, to generate a set of labels as standard result. Total of 34,399 entities are labeled. We get the result in a single file, with each line representing a label of a mention to an entity in the following format:

file_id   mention_start_pos   entity_URL   mention

Figure 4 shows recall-windowsize and precision-windowsize plots. Recall refers to the proportion of the entities labeled correctly in standard result. Precision refers to the proportion of the entities labeled correctly in all entities labeled. For example,when the window size is 3,our approach labels a total of 116,501 entities, of which 16,335 are correctly labeled. The recall is 0.4748(16335/34399) and the precision is 0.1402(16335/116501). It's easy to note that both recall and precision are increasing along with the window size,for the bigger the window is, the more concrete the concept will be. For example, the concrete concept "online travel agencies" will be detected when the window size is 3 or more. However,when we set the window size is 2,this concrete concept will be ignored. Instead,we will get two common concepts,"online" and "travel agencies".

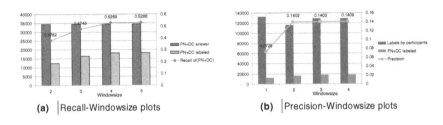

   (a) |Recall-Windowsize plots   **(b)** |Precision-Windowsize plots

Fig. 4. The result of testing

5   Conclusion

In this report, we present ELS, the system for labeling entities and linking with the Wikipedia URLs. In the ELS system, the MySQL database is used to maintain all the given entities. Specially, to improve the query efficiency, the given entities are inserted into 27 different tables according their first letters. For the input test corpus, we convert each file into a list of tokens and then label the entities by matching them with the entities in the database. Based on the list of tokens, the windows model is designed to make sure the labeled concepts are concrete.

For each labeled entity, several different Wikipedia URLs may be related. In the report, the edit distance is used to compute the distance between the labeled entity and each related Wikipedia URL. We select the URL which has minimum distance with the labeled entity as the final link.

Acknowledgments. This report is supported by the 863 Program under (Grant No. 2012AA011004), the National Natural Science Foundation of China (Grant No. 61003060) and the Fundamental Research Funds for the Central Universities (Grant No. N110404010).

References

1. Wikipedia, the free encyclopedia, `http://en.wikipedia.org/wiki/Wiki`
2. MySQL, `http://www.mysql.com/`.
3. Wagner, R.A., Fischer, M.J.: The string-to-string correction problem. Journal of the ACM 21(1), 168–173 (1974)
4. OpenNLp, `http://opennlp.apache.org/`

Combining POS Tagging, Lucene Search and Similarity Metrics for Entity Linking

Shujuan Zhao[1], Chune Li[1], Shuai Ma[1,*], Tiejun Ma[2], and Dianfu Ma[1]

[1] SKLSDE Lab, Beihang University, China
[2] University of Southampton, UK
{zhaosj@act.,lichune@act.,mashuai@,madf@}buaa.edu.cn,
tiejun.ma@soton.ac.uk

Abstract. Entity linking is to detect proper nouns or concrete concepts (a.k.a mentions) from documents, and to map them to the corresponding entries in a given knowledge base. In this paper, we propose an entity linking framework POSLS consisting of three components: mention detection, candidate selection and entity disambiguation. First, we use part of speech tagging and English syntactic rules to detect mentions. We then choose candidates with Lucene search. Finally, we identify the best matchings with a similarity based disambiguation method. Experimental results show that our approach has an acceptable accuracy.

Keywords: Entity Linking, POS Tagging, Lucene Search, Similarity Metrics, Mention Detection.

1 Introduction

Entity linking is to detect proper nouns or concrete concepts (a.k.a mentions) from documents, and to map them to the corresponding entries in a given knowledge base (KB) [1]. The research has attracted a lot of interests since its invention, due to the rapid expansion of Web information that leads to a great need of extracting useful knowledge from the Web. Moreover, the structured Web knowledge, i.e. Wikipedia[1], is increasingly becoming mature, which makes it possible to dig out more detailed information. For example, question answering tasks first find the expansion terms or synonyms of questions by linking user queries to Wikipedia, and then search answers with these synonyms [2]. However, linking entities manually is tedious and requires a lot of efforts. Our goal is to link mentions automatically in documents to Wikipedia URLs to significantly reduce manual efforts. Thus we propose an entity linking framework POSLS, which combines Part-Of-Speech tagging [3], Lucene search [4] and similarity metrics [5] for entity linking, and we correlate a target mention with its corresponding unique

---

* Contact author, and Shuai is supported in part by NGFR 973 grant 2014CB340300 and 863 grant 2011AA01A202, SKLSDE Lab grant SKLSDE-2012ZX-08, and the Fundamental Research Funds for the Universities.
[1] http://www.wikipedia.org/

X. Lin et al. (Eds.): WISE 2013, Part I, LNCS 8180, pp. 503–509, 2013.

KB entry. Specifically, the KB here contains 2,860,422 entries, and each entry is associated with two columns: entity names (in forms of Wikipedia URLs) and the collected mentions associated with the entities.

Given a corpus of plain texts and a KB as inputs, our approach first utilizes the Stanford's POS tagging technique [3] to analyze categories (verb, noun or adjective etc.) of every word, after which proper nouns are detected. Moreover, most of the concrete concepts could be identified based on the relationships of adjacent words. Next, the candidate URLs (entities) corresponding to a mention are to be detected. In order to speed up the search process, we treat each entry of the KB as a document and index them in Lucene search engine [4]. After querying the mention using Lucene, we further decide the best matching URLs, by comparing the similarity between the detected mentions and the given entity mentions in the KB for candidate URLs with similarity metrics, e.g., q-grams and edit distance (see [5] for a survey). Meanwhile, we also utilize history information for the disambiguation of some proper nouns in the process.

In conclusion, our main contributions are (1) an entity linking framework combining existing techniques, such as POS Tagging, Lucene search and similarity metrics, (2) a set of English syntactic rules for entity detection, and (3) an entity disambiguating method based on similarity metrics.

Organization. The paper is organized as follows. Section 2 depicts the related work. Section 3 describes how our framework POSLS works for entity linking. Section 4 contains an experimental analysis, followed by the conclusion section.

2   Related Work

The challenges of entity linking lie in mention detection and entity disambiguation. As to detecting mentions, Medelyan et al. [6] proposed an n-gram method, using a sliding window on the input article and comparing every n-gram with stop words omitted (stop words appear frequently and have no special meanings, such as "a" and "the"). Mihalcea and Csomai [7] constructed a controlled vocabulary composed of Wikipedia article titles and surface forms (anchor texts that refer to other Wikipedia links), to which is referred in mention detection. However, it is costly and time-consuming to prepare a vocabulary bank and use n-gram searching when the input document is large. Mendes et al. [8] firstly used LingPipe Extract Dictionary-Based Chunker [9] and then exploited POS tagger of LingPipe to get rid of mentions that were made of verbs, adjectives, adverbs and prepositions. Our approach is similar, but we first make the POS analysis, and then take advantage of English syntactic rules to find mentions that are primarily nominal phrases.

For entity disambiguation, the key is to compute the relevance with certain similarity metrics to get the top match results. [10] built a vector space based on a bag-of-word model and made use of the cosine similarity, and [11] further used the category feature of Wikipedia attributes. Other better methods may consider overlap (Jaccard Cofficient) between the first paragraph of the Wikipedia article and input document [7] and the edit distance between titles and mentions [12], or apply intricate interlinks of articles to compute the relevance score of URLs [13]. Generally, pure similarity comparing methods get an ordinary performance in specific domains. Nowadays, similarity metrics combining machine learning [12, 14] or graph model [13] usually obtain a good performance. However, preparing a representative training set is costly and constructing graph is time-consuming. Here we utilize similarity metrics [5] and previous detected mentions in a document to find the best matchings.

3   POSLS: An Entity Linking Framework

Given a corpus of documents and a knowledge base with Wikipedia URLs and their mapping mentions, our entity linking framework POSLS automatically detects mentions (proper nouns and concrete concepts) in these documents, and links them to best matching KB entries. The final output contains documents IDs, offsets of the detected mentions in the documents, detected mentions and their matching URLs in KB.

Table 1. English Syntactic Rules for Concrete Mentions

No.	Rules	Meanings	Explanations
R_1	(NN| NNP| NNS| NNPS)$^+$	mentions composed of nouns	noun phrases: one or more singular, plural or proper nouns
R_2	(JJ| JJS)$^+ \cdot R_1$	mentions starting with adjectives and ending with nouns	one or more adjectives followed by R_1
R_3	$R_1 \cdot$ 'of' $\cdot$ S$^* \cdot R_1$	mentions with "of", such as "History of China"	noun phrases followed by "of", an arbitrary string and R_1
R_4	$R_1 \cdot$ T $\cdot R_1$	mentions with the genitive marker, e.g., "Stanford's POS tagging technique"	noun phrases followed by T and R_1

Symbols *, $^+$, | and $\cdot$ denote any number of occurrences, one or more occurrences, alternation and concatenation, respectively. NN: singular noun or mass; NNP: proper singular noun; NNS: plural noun; NNPS: proper plural noun; JJ: adjective; JJS: superlative adjective; S: characters; T: the generative maker 's or '.

3.1   Detecting Mentions

The first step is to detect possible mentions appeared in the input documents. We utilize the Stanford's POS tagging technique [3], which further uses the Penn Treebank Tag set [15] to get proper nouns and acquire concrete concepts, in terms of a set of English syntactic rules expressed in regular expressions, shown in Table 1. The rules are reasonable for mentions that are nominal phrases.

The four rules could embody necessary possible concepts appeared in English articles. Experiments also show that the method could detect mentions with a high recall. Besides, we ignore common single words, such as "trip" and "school", from a list of 1500 most frequently used nouns[2].

3.2 Searching Candidates

The second step is to find a list of URL candidates from the given KB for each mention. Ambiguity is common in English because of polysemy, difference of contexts and morphological diversity (acronym, abbreviation and alterable order) [16]. For instance, "tree" may refer to "plant tree" or "tree data structure"; although "China" could be a country name, it could also refer to "the history of China" or "the culture of China" in different contexts. In the meanwhile, the acronym "KB" might have various linkings, which could refer to "Knowledge Base", "Kilobyte", or even a bank of Iceland – "Kaupthing Bank".

Realizing that the KB entries are excessive, we make use of Apache Lucene, a fast retrieval software library, to select matching URL candidates. We first tokenize URLs and entities in the KB, and build an inverted index for each token by treating each line as a small document. Different weights are set to tokens appeared in left and right columns under the assumption that there are almost no errors in the URL fields. Detected mentions in first step are sent to Lucene as queries, and the software ranks the given results based on the TF/IDF relevance between the mention and each line. We choose the top 40 as candidates.

3.3 Entity Disambiguation

The last step is to determine the best matching URL from the candidate set for each detected mention. Considering that the KB has provided mapping pairs between the URLs and entity mentions, we compare the similarities between the detected mentions and the provided entity mentions in the KB. Due to the diversity of expressions, there may be more than one collected entities, separated by "##", that corresponds to the same URL. We utilize similarity metrics, e.g., q-grams, edit distance and Jaro-Winkle [5], for entity disambiguation.

The method works as follows. For each candidate URL, we first get its provided mapping mentions in the KB. We then compare them with the detected mention based on similarity metrics. Meanwhile, we record the current most similar URL and its similarity value, which are updated constantly. When a URL has a similarity between its some given mention and the detected mention that is larger than the threshold, it becomes a possible mapping. Then we compare the similarity value with the current stored value. If the new value is higher, we update the most similar result and maximum similarity accordingly. In addition, we notice that one entity mention may appear in several KB entries, caused by the morphology of English. Therefore, when two URLs have the same mapping entity mention, we further compute the similarity of URLs and the detected

---

[2] http://www.talkenglish.com/Vocabulary/Top-1500-Nouns.aspx

mention. The one with a higher similarity becomes the most similar URL. Finally, the most similar URL is treated as the best matching. Moreover, we make use of history information to match some proper nouns which are parts of the other nouns. For these proper nouns, we just copy their corresponding complete mapping results directly. Though we mentioned that similarity metrics generally have an ordinary performance in entity linking, it is not true for the situation that the KB already gives the matching entries of URLs and entity mentions.

4   Experimental Study

The testing dataset is a corpus of domain-independent articles for entity linking, provided by the WISE challenge 2013. With the given correct answers, we evaluate the performance of our English syntactic rules and POSLS framework.

4.1   English Syntactic Rules Evaluation

By comparing with the correct answers, we tested our mention detection method with four datasets, each consisting of 100 or 200 articles. The results have a high average recall of 0.749, as shown in Table 2.

Table 2. English Syntactic Rules Results Evaluation

Group	# of articles	Recall
1	100	0.75603
2	100	0.73036
3	200	0.75294
4	200	0.76595

4.2   The POSLS Framework Evaluation

The POSLS method is implemented in JAVA, and all experiments are conducted on a dual core computer with 4G memory. We conducted a series of experiments on several data sets with the edit distance, q-gram and jaro-winkler metrics. The results are shown in Tables 3 and 4 with 100 and 600 testing articles, respectively. The results are computed by using exact equality of the detected mentions and standard answers (redirecting URLs are treated as correct for they refer to the same Wikipedia articles). In addition, if we consider those correct linkings (not listed in given answers), the following results could be much better.

From the experiment results, we can see that the edit distance similarity has the best performance in precision and recall, while the q-gram metric has the the best performance in efficiency. In general, our POSLS framework has an acceptable precision and recall.

Table 3. POSLS Results Evaluation with Different Similarity Metrics (100)

Similarity Functions	Precision	Recall	F-measure	Time
Edit Distance	0.2972973	0.5306122	0.3810793	84959ms
Q-gram	0.2701950	0.4948080	0.3495496	86345ms
Jaro-Winkler	0.2116380	0.4740260	0.2926271	82861ms

Table 4. POSLS Results Evaluation with Different Similarity Metrics (600)

Similarity Functions	Precision	Recall	F-measure	Time
Edit Distance	0.2720268	0.4491751	0.3388446	454381ms
Q-gram	0.2418758	0.4110738	0.3045528	466615ms
Jaro-Winkler	0.1916144	0.4035235	0.2598420	421335ms

5   Conclusion

In this paper, we propose an entity linking framework POSLS which combines POS tagging, Lucene search and similarity metrics. First, we built a set of English syntactic rules according to POS tags to detect proper nouns and concrete concepts. Then a group of candidates were selected, using Lucene search. Finally, we introduced a similarity based entity disambiguation method. We also experimentally verified the effectiveness and accuracy of the POSLS framework.

There is much to be done in the future. More sophisticated English syntactic rules are to be developed to improve the mention detection quality. Further, we are exploring new techniques to improve the efficiency and accuracy.

References

1. Dredze, M., McNamee, P., Rao, D., Gerber, A., Finin, T.: Entity disambiguation for knowledge base population. In: COLING (2010)
2. MacKinnon, I., Vechtomova, O.: Improving complex interactive question answering with wikipedia anchor text. In: Macdonald, C., Ounis, I., Plachouras, V., Ruthven, I., White, R.W. (eds.) ECIR 2008. LNCS, vol. 4956, pp. 438–445. Springer, Heidelberg (2008)
3. Toutanova, K., Klein, D., Manning, C.D., Singer, Y.: Feature-rich part-of-speech tagging with a cyclic dependency network. In: HLT-NAACL (2003)
4. Lucene Search, http://lucene.apache.org/
5. Elmagarmid, A.K., Ipeirotis, P.G., Verykios, V.S.: Duplicate record detection: A survey. IEEE TKDE 19(1), 1–16 (2007)
6. Medelyan, O., Witten, I.H., Milne, D.: Topic indexing with wikipedia. In: The Wikipedia and AI Workshop at AAAI 2008 (2008)
7. Mihalcea, R., Csomai, A.: Wikify!: linking documents to encyclopedic knowledge. In: CIKM (2007)
8. Mendes, P.N., Jakob, M., García-Silva, A., Bizer, C.: Dbpedia spotlight: shedding light on the web of documents. In: I-SEMANTICS (2011)

9. Carpenter, B., Baldwin, B.: Lingpipe (2008)
10. Zhang, W., Su, J., Tan, C.L., Wang, W.T.: Entity linking leveraging automatically generated annotation. In: COLING (2010)
11. Cucerzan, S.: Large-scale named entity disambiguation based on wikipedia data. In: EMNLP-CoNLL (2007)
12. Zhang, W., Su, J., Tan, C.L., Cao, Y., Lin, C.-Y.: A lazy learning model for entity linking using query-specific information. In: COLING (2012)
13. Han, X., Sun, L., Zhao, J.: Collective entity linking in web text: a graph-based method. In: SIGIR (2011)
14. Zheng, Z., Li, F., Huang, M., Zhu, X.: Learning to link entities with knowledge base. In: HLT-NAACL (2010)
15. Marcus, M.P., Marcinkiewicz, M.A., Santorini, B.: Building a large annotated corpus of English: The Penn Treebank. Computational Linguistics 19(2), 313–330 (1993)
16. Kortmann, B., Schneider, E.W., Burridge, K., Mesthrie, R., Upton, C.: A handbook of varieties of English A Multimedia Reference Tool. Phonology, vol. 1, Morphology and Syntax, vol. 2. Mouton de Gruyter (2004)

Predicting Microblog User's Age
Based on Text Information

Ye Li[1,2], Tao Liu[1,2], Hongyan Liu[3,*], Jun He[1,2,*], and Xiaoyong Du[1,2]

[1] Key Labs of Data Engineering and Knowledge Engineering, Ministry of Education, China
[2] School of Information, Renmin University of China
[3] School of Economics and Management, Tsinghua University
{echo_li,liutao.info,hejun,duyong}@ruc.edu.cn,
hyliu@tsinghua.edu.cn

Abstract. User age information plays a crucial role in many real applications such as precise marketing, directional promotion and personalized recommendation. In this paper, we focus on predicting user age range in Sina Weibo. To protect user privacy, we only have user basic profile information and user published messages (tweets), which are all mapped to integers. From these meaningless integers, we have to seek out underlying features or structures. Through analysis, we extract significant features related to age. In order to evaluate the correlation between user basic information and age ranges, we choose mutual information as measurement. To handle the problem of high dimensions and data sparsity caused by traditional word vector model of tweet contents, we propose aggregated tweet features corresponding to different age ranges. Using these features, we compared many classification algorithms. Finally, the model based on decision tree can achieve best prediction accuracy up to 83%.

Keywords: microblog, feature extraction, age prediction, classification.

1 Introduction

Sina Weibo is the largest micro-blog service provider in China which has more than 500 million registered users and 46.2 million active users per day. Users provide some basic information when registering. Users build up social network through following actions. In Sina Weibo, a user can follow any other user he or she is interested in and other users can also follow him or her without his or her permission. Users can publish their own tweets or retweet from other users they follow.

User demographic prediction has been an interesting research problem in Micro-blog. Among the many demographic characteristics, age prediction arises considerable attention. It is proved that with correct prediction we can do better on precise marketing, directional promotion, personalized recommendation and so on. Thus help companies achieve profitable growth at lower cost. There are many such researches on Twitter [1] which is the most famous micro-blog service provider around the

---

* Corresponding authors.

X. Lin et al. (Eds.): WISE 2013, Part I, LNCS 8180, pp. 510–515, 2013.

world, but few on Sina Weibo, and there are also many differences between Twitter and Sina Weibo which is more in line with Chinese people's habits.

In this paper, we focus on the prediction of users' age ranges in Sina Weibo with user basic information and user published messages. To protect users' privacy, all the information we have has been mapped to meaningless integers. On the whole it is a classical classification problem. To predict user age range, we first extract significant features from a mass of integers. For assessing correlation between user basic information and age ranges, we choose mutual information as measurement. Meanwhile, considering the high dimension and data sparsity caused by traditional word vector model when processing user published messages, we propose new aggregated features corresponding to four different age ranges. They well capture the information we need and avoid the problems caused by word vector model.

Using these features, we compare many traditional classifiers, such as, *decision tree, naïve bayes, SVM* and some complex classifier, like *Deep Belief Network* [2] and so on. Finally, we choose decision tree as the prediction model which can achieve the better performance than others and the accuracy is up to 83%.

The rest of the paper is organized as follows. Section 2 defines the task and describes the dataset. In section 3, we show the feature extraction and selection procedure. Section 4 shows all the models we have tried and experiment details. In Section 5 we conclude our work.

2   Dataset and Task Description

The dataset we use in this paper is from *WISE challenge* 2013 *Track* 2[1]. The information we have can be divided into three parts - user basic information, user published message and user birthday. User basic information includes features such as anonymized user ID, tags, job, education, user self-description and gender. To protect user privacy, values of all these features are mapped into integers without real meaning. User basic information covers 1126049 users, among which 980392 users have published message (for simplicity, we call each message a tweet). Users' tweets contain tweet timestamps and tweet contents with each term mapped to an integer. In the training set, all users have their birthday information.

The target of our work is to predict user's age range according to user basic information and users' tweets. There are four ranges represented by an integer from one to four. Table 1 shows the specific age scope per range.

Table 1. Age Ranges

Range	Scope
1	[0 ~18]
2	(18 ~ 24]
3	(24 ~ 35]
4	> 35

[1] http://www.wise2013.org/wise2013challenge.html#T2

Since the training dataset contains user birthday, we can easily get user age through simple calculation. It turns out to be a supervised classification problem, where each age range corresponds to a class label. To simplify expression, we will use *Range* to denote age range in this paper. Since all the information is preprocessed and mapped to integers, we have to find the underlying features or structures hidden in the numbers.

3   Feature Extraction

In this section we introduce how to construct features and how to extract useful features from the provided dataset. The features are grouped into two categories - ones related to user basic information and the other related to tweet contents.

3.1   User Basic Information Features

In this section, we make use of the attributes (features) in the given dataset of user basic information to construct useful features. Attributes in basic information part are listed in Table 2.

Table 2. User Basic Information

Attributes	Explanation
Tags	User labels, represented by integers separated by blank space
Jobs	Occupational information, represented by integers separated by blank space
Education	Educational information, represented by integers separated by blank space
Description	Self-introduction, represented by integers separated by blank space
Gender	User gender, 1 for male and 2 for female

Based on the attributes shown in Table 2, we extract 25 features shown below with detailed explanations.

- **Existence of tags:** binary value of 0 or 1, indicating whether user basic information contains tags.
- **Length of tags:** the number of tags.
- **Tag terms' correlation with each age range:** since we will train a prediction model through supervised learning process, we can use some kind of assessment to evaluate the correlation between user tag words (terms mapped to integers) and each range. Here we use mutual information to evaluate. *Mutual Information (MI)* is an useful measure in Information Theory. Equations (1) and (2) show the definition of mutual information. Here we use y to represent Range, $y \in \{1, 2, 3, 4\}$, t_i to represent term in tags. $H(x)$ means the joint entropy. The probability of $p(x, y)$ can be got using maximum likelihood estimation.

$$H_{(tag, range)} = \sum_{t_i \in tag} MI(t_i, y) = \sum_j \left(H(t_j) + H(y) - H(t_j, y) \right) \tag{1}$$

$$H(x, y) = p(x, y) log p(x, y) \tag{2}$$

By evaluating correlation using mutual information, we construct four features related to four different ranges based on tag information.

- **Existence of jobs:** binary value of 0 or 1, indicating whether user basic information contains job information.
- **Length of jobs:** the number of job terms
- **Job terms' correlation with each age range:** similar to tags, four features are constructed to express correlation between each job term and each range.
- **Existence of education:** binary value of 0 or 1, indicating whether user basic information contains education information.
- **Length of education:** the number of terms used in the education information.
- **Education terms' correlation with age range:** similar to tags, four features are constructed to express correlation between education term and each range.
- **Existence of description:** binary value of 0 or 1, indicating whether user basic information contains self description.
- **Length of description:** the number of terms used in description.
- **Description terms' correlation with per age range:** similar to tags, four features are constructed to express correlation between each description term and each range.
- **Gender:** 1 indicates male and 2 indicates female.

3.2   User Tweet Features

This section shows features we extract from user tweets. User tweets have two different kinds of attributes which are displayed in Table 3.

Table 3. User Tweet Information

Attribute	Explanation
Timestamp	Time at which user published tweet
Tweet Content	each term is mapped to integers

Based on the attributes shown in Table 3, we extract 9 features as follows.

- **User tweet quantity:** the number of tweets a user published.
- **User retweet ratio:** By examining the tweet content of different users, we find a frequently occurred integer "92725" in tweets. Although the total percentage of the tweets containing "92725" is not very large, it surprises us that it usually appears at the very beginning of the tweet, accounting for 92.94% of the tweets containing this number. Associating with Sina Weibo's retweet mechanism, we make a reasonable guess that integer "92725" may map to user's retweet action. Based on this assumption, for each user who published tweets, we have the ratio of the number of tweets with integer "92725" to the total number of tweets , which we call retweet ratio.
- **Tweet timestamp:** time point when user published one tweet.
- **Average tweet length:** From the intuitive point of view, user average tweet length can reflect user's obsession and the willingness to express oneself on Weibo, which may reflect one's age.

- **Average time gap of tweet publication:** It is the average time gap between two sequential tweet timestamps. Shorter time gap means more frequent tweet publication actions.

Till now, we extract features only considering tweet statistical information. Tweet content is not included yet. The simple way to extracting features from tweet content is to build the bag-of-words model, representing users with the tweet word vectors. However, the number of unique integers in tweet contents is more than 100 thousand (147779 in total). If we use word vector model, the vector dimension is so high that it is out of processing ability. Meanwhile, we experience severe data sparse problem.

In order to avoid the problems we mentioned above and make full use of the tweet contents at the same time, we propose four aggregated features related to four distinct age ranges. We find the most representative terms among four ranges, which is only a small subset of the whole term set. Meanwhile, the aggregated features decrease the high dimensions while containing necessary information.

To find the representative tweet terms per age range, we turn to frequent itemset mining algorithm, *FP-Growth* [3], to find the frequent terms occurred in tweets of each age range users. As a result, four different representative term sets are constructed, which we call Representative term sets. Based on these sets, we obtain four features for each user.

- **Number of representative tweet terms in each age range:** It is the number of user tweet terms included in the representative term set for each age range.

Above are all 34 features we extract from the given dataset. However, we still need to do feature selection to eliminate irrelevant features. Since we do not want to spend too much time on it, we only take some basic statistical experiments to verify whether the features really have some correlation with user age. By experiments, we at least wipe out tweet timestamp and keep the rest 33 features for classification.

4   Predicting User Age Range

We choose several different classification methods to construct our prediction models. We also generate different training and test datasets with different sizes to evaluate the performance of different classifiers. The methods we have taken are listed in Table 4, which shows the result on training dataset of 800000 users and test dataset of 80000 users. Among them, *Deep Belief Network* (*DBN*) is one of the deep learning algorithms that stand out in machine learning recently. Although *DBN* has an outstanding accuracy on a small training set of 5000 users, its accuracy decreases sharply as training examples increase. Thus even *DBN* has a great ability on feature learning, it is not suitable in this situation.

Table 4. Method Comparasion

Methods	Explanation	Precision
C4.5	*Decision Tree Classification Algorithm*	0.83
SVM	*Support Vector Machine*	0.65
NB	*Naive Bayes*	0.60
DBN	*Deep Belief Network*	0.63

Based on this result, we finally choose decision tree as classification method, as others have similar or poorer effect compared with decision tree. Here we select C4.5 decision tree algorithm [4] for prediction.

To compare the effects of different size of training dataset, we construct different training dataset by randomly selection from the whole training dataset. The number of users in each test dataset is 10% of the corresponding training dataset. As the size of training set increases, the accuracy of the model increases, too. Table 5 shows the results of experiments. Up to now, the best accuracy we can achieve is 83%.

Table 5. Prediction Model's Performance

Methods	Training sets size	Precision
C4.5	20000	0.64
C4.5	50000	0.75
C4.5	200000	0.77
C4.5	500000	0.79
C4.5	600000	0.80
C4.5	800000	0.83

5   Conclusions

In this paper, we build a model to predict user age range by extracting user features from dataset with limited semantics. To tackle the high dimension and data sparsity caused by word vector model when featuring user messages, we propose new features corresponding to different age ranges which make full use of tweet contents and avoid the severe problems mentioned above. By contrasting various models' performance, we construct a decision tree model using C4.5 algorithm and achieve accuracy of 83%.

Acknowledgements. This work was supported by the 973 program of China under Grant No. 012CB316205, the National Social Science Fund of China under Grant No. 12&ZD220, the NSFC under Grant Nos. 71272029, 61033010 and 71110107027.

References

1. Rao, D., Yarowsky, D., Shreevats, A., Gupta, M.: Classifying Latent User Attributes in Twitter. In: Proceedings of the 2nd International Workshop on Search and Mining User-generated Contents (2010)
2. Bengio, Y.: Learning Deep Architectures for AI. Foundations and Trends in Machine Learning (2009)
3. Han, J., Pei, J.: Mining frequent patterns by pattern-growth: methodology and implications. ACM SIGKDD Explorations Newsletter, 2(2) (December 2000)
4. Quinlan, J.R.: C4.5: Programs for Machine Learning. Morgan Kaufmann (1993)

Predicting Users' Age Range
in Micro-blog Network

Chengyu Wang, Bing Xiao, Xiang Li, Jiawen Zhu,
Xiaofeng He, and Rong Zhang

Software Engineering Institute, East China Normal University, Shanghai, China
{chengyuwang,bingxiao,xiangli,jiawenzhu}@ecnu.cn,
{xfhe,rzhang}@sei.ecnu.edu.cn

Abstract. In this report, we present our work on WISE 2013 Challenge
Track II to predict the age range of Weibo users. In this challenge, a
dataset consisting of Sina Weibo user information was presented. The
goal of the challenge is to predict users age range. With personal information and original tweets for over one million users as training data, we
analyze and process the dataset, and experiment a series of prediction
methods including SVM, decision tree etc. The result shows that ensemble classifiers based on AdaBoost achieves the best prediction results in
this challenge.

Keywords: WISE challenge, classification, AdaBoost.

1   Introduction

Sina Weibo (http://weibo.com) is one of the most popular micro-blog services
in China[1]. In this challenge, we are given a dataset consisting of Weibo users
with personal information. The goal of the challenge is to predict the age range
of a Weibo user.

Parsing the original data is needed firstly. Files of data where Chinese words
have been mapped to numbers were too large and interspersed, so we remap the
numbers in every attributes together. As thus we can use these unique numbers
directly rather than separate by the attribute name. Feature selection begins
after finding that the number of features is over 300 thousand and most of the
words are rarely used. The lengths of job, education and other attributes of
different users are regarded as features here. Entropies of different attributes in
each age class can help in the feature selection process. The remainder of this
paper proposes several prediction methods and models that have been applied
during the experiment. Three categories of prediction approaches are introduced
including building classifiers, classifying via regression and ensemble classifiers.
The result suggests that adaptive boosting methods can greatly improve the
performance and achieve higher accuracy in prediction. The conclusion is given
in the last section.

X. Lin et al. (Eds.): WISE 2013, Part I, LNCS 8180, pp. 516–521, 2013.

2   Dataset Description and Data Preprocessing

The WISE 2013 Challenge Track II Dataset includes five data files with trainID-BIRTH.txt, trainInfos.txt and traintweets.txt being training set, while testInfos.txt and testtweets.txt being the testing set.

The data formats of each data file are listed:

- trainIDBIRTH.txt: userid birthdate
- trainInfos.txt and testInfos.txt: userid tags jobs education description
- traintweets.txt and testtweets.txt: userid timestamp1,tweet1 timestamp2, tweet2 timestamp3,tweet3...

In the dataset, only original tweets are preserved and segmented. Note that contents with Chinese are preprocessed by mapping each Chinese character to a unique number. So there are only alphanumeric contents in the dataset that has been optimized and preprocessed. The size of training data is 1126049. The size of testing data is 17519.

In trainIDBIRTH.txt, users' IDs and birthdates are listed, where we can get the age of each user first. In trainInfos.txt, each line is corresponding to one user and his or her personal information(user ID, tags, jobs, personal description, age, gender and education) is included.

We write Java programs to parse the original data files, namely trainID-BIRTH.txt, trainInfo.txt and traintweets.txt. The age is calculated using the birthdate of the corresponding user. The user ages are split into following four age ranges.

- Range 1: age younger than or equal to 18 years old;
- Range 2: age older than 18 but younger than or equal to 24 yesrs old;
- Range 3: age older than 24 but younger than or equal to 35 yesrs old;
- Range 4: age older than 35 years old.

The class is equivalent of the age range of the user.

As for text, different elements from different attributes can have the same number since contents in Chinese are preprocessed by mapping each Chinese word to a unique number. However, the same number from different attributes should not be treated as the same. We remap these numbers into a new vector space.

After the remapping process, we generate the following statistics.

3   Feature Creation and Selection

3.1   Feature Creation

We first use the bag-of-words model to generate features. It is commonly used in methods of document classification where the occurrence of each word is used as a feature for training a classifier. We can apply this method to different attributes mentioned above and all the words have already been remapped.

However, this method should not be completely applied in our problem because of the following reasons:

Table 1. Remapping Data

Attribute	Range	Total Number
tags	0-58944	58945
jobs	58945-80747	21803
education	80748-94027	13327
description	94028-218885	124881
tweet	21886-391821	172936

1. The number of features is over 300 thousand thus is already too large to handle. The training process can be extremely time-consuming and memory-consuming.
2. According to the statistics, over 90 percent of the words are used by only a small portion of the users and only a little part of the vocabulary are frequently used.

According to our dataset, the tweets posted by different micro-blog users may have various lengths. It proves true for personal information provided by users, too. For example, younger users, especially teenagers, have a tendency to provide little information in their jobs column since most of them do not have a job. We retrieve the lengths of job, education and other attributes of different users and regard them as features.

3.2   Feature Selection

As mentioned above, only a small part of words are frequently used. We employ a larger threshold for tweet words while for others, we suggest a smaller threshold. Next, we calculate the number of word occurrences in data from four classes (age ranges) for every word. Entropy is defined as:

$$H(x) = -\sum_{i=1}^{n} p(x_i) \log p(x_i) \tag{1}$$

Then, we calculate the entropies of different tags. Similarly, we get entropies of jobs, education, etc. as well. According to information theory, lower entropy suggests that only certain group of users have a high probability to use the corresponding word. Thus, we use features that have low entropies.

We also apply forward search algorithm to further reduce the number of features in order to find a good feature subset. Also, Principle Components Analysis (PCA) can reduce dimensionality but our experiments show that it will make the performance of our classifier worse.

4   Age Range Prediction Methodology

4.1   Classification

In our age range prediction approach, we define four classes {1, 2, 3, 4} where class 1 is defined as age range 1 (less than 18) and class 2 is defined as age range

2 (larger than or equal to 18 but less than 24) and so on. The object is to learn a model f such that f: X ⇒ {1, 2, 3, 4}. In this section, we introduce a series of classifiers we used, including Naive Bayes and Support Vector Machine with different kernels.

Naive Bayes classifiers[2] are simple classifiers based on Bayes' theorem. Although strong independence assumptions almost never hold true in the real world, Naive Bayes classifiers can be trained efficiently in supervised learning.

A support vector machine[3] constructs a hyper plane or set of hyper planes in a high- or infinite-dimensional space, making the separation easier in that space. The mappings used by SVM schemes are designed to ensure that dot products may be computed easily in terms of the variables in the original space, by defining them in terms of a kernel function selected to suit the problem. We use two kinds of kernel functions: polynomial kernel and Gaussian radial basis function. The formulas can be written as:

Polynomial kernel:

$$k(x_i, x_j) = (x_i^T . x_j + 1)^d \tag{2}$$

Gaussian radial basis function:

$$k(x_i, x_j) = exp(-\gamma |x_i - x_j|^2) \tag{3}$$

Since we wish to solve a multiclass classification problem, we reduce the single multiclass problem into multiple binary classification problems. Using the one-versus-one approach, classification is done by a max-wins voting strategy, in which every classifier assigns the instance to one of the two classes, then the vote for the assigned class is increased by one vote, and finally the class with the most votes determines the instance classification.

4.2   Classification via Regression

Because we can use a regression model to predict the age of micro-blog users, we can solve the classification problem in a regression approach. The problem can be solved in two steps:

Step1: build a model f such that f: X ⇒ Y where Y is the predicted age.
Step2: calculate the class label L such that:

$$Y \in [0, 18] \Rightarrow L = 1$$
$$Y \in (18, 24] \Rightarrow L = 2$$
$$Y \in (24, 35] \Rightarrow L = 3$$
$$Y \in (35, +\infty) \Rightarrow L = 4$$

We again use SVM to perform the regression task. Same to SVM classifiers, a non-linear function is produced by linear learning machine mapping into high dimensional kernel induced feature space. The capacity of the system is controlled by parameters that do not depend on the dimensionality of feature space.

4.3 Ensemble Classifiers

We use machine learning meta-algorithm such as bagging and boosting to train weak classifiers and then add them to a final strong classifier. The combination of weak classifiers can improve the performance of our classification model.

A random forest[4] is a classifier consisting of a collection of tree-structured classifiers. It constructs a multitude of decision trees at training time and outputs the class that is the majority of the classes output by individual trees. The method combines the bagging method and the random selection of features in order to construct a collection of decision trees with controlled variation.

The Adaboost[5] algorithm maintains a set of weights over the original training set and adjusts these weights after each classifier is learned. On each round, the weights of each incorrectly classified example are increased, and the weights of each correctly classified example are decreased.The new classifier focuses on the examples which have been classfied incorrectly.

We employ AdaBoost in conjunction with Decision Tree learning algorithms to improve the performance.

5 Experiments and Results

In this section, we report our results regarding the performance of Naive Bayes Model, Support Vector Machine with polynomial kernel, Support Vector Machine with Gaussian radial basis kernel, Classification via Support Vector Regression, Random Forest Model and AdaBoost Decision Tree.

We employ accuracy to evaluate the performance of our prediction models. Recall that for classification tasks, the terms true positives, true negatives, false positives, and false negatives compare the results of the classifier under test with trusted external judgments. The terms positive and negative refer to the classifier's prediction (expectation), and the terms true and false refer to whether that prediction corresponds to the external judgment (observation). The term accuracy is defined as:

$$Accurary = \frac{tp + tn}{tp + tn + fp + fn} \tag{4}$$

The accuracies of different models are shown in the table.

Table 2. Accuracies

Model	Accuracy(%)
Naive Bayes	58.6
SVM with polynomial kernel	75.43
SVM with RBF kernel	78.65
SVM Regression	76.42
Random Forests	65.25
AdaBoost Decision Tree	**83.2**

From the statistics, we observe that simple model such as Nave Bayes is least accurate. This is because the Nave Bayes model has a relatively large bias against the real model but the training time is the shortest among all models.

Support Vector Machine can efficiently perform non-linear classification thus yield a much better result than Nave Bayes model. It is also suggested that using Gaussian radial basis kernel to map inputs into high-dimensional feature spaces has a slightly better performance than polynomial kernel. Using Support Vector Machine for Regression can improve performance, too. The main drawbacks of Support Vector Machine are that the training process consumes much larger memory space and longer time, especially when the parameter C is relatively large. The trained models have a smaller bias but a larger variance on different testing sets.

Since single strong classifiers do not enjoy a satisfying accuracy, we continue to use ensemble classifiers. Random Forests are applied to construct trees considering randomly selected features. Although the model enjoys a good accuracy in training, it has a tendency to over-fit when different testing tests are used. However, AdaBoost Decision Tree works well and is less susceptible to the overfitting problem than other learning algorithms in our experiments. Although a single Decision Tree model does not produce a high accuracy, it is still useful in the final linear combination of classifiers.

6   Conclusions

In this paper, we introduced several approaches based on machine learning techniques to predict Micro-blog users' age range in WISE 2013 Challenge. The dataset provided by the WISE 2013 Challenge Committee was optimized and processed. A series of features were generated from the dataset. We presented and analyzed prediction models and it was suggested that ensemble classifiers, especially using adaptive boosting methods, can greatly improve the performance and achieve higher accuracy in prediction. Therefore, we can come to the conclusion that the approaches we presented were valid and effective.

References

1. Wu, X., Wang, J.: How about micro-blogging service in China: analysis and mining on sina micro-blog. In: Proceedings of 1st International Symposium on From Digital Footprints to Social and Community Intelligence. ACM (2011)
2. McCallum, A., Nigam, K.: A comparison of event models for naive bayes text classification. In: AAAI 1998 Workshop on Learning for Text Categorization, vol. 752 (1998)
3. Gunn, S.R.: Support vector machines for classification and regression. ISIS technical report 14 (1998)
4. Breiman, L.: Random forests. Machine Learning 45(1), 5–32 (2001)
5. Dietterich, T.G.: An experimental comparison of three methods for constructing ensembles of decision trees: Bagging, boosting, and randomization. Machine Learning 40(2), 139–157 (2000)

Author Index